ADVANCES AND CURRENT TRENDS IN BIOMECHANICS

Advances and Current Trends in Biomechanics

Edited by

Jorge Belinha

Department of Mechanical Engineering, School of Engineering, Polytechnic of Porto, Portugal

José Carlos Reis Campos

Faculty of Dental Medicine, University of Porto, Portugal

Elza Fonseca

Department of Mechanical Engineering, School of Engineering, Polytechnic of Porto, Portugal

Maria Helena Figueiral Silva

Faculty of Dental Medicine, University of Porto, Portugal

Maria Arcelina Marques

Department of Physics, School of Engineering, Polytechnic of Porto, Portugal

Maria Fernanda Gentil Costa

School of Health, Polytechnic of Porto, Portugal

Susana Oliveira

Faculty of Dental Medicine, University of Porto, Portugal

CRC Press
Taylor & Francis Group
Boca Raton London New York

CRC Press is an imprint of the
Taylor & Francis Group, an **informa** business
A BALKEMA BOOK

CRC Press/Balkema is an imprint of the Taylor & Francis Group, an informa business

Typeset by Integra Software Services Pvt. Ltd., Pondicherry, India

Library of Congress Cataloging-in-Publication Data

A catalog record has been requested for this book

Published by: CRC Press/Balkema
Schipholweg 107C, 2316 XC Leiden, The Netherlands
e-mail: enquiries@taylorandfrancis.com
www.routledge.com – www.taylorandfrancis.com

ISBN: 978-1-032-10806-3 (Hbk)
ISBN: 978-1-032-10807-0 (Pbk)
ISBN: 978-1-003-21715-2 (eBook)
DOI: 10.1201/9781003217152

Advances and Current Trends in Biomechanics – Belinha et al. (Eds)
© 2022 Copyright the Editor(s), ISBN: 978-1-032-10806-3

Table of contents

 ISBN: 978-1-032-10806-3

Preface

This book contains the selection of the most relevant papers presented at the 9th National Biomechanics Congress (CNB 2021) that took place in Porto, Portugal, on the 19th and 20th of February 2021.

The National Congress of Biomechanics (CNB) is a scientific meeting organized in Portugal, in the area of Biomechanics, under the auspices of the Portuguese Biomechanical Society (SPB).

The CNB promotes a forum for scientific discussion among researchers in the various areas of Biomechanics, through partnerships and research with common interests. The CNB also encourages the participation of M.Sc. and Ph.D. students aiming to promote the growth of Biomechanics in Portugal.

It should be noted that the diversity of the themes addressed in this within the scope of Biomechanics, showing the multidisciplinary of this area of knowledge in different topics, such as biofabrication; biomaterials; cardiovascular biomechanics, biofluids and hemodynamics; biomechanics of the injury/impact; biomechanics of rehabilitation; sports biomechanics; biomechanics of the skull and spine; biomechanics of the musculoskeletal system; orofacial biomechanics; orthopedic biomechanics; experimental and numerical biomechanics; tissue engineering; and biomedical devices.

The editorial team of this book would like to acknowledge all authors who shared their highly qualified works and members of the scientific committee that assisted in the review process, enabling the improvement of the quality of the submitted works.

The topics addressed in this book reveal the advances in Biomechanics and future research lines and trends.

The editorial team

Jorge Belinha - ISEP-IPP | INEGI
José Carlos Reis Campos - FMDUP | INEGI
Elza Fonseca - ISEP-IPP | INEGI
Maria Helena Figueiral de Silva - FMDUP | INEGI
Maria Fernanda Gentil Costa - ESS-IPP | INEGI
Susana Oliveira - FMDUP
Maria Arcelina Marques - ISEP-IPP | INEGI

 ISBN: 978-1-032-10806-3

Organizing committee

Chairs:

President:
Jorge Belinha - ISEP-IPP | INEGI

Vice-president:
José Carlos Reis Campos - FMDUP | INEGI

Elza Fonseca - ISEP-IPP | INEGI
Maria Helena Figueiral de Silva - FMDUP | INEGI
Maria Fernanda Gentil Costa - ESS-IPP | INEGI
Susana Oliveira - FMDUP
Maria Arcelina Marques - ISEP-IPP | INEGI

Local organizers:

Marco Parente - FEUP | INEGI
Madalena Peyroteo Gomes - INEGI
Ana Catarina Guerra - INEGI
Maria Inês Barbosa - INEGI
Ana Isabel Pais - INEGI
Frederico Jacob - CIETI | ISEP
Pedro Martins - CIETI | ISEP

Advances and Current Trends in Biomechanics – Belinha et al. (Eds)
© 2022 Copyright the Editor(s), ISBN: 978-1-032-10806-3

Scientific committee

Adélia Sequeira | *Universidade de Lisboa*
Amílcar Ramalho | *Universidade de Coimbra - CEMMPRE*
António Completo | *Universidade de Aveiro*
António Figueiredo | *Universidade de Coimbra*
António Ramos| *Universidade de Aveiro*
António Torres Marques | *Universidade do Porto*
António Veloso | *Universidade de Lisboa*
Arcelina Marques | *Instituto Superior de Engenharia do Porto – INEGI*
Aurélio Faria | *Universidade da Beira Interior*
Carolina Vila-Chã | *Instituto Politécnico da Guarda - CIDESD*
Cristina Manuela Peixoto dos Santos | *Universidade do Minho – CMEMs*
Daniel Marinho | *Universidade da Beira Interior - CIDESD*
Daniela Vaz | *CQC-FCTUC e ESSLei - Instituto Politécnico de Leiria*
Denis Coelho | *School of Engineering, Jönköping University*
Eduardo Campos | *Bysteel, UK*
Elza Fonseca | *Instituto Superior de Engenharia do Porto – INEGI*
Fernando Pinho | *FEUP*
Fernando M.F. Simões | *Instituto Superior Técnico – CERIS*
Filipa João | *Universidade de Lisboa*
Filipe Samuel Silva | *Universidade do Minho*
Héctor Jaramillo Suárez | *Universidad Autónoma de Occidente, Cali, Colombia*
Henrique Almeida | *Instituto Politécnico de Leiria*
Henrique Xavier Campos | *HXC Architecture | Montpellier France*
Henrique Pereira Neiva | *Universidade da Beira Interior - CIDESD*
Joana Reis | *Escola de Ciências e Tecnologia - Universidade de Évora*
João Abrantes | *Universidade Lusófona de Humanidades e Tecnologias - MovLab/CICANT*
João Folgado | *Instituto Superior Técnico - Universidade de Lisboa*
João Paulo Vilas-Boas | *FADEUP-CIFI2D/LABIOMEP-UP*
João Manuel R. S. Tavares | *FEUP-INEGI*
João Sampaio Fernandes | *FMDUP*
Joaquim Gabriel Mendes | *FEUP-INEGI*
Jorge Ambrósio | *Universidade de Lisboa*
Jorge Belinha | *Instituto Superior de Engenharia do Porto - INEGI*
Jorge Ferreira | *Universidade de Aveiro*
Jorge Laíns | *Centro de Medicina de Reabilitação da Região Centro*
José António Simões | *Universidade de Aveiro e ESAD.IDEA*
José Luís Alves | *Universidade do Minho*
José Manuel Casanova | *Universidade de Coimbra*

José Mário Rocha | *FMDUP*
José Reis Campos | *Universidade do Porto*
José Xavier | *Universidade NOVA de Lisboa (FCT NOVA)*
Leandro Machado | *Universidade do Porto*
Lídia Carvalho | *INESC-TEC*
Luciano Menegaldo | *Universidade Federal do Rio de Janeiro*
Luís Roseiro | *Instituto Politécnico de Coimbra – ISEC*
Luísa Sousa | *FEUP-INEGI*
Marco Parente | *Universidade do Porto*
Marco Santos | *Universidade de Aveiro*
Margarida S. Fernandes | *FMDUP*
Maria Augusta Neto | *Universidade de Coimbra*
Maria Helena Figueiral de Silva | *FMDUP-INEGI*
Maria Fernanda Gentil Costa | *Escola Superior de Saúde - P.Porto*
Mário Augusto Vaz | *FEUP-INEGI*
Mário Costa | *Instituto Politécnico da Guarda*
Miguel Lourenço | *Instituto Politécnico da Guarda*
Miguel Tavares da Silva | *Instituto Superior Técnico - Universidade de Lisboa*
Mónica Abrantes Oliveira | *TEMA-Universidade de Aveiro*
Nuno Dourado | *Universidade do Minho*
Paulo Fernandes | *Instituto Superior Técnico*
Paulo Flores | *Universidade do Minho*
Paulo Piloto | *Instituto Politécnico de Bragança*
Paulo Roriz | *ISMAI, CIDESD, LABIOMEP*
Pedro Martins | *Universidade do Porto*
Pedro Morouço | *Instituto Politécnico de Leiria*
Renato Natal Jorge | *FEUP-INEGI*
Rita Santos Rocha | *Instituto Politécnico de Santarém – CIPER/ULisboa*
Ronaldo Gabriel | *Universidade de Trás-os-Montes e Alto Douro*
Rubim Santos| *Escola Superior de Saúde - P.Porto*
Rui A. Lima | *Universidade do Minho*
Rui Miranda Guedes | *FEUP-INEGI*
Rui B. Ruben | *ESTG-CDRsp, Politécnico de Leiria*
Susana Oliveira | *FMDUP*
Vera Moniz-Pereira | *Universidade de Lisboa*
Vicente Silva Campos | *CHU Lisboa Central*

Acknowledgments

The editorial team truly acknowledge the scientific and institutional support of the following partners:

Instituto Superior de Engenharia do Porto
Faculdade de Medicina Dentária da Universidade do Porto
Fundação para a Ciência e a Tecnologia
Instituto de Ciência e Inovação em Engenharia Mecânica e Engenharia Industrial
Instituto Politécnico do Porto
Universidade do Porto
Laboratório de Biomecânica da Universidade do Porto
Sociedade Portuguesa de Biomecânica
Sociedade Brasileira de Biomecânica
Federación Iberoamericana de Ingeniería Mecánica

Advances and Current Trends in Biomechanics – Belinha et al. (Eds)
© 2022 Copyright the Author(s), ISBN: 978-1-032-10806-3

New biomechanism: A Bioactive Kinetic Screw

C.A. Andreucci & R.N. Jorge
Faculty of Engineering of the University of Porto (FEUP), Portugal

E.M.M. Fonseca
Mechanical Engineering Department, School of Engineering, Polytechnic of Porto (ISEP), Portugal

ABSTRACT: A new biomechanism, Bioactive Kinetic Screw (BKS), for bone screws created by the first author will be presented through a bone implant screw, in which the bone particles, cells and molecules removed during bone drilling will be used as an autogenous homogeneous transplant in the same implant site, aiming to obtain primary and secondary bone stability simplifying the surgical procedure (one step) and improving healing process. With this new biomechanism we will have the possibility to be proven, of obtaining an intermediate phase of healing between primary and secondary bone stability, using the natural process of inflammation, healing, repair and in the final stage of osseointegration, we could have a bone remodelling through the implant screw.

1 INTRODUCTION

Develop a biomechanism for bone fixation that involves desirable biological and mechanical characteristics in a biomechanical rehabilitation in a safe and practical way, are the founding principles of all the specialties involved in these studies; from researches with molecules, proteins and genetic therapies to the functional biomechanical restoration of limbs and organs (Adell et al. 1981).

The successful close interaction of implant surfaces with their biological environment is one of the most crucial issues in biomechanics of implantology (Albrektsson and Johansson 2001). Bone is a strong and lightweight composite multi-scale anisotropic material which presents a hierarchy of microstructures. Besides its multi-scale nature, bone can adapt its structure through remodelling phenomena, after a primary mechanical stability, which induces changes in its structure and biomechanical properties to accommodate the presence of an implant (secondary stability).

Despite the difficulties, the industrial design of implants has often been driven by an aggressive 'copycat' marketing approach rather than by fundamental advances in biomechanics (Abraham 2014). The difficulty also comes from the multi-factorial determinants of the problem, given by the implant properties, by the surgical protocol (which is not standardized) and by the patient's bone quality and its mechanical behaviour. The implant osseointegration is far from being understood, and measuring peri-prosthetic bone properties remains a challenge.

Implant surface treatment and coatings (surface topography) are responsible for increase the surface area of the bone-implant contact (BIC), decrease the stresses, and enhance adhesion qualities to the BIC interface at initial healing. Today we have improvements in biomechanical characteristics that optimize primary and secondary stability, mainly related to the screw threads (primary stability) and the surface of the screws (secondary stability) (Abuhussein et al. 2010).

DOI: 10.1201/9781003217152-1

Using Finite Element Analysis (FEA), some authors (Chang et al. 2012) evaluated the pattern of micro motion within implants and surrounding bone with different thread designs under immediate loading. The results revealed that all micro motion was located near the interface of cortical and cancellous bone and the square thread profile had the most favourable micro motion value. Multiple studies (Abuhussein et al. 2010) demonstrated that implants with smaller pitch showed the greater surface area and better stress distribution particularly in low-density bone.

It has been a common clinical observation that bone is often lost to the first thread after loading, regardless of the manufacturer type or design the bone loss often stops at the first thread because changes the shear load created by the crest module to a component of compressive loading (Abuhussein et al. 2010).

The bone response, which means rate, quantity and quality, are related to implant surface properties, the composition and charges are critical for protein adsorption and cell attachment. Hydrophilic surfaces seem to favour the interactions with biological fluids and cells when compared to the hydrophobic ones and hydrophilicity is affected by the surface chemical composition (Adell et al. 1981). Anodized surfaces interfere positively in bone response with higher values for biomechanical and histomorphometric tests when compared to machined surfaces. Calcium phosphate (CaP) coatings, mainly composed by hydroxyapatite, have been used as a biocompatible, osteoconductive and resorbable blasting materials. The idea behind the clinical use of hydroxyapatite is to use a compound with a similar chemical composition as the mineral phase of the bone to avoid connective tissue encapsulation and promote peri-implant bone apposition.

In some studies (Su et al. 2013) nano-hydroxyapatite/polyamide 66/glass fiber (n-HA/PA66/GF) was used as bioactive bone screws in comparison with metallic screws to investigate the in vivo biocompatibility, internal fixation properties and osteogenesis. The n-HA/PA66/GF screws exhibited good biocompatibility, high mechanical strength and extensive osteogenesis in the host bone; the maximum push-out load of the bioactive screws was greater than that of the metallic screws.

Bioactive materials, able to induce hydroxyapatite precipitation in contact with body fluids, are of great interest because of their bone bonding capacity. A comparison among bioactive materials with different surface features (chemistry, charge, morphology) was performed to verify case by case the mechanism of action and the relationship with kinetics and type of precipitated hydroxyapatite over time. The results suggest that the OH groups on the surface have several effects on the surface chemical behaviour: a detailed description of their abundance and chemical acidic/basic reactivity can be helpful for a better understanding of surface chemistry. Abundance of OH groups mainly affect hydrophilicity of surfaces, while the isoelectric point, surface charge and ions adsorption from the solution mainly depends on the OH acidic/basic strength.

The main outcome of the present research is a detailed understanding of the bone implant screwing processes, contributing for the application of optimum implant screw insertion parameters that will minimize the bone damage due the invasive procedure, trying to obtain primary and secondary stability optimizing healing process.

The results on this new model BKS contribute to a comprehensive understanding of thermo mechanical aspects of bone drilling and screwing process and associated predictive models creating a new simplified protocol for implant screws surgery. The new developed model and protocol if succeeds in results, can be used in any bone surgery when implant screw integration is needed. The drilling and the new implant screwing process under different aspects provide an important tool that will enable for the health professionals to plan an accurate and safe surgery with better post-surgery patient recovery.

2 MATERIALS AND METHODS

The use of screws for bone implants and fixations has been done in a safe and systematic way since 1985 (Albrektsson and Albrektsson 1987). Biomechanics has not improved real changes since then, with the focus given more over the last four decades to the surfaces and threads of the

Figure 1. New Bioactive Kinetic Screw (BKS) made by the first author.

screws. The innovation of a new biomechanism (BKS) indicated in Figure 1, using biomechanical concepts already well known and structured in the same implant screw, will be described.

In the new biomechanism the particles removed during bone drilling and screwing at the same time due to the new characteristics, will use the chips (particulate bone, cells and molecules) to fill the through hole by the drill flutes in the new screw.

The screw was created based on the concepts of using drill bits for bone drilling and screws for fixing bone implants. Bearing in mind that the traditional drill flutes eliminate the particulate material to be removed after cutting, the author created a limit for these flutes that are found in a through hole connecting and limiting these cutting grooves avoiding the removal of particles coming from the drilling. This drilling particulate material will be compacted in and through the screw, increasing the bone contact surface and creating a bone bridge of particulate material through the screw (Elsayed 2019).

Developing a screw that will also be the drill bit for bone implants fixation, creates the opportunity to decrease the drilling stages in the surgical protocols, using and not discarding the bone from the site where the implant will be screwed, taking advantage of the natural characteristics of the particulate bone, blood, molecules and cells of the implanted site, stimulating the biological processes of bone healing. The advantages obtained with the simplification of surgical protocols for implants and the optimization of immediate and long-term results, justify the research of a new biomechanism and its practical results in genetic, molecular and cellular levels in bone healing for biomechanical efficiency (Degirmenci and Atala 2019).

The new biomechanism can be applied to any biomaterial for implants screws, any kind of surfaces, coatings and threads, which provides more possibilities for practical applications and optimization of results. The characteristics that were considered today to optimize the success rate in situations of low quality and bone quantity were chosen, where the risks of implant failures are lower when related to implant design. Commercially pure titanium will be used as standard biomaterial for being the oldest, accepted and well-studied short and long term results for dental implants; remembering that the new biomechanism can be applied to any biomaterial for bone screws, for example nano-hydroxyapatite/polyamide 66/glass fiber (n-HA/PA66/GF), (Su et al. 2013) any kind of surfaces, coatings and threads (Abuhussein et al. 2010), which provides more possibilities for practical applications and optimization of results.

3 RESULTS AND DISCUSSION

The studies are in their manufacturing phase of the new biomechanism in the form of a dental implant to perform the laboratory tests that will be used to validate the computational models.

The resources and financing for the research have already been approved and are in their final phase for the beginning of data collection.

The possibility of obtaining bone healing through the screw can become a new reality for bone implants, where we will have for the first time direct bone to bone contact through the screw.

In every bone drilling for insertion of an implant screw, there is a bone loss with almost all the measurements of the screw used. Clearly taking advantage of this particulate bone, which is normally discarded during drilling, is in itself already a great benefit for the patient and the surgeon.

The simplification of the surgical technique is also a great natural benefit of the new biomechanism, since with a lower number of intermediate drills and the direct self-drilling of the implant in cases of low bone quality (trabecular bones) we can decrease the time and the risks related to the surgical technique (Steigenga 2003).

In every bone drilling for insertion of an implant screw, there is a bone loss with almost all the measurements of the screw used. Clearly taking advantage of this particulate bone, which is normally discarded during drilling, is in itself already a great benefit for the patient and the surgeon.

The natural presence of bone through the screw and direct bone contact through the hole present in the new biomechanism as seeing on Figure 1, will enable the new formation of blood vessels and better bone quality around the implant, to be tested (Elsayed 2019).

4 CONCLUSIONS

The new BKS created by the first author allows the use of the bone, blood, cells, and molecules removed in the bone drilling to be introduced through the screw using an innovative insertion and compacting biomechanism, based on biomechanical concepts already proven and used since the initial use of screws for fixations and bone implants. The innovation is based on the system created by associating different biomechanical concepts in the same screw, optimizing efficiency in molecular, cellular, and macroscopic levels. Determining the design of a bone implant for all the practical clinical conditions that present itself to achieve osseointegration is not yet fully possible. In cases of low quality and bone quantity, we find the greatest number of implant failures. With this factor in mind, we opted to use anodized surface with calcium phosphate coating (CaP), square threads and small pitch in the new biomechanism to be tested.

All these features can be modified according to individual surgery plan. More studies should confirm these properties and their consequences for bone healing of screws.

REFERENCES

Abuhussein, H., Pagni, G., Rebaudi, A., Wang, H. L. 2010. The effect of thread pattern upon implant osseointegration. *Clinical Oral Implants Research*. 21: 129–136.

Abraham, C. M. 2014. A Brief Historical Perspective on Dental Implants, Their Surface Coatings and Treatments. *The Open Dentistry Journal*, 8(Suppl 1-M2): 50–55.

Adell, R., Lekholm, U., Rockler, B., Branemark, P. 1981. A 15 year study of osseointegrated im-plants in the treatment of the edentulous jaw. International J. Oral Surgery, 10(6): 387–416.

Albrektsson, T., Johansson, C. 2001. Osteoinduction, osteoconduction and osseointegration. *Europe Spine J*. 10(Suppl 2): S96–S101.

Albrektsson, T., Albrektsson, B. 1987. Osseointegration of bone implants: A review of an alternative mode of fixation. *Acta Orthopaedica Scandinavica*, 58(5): 567–577.

Chang, P. K., Chen, Y. C., Huang, C. C., Lu, W. H., Chen, Y. C., et al. 2012. Distribution of micromotion in implants and alveolar bone with different thread profiles in immediate loading: a finite element study. *International J Oral Maxillofacial Implants*, 27: 96–101.

Degirmenci, K., Atala, M. H. 2019. The impact of different dental implant surface properties to osseointegration success of dental implants: A systematic review. *Clinical Oral Implants Research*, 30(19): 280.

Elsayed, M.D. 2019. Biomechanical Factors That Influence the Bone-Implant-Interface. *Research Rep Oral Maxillofacial Surgery*. 3: 1–14.

Steigenga, J. T., al-Shammari, K. F., Nociti, F. H., Misch, C. E., Wang, H. L. 2003. Dental implant design and its relationship to long-term implant success. *Implant Dentistry*. 12: 306–317.

Su, B., Peng, X., Jiang, D., Wu, J., Qiao, B., Li, W., et al. 2013. In Vitro and In Vivo Evaluations of Nano-Hydroxyapatite/Polyamide 66/Glass Fiber (n-HA/PA66/GF) as a Novel Bioactive Bone Screw. *PloS one*, 8(7): 1–9.

Advances and Current Trends in Biomechanics – Belinha et al. (Eds)

Soft materials for exoskeleton development

A. André
INEGI, Porto, Portugal

P. Martins
INEGI, LAETA, FEUP, Porto, Portugal

ABSTRACT: Nowadays, people have the need to be independent from others during the performance of their daily-living activities. To help them achieving some of that freedom, when mobility disorders are present, exoskeletons started to be developed and used. Nevertheless, these solutions were always built considering rigid, uncomfortable and heavy materials. Precisely due to the nature of the materials used, a need of an alternative was imperative. Currently, the new materials in development, for rehabilitation and aided locomotion applications, must follow some requirements in contrast to the traditional ones. Having an affordable price, being light weighted and soft to the touch are some of the of the ideal requirements. Considering this situation, the current paper intends to highlight new materials able to be a real alternative to conventional ones.

1 INTRODUCTION

The number of paralyzed people or with reduced mobility is increasing due to many factors (Bing, 2017), where aging (Chen, 2020), diseases (Moreno, 2020) and accidents (Nas, 2015) are being the biggest contributors for this situation.

In countries where the life expectancy is higher, such as Germany, Norway or even Hong Kong, the age aspect is even more relevant, since there is a direct relation between aging and the increased mobility problems (Levesque, 2020). Plus, data from United Nations (ONU) estimates that the percentage of people over 65 years old will be around 20% of global population in thirty years (Chen, 2015).

Neurodegenerative pathologies, such as Parkinson disease (Shah, 2020) and amyotrophic lateral sclerosis (ALS) (Gruis, 2011), are also responsible for mobility impairments. These diseases typically cause the progressive loss of structure and function of the central nervous system (CNS), which is responsible for the sensory information and the commands to the muscles. This fact has a huge impact on locomotion abilities. Nowadays, numbers suggest that about 50 million people worldwide have some type of neurodegenerative disease (Bellou, 2020).

Injuries, mostly in spinal cord or brain, caused by motorized accidents, falls or sporting practices (NSCISC, 2019) can also affect mobility. Every year, about 40 million people suffer from this cause of mobility impairments (Yip, 2012).

When people are in this situation, simple activities, such as getting out of bed or walking around, could be painful and stressful, because something simple and familiar is no longer easily achievable. As a result, in an attempt to fight this social problem, there have been societal efforts to help these people in several different ways. One solution could be the development of exoskeletons by engineers and researchers. The first prototypes, although useful, had many disadvantages mainly related to the nature of the materials used for the construction.

DOI: 10.1201/9781003217152-2

As primary choice, rigid materials, like aluminum (Vertechy, 2009), were used, which resulted in some discomfort and additional weight to the users (Gonzalez-Vargas, 2017). Also, the degrees of freedom of the solution provided were not enough for going along with the natural human movements.

Answering some of these inconveniences, new materials started to be used, with a special focus on soft materials.

2 SOFT MATERIALS

Soft materials, unlike rigid and metallic alternatives, are mainly polymers and polymer-based composites with the ability to change their shape or size when externally stimulated by an input (Liu, 2019). These initial triggers could be a thermal source (Kim, 2012), light (visible or infrared spectrum) (Luo, 2015), magnetic field (Breger, 2015) or even an electric field (Correia, 2019), Figure 1.

Thermally stimulated soft materials, such as silicone-based elastomers (Kang, 2019), synthetic hydrogels (El-Atab, 2020) or shape memory materials (Renata, 2017), can be activated by an external thermal source, including infrared (IR), thermal radiation and Joule heating (Bilodeau, 2018). Typically, they are low cost, biodegradable, show low density and temperature actuation. However, the low efficiency and high-power consumption are some of the problems these materials face (Bar-Cohen, 2019).

Photo-responsive materials are based on photochromic molecules with the ability to capture optical signals (visible light or near-infrared) and convert them into different property modifications (Florea, 2012). These molecules, such as spiropyran (Florea, 2012), can be added into gels, polymers, and fluids (El-Atab, 2020). The result is environmental-friendly materials with the possibility of remote control. As disadvantages, high intensities of light are required, as well as limited deformations and slow actuation speed (Bar-Cohen, 2019).

Materials stimulated by magnetic fields use magnetic particles and fillers incorporated in many different soft compounds, such as polymers, gels, papers or fluids (Hines, 2017), since magnetic fields are able to penetrate the majority of these materials. This result into a magnetization profile with variable magnitude and direction (Garstecki, 2009). Using this stimulation method, the associated materials have a linear effect with quick response. However, to activate them, large dimension magnetic coils are required and the fabrication process of these materials are also very complex (Bar-Cohen, 2019).

Finally, the materials stimulated by electrical fields, such as polyvinylidene fluoride (PVDF), are able to convert an electric input into a mechanical response output (Correia, 2019). Some advantages and disadvantages of these materials will be mentioned in next section.

Besides the different stimulation method, all these different materials might be used during the conception and development of an exosuit as actuators.

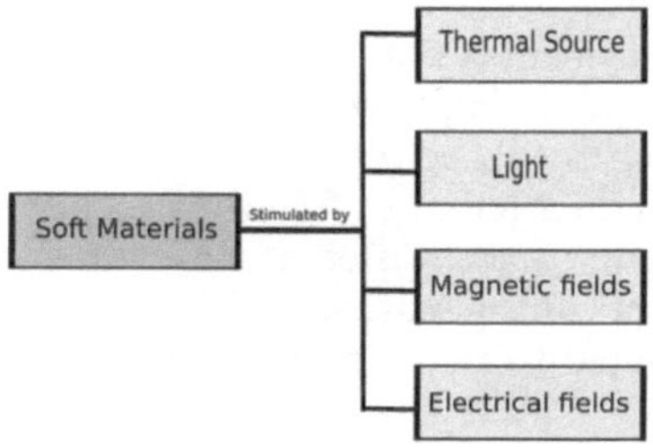

Figure 1. Soft materials stimulation methods.

Figure 2. Example of a PVDF film, a field-activated EAP.

3 ELECTROACTIVE POLYMERS

The materials stimulated by electrical fields are the most convenient and practical to use, not only in exosuit solutions, but also in many other applications. These materials are typically known as electroactive polymers (EAP) and artificial muscles, due to their ability to emulate biological muscles (Bar-Cohen, 2019).

As polymers, these materials have multiple advantageous properties, such as mechanical flexibility, low density, as well as being easy to process and mass produce (Bar-Cohen, 2019). Besides, the ability that some EAPs exhibit in converting mechanical strain into an electrical signal is also appreciated. This reverse effect allows EAP might be used as sensors and energy harvesters as well (Bar-Cohen, 2019).

Based on the mechanism that activates EAP materials, they can be divided into two major groups, the ionic and field-activated polymers. Ionic EAPs, in which the stimulation process involves diffusion of ions, can produce large bending displacements, require low voltage and have natural bi-directional actuation. As disadvantages, these types of soft materials show slow response and induce relatively low actuation force. An example of ionic EAP material is polypyrrole, a polymer that is obtained by oxidative polymerization of pyrrole (Maziz, 2017).

Field-activated EAPs, driven by Coulomb forces, show a rapid response, can hold strain under DC activation and induce relatively large actuation forces. However, actuating them requires high field strength, which could result in high voltages (Bar-Cohen, 2019). One of the most widely known field-activated EAP is the poly(vinylidene fluoride) (PVDF) (Sessler, 1981), Figure 2.

In literature, it is possible to find several studies showing promising results using EAPs. Senders et al (Senders, 2010) and Ganet et al (Ganet, 2016) are some examples of the utility of these materials for medical applications. Correia et al (Correia, 2019) and Maziz et al (Maziz, 2017) studied these materials for exosuit development.

4 DISCUSSION AND CONCLUSION

Soft materials have arrived to fill in an existent need left by traditional materials used in exoskeleton development. Compared with aluminum (for example), they originated friendlier touch sensations when in contact with the human body. Nevertheless, these soft materials are not suitable to replace entirely traditional solutions, since they do not produce enough force to sustain limbs or even the entire body.

Still, the results achieved in literature using soft materials show their potential as an alternative to more conventional ones. To understand how adequate these solutions could be, will require further research and development.

ACKNOWLEDGMENTS

The authors gratefully acknowledge funding from FCT and FSE, Portugal, under grant SFRH/BD/147807/2019, and project MImBI - PTDC/EME-APL/29875/2017 financed through FEDER and FCT. This work was supported by FCT, through INEGI, under LAETA, project UIDB/50022/2020.

REFERENCES

Bar-Cohen, Y. e. a., 2019. Electroactive polymer (EAP) actuators—background review. *Mechanics of Soft Materials.*

Bellou, E. e. a., 2020. Polygenic risk and pleiotropy in neurodegenerative diseases. *Neurobiology of Disease*, Volume 142.

Bilodeau, R. e. a., 2018. All-soft material system for strong soft actuators. *2018 IEEE International Conference on Soft Robotics (RoboSoft)*, pp. 288–294.

Bing, C. e. a., 2017. A wearable exoskeleton suit for motion assistance to paralysed patients. *Journal of Orthopaedic Translation*, Volume 11, pp. 7–18.

Breger, J. e. a., 2015. Self-Folding Thermo-Magnetically Responsive Soft Microgrippers. *ACS Applied Materials & Inerfaces*, pp. 3398–3405.

Chen, B. e. a., 2015. Recent developments and challenges of lower extremity exoskeletons. *Journal of Orthopaedic Translation*, pp. 26–37.

Chen, B. e. a., 2020. State-of-the-art research in robotic hip exoskeletons: A general review. *Journal of Orthopaedic Translation*, Volume 20, pp. 4–13.

Correia, D. M. e. a., 2019. Ionic Liquid Cation Size-Dependent Electromechanical Response of Ionic Liquid/Poly(vinylidene fluoride)-Based Soft Actuators. *The Journal of Physical Chemistry*, pp. 12744–12752.

El-Atab, N. e. a., 2020. Soft Actuators for Soft Robotic Applications: A Review. *Advanced Intelligent Systems.*

Florea, L. e. a., 2012. Photo-Responsive Polymeric Structures Based. *MAcromolecular Materials and Engineering*, pp. 1148–1159.

Ganet, F. e. a., 2016. Development of a smart guide wire using an electrostrictive polymer: option for steerable orientation and force feedback. *Sci Rep.*

Garstecki, P. e. a., 2009. Propulsion of flexible polymer structures. *Journal of Physics: COndensed Matter.*

Gonzalez-Vargas, J. e. a., 2017. *Wearable Robotics: Challenges and Trends.* s.l.:Springer.

Gruis, K. e. a., 2011. Amyotrophic lateral sclerosis patients' self-reported satisfaction with assistive technology. *Muscle&Nerve*, Volume 43, pp. 643–647.

Hines, L. e. a., 2017. Soft Actuators for Small-Scale Robotics. *Advanced Materials.*

Kang, D. J. e. a., 2019. Programmable soft robotics based on nano-textured thermo-responsive actuators. *Nanoscale.*

Kim, J. e. a., 2012. Designing Responsive Buckled Surfaces by Halftone Gel Lithography. *Science*, Volume 335.

Levesque, L. e. a., 2020. Study of human-machine physical interface for wearable mobility assist devices. *Medical Engineering and Physics*, pp. 33–43.

Liu, J. e. a., 2019. Responsible and Foldable Soft Materials. *Trends in Chemistry*, pp. 107–122.

Luo, R. e. a., 2015. Gradient Porous Elastic Hydrogels with Shape-Memory Property and Anisotropic Responses for Programmable Locomotion. *Advanced Functional Materials*, Volume 25, pp. 7272–7279.

Maziz, A. e. a., 2017. Knitting and weaving artificial muscles. *Sci. Adv.*

Moreno, J. e. a., 2020. Hybrid Wearable Robotic Exoskeletons for Human Walking. Em: *Wearable Robotics Systems and Applications.* s.l.:s.n., pp. 347–364.

Nas, K. e. a., 2015. Rehabilitation pf Spinal Cord Injuries. *World Journal of Orthopedics*, pp. 8–16.

NSCISC, N. S. C. I. S. C., 2019. *Spinal Cord Injury Facts and Figures at a Glance*, s.l.: s.n.

Renata, C. e. a., 2017. Shape change/memory actuators based on shape memory materials. *J Mech Sci Technol*, pp. 4863–4873.

Senders, C. e. a., 2010. Force requirements for artificial muscle to create an eyelid blink with eyelid sling. *Arch Facial Plast Surg.*

Sessler, G., 1981. Piezoelectricity in polyvinylidenefluoride. *The Journal of the Acoustical Society of America.*

Shah, V. e. a., 2020. Quantity and quality of gait and turning in people with multiple sclerosis, Parkinson's disease and matched controls during daily living. *Journal of Neurology*, pp. 1188–1196.

Vertechy, R. e. a., 2009. Development of a New Exoskeleton for Upper Limb Rehabilitation. *2009 IEEE 11th International Conference on Rehabilitation Robotics*, pp. 188–193.

Yip, P. e. a., 2012. Spinal cord trauma and the molecular point of no return. *Molecular Neurodegeneration*, Volume 7.

Advances and Current Trends in Biomechanics – Belinha et al. (Eds)

CBCT in the upper airways evaluation

F. Cavadas
Faculty of Dental Medicine, Universidade Católica Portuguesa, Viseu, Portugal

J. Fonseca
Integrated Dental Medicine Center of Coimbra, Portugal

T. Marques & P. Fonseca
Centre of Interdisciplinary Research in Health, Faculty of Dental Medicine, Universidade Católica Portuguesa, Viseu, Portugal

ABSTRACT: Cone Beam Computer Tomography (CBCT) is an auxiliary diagnostic tool that is increasingly used in dentistry and, therefore, would be an asset if it were subjected to airway segmentation, increasing the probability of detecting any anomaly and, consequently, assist in the screening of patients with Obstructive Sleep Apnea Syndrome (OSAS). The main aim of this study is to determine the usefulness of CBCT in the early detection of oropharyngeal constriction, since it is an anatomical factor predisposing to OSAS. Considerable percentages were verified associated to the risk of OSAS, more specifically, 14.3% to the high risk and 28.6% to the moderate risk.

1 INTRODUCTION

The Cone Beam Computerized Tomography (CBCT) is an auxiliary means of diagnosis or of the surgical-prosthetic treatment plan, which is increasingly used in dentistry. Therefore, it would be an asset if it were subjected to airway segmentation, increasing the probability of detecting some anomaly and, consequently, assisting in the screening of patients with Obstructive Sleep Apnea Syndrome (OSAS). Given that the literature has pointed out CBCT as an important source of airway data despite being sometimes neglected. (Zimmerman et al, 2017)

Obstructive Sleep Apnea Syndrome is defined as the repeated occurrence of episodes of total (apnea) or partial (hypopnea) obstruction of the upper airways during sleep. When left untreated, it can cause several comorbidities, associated for example with cardiovascular diseases, neurovascular diseases, increased risk of work and traffic accidents in the face of excessive sleepiness, decreased concentration and cognitive functions. (Daurat et al, 2016)

The airway is divided into 3 parts: nasopharynx, oropharynx and hypopharynx. One of the regions relevant to the pathophysiology of OSAS is the oropharynx which runs from the uvula to the epiglottis. The obstruction can appear at different levels of the oropharynx, being more frequent in the retropalatal, retroglossal region, lateral walls and epiglottis. (Salman 2019)

Therefore, the aim of this study is to determine the CBCT utility in the early detection of oropharynx constriction once it is an OSAS anatomic predisposing factor.

DOI: 10.1201/9781003217152-3

2 MATERIAL AND METHODS

After study approvement (opinion nº 27/2019 of 6/02/2020) by the Ethics Committee for Health (CES) of the Universidade Católica Portuguesa, twenty-eight CBCTs performed at the Clínica Dentária Universitária in Viseu were analyzed. The CBCT data used were obtained through Planmeca® ProMax 3D (Planmeca, Helsinki, Finland) having been exposed to 90kV, 8mA and 13.713s. The software used for the analysis is Romexis (version 5.1.O.R Planmeca®). The images are displayed according to the size 502x502x436mm and voxel 400µm. The measurement is made automatically by the image software whose reliability is proven by the existing literature. (Kamaruddin et al, 2019)

For airway measurement, the 3 anatomical planes, sagittal, coronal and axial are oriented according to the direction of the airways. (Figure 1)

After activating the "Extract Airways" tool, the first point is marked at the level of the Posterior Nasal Spine (PNE) and its marking is continued to follow the curvature of the airways, the last point being marked at the middle point of the 4th cervical vertebra. (Figure 2)

Before calculating its dimensions (Figure 3), the "Airways tool" option is selected and the variable that defines the image limit is adjusted. The total volume and the minimum airway area are calculated. (Figure 4)

From the narrowest area of the airway, the antero-posterior and laterolateral dimensions are measured manually with the aid of the "linear measurement" tool integrated in the software. (Figure 5) When necessary, the segmentation is readjusted so that the entire airway is included.

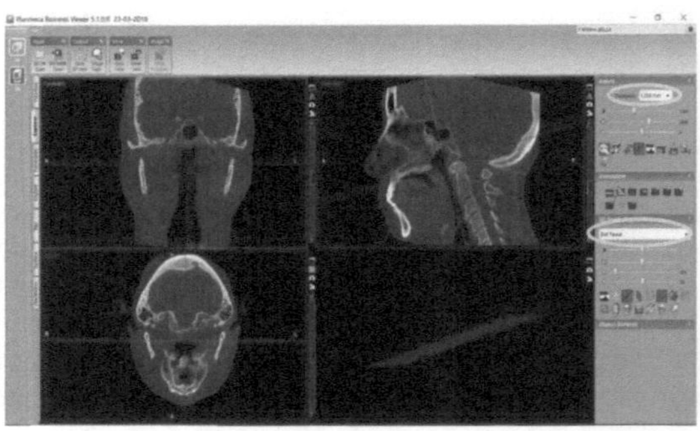

Figure 1. Three anatomical planes (sagittal, coronal and axial).

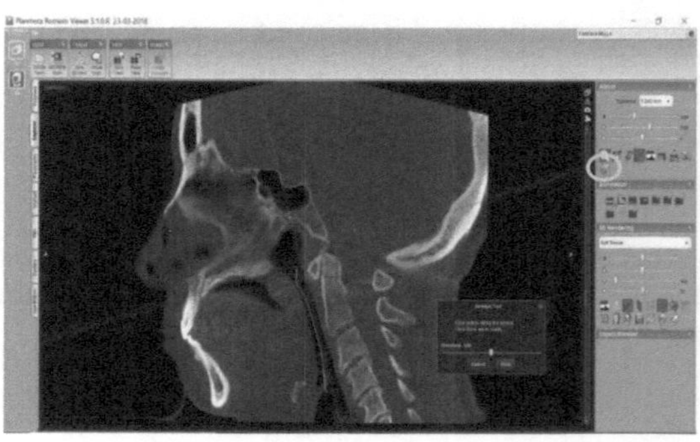

Figure 2. Definition of anatomical airway limits.

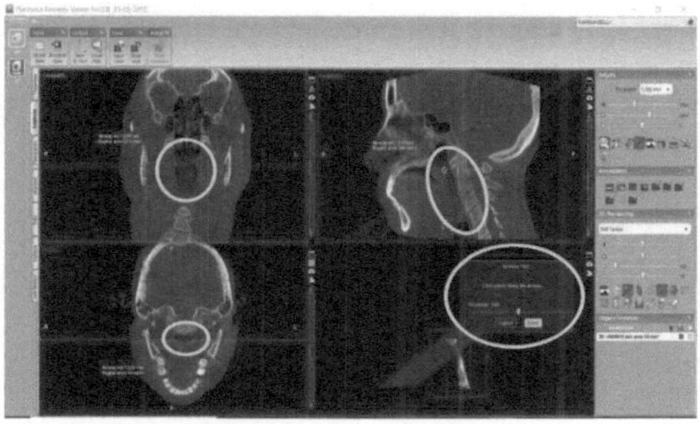

Figure 3. Airway values before adjusting your limits.

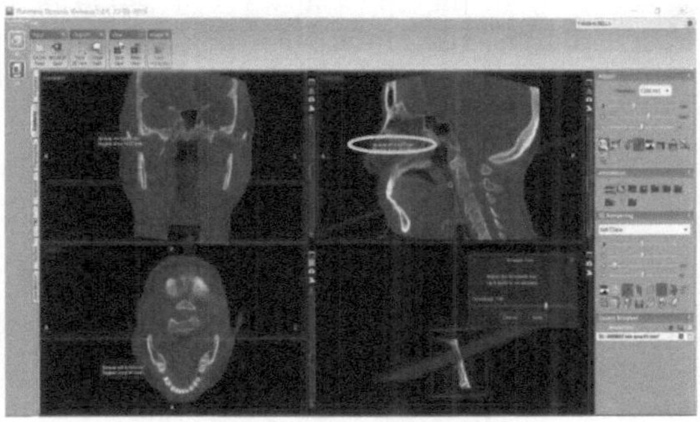

Figure 4. Calculation of total volume and minimum airway area.

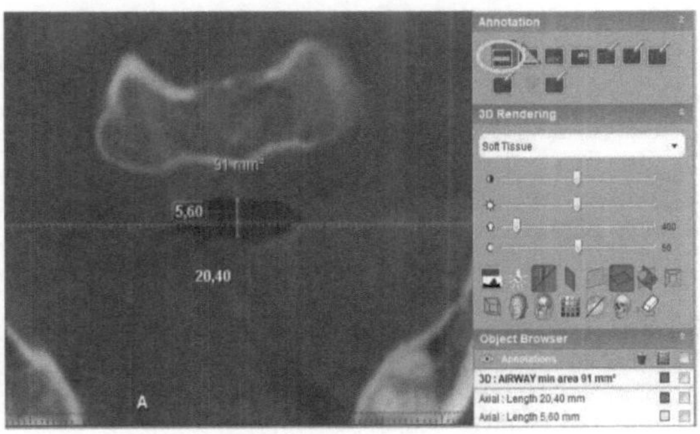

Figure 5. Calculation of anteroposterior and laterolateral dimensions.

3 RESULTS

Regarding the most constricted area of the upper airways, the average was 142.59mm^2. Four of the evaluated CBCT (14.3%) have measurements <52mm^2. Eight (28.6%) have between 52 and 100mm^2 and 16 (57.1%) have an airway >100mm^2.

Regarding the total volume, we found that in our sample the median was 16.9cm^3 and that there are disparate values since the minimum volume found was 8.8cm^3 and the maximum was 46.9cm^3. Finally, when measuring the anteroposterior (A-P) and laterolateral dimensions of the most constricted area, it was found that, in general, the constricted area had greater length at the laterolateral level than at the A-P level.

It was also found that the constriction was more frequent at the retropalatal level (53.6%; n=15), followed by the retroglossal level (35.7%; n=10) and, finally, at the epiglottis level (10.7%; n=3).

4 DISCUSSION

The value of oropharyngeal constriction was associated with the risk of developing OSAS by Steffy et al in 2018. They considered a high risk to be less than 52mm^2, a moderate risk between 52 and 100mm^2 and a low risk to be more than 100mm^2. (Steffy et al, 2018) Although the greater number of our patients (57.1%) has a low risk of OSAS, there is still a significant percentage with high (14.3%) and moderate risk (28.6%).

In relation to the total volume although it is suggestive that a smaller volume will not be so favorable for a correct functioning of the upper airways, it would be interesting to quantify from what values may exist a compromise in the physiology of the airway. To this end, there is an urgent need to standardize the protocol for airway measurement. (Zimmerman et al, 2017)

5 CONCLUSIONS

The CBCT allows the calculation of the total volume of the airways, as well as its area in the different anatomical planes (sagittal, coronal and transversal). It also allows the identification of the regions of greatest anteroposterior and laterolateral constriction of the airways. We can conclude that the use of this protocol of CBCT analyze, the dentist can help in the screening of patients with suspected SAOS.

REFERENCES

Cavadas, A.F.M. 2020. *O Contributo Do Médico Dentista para o Diagnóstico Precoce da Síndrome da Apneia Obstrutiva do Sono*. (MSc Thesis, Universidade Católica Portuguesa).

Daurat, A., Sarhane, M., Tiberge, M. 2016. *Obstructive sleep apnea syndrome and cognition: a review*. Neurophysiol Clin, 46(3), 201–215.

Salman SO. 2019. *Modern Management of Obstructive Sleep Apnea* (1st ed.), Springer Nature Switzerland AG, Springer International Publishing, 129 p.

Steffy, D.D., Tang, C.S. 2018. *Radiographic Evaluation of Sleep- Disordered Breathing*. Radiol Clin NA, 56(1), 177–185.

Kamaruddin, N., Daud, F., Yusof, A., Aziz, M.E., Rajion, Z.A., 2019. *Comparison of automatic airway analysis function of Invivo5 and Romexis software*. Peer J, 7(1):1–10.

Zimmerman JN, Lee J, Pliska BT. 2017. Reliability of upper pharyngeal airway assessment using dental CBCT: A systematic review. Eur J Orthod.: 39(5):489–96.

Advances and Current Trends in Biomechanics – Belinha et al. (Eds)
© 2022 Copyright the Author(s), ISBN: 978-1-032-10806-3

Stress and strain analysis in zirconium triangular dental implants

R. Couto, D. Martins, A.R. Carreiras & E.M.M. Fonseca
Mechanical Engineering Department, School of Engineering, Polytechnic of Porto (ISEP), Portugal

C. Andreucci & R.N. Jorge
Faculty of Engineering of the University of Porto (FEUP), Portugal

ABSTRACT: The main objective of this work is to study the influence of the internal diameter size in the triangular dental zirconium (Zr-7Cu-5Sn) implant. Dental implant affects bone tissue, in response to different mechanical stimuli, and it depends on some parameters (material stiffness, thread shape, thread pitch, length, diameter). Different researches, have been developed new models to describe the bone remodelling process, considering different bone stimuli (stress, strain energy, damage, cell biology, metabolic and external factors). The results about the chosen dental implant diameter, using a metallic zirconium material implant, give additional information about the transferred stress and strain energy density to the bone. In this way, it will be possible to guarantee the implant stability and sustain loads during or after osteointegration, using a material that has benefits to the biomechanical behaviour in general, highlighting the low elastic modulus.

1 INTRODUCTION

For the osteointegration improvement, a low elastic modulus in the screw and dental implant is important to mimicking the bone substitutes and allow a balanced stress distribution between bone-implant (Herekar et al. 2014).

The main goal of this work is to verify the size effect of the internal diameter on stress and strain distribution, in a zirconium dental implant with a triangular implant, that has a low elastic modulus, using the finite element method.

Different diameters allow different spectrums of stress and strain, which mean different biomechanical responses. In terms of stress, stress shielding effect can cause bone loss, that consequently allow a failure of the bone-implant construct (Herekar et al. 2014). Stress shielding reduces the bone activity, and according to the article (Uhthoff, Jaworski, 1978), it causes bone loss. Thus, it is important to investigate which the favourable diameter for stress distribution, avoiding that bone loss.

The zirconium metallic alloys, mentioned in (Kim et al. 2019), have excellent biocompatibility, comparing with common titanium alloys. That zirconium alloys developed by (Kim et al. 2019) have elastic modulus surrounding 14-18GPa, that is lower than so many common titanium alloys, that have elastic modulus surrounding 105GPa (Haase and Rouhi, 2013). This is an excellent characteristic of that zirconium alloys, that allows to decrease the stress shielding effect of bone-implant construct.

The thread implant profile is one of the major contributors to initial implant stability. Implant stability will promote successful osteointegration and will lead to stress distribution over wide area at low level (Haase and Rouhi, 2013). The present study only uses a triangular shape implant, that some articles referred to be the best shape for stress distribution (Haase and Rouhi, 2013).

DOI: 10.1201/9781003217152-4

In the past decades, there is a lot of studies that investigates the effect of diameter on biomechanical behaviour (Eazhil et al. 2016), (Saluja et al. 2012), (Zeinab et al. 2014), (Lung et al. 2019). But that studies never investigated the effect of diameter, when Zr-7Cu-5Sn, a new biomaterial developed by (Kim et al. 2019), is used.

With the presented work, it is intended to analysis the stress and strain level developed on a new zirconium (Zr-7Cu-5Sn) dental implant using two different internal diameters. The results allow to conclude about the mechanical stimuli effect when the internal diameter was changed in the zirconium material.

2 MATERIALS AND METHODS

To quantify the level of mechanical stimuli transferred between the implant and neighbouring bone, an average value of the von Mises stresses (σ_{bi} and σ_{tj}, stress within bone and in adjacent triangular implant thread) were measured along a path. ε_{bi} and ε_{tj} were also measured along a path, at the same points defined, and it represents the strain in bone and in adjacent triangular implant thread, respectively.

Figure 1 shows the measured points to obtain the level of stress and strain. von Mises stresses measured allowed the stress transfer parameter (STPs) calculation, as proposed by (Haase and Rouhi, 2013), according to the Equation 1.

$$\text{STPs} = \frac{\sigma_b}{\sigma_t} + \sum\nolimits_{i=j=2}^{i=j=4} \frac{\sigma_{bi}}{\sigma_{tj}} \tag{1}$$

Total mechanical strains measured allowed the strain energy density *parameter (SEDTPs)* calculation, also proposed by (Haase and Rouhi, 2013), according to the Equation 2.

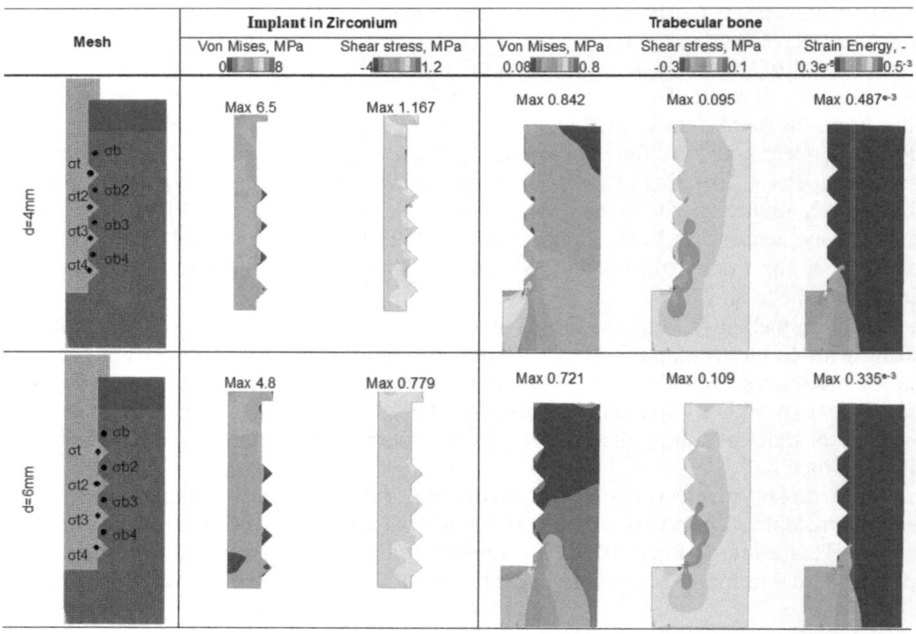

Figure 1. Numerical results in triangular dental implant and trabecular bone (d=4 and 6mm, Zr-7Cu-5Sn).

$$\text{SEDTPs} = \frac{\sigma_b \varepsilon_b}{\sigma_t \varepsilon_t} + \Sigma_{i=j=2}^{i=j=4} \frac{\sigma_{bi} \varepsilon_{bi}}{\sigma_{tj} \varepsilon_{tj}} \quad (2)$$

To obtain the level of stress and strain, two-dimensional axisymmetric computational models of a bone-implant were constructed using the finite element method with ANSYS®. The contact between all different model parts (cortical, trabecular and implant) were assumed to be perfectly bonded. All materials were assumed linear elastic, homogeneous, and isotropic. The corresponding elastic properties such as Young's modulus (*E*) and Poisson ratio (υ) were determined from the literature (Kim et al. 2019), (Haase and Rouhi, 2013). The model consists of a homogeneous cortical (*E*=20GPa) and a trabecular bone (*E*=1GPa), both with a Poisson ratio of 0.35 (Haase and Rouhi, 2013). Implants were modelled as a zirconium alloy (Zr-7Cu-5Sn) with E=14.4GPa (Kim et al. 2019) and Poisson ratio 0.30. The present work compares two different internal diameters equal to 4 and 6mm, with a length 11mm and a constant number of 4 threads.

3 RESULTS AND DISCUSSION

Figure 1 shows the obtained von Mises stress, shear stress and strain energy, presented in the implant and trabecular bone in all models in study.

Analysing the von Mises stress pattern, the implant with lower diameter has a higher peak of stress. The threads have the lowest stresses for both diameters. So, apparently, the threaded part allows a uniform stress distribution. The trabecular bone for a diameter of 4mm has higher stresses, comparing with the trabecular bone for a diameter of 6mm. For both diameters, the peak of stress occurs at the bottom right corner of the threaded zone in trabecular bone.

The strain energy density around the implant tip is reduced with a higher diameter. The lower diameter exhibited higher strain energy density than higher diameter. For both diameters, bellow the implant, the strain pattern shows higher values, comparing with the threaded zone around the implant.

The shear stress level represents the interaction effect in the implant interface. The maximum shear stress for both models is concentrated between the distal thread implant to the bone interface. Shear stress is found at the bone interface and with high effect localized proximally. Shear stress presents a high risk for failure at the interface.

Figure 2 represents the calculation of the stress transfer parameter and strain energy density transfer parameter according to the different diameter implant size.

As conclusion, decreasing the internal diameter of the implant results in a 2.3% increase of the total stress transferred between the implant and neighbouring bone. The load transferred to bone increases with the smaller implant diameter. Implants with a reduced diameter result in less stress shielding, preventing the bone loss. Despite the higher stresses for a diameter of 4mm, it has a higher STPs ratio, that means, quantitatively, that a lower diameter increases the load transferred to bone. These results could be compared with the studies (Haase and Rouhi, 2013) and

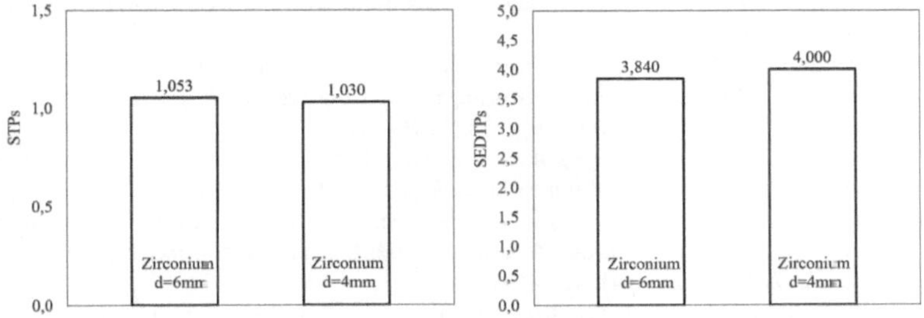

Figure 2. Stress transfer parameter (STPs) and strain energy density transfer parameter (SEDTPs).

(Lung et al. 2019), that achieved the same findings using titanium, which allows to conclude that zirconium has the same response, comparing with titanium, in terms of stress distribution, decreasing the implant diameter. The investigation (Saluja et al. 2012) refers that a wider diameter increases the contact surface between the implant and the bone, and consequently it allows a lower stress pattern surrounding the implant, agreement with the presented results.

The study (Eazhil et al. 2016) mentions that a wider diameter gives more stability and fracture resistance, which allows to conclude that the implant diameter should not be overly low. Authors (Zeinab et al. 2014) concluded that the optimal diameter is 3.5mm. The SEDTPs ratio is higher for a diameter of 4mm, comparing with a diameter of 6mm, which tells that a lower diameter is favourable for the distribution of that mechanical stimulus. High concentration of the strain energy density in the implant may cause damage and sub-sequent bone absorption around the implant. So, a higher distribution of strain energy density increases bone mass and strength, and a lower distribution of the same stimulus causes bone loss and reduces bone strength (Korabi et al. 2017). The study by (Haase and Rouhi, 2013) achieved the same results, in terms of strain energy density distribution. They show that increasing a titanium dental implant diameter, the SEDTPs value decrease. In terms of strain values in bone, the investigation produced by (Ding et al. 2009) reached the same results. They mention that a wider diameter promotes lower strain values in bone, around the implant.

4 CONCLUSIONS

A smaller diameter is favourable for stress and strain energy density distribution, and it allows to conclude that its use prevents the bone loss, not affecting bone mass and strength. However, a wider diameter gives more stability and fracture resistance. Due to this, it is recommended to use an optimal diameter value. If it was too low, it will have a good performance in preventing bone loss, but it will give a weak stability and a low fracture resistance. If it was too high, it will have a strong stability and a high fracture resistance, but it will cause bone loss, decreasing the bone mass and strength. Comparing with other studies, and looking for the behaviour of both diameters, considering the stimuli referred, we recommend using an optimal diameter of 4mm for this material, finding the perfect balance in the biomechanical performance of the implant, considering the two important mechanical stimuli. A good biomechanical performance is essential to get a desirable osteointegration. This biomaterial, Zr-7Cu-5Sn, is a metallic zirconium alloy that has a low elastic modulus, comparing with other common biomaterials. Due to its elastic properties, this biomaterial favours a better distribution of mechanical stimuli, that takes lots of advantages in a biomechanical behaviour of a bone-implant construct. The use of that new biomaterial, Zr-7Cu-5Sn, in implants with a diameter of 4mm, and with that triangular shape, allows a great biomechanical performance, giving a balanced behaviour between bone loss and stability or fracture resistance.

REFERENCES

Ding X., Zhu X.H., Liao S.H., Zhang X.H., Chen H. 2009. Implant-bone interface stress distribution in immediately loaded implants of different diameters: a three-dimensional finite element analysis. *Journal of Prosthodontics*, 18(5): 393–402.

Eazhil R., Swaminathan S.V., Gunaseelan M., Kannan G.V., Alagesan C. 2016. Impact of implant diameter and length on stress distribution in osseointegrated implants: A 3D FEA study. *J. International Society of Preventive and Community Dentistry*, 6(6): 590–596.

Haase, K., Rouhi, G. 2013. Prediction of stress shielding around an orthopedic screw: Using stress and strain energy density as mechanical stimuli. *Computers in Biology and Medicine*, 43(11): 1748–1757.

Herekar, M.G, Patil, V.N., Mulani, S.S., Sethi, M., Padhye, O. 2014. The influence of thread geometry on biomechanical load transfer to bone: A finite element analysis comparing two implant thread designs. *Dental Research Journal*, 11(4): 489–494.

Korabi, R., Yona, K.S., Rittel, D. 2017. On stress/strain shielding and the material stiffness paradigm for dental implants. *Clinical Implant Dentistry and Related Research*, 19(5): 935–943.

Kim, M., An, S., Huh, C., Kim, C. 2019. Development of Zirconium-Based Alloys with Low Elastic Modulus for Dental Implant Materials. *Applied Sciences*, 9(24): 5281.
Lung, H., Hsu, J.T., Wu, A.Y.J., Huang, H.L. 2019. Biomechanical Effects of Diameters of Implant Body and Implant Platform in Bone Strain around an Immediately Loaded Dental Implant with Platform Switching Concept. *Applied Sciences*, 9(10): 1998.
Saluja B., Alam M., Ravindranath T., Mubeen A., Adya N., Bhardwaj J., Dhiraj A. 2012. Effect of length and diameter on stress distribution pattern of INDIDENT dental implants by finite element analysis. *Journal of Dental Implants*, 2(1): 19–25.
Uhthoff, H.K., Jaworski, Z.F.G. 1978. Bone loss in response to long-term immobilisation. *The Journal of Bone and Joint Surgery*, 60-B(3): 420–429.
Zeinab, A., Reza, T., Kambiz, G.O. 2014. Optimum Selection of the Dental Implants According to Length and Diameter Parameters by FE Method in the Anterior Position. *International Journal of Bioscience, Biochemistry and Bioinformatics*, 4(4): 265–269.

Advances and Current Trends in Biomechanics – Belinha et al. (Eds)
© 2022 Copyright the Author(s), ISBN: 978-1-032-10806-3

The diameter effect on titanium dental implant analysis using four different thread profile shapes

D. Martins, R. Couto, A.R. Carreiras & E.M.M. Fonseca
Mechanical Engineering Department, School of Engineering, Polytechnic of Porto (ISEP), Portugal

C. Andreucci & R.N. Jorge
Faculty of Engineering of the University of Porto (FEUP), Portugal

ABSTRACT: The main goal of this work is to study the influence on mechanical bone behaviour using different thread profiles shapes considered in titanium dental implants. The biomechanical behaviour of bone and thread implant profile play an important role on the study of stresses and strains at bone-implant interface. The success of dental surgeries depends on the mechanical stimulus (stress, strain, strain rate, strain energy, density...) between the bone and the implant. In this work, four different threads profiles shapes were considered (Model 1 - Plateau typeA, Model 2 - Plateau typeB, Model 3 - Triangular, Model 4 - Rectangular) and for each one two internal diameters equal to 4 and 6mm. The results about the chosen thread profile shape and the diameter give additional information about the transferred stresses between bone-implant to guarantee the stability of the implant and sustain loads during or after osseointegration.

1 INTRODUCTION

In dental medicine, osseointegration might be affected by many factors, implant, screw design, surface treatment, bone quality, surgical technique, post-operative care (Lee et al. 2010).

In particular, the implant design is responsible for offering the better bone surface contact, and that way, allows to determine the initial stability and the implant capability to withstand loading. The implant, screw diameter and thread configuration are design related factors that have been modified over time (Orsini et al. 2012). According to different investigations the thread profile design of an implant is one of the dominant problems and affect the developed stress levels into the bone resorption (Lee et al. 2010), (Haase et al. 2013). It was observed that decreasing stress leads to bone loss while increasing stress leads to bone formation. However, it was also demonstrated that bone resorption occurs under extreme stresses. Therefore, implant threads must be designed to maximize the transmission of favourable stresses and minimize the extreme unwanted stresses (Abuhussein et al. 2010).

There are many thread profiles shapes used in dental implants, and different researchers have reported different conclusions. In the present study, it was analysed four different implant geometries with two different internal diameters equal to 4 and 6 mm. The dental implant usually varies in a diameter between 3 to 7mm, dependent on surgical and prosthetic prerequisites (Velmurugan et al. 2017). This dimension is designed to gain maximum stability between bone-implant. From a biomechanical point of view of wider implants permits engagement of a maximal measure of bone, and stress distribution is, in theory improved (Velmurugan et al. 2017). The stress distribution varies with respect to implant design.

DOI: 10.1201/9781003217152-5

Many publications present different numerical simulations used to test different geometries combined with different materials. Nevertheless, in this paper, it was only analysed a titanium alloy for the implant material. Dental implants can be fabricated in titanium alloys due to its good mechanical properties, low density, and good bone-contact biocompatibility. This type of implants demonstrates high success rates, and it is very unlikely that they will be associated with complications or failure (Nicholson, 2020). However, the number of the involved variables are considerable, material properties, dimensions, analysis type, applied boundary conditions, etc. All these findings have great significance in the biomechanical behaviour of the transferred loads between bone-implant. Nevertheless, more information is obtained about the better performance, dependent on the chosen thread profile shape. The success of dental implants and screws depends on the way stresses are transferred to the surrounding bone-implant (Herekar et al. 2014), (Geng et al. 2001).

In this work, the finite element method was the selected technique to obtain the solution in all different the models. This tool has been widely used in the literature to evaluate the design and function of dental implants besides, it has been proven to be a good theoretical method for analysing dental implant systems (Holmgren et al. 1998). Our work gives additional information about the developed stresses on a titanium dental implant using lesser expensive computational models (two-dimensional axisymmetric planes) in all four thread profiles shapes with two different internal diameters.

2 MATERIALS AND METHODS

Two-dimensional axisymmetric computational models of a bone-implant were constructed using the finite element method with ANSYS®. The numerical model was assumed as a state of ideal osseointegration, where the implant and the cortical and trabecular bone were assumed to be perfectly bonded. All materials were assumed linear elastic, homogeneous, and isotropic. The corresponding elastic properties such as Young's modulus (E) and Poisson ratio (υ) were determined from the literature (Haase et al. 2013). The model consists of a homogeneous cortical bone (E=20GPa) and a trabecular bone (E=1GPa) with a Poisson ratio of 0.35 (Haase et al. 2013). Implants were modelled as a titanium alloy (Ti–6Al–4V) with E=105GPa (Orsini et al. 2012) and Poisson ratio of 0.35. The analysis was based in two different implant inner diameters, 4mm and 6mm, each one in four different threads profiles shapes (Model 1 - Plateau typeA, Model 2 - Plateau typeB, Model 3 - Triangular, Model 4 - Rectangular).

Figure 1 shows all two-dimensional computational models in study and the respectively used mesh with elements length equal to 0.25mm. The bottom edge of the model on Y-axis was assumed to be fixed. An average masticatory axial force of 80N was determined from the literature and included in the screw head (Haase et al. 2013).

3 RESULTS AND DISCUSSION

Figure 2 shows the von Mises stress in all models, presenting the dental implant and the trabecular bone, respectively. Close to each model is presented the maximum value, according to the presented range stress level.

Figure 3 shows the von Mises stress level throughout to the implant length, representing the obtained values near of the cortical and trabecular bone (from 1 to 6).

The comparison between all 4 models with different thread profiles shapes and 2 different internal diameters on titanium dental implant revealed that the lower stress distribution in trabecular bone appears in Model 4 (d=6mm) and the higher value occurs in Model 3 (d=4mm). For each internal implant diameter, Model 4 (rectangular profile) revealed the lower stress distribution.

Models 2 and 3 and lesser diameter allow great level of stresses through the trabecular bone. Beyond the thread profiles, the dental implant diameter also induces different level of

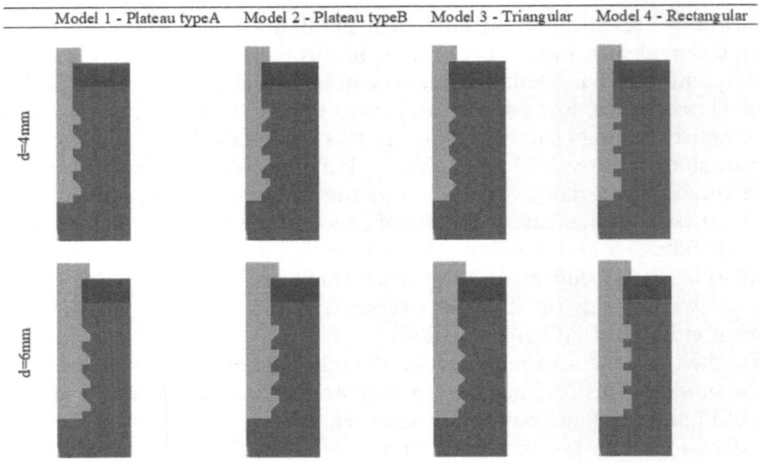

Figure 1. Meshes in all models (d=4 and 6mm, Ti). In all meshes the dental implant is represented in blue colour, the cortical bone is purple and the trabecular bone in red colour.

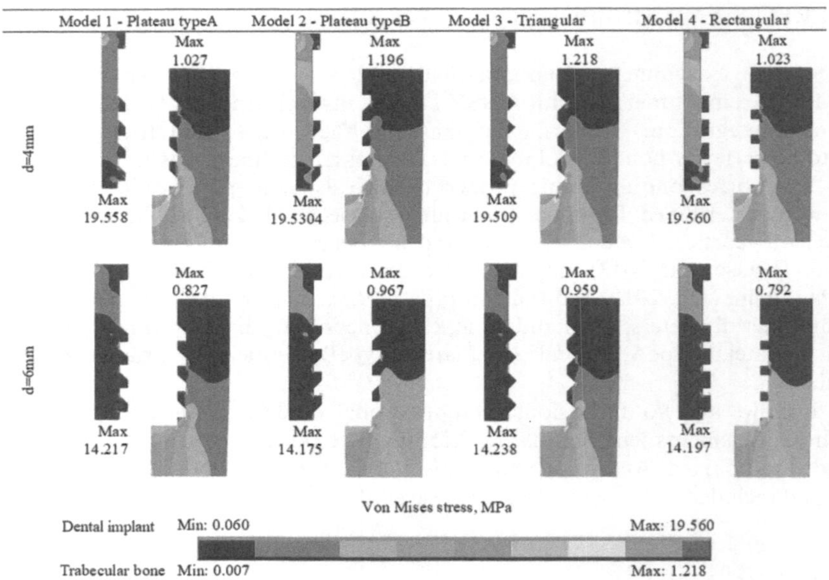

Figure 2. Meshes and von Mises stresses in all models (d=4 and 6mm, Ti).

stress throughout the bone, in general with higher diameter the stress decreases into the bone, as mentioned in (Velmurugan et al. 2017).

The level of stress dissipates along the implant length but culminates in the most distal implant thread (Haase et al. 2013) and pass to the adjacent trabecular bone. The same conclusion was observed in our models, where Models 2 and 3 were the most critical with visible high stress concentration at distal near of the thread profile corner. The rectangular or square dental profile (Model 4) allows obtain less stresses and shows better performance compared with the other profiles in trabecular bone, as demonstrated in (Karaman et al. 2018).

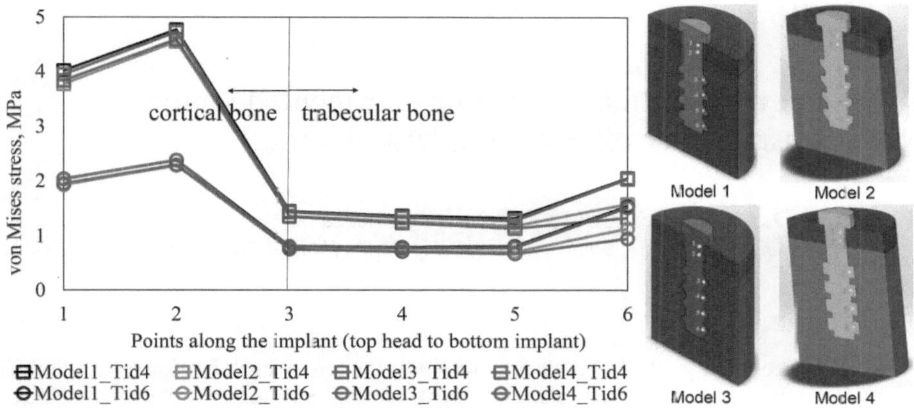

Figure 3. Von Mises stress of each model (d=4 and 6mm, Ti) in 6 points throughout to the implant length.

4 CONCLUSIONS

It is well accepted that bone loss can be reduced by a greater transfer of mechanical stimuli from an implant. Although the specific mechanical stimuli to bone formation are unknown, bone remodelling theorists propose both stress and strain energy density as contributing. In our study, the numerical method was used to simulate the various implant profile shapes and different diameters. Through some studies, including our work, using the finite element method it was demonstrated that increase the implant diameter lets to a decrease of the stress in the cortical bone. And this allowed to better understand how load is transferred from the implant to the adjacent bone.

REFERENCES

Abuhussein, H., Pagni, G., Rebaudi, A., Wang, H.L. 2010. The effect of thread pattern upon implant osseointegration. *Clinical Oral Implants Research*, 21(2): 129–136.

Geng, J.P, Keson, B.C., Liu, G.R. 2001. Application of finite element analysis in implant dentistry: A review of the literature. *The Journal of Prosthetic Dentistry*, 85(6): 585–598.

Haase, K., Rouhi, G. 2013. Prediction of stress shielding around an orthopedic screw: Using stress and strain energy density as mechanical stimuli. *Computers in Biology and Medicine*, 43(11): 1748–1757.

Herekar, M.G, Patil, V.N., Mulani, S.S., Sethi, M., Padhye, O. 2014. The influence of thread geometry on biomechanical load transfer to bone: A finite element analysis comparing two implant thread designs. *Dental Research Journal*, 11(4): 489–494.

Holmgren, E.P., Seckinger, R.J., Kilgren, L.M., Mante, F. 1998. Evaluating parameters of osseointegrated dental implants using finite element analysis-a two-dimensional comparative study examining the effects of implant diameter, implant shape, and load direction. *Journal of Oral Implantology*, 24(2): 80–88.

Karaman, H., Yuce, C., Karpat, F., Dhanasekaran, L., Khandaker, M. 2018. Structural Analysis of Dental Implants with Various Micro Grove Profiles. *International Research Journal of Advanced Engineering and Science*, 3(1): 101–104.

Lee, C.C., Lin, S.C., Kang, M.J., Wu, S.W., Fu, P.Y. 2010. Effects of implant threads on the contact area and stress distribution of marginal bone. *Journal of Dental Sciences*, 5(3): 156–165.

Nicholson, J.W. 2020. Titanium Alloys for Dental Implants: A Review. *Prosthesis*, 2(2). 100–116.

Orsini, E., Giavaresi, G., Trirè, A., Ottani, V., Salgarello, S. 2012. Dental Implant Thread Pitch and Its Influence on the Osseointegration Process: An In Vivo Comparison Study. *The International Journal of Oral & Maxillofacial Implants*, 27(2): 383–92.

Velmurugan, D., Santha, A.M., Sarate, S.G. 2017. Dental implant materials, implant design, and role of FEA- a brief review. *Journal of Evolution of Medical and Dental Sciences*, 6(44): 3487–3492.

Advances and Current Trends in Biomechanics – Belinha et al. (Eds)

Stress level developed in cemented or uncemented femoral component using three different implant materials

A.R. Carreiras, D. Martins, R. Couto & E.M.M. Fonseca
Mechanical Engineering Department, School of Engineering, Polytechnic of Porto (ISEP), Portugal

ABSTRACT: The main goal of this work is to study the stress level developed in the femoral component (cemented or uncemented) using different material and implant diameters. The stress analysis is obtained along bone-cement or bone-implant interface. Two different models were considered (Model 1 – cemented femoral component, Model 2 – uncemented femoral component) and for each one three implant diameters equal to 6, 10 and 14 mm and three different materials (cobalt-chromium, titanium, iso-elastic titanium alloy). The long-term stability of femoral components cemented or uncemented critically depend on bone, cement and implant elements, and their correspondent interfaces. The interface deterioration as the most immediate cause for loosening failure of implants. The results about the developed stress through the interface length in these models play an important role to analysis the implant stability.

1 INTRODUCTION

The implant material and dimension are one of the dominant problems and affect the stress level developed into the bone resorption (Haase and Rouhi, 2013). The interface deterioration is the most immediate cause for loosening failure of implants (Bousnane et al., 2018). Several researchers have shown that numerical and experimental simulations are used to test different femoral components combined with different shapes and materials (Ridzwan et al., 2006), (Fernandes et al., 2016). They concluded that the implant long-term stability is dependent on the bond between bone-cement (Bousnane et al., 2018). In other investigation, it is referred that increasing the cement thickness predicts to significantly reduce stress in the cement mantle of a femoral implant, (Lee et al., 1993).

Bone cement has been widely used in orthopaedic situations. As reported by the authors (Postawa and Szarek, 2007) the interface between bone and cement is one of the weak-link zones, because cement is unable to bond to living bone and produces shrinkage during polymerisation and influence prothesis loosening. Minimizing gaps between the bone and the implant is important for long-term fixation (Moritaa, et al., 1998), (Postawa and Szarek, 2007).

Lee, C (2005) describes the bone cement mechanical properties addressing the relative importance in short and long-term properties. Bone cement has mechanical properties that influence the function of a replacement hip joint. In compression it is strong, and it is weak in tension, (Lee, C. 2005). Bone cement also has a low bending modulus of elasticity, (Lee, C. 2005).

The objective of this work is to analyse the stress level developed between the interface cement-bone and implant-bone, from proximal end to distal end length and to enlarge the study from the authors of this work (Carreiras et al.,2020). The used methodology allows to obtain results from von Mises and shear stress, with the finite element method with lesser expensive computational models (two dimensional axisymmetric elements).

DOI: 10.1201/9781003217152-6

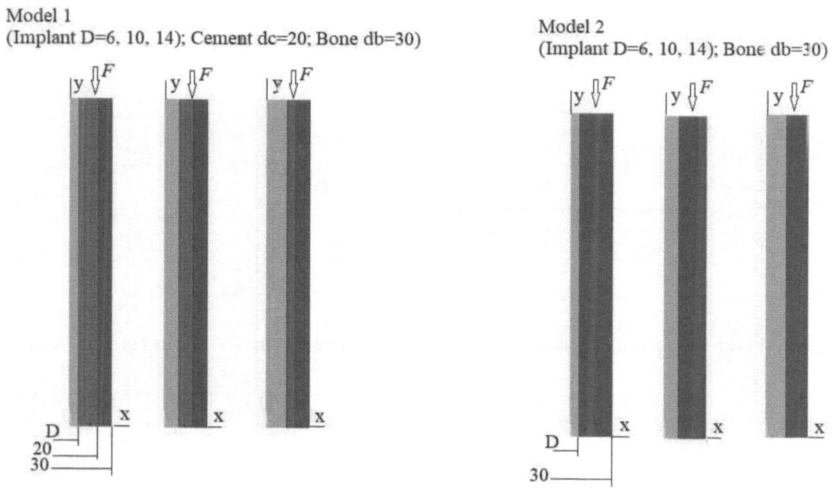

Figure 1. Cemented (Model 1) and uncemented femoral component (Model 2).

2 MATERIALS AND METHODS

The finite element method was used to simulate an implant-cement-bone (Model 1) and implant-bone (Model 2) perfectly bonded. Two-dimensional and axisymmetric models were constructed using ANSYS®, Figure 1. The numerical model is structural and linear, using axisymmetric plane finite element with 8 nodes, PLANE183. An average compression applied load F=700N was determined from the literature and included in the model (Ridzwan et al., 2006).

The outside diameter of the bone is equal to 30mm and the solid implant diameter is considered between 6, 10 and 14mm. In cemented femoral component (Model 1) the cement diameter is equal to 20mm. According to the dimensions, three thicknesses of cement were considered 7, 5 and 3mm. In their published research, (Kwak et al., 1979) recommended a cement layer between 3 and 4mm that will place least stress on the implant.

The elastic properties such as Young's modulus (E) and Poisson ratio (υ) were determined from the literature, (Lee et al., 1993). The model consists of a homogeneous cortical bone (E_b=20GPa), cement bone (E_c=2GPa), titanium alloy (E_i=115GPa), iso-elastic titanium alloy (E_i=40GPa) and cobalt-chromium (E_i=200GPa) with Poisson of 0.3.

3 RESULTS AND DISCUSSION

The obtained numerical results, using finite element analysis, allow to determine the von Mises and shear stress developed in all models. Figure 2 and Figure 3 show the stress distribution between cement-bone and implant-bone interface.

A comparison between all models, it was demonstrated that the higher von Mises and shear stress are near the proximal end and decreases to distal end. In particular, von Mises stress at the cement-bone or implant-bone interface have a higher concentration near at the proximal end and decrease to distal end.

In Model 1, smaller implant diameters produced higher stress and in Model 2 higher stiff materials produce higher stress level. Figure 4 represents the level of von-Mises in Model 1 (Titanium_D6) in comparison with the other models.

Iso-elastic titanium and higher implant diameter are the combination that produce lower and constant stress through all length in Model 2, Figure 3. The cement in the femoral component allows to decrease the stress level. As a result, flexible implant produces less stress level, therefore

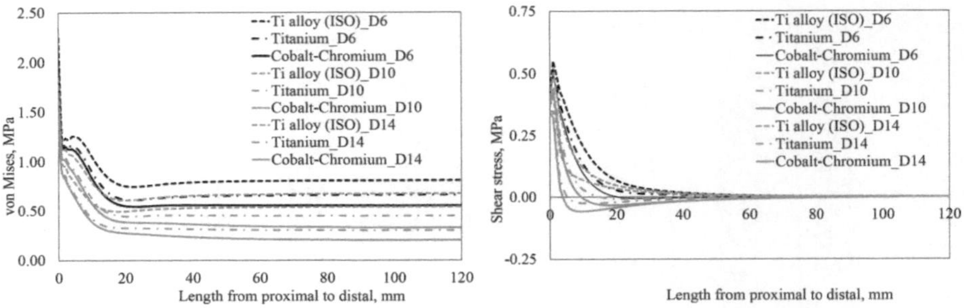

Figure 2. Cemented component (Model 1), von Mises and shear stress (MPa) in cement-bone interface.

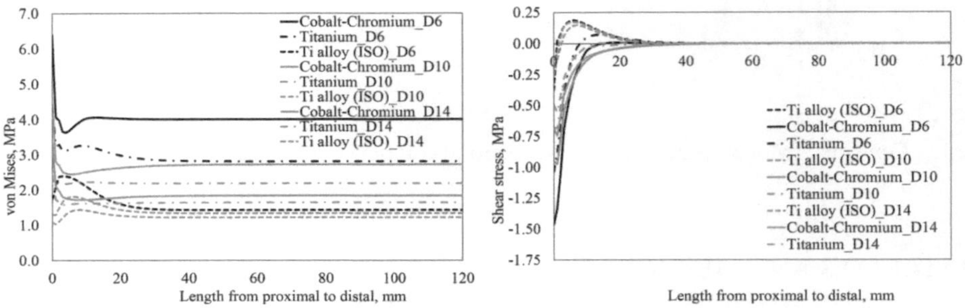

Figure 3. Uncemented component (Model 2), von Mises and shear stress (MPa) in implant-bone interface.

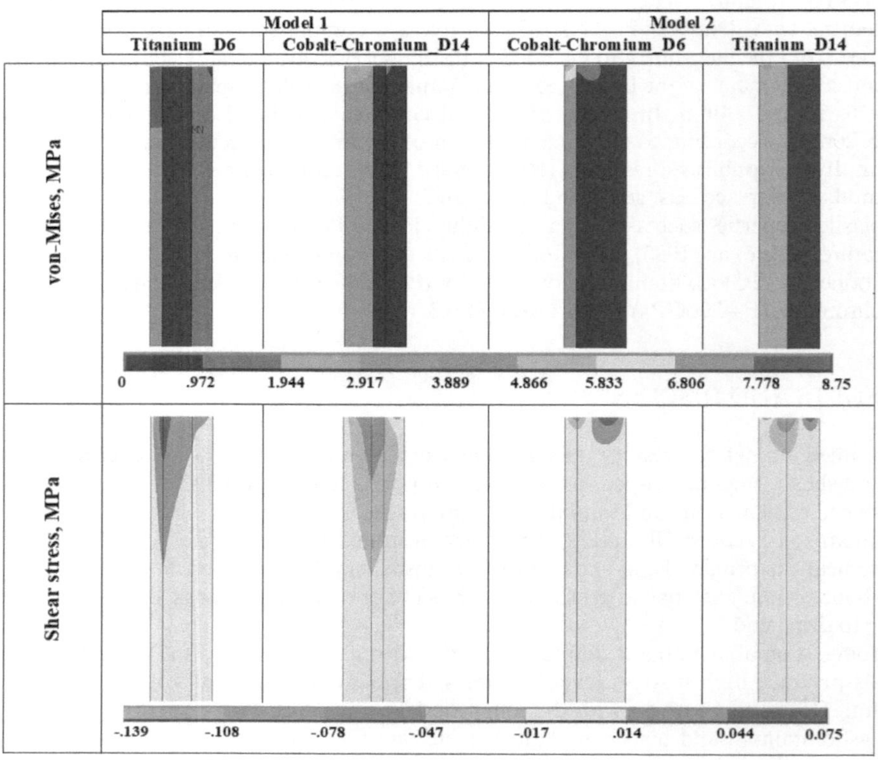

Figure 4. Stress level developed in some models.

less stress shielding in the bone, as reported by (Carreiras et al., 2020). One problem with flexible uncemented implants is the increased of relative interface motions. However, it is crucial to combine the effect with interface shear stress.

About shear stress, the results could represent the interaction effect in the implant material interface. The maximum shear stress for both models is concentrated between the implant to the bone interface, decreased gradually towards the lateral bone side. Shear stress at the cement-bone or implant-bone interface is found and with high effect localized proximally as reported by other authors (Bousnane et al., 2018). Shear stress presents a high risk for failure at the interface. Figure 4 shows this pronounced effect in Model 1 (Titanium_D6) as the most critical.

Model 2 represents localized high shear stress at proximal end where the influence of stiffer material and lesser implant diameter represents the most critical peak, Figure 3. Iso-elastic titanium and the combination with smaller implant diameter allows lesser shear stress.

In Model 1, due the cement effect, the peak of shear stress is very similar between materials and used implant diameters, and their effect is less when compared with Model 2. According to the research from (Lee et al., 1993), increasing the cement thickness, tensile stress is reduced by fifty percent and shear stress at least twelve percent. This type of observations was verified in our models when the same implant (diameter and material) was used in Model 1 and 2. The level of stress is more reduced in Model 1.

4 CONCLUSIONS

The maximum stresses occur at or very near the implant cement interface, indicating the probability of debonding, as an initialling mechanism in the implant loosening process. The level of shear stresses is in the same range as the von Mises stress, whereby it is important to consider due the probable failure models. Flexible implants reduce stress shielding and bone resorption, however they increase interface stresses, as the previous obtained results. In all models, von Mises stresses were severely concentrated around the implant, being the concentration in the Model 1 was higher than in the Model 2, suggesting bone absorption around the implant.

REFERENCES

Bousnane, T., Benbarek, S., Sahli, A., Serier, B., Bouiadjra, B.A.B. 2018. Damage of the Bone-Cement Interface in Finite Element Analyses of Cemented Orthopaedic Implants. *Periodica Polytechnica Mechanical Engineering*, 62(2): 173–178.

Carreiras, A.R., Fonseca, E.M.M., Martins, D., Couto, R. 2020. The axisymmetric computational study of a femoral component to analysis the effect of titanium alloy and diameter variation. *Journal of Computational Applied Mechanics*, 51(2): 403–410.

Fernandes, M.G., Alves, J.L., Fonseca, E.M.M. 2016. Diaphyseal femoral fracture: 3D biomodel and intramedullary nail created by additive manufacturing. *International Journal of Materials Engineering Innovation*, 7(2): 130–142.

Haase, K., Rouhi, G. 2013. Prediction of stress shielding around an orthopedic screw: Using stress and strain energy density as mechanical stimuli. *Computers in Biology and Medicine*, 43(11): 1748–1757.

Lee, C. 2005. The Mechanical Properties of PMMA Bone Cement. In: *The Well-Cemented Total Hip Arthroplasty*. Springer, Berlin, Heidelberg.

Lee, I.Y., Skinner, H.B., Keyak, J.H. 1993. Effects of Variation of Cement Thickness on Bone and Cement Stress at the Tip of a Femoral Implant. *The Iowa Orthopaedic Journal*, 13: 155–159.

Moritaa, S., Furuyaa, K., Ishiharab, K., Nakabayashib, N. 1998. Performance of adhesive bone cement containing hydroxyapatite particles, *Biomaterials*, 19(17): 1601–1606.

Postawa, P., Szarek, A. 2007. Analysis of changes in bone cement damping factor and its effect on bone load. *Journal of Achievements in Materials and Manufacturing Engineering*, 23(1): 35–38.

Ridzwan, M.I.Z., Shuib, S., Hassan, A.Y., Shokri, A.A. 2006. Effects of Increasing Load Transferred in Femur to the Bone-Implant Interface. *Journal of Applied Sciences*, 6(1): 183–189.

Kwak, B.M., Lim, O.K., Kim, Y.Y., Rim, K. 1979. An Investigation of the Effect of Cement Thickness on an Implant by Finite Element Stress Analysis. *International Orthopaedics*, 2: 315–319.

Advances and Current Trends in Biomechanics – Belinha et al. (Eds)

Development of prosthetics for upper extremities using affordable materials

D. Vieira & A.G. Magalhães
Mechanical Engineering Department, School of Engineering, Polytechnic of Porto (ISEP), Portugal

ABSTRACT: This document briefly describes the research work and subsequent development of prostheses for the upper extremities – hands and wrists – using low-cost material, efficient manufacturing methods, allowing a high adaptability, and reducing production time.

To increase the anatomical realism, the development of an outer coating film was also considered. The importance of these specifications relies on the targeted audience – children and pre-adolescents – which also brings the necessity of an easy adaptation or replacement of the product (partial or total), in case of damage or growth, within a short execution time. Hence the need to carefully choose the right materials and the manufacturing methodology.

1 INTRODUCTION

A relatively large part of the population, at some point in their lives, is faced with the misfortune of losing a lower or upper limb of their bodies, with dramatic consequences from the point of view of their social integration.

The loss of limbs can occur in several ways and the causes can be dividing divided into three major groups: congenital diseases formed mostly in embryonic development; traumatic experiences caused by accidents and the need to amputate a limb due to a disease or infection (sepsis) (Novaes, Edgard. 2012). If the malformation is congenital, which is estimated to occur in 2% of cases, the use of prostheses is widely accepted as an effective solution to restore some normality (Leal, Nuno. 2011). However, we can point out some barriers to the adoption of this methodology: flexibility, cost and development and delivery time. It is estimated that the time between delivery of a prosthesis can vary between 1 and 22 months.

The high cost of a prosthesis and the need for constant adaptation to the child's natural growth is, perhaps, the main reason for the use of a prosthesis to take place at a more advanced stage. This makes it impossible for the child to learn several simple movements and abilities like drawing and writing. According to information published by the Portuguese Society of Physical Medicine and Rehabilitation (Afonso, Carla. et al. 2009), it appears that 53% of children with this type of disability rarely use prostheses, 32% use them constantly and the remaining 15% use prostheses just to practice recreational activities. In addition, analyzing the ages when they initiate the use of prostheses, it appears that 64% of children started between 10 and 16 years old, 28% between 5 and 10 years old and only a small percentage of 8% started to use prostheses with less than 5 years. This reality impairs the child's psychosomatic development and makes it imperative to develop easily accessible prostheses.

DOI: 10.1201/9781003217152-7

2 MATERIALS AND METHODS

As general requirements of this project, the following were considered relevant: easy adaptability, easy placement, and low cost of manufacturing. The need to guarantee a low final cost motivated the search for materials and the finding of constructive solutions compatible with an easily automated production.

In this sense, the authors opted for the use of PLA (polylactic acid) and TPU (thermoplastic polyurethane).

PLA is a thermoplastic that can be melted, and its shape can be changed without its mechanical properties being strongly altered. It is derived from reusable resources and decomposes slowly when exposed to the environment. It's not a Newtonian fluid (viscosity decreases with increasing pressure) and the mechanical properties are good compared to traditional polymers (Martins, Tomás Sousa. 2017).

TPU is a thermoplastic that is characterized as the bridge between elastic and plastic regime. It is an easily deformable material, with a wide range of applications, resistant to abrasion, flexible and food safe (Fiorio, Rudinei. 2007).

To produce the pieces, 3D printing was chosen as it allows for personalized production, while reducing development and manufacturing times.The first phase of project design, a dimensional analysis of a hand was carried out using scanner techniques, Figure 1 (A). After studying the adaptability, the first prototype, Figure 1 (B), was designed, defining the kinematics of the main movements, angles that each finger must perform and other additional movements. After that step, a three-dimensional prototype was created in SolidWorks and the conversion was made into a readable file (G.Code) by the 3D printer.

Finally, the possibility of developing an outer skin with an anatomical configuration more like that of a hand was studied in the sense of being able to serve as the outer covering of the prosthesis, Figure 2.

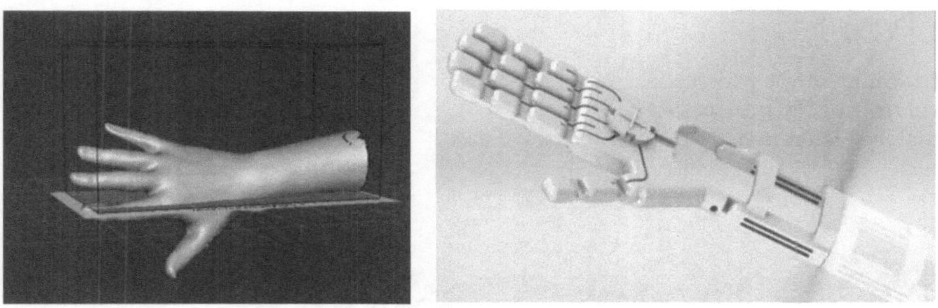

Figure 1. (A) Dimensional survey on a scanner. (B) First prototype developed.

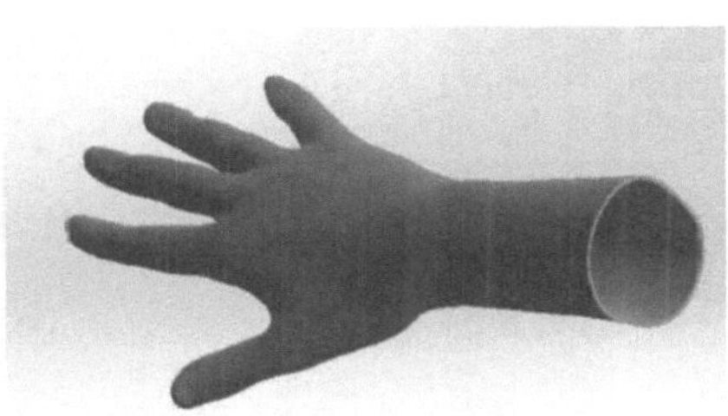

Figure 2. Exterior coating membrane to produce greater anatomical resemblance.

3 RESULTS AND DISCUSSION

The creation of prototypes proved to be fundamental for assessing the feasibility of a prosthesis being produced by a printing process and, simultaneously, allowing the correction of some geometric and functional details.

To date, three prototypes have been made to achieve the final result.

Each prototype consists of a set of rigid parts produced in PLA, connectors act as the ligaments between the phalanges that allow the flexing of the fingers; a set of cables for flexing the fingers; and possibly some metallic components such as screws and nuts for fixing and connecting parts.

The first prototype observed in Figure 1 (B) was equipped with all the components mentioned above. It also contained a test mechanism that allowed the flexion of the fingers necessary to hold the writing material.

After the printing of the parts and the assembly, some errors were detected that should be highlighted:

-The cable tensioning mechanism is not functional;
-Execution of the holes for the cables to activate the fingers;
-Dimensional tolerances;
-Non-efficient nylon cables;
-Dimensioning of the prototype;
-Lack of space to fit the stump.

The second prototype illustrated in Figure 3 presents a totally innovative design and geometry. To correct the errors seen in the first prototype, the tension mechanism, as well as excess material, were removed, allowing a better fit of the stump. In addition to these more notorious changes, the nylon cable was replaced by a steel cable and the TPU connectors were improved making them more malleable.

Although the corrections proved to be positive, there are still some factors to improve, among them:

-Dimensioning of the prosthesis;
-The comfort of the prosthesis;
-The material and diameter of the cable to be used.

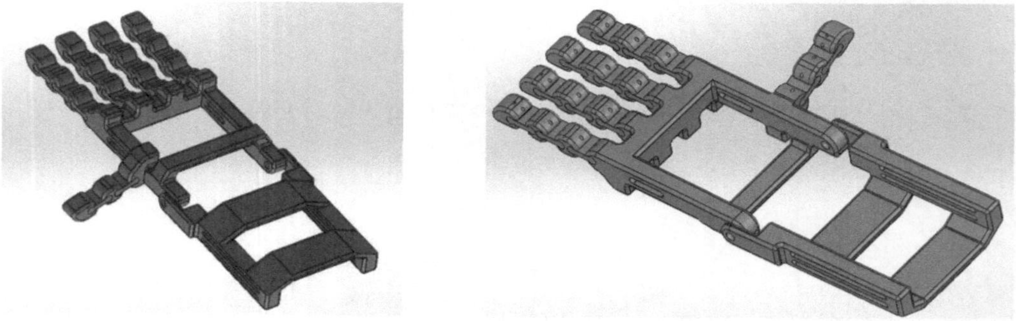

Figure 3. Representative assembly of the second prototype.

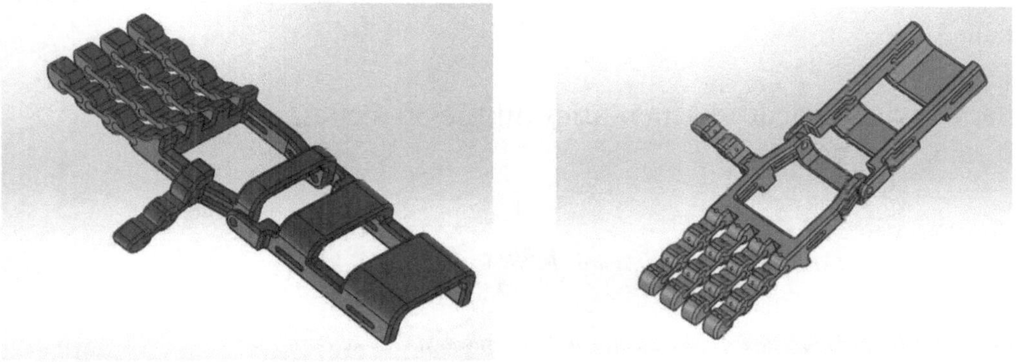

Figure 4. Representative assembly of the third prototype.

These factors were later improved with the production of the third and current prototype, Figure 4. In this, the dimensioning was corrected, comparing the final dimensions by hand of a 10-year-old child, however, it is not yet the final product.

4 CONCLUSIONS

The developed and promoted work that is still in progress, allows us to conclude that the printing technique is an excellent opportunity for the manufacture of prostheses for children, where the need for speed of execution and the fast response to constant functional adaptations due to growth can be fundamental.

Changes are still being made, always seeking to optimize the mechanism, comfort and design that best suits the intended age group. In this sense, it is a valuable contribution to social inclusion and the promotion of an active and healthy life.

REFERENCES

Afonso, Carla., Coelho, João.P., Cadete, Anal., Lopes, A., Vasconcelos, Ana., Batalha, Isabel. 2009. Amputações e Malformações Congénitas do Membro Superior na População Pediátrica - Revisão de 27 Anos. Revista da Sociedade Portuguesa de Medicina Física e de Reabilitação, 17(1).

Fiorio, Rudinei. 2007. Efeitos da Incorporação de Oligouretanos de Baixa Massa Molar nas Propriedades Térmicas, Mecânicas e Reológicas do Poliuretano Termoplástico. Dissertação para a obtenção do título de Mestre em Engenharia.

Leal,Nuno. 2011. Desenvolvimento do processo de fabrico de próteses humanas em silicone para substituição de órgãos em tecidos moles. Relatório do Projeto Final / Dissertação do MIEM, FEUP.

Martins, Tomás Sousa. 2017. Influência dos parâmetros de fabrico nas propriedades mecânicas de peças obtidas por impressão 3D com um único material. / Dissertação para obtenção do Grau de Mestre em Engenharia Mecânica.

Novaes, Edgard. 2012. Deformidades congénitas dos membros superiores. Parte I: falhas de formação. São Paulo. Rev. Brasileira de Ortopedia, 47(5).

Advances and Current Trends in Biomechanics – Belinha et al. (Eds)
© 2022 Copyright the Author(s), ISBN: 978-1-032-10806-3

Gender differences in kinematics on the 30-s chair stand test in knee osteoarthritis

Vitor Ferreira
School of Health Sciences, University of Aveiro (ESSUA), Portugal

Leandro Machado
CIF2D, LABIOMEP, Faculdade de Desporto da Universidade do Porto, Portugal

Paulo Roriz
Research Center in Sports Sciences, Health Sciences and Human Development, CIDESD, University Institute of Maia, ISMAI, Portugal
LABIOMEP, Portugal

ABSTRACT: The sit-to-stand movement is a fundamental activity of daily living and required for an upright posture. Understanding gender differences in the execution of this task may allow additional effective interventions. The purpose of this study was to investigate the biomechanical gender differences in kinematics on the 30-s chair stand test performance in patients with knee osteoarthritis. A 3D 12-camera motion analysis system was used to capture the body motion during the test. Thirty participants with medial knee osteoarthritis were recruited: 15 men and 15 women. A higher valgus angle was found in female participants with statistical significance ($p = 0.013$), but only in the less painful knee. Likewise, a closer approach of the knees during the movement was notorious. The predominance of upright position with the knee in valgus has implications in knee mechanics. which may allow the definition of different gender rehabilitation strategies.

1 INTRODUCTION

Osteoarthritis (OA) is the most common cause of musculoskeletal pain and disability in the knee (Silverwood et al., 2015). It affects daily activities such as climbing stairs, getting out of a chair and walking (Millor et al., 2013). Like walking, the sit-to-stand movement is essential for independent living in the elderly. Therefore, the biomechanical analysis of this task can contribute for the implementation of dedicated rehabilitation programs. The sit-to-stand performance is extensively used for assessing lower-extremity function and strength as well as balance control in older adults. It measures lower body strength and relates it to some of the most demanding activities of daily living (e.g., climbing stairs, getting out of a chair, bathing or rising from the horizontal to the upright position). The 30-seconds (30-s) chair stand test is the test that best reproduces this activity. To perform the test the participant is asked to stand up from a chair with arms crossed on the chest, and then return to the sitting position as many times and as safely as possible within a 30 seconds period of time (Sagawa et al., 2017).

Gender differences in the execution of different movements or activities are expected to occur. For example, in walking healthy females and those with knee OA, compared to the male counterparts, seem to exhibit significant greater knee abduction and hip adduction angles (Phinyomark et al., 2016). Furthermore, published literature on risk factors for the onset of knee OA reported a pooled odds ratio of 1.84 (95% confidence interval, 1.32–2.55)

DOI: 10.1201/9781003217152-8

for women compared with men (Buckwalter and Lappin, 2000). However, the reasons for higher risk of knee OA in women are not well understood. Biomechanical singularities in the execution of daily activities may explain the increased risk in the development of OA, partly because the pathology is induced by mechanical damage to soft tissues.

To the best of our knowledge there are no studies reporting gender-related differences in the sit-to-stand and stand-to-sit movement. Understanding them may contribute for more clinical and exercise effective interventions. The purpose of this study was to investigate the gender differences in the kinematics on the 30-s chair stand test performance in patients with knee osteoarthritis.

2 MATERIALS AND METHODS

Thirty participants with symptomatic medial knee osteoarthritis were recruited: 15 men (mean age 60.5 ± 7.4 years, mean body mass of 81.1 ± 13.6 kg, mean height of 1.68 ± 0.05 m, and mean body mass index of 28.8 ± 4.4 kg/m^2) and 15 women (mean age 61.2 ± 9.4 years, mean body mass of 73.1 ± 11.0 kg, mean height of 1.57 ± 0.10 m, and mean body mass index of 30.1 ± 6.0 kg/m^2). Clinical and radiological inclusion criteria followed those established by the American College of Rheumatology. Only patients with grade 2 and 3 on the Kellgren–Lawrence grading scale were included. The exclusion criteria were the presence of a joint prosthesis or a history of lower limb, neurological or orthopaedic disorders, other than the presence of knee OA, that could affect the patient gait or balance. The study was approved by hospital and university ethics committees.

A 3D 12-camera motion analysis system (Qualisys Oqus, Gothenburg, Sweden) was used to capture the body motion during the 30-s chair stand test. A total of 42 reflective markers were used to locate the pelvis and the lower limbs as described in a previous study (Ferreira et al., 2020). Two force plates (Bertec, Columbus, USA) embedded in the floor were used to capture the ground reaction forces for each foot. The motion and force plate data were synchronized and sampled at 200 Hz and 1000 Hz, respectively. The subjects sat on an armless chair with the knees at 90° and the upper limbs crossed in the chest. Starting from the seated position, the subjects stand completely up, until full extension of the hips and knees (Figure 1). The 30-s chair stand test was divided in two phases, the 'stand-up' and 'sit-down', for the movement of sit-to-stand and stand-to-sit, respectively. The mean of the first three and of the last three repetitions was used to further analysis as previously described (Ferreira et al., 2019).

The joint kinematics and kinetics were generated using V3D (Version 6.00.12, C-Motion, Rockville, USA). The raw data were smoothed using a fourth-order Butterworth lowpass digital filter of 6 Hz. The peak values of the knee joint angles were extracted from the waveforms for the three anatomical planes and expressed in degrees (°). The distance between knees centres was calculated by measuring it in the frontal plane. Likewise, the flexion angle was calculated at the minimum distance between the knees.

The T-test for independent samples was used to analyse the difference between gender (women and men), the knees (painful or non-painful) and other biomechanical parameters.

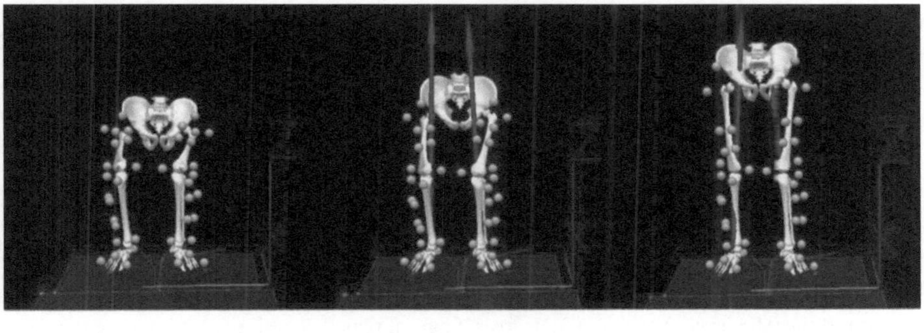

Figure 1. An animation of the sequence in the 30s chair stand test in Visual 3D (from left to right, sitting to standing position).

Tests were completed with SPSS 25.0 (IBM, Chicago, IL). The 95% confidence interval (α = 0.05) was adopted.

3 RESULTS

The descriptive statistics (mean and ±1 standard deviation) for kinematics are presented in Table 1. For the less painful knee, male participants have a peak of adduction of -3.8° (± 4.6), while female participants have a peak of adduction of -7.8° (± 5.3) in the frontal plane. This difference in the frontal plane, has statistical significance (p = 0.013). Similarly, in the horizontal plane, a higher value of external rotation was found in female participants (-15.3° ± 11.4) than in male participants (-7.8° ± 5.7).

Additionally, significant differences (p = 0.001) for gender were found for the minimum distance between the knees, either for stand to sit and sit to stand, with males exhibiting higher distances (0.27 ± 0.3 m) than females (0.22 ± 0.4 m). However, for knee flexion angle at minimum distance, the differences for gender were not significant.

4 DISCUSSION

The primary purpose of this study was to examine the effects of gender on 30s chair stand test kinematics in individuals with knee OA. There were gender-related differences in sit-to-stand movement, mainly in the frontal and horizontal plane. It was possible to observe in the female participants a more pronounced angle in adduction than in males, particularly, in the less painful knee. This most pronounced valgus angle between the participants, can partially explained by the physiological difference in the anatomy of the lower limbs between genders. Especially, the shape of the pelvis, which accentuates the angle in valgus for the knees (Shultz et al., 2008). This angle difference, especially in the static position, is also visible during the sit-to-stand movement and may contribute to the increased risk of developing knee OA in women. In the long-term the excessive compressive force may exacerbate the condition and, consequently, increase the difficulty to perform such a simple movement.

When we tried to verify, if the participants put their knees close or away from each other during the test, we found that the female participants brought their knees closer, both in the movement of sit-to-stand and stand-to-sit. There were no differences in the flexion angle at the minimum distance between the knees. This adaptation to the movement will probably be

Table 1. Kinematics of the knee in 30-s chair stand test. The results are for the mean of peak values.

Knee	Kinematics	Group	Mean (SD)	p value
Most painful knee	Sagittal plane	Men	9.1 (±5.0)	0.150
		Women	5.2 (±8.8)	
	Frontal plane	Men	-3.7 (±6.7)	0.120
		Women	-7.4 (±5.9)	
	Horizontal plane	Men	-8.0 (±8.1)	0.130
		Women	-12.8 (±8.8)	
Less painful knee	Sagittal plane	Men	7.3 (±6.4)	0.738
		Women	6.4 (±7.9)	
	Frontal plane	Men	-3.8 (±4.6)	0.013*
		Women	8.6 (±5.3)	
	Horizontal plane	Men	-7.8 (5.7)	0.029*
		Women	-15.3 (±11.4)	

* Significant at the 0.05 level (2-tailed). (°): degrees. Sagittal plane: negative values represent extension. Frontal plane: negative values represent adduction. Horizontal plane: negative values represent external rotation.

a consequence of the difference in the biomechanics in the execution of the movement. Partly, will also be related to the different anatomy between the genders, but also with some difference in the capacity of production of force by the muscular system. In a previous study (Bouchouras et al., 2015), was found that women with knee OA displayed significantly lower vastus lateralis coupled with a higher biceps femoris electromyographic activity, and higher agonist–antagonist co-contraction and co-activation than asymptomatic women. This decrease in the role of the vastus lateralis in movement may be a reason for the decrease in the ability to keep the knees together in a normal position. Understanding these changes has an important place in the knee OA rehabilitation programs. For example, it suggests the strengthening of the vast vastus lateralis. As such, in the rehabilitation process, these differences should be taken into account and different strategies adopted according to gender.

Some limitations to the current study should be considered. We did not collect electromyography data. Therefore, muscle activation patterns were not included in the analysis. This analysis would be an important for in deep analysis of muscle assessment during the task. A second limitation is the lack of a control group, including participants with the same characteristics, but without a previous diagnosis of knee OA. The participation of these elements would allow a more accurate comparison of the differences.

5 CONCLUSION

This study contributed for better understanding of the functional changes in patients with moderate OA, in particular the gender-related biomechanics differences during the 30s chair stand test. Apparently, female participants perform the movement of sit-to-stand and stand-to-sit with a more evident valgus on the knee, mainly in the less painful knee, which corresponds to a greater approximation of the knees during the performance of the movements. The knowledge of these differences will be important for the definition of different rehabilitation strategies depending on the gender.

REFERENCES

Bouchouras, G., Patsika, G., Hatzitaki, V., Kellis, E., 2015. Kinematics and knee muscle activation during sit-to-stand movement in women with knee osteoarthritis. Clin Biomech (Bristol, Avon) 30, 599–607. https://doi.org/10.1016/j.clinbiomech.2015.03.025

Buckwalter, J.A., Lappin, D.R., 2000. The disproportionate impact of chronic arthralgia and arthritis among women, in: Clinical Orthopaedics and Related Research. https://doi.org/10.1097/00003086-200003000-00018

Ferreira, V., Machado, L., Roriz, P., 2020. The effects on lower limbs kinematics of different lateral wedge insoles. J. Mech. Eng. Biomech. 4, 77–82. https://doi.org/10.24243/jmeb/4.4.233

Ferreira, V., Machado, L., Vilaça, A., Xará-Leite, F., Roriz, P., 2019. Biomechanics performance in 30-s chair stand test in patients with medial knee osteoarthritis. Int. J. Hum. Factors Ergon. 6, 319–330. https://doi.org/10.1504/IJHFE.2019.105359

Millor, N., Lecumberri, P., Gomez, M., Martinez-Ramirez, A., Izquierdo, M., 2013. An evaluation of the 30-s chair stand test in older adults: frailty detection based on kinematic parameters from a single inertial unit. J Neuroeng Rehabil 10, 86. https://doi.org/10.1186/1743-0003-10-86

Phinyomark, A., Osis, S.T., Hettinga, B.A., Kobsar, D., Ferber, R., 2016. Gender differences in gait kinematics for patients with knee osteoarthritis. BMC Musculoskelet. Disord. 17. https://doi.org/10.1186/s12891-016-1013-z

Sagawa, Y., Bonnefoy-Mazure, A., Armand, S., Lubbeke, A., Hoffmeyer, P., Suva, D., Turcot, K., 2017. Variable compensation during the sit-to-stand task among individuals with severe knee osteoarthritis. Ann Phys Rehabil Med 60, 312–318. https://doi.org/10.1016/j.rehab.2017.03.007

Shultz, S.J., Nguyen, A.D., Schmitz, R.J., 2008. Differences in lower extremity anatomical and postural characteristics in males and females between maturation groups. J. Orthop. Sports Phys. Ther. https://doi.org/10.2519/jospt.2008.2645

Silverwood, V., Blagojevic-Bucknall, M., Jinks, C., Jordan, J.L., Protheroe, J., Jordan, K.P., 2015. Current evidence on risk factors for knee osteoarthritis in older adults: a systematic review and meta-analysis. Osteoarthr. Cartil. 23, 507–515. https://doi.org/10.1016/j.joca.2014.11.019

Advances and Current Trends in Biomechanics – Belinha et al. (Eds)
© 2022 Copyright the Author(s), ISBN: 978-1-032-10806-3

Variability of spatiotemporal gait parameters in patients with knee osteoarthritis

Vitor Ferreira
School of Health Sciences, University of Aveiro (ESSUA), Portugal

Leandro Machado
CIF2D, LABIOMEP, Faculdade de Desporto da Universidade do Porto, Portugal

Paulo Roriz
Research Center in Sports Sciences, Health Sciences and Human Development, CIDESD, University Institute of Maia, ISMAI, Portugal
LABIOMEP, Portugal

ABSTRACT: Assessment of spatiotemporal gait parameters provides objective data for the study of gait. On the other hand, the study of gait variability may be more sensitive in quantifying changes in the motor control. The purpose of this study was to investigate the differences in variability of spatiotemporal gait parameters. Kinematic data were recorded using 3D motion cameras, from participants walking along a 10-meter walkway at a self-preferred walking speed. Two groups (patients with medial knee osteoarthritis; healthy young adults) of gender-matched participants were recruited. Mean values and coefficient of variation for spatiotemporal parameters were calculated. Several statistical differences were found in gait parameters between groups. However, when the coefficient of variability between groups was compared, no statistically significant differences were found, except for the left stance time. Although there were several significant differences in spatiotemporal gait parameters, the assessment of variability did not reflected changes in motor control on gait in patients with knee osteoarthritis.

1 INTRODUCTION

Gait is frequently used to quantify physical function, quality of life and health status. On the other hand, it is often cited as a major cause for falls. Several events are described during the walking cycle. After the heel strike of one member and before the toe-off of the contralateral member, there is a double support period that can be used to alter gait and correct deviations, leading to an improved control of gait (Williams and Martin, 2019). A large double support period increases stability and leads to a lower walking speed and a larger walking period.

Certainly, spatiotemporal gait parameters provide objective data for assessment of gait global performance. However, it may not be sufficient to find reliable predictors for falls (Mortaza et al., 2014). On the other hand, the variability of these parameters can be more adequate in quantifying changes in the motor control of gait for clinically relevant conditions. For instance, variability of stance time or swing time was significantly higher in fallers, compared to nonfallers. The results suggested that variability of temporal and spatial parameters of gait was the only functional component that differentiates fallers from nonfallers, and that should be used in clinical screening programs for assessing the risk of falling (König et al., 2014).

DOI: 10.1201/9781003217152-9

Studying gait variability can also contribute to quantify age-related changes in the motor control of gait. In general, high variability is related to major attention involvement, whereas low variability reflects automatic processes that require minimal attention (Dhawale et al., 2017). On the other hand, understanding variations in chronic musculoskeletal diseases can improve rehabilitation programs.

Osteoarthritis (OA) of the knee is a major cause of pain and locomotor disability worldwide. Moreover, resting and movement pain are two distinct characteristics of OA pain (Fu et al., 2018). Consequently, patients suffering from late- to end- stage OA become limited in their mobility by the presence of persistent pain on movement. The impact of these structural and sensory changes on the variability of movement during gait requires further research. Some studies report that pain with acute characteristics increases variability, allegedly because alternative motor solutions are explored to reduce pain, while chronic pain was associated with reduced motor variability, suggesting to reflect an attempt to avoid painful movements.

Understanding the effect of aging and cumulatively the effect of a chronic disease that limits walking can provide important information for clinicians. It is also important, to fine-tune rehabilitation strategies to improve independence and gait efficiency in older populations. Therefore, the purpose of this study was to investigate the differences in gait variability, during walking at a self-preferred walking speed, in patients with medial knee osteoarthritis, compared with healthy young adults.

2 MATERIALS AND METHODS

Two groups of gender-matched participants were recruited: 16 healthy young adults (mean age 22.1 ± 6.0 years, women 50%) and 16 medial knee OA patients (mean age 62.8 ± 7.2 years, women 50%) with a Kellgren and Lawrence grade scale of 2 or 3.

The inclusion criteria included participants without any physical disability, aged from 20 to 30 years old, for the group of healthy young adults, and aged from 50 to 75 years old for the group with knee OA. The participants were excluded based on any neurological or musculoskeletal disorders that could impair assessment of gait or balance.

The study was conducted in accordance with the Declaration of Helsinki and was approved by hospital and university ethics committees.

Kinematic data were recorded using twelve cameras (Qualisys Oqus, Qualisys, Gothenburg, Sweden), sampling at 200 Hz, from participants walking along a 10-meter walkway at a self-preferred walking speed. Ground reaction forces (GRF) data were collected at 1000 Hz. The Calibrated Anatomical System Technique (CAST) was employed to characterize body segments motion during the walking trials as described in a previous study (Ferreira et al., 2020).

Participants were given time to adapt to the procedures and allowed some time to practice until they felt comfortable walking at a normal speed. Three trials of each participant were recorded to eliminate variation in walking speed and to enhance the reliability and accuracy of the average. Data were analysed with the Qualisys Track Manager software (Qualisys, Gothenburg, Sweden) to identify, edit and process gap markers trajectories. Data were filtered using a 4th-order low-pass Butterworth filter with a 6 Hz cut-off frequency. Post-processing calculation of all gait parameters was conducted using Visual3D software (C-Motion, Rockville, USA). Gait events were found using the vertical GRF with a 20 N threshold. Several gait spatiotemporal parameters were analysed, and the coefficient of variation (CV) was calculated by the following equation: $CV=SD/M\times100$ (where SD is the standard deviation and M is the mean). The mean values of gait spatiotemporal parameters and CV variation were exported to SPSS 25.0 (IBM, Chicago, USA) and an independent T-Test was run on the data with a 95% confidence interval ($\alpha = 0.05$).

3 RESULTS

As shown in **Table 1**, several significant statistical differences were found ($p \leq 0.05$) in spatiotemporal gait parameters between groups. Despite innumerable statistical differences between

Table 1. Spatiotemporal and coefficient of variability gait parameters.

		Spatiotemporal parameters		Coefficient of variability	
Variable	Group	Mean (SD)	p value	Mean (%) (SD)	p value
Cycle time	Control	1.06 (±0.07)	0.714	2.29 (±3.68)	0.158
	OA	1.08 (±0.17)		5.79 (±8.80)	
Left stance time	Control	0.65 (±0.04)	**0.016***	2.49 (±2.33)	**0.045***
	OA	0.73 (±0.12)		5.27 (±4.70)	
Left step length	Control	0.70 (±0.05)	**0.009***	2.64 (±1.87)	0.382
	OA	0.63 (±0.10)		4.28 (±7.12)	
Left step time	Control	0.52 (±0.04)	**0.022***	6.81 (±12.74)	0.428
	OA	0.57 (±0.06)		4.15 (±3.70)	
Left stride length	Control	1.41 (±0.10)	**0.003***	1.83 (±1.38)	0.085
	OA	1.25 (±0.19)		8.39 (±14.15)	
Right stance time	Control	0.65 (±0.06)	0.095	7.55 (±11.31)	0.829
	OA	0.70 (±0.11)		8.84 (±20.63)	
Right step length	Control	0.71 (±0.05)	**0.011***	2.52 (±1.54)	0.317
	OA	0.65 (±0.07)		4.35 (±7.04)	
Right step time	Control	0.55 (±0.05)	0.343	5.47 (±7.97)	0.493
	OA	0.57 (±0.06)		3.98 (±3.15)	
Right stride length	Control	1.40 (±0.11)	**0.003***	2.99 (±4.99)	0.151
	OA	1.23 (±0.18)		8.44 (±13.69)	
Speed	Control	1.33 (±0.11	**0.000***	2.26 (±1.19)	0.815
	OA	1.14 (±0.16)		2.41 (±2.16)	
Stride length	Control	1.41 (±0.11)	**0.003***	2.62 (±3.58)	0.116
	OA	1.22 (±0.20)		6.87 (±9.68)	
Stride width	Control	0.12 (±0.02)	0.976	7.84 (±3.66)	0.609
	OA	0.12(±0.03)		8.87 (±7.10)	
Mean step length	Control	0.71 (±0.05)	**0.006***	1.66 (±1.13)	0.388
	OA	0.64 (±0.07)		2.20 (±2.15)	
Mean step time	Control	0.53 (±0.03)	**0.050***	1.35 (±0.71)	0.750
	OA	0.57 (±0.06)		1.46(±1.12)	

* Significant at the 0.05 level (2-tailed).

the spatiotemporal parameters that characterize the gait cycle, when the CV of these parameters was compared, no statistically significant differences were found ($p > 0.05$) between groups, except for the left stance time ($p = 0.045$). In the control group the CV of left stance time was 2.5 ± 2.3 % and in the OA group was 5.3 ± 4.7 %. Additionally, both groups exhibited higher variability in spatiotemporal parameters for the double limb support time. On average, participants of the control group (21.0 ± 11.5 %) exhibit slightly higher variability than OA participants (15.6 ± 11.7 %). In the CV of the right initial double limb support time, participants of the control group present 13.0 ± 15.3 % and participants of the OA group 19.0 ± 19.9 %. In the CV of the right terminal double limb support time, participants of the control group present 12.6 ± 18.5 % and participants of the OA group 22.0 ± 16.5 %.

4 DISCUSSION

The present study tested the hypothesis that gait variability would be reflected in the gait pattern of patients with knee OA. However, compared to a group of healthy young adults, the results did not confirm the initial hypothesis. The existing knowledge of gait variability in knee OA patients is limited. Only some studies examined the variability assessed by spatiotemporal measures in knee OA patients. The double support time was the variable reported with the most significant changes (Svoboda et al., 2017). One study assessed the temporal variability in patients with severe knee OA awaiting total knee replacement and observed a CV of 22.8 (± 2.2) % on the double support time. In our study, a similar trend was found (15.6 % ± 11.7). Nevertheless, in the previous study, patients were recruited in an advanced stage, while those from the present studied were in a moderate stage of the disease. On the other hand, the CV is more associated with changes in neurological conditions, and because OA is essentially an orthopaedic condition, changes in motor control may not be detectable. The present study found only a statistically significant difference in CV in the left stance time. To the best of our knowledge there are no studies that specifically compare the left with the right side in patients with knee OA. Only, the affected with the unaffected knee (Kiss et al., 2012). Our study included 9 patients with left knee OA and 7 with right knee OA. The significant difference found only on the left side can be explained due to specific characteristics of the participants, such as member dominance. This issue was not analysed in the present study and due to the apparent lack of information in the literature, it may be a study subject in future research.

Some significant differences were found in several spatiotemporal parameters (Table 1). These results are very similar to those found in a previous systematic review (Kiss et al., 2012), and may be related to the demand that these patients need a more pronounced stance phase, and particularly a longer double limb support time, to ensure a better stability during gait. In a previous study (Callisaya et al., 2010), arthritis was only associated with greater step time, has observed in our study ($p = 0.050$). A possible explanation may be related to impairments commonly associated with arthritis such as pain or decreased strength interfering with timing of steps (Bogen et al., 2020). Future research should confirm possible changes in knee OA gait variability based on disease severity criteria, such as the Kellgren and Lawrence scale.

The strengths of this study were the use of two groups of gender-matched participants, the use of sophisticated measurements by trained staff, using standardised protocols, and the range of measures of gait spatiotemporal and variability parameters studied. Our findings are, however, limited by the lack of a control group matched with the same age of the knee OA patients, but without a clinical history of knee OA. In addition, the total number of participants is relatively small.

5 CONCLUSION

The comparison of several parameters shows significant changes in gait automatisms among patients with knee OA comparing to healthy young adults. Particularly, an increase in the stance and double stance phase time may be related to the need of increasing gait stability. Although the differences found in gait parameters between healthy young adults and OA patients, the age difference between groups could explain the differences. A fact that supports this conclusion is related to the comparison of the variability between the groups, suggesting no particular changes in motor control of gait in patients with moderate knee OA.

REFERENCES

Bogen, B., Moe-Nilssen, R., Aaslund, M.K., Ranhoff, A.H., 2020. Muscle Strength as a Predictor of Gait Variability after Two Years in Community-Living Older Adults. J. frailty aging 9, 23–29. https://doi.org/10.14283/jfa.2019.24

Callisaya, M.L., Blizzard, L., Schmidt, M.D., McGinley, J.L., Srikanth, V.K., 2010. Ageing and gait variability—a population-based study of older people. Age Ageing 39, 191–197. https://doi.org/10.1093/ageing/afp250

Dhawale, A.K., Smith, M.A., Ölveczky, B.P., 2017. The Role of Variability in Motor Learning. Annu. Rev. Neurosci. https://doi.org/10.1146/annurev-neuro-072116-031548

Ferreira, V., Machado, L., Simões, R., Roriz, P., 2020. Intra-session reliability of different angles of lateral wedge insoles in the biomechanical parameters of the knee. Phys. Ther. Rehabil. 7, 10. https://doi.org/10.7243/2055-2386-7-10

Fu, K., Robbins, S.R., McDougall, J.J., 2018. Osteoarthritis: The genesis of pain. Rheumatol. (United Kingdom). https://doi.org/10.1093/rheumatology/kex419

Kiss, R.M., Bejek, Z., Szendrői, M., 2012. Variability of gait parameters in patients with total knee arthroplasty. Knee Surgery, Sport. Traumatol. Arthrosc. https://doi.org/10.1007/s00167-012-1965-y

König, N., Taylor, W.R., Armbrecht, G., Dietzel, R., Singh, N.B., 2014. Identification of functional parameters for the classification of older female fallers and prediction of "first-time" fallers. J. R. Soc. Interface. https://doi.org/10.1098/rsif.2014.0353

Mortaza, N., Abu Osman, N.A., Mehdikhani, N., 2014. Are the spatio-temporal parameters of gait capable of distinguishing a faller from a non-faller elderly? Eur. J. Phys. Rehabil. Med.

Svoboda, Z., Bizovska, L., Janura, M., Kubonova, E., Janurova, K., Vuillerme, N., 2017. Variability of spatial temporal gait parameters and center of pressure displacements during gait in elderly fallers and nonfallers: A 6-month prospective study. PLoS One. https://doi.org/10.1371/journal.pone.0171997

Williams, D.S., Martin, A.E., 2019. Gait modification when decreasing double support percentage. J. Biomech. https://doi.org/10.1016/j.jbiomech.2019.05.028

Advances and Current Trends in Biomechanics – Belinha et al. (Eds)
© 2022 Copyright the Author(s), ISBN: 978-1-032-10806-3

3D bioprinting: Parameters optimization for agarose

A.M. Teixeira & A. André
Institute of Science and Innovation in Mechanical and Industrial Engineering, Porto, Portugal

P. Martins
INEGI, LAETA, FEUP, Porto, Portugal

ABSTRACT: Agarose is a biocompatible hydrogel that has promising features as soft tissue mimicking material and scaffold for tissue engineering. Those structures might be developed with a parametric design using 3D bioprinting, which is a technique that can be used with a variety of biomaterials, including hydrogels. The aim of this work is to evaluate the printability of agarose, investigating and optimizing the printing parameters. To accomplish that, an alternating line was designed and printed in order to analyze its thickness and to calculate the parameter optimization index. The optimum printing parameters for a concentration of agarose of 1.5% (w/w) were obtained and corresponded to 40 kPa of pressure, a print speed of 10 mm/s and a needle gauge of 25G.

1 INTRODUCTION

3D bioprinting is a promising and growing approach to develop idealized tissues and organs, combining cells into scaffolds (Mandrycky *et al.*, 2016). This technique allows the production of custom designs with a complex internal morphology and perform controlled material extrusion, in a layer-by-layer fashion, to achieve the desired properties (Chae *et al.*, 2018; Mohseni *et al.*, 2018). There are different 3D bioprinting techniques that allow to print biomaterials with or without cells, such as inkjet, laser-assisted and extrusion bioprinting (Pereira and Bartolo, 2015). The choice of the technique depends on the requirements of the material and the purpose of the print (i.e. target tissue). Biomaterials such as hydrogels are commonly used to mimic and regenerate human tissues, especially soft tissues such as skin, vessels, neural tissue, cartilage, adipose tissue, skeletal muscle, among others. They are an attractive material since they can mimic the natural extracellular matrix and present good biocompatibility, low inflammation, and suitable mechanical properties which can be tailored, for example, by using hydrogels blends or crosslinking agents (Mandrycky *et al.*, 2016; O'Halloran *et al.*, 2018). Among these materials, gelatin, alginate, chitosan, collagen and agar are some of the most common natural polymers used in biofabrication (Ratheesh *et al.*, 2017). Agarose is a hydrogel not commonly reported in literature either for tissue engineering applications or for 3D bioprinting. However, it was already stated that agarose has promising properties, such as good biocompatibility and physiochemical properties, presents thermoreversible gelation behavior and covers the properties of different tissues (Zarrintaj *et al.*, 2018), including the mechanical properties of breast tissues (Teixeira and Martins, 2020). Therefore, agarose seems to be a promising material to produce scaffolds for soft tissues applications. Since those scaffolds are often developed using 3D bioprinting techniques, in this study the authors focused on this technology and in the printability of agarose as the first step to produce 3D bioprinted agarose scaffolds.

DOI: 10.1201/9781003217152-10

2 MATERIALS AND METHODS

To evaluate the printability of agarose, a concentration of 1.5% (w/w) was used. The agarose powder was dissolved in distilled water, poured into the printer cartridge, and kept at 40°C until printing. The experiment was performed using an INKREDIBLE+ bioprinter (CellInk, Sweden), at room temperature (~25°C). The print speed was defined at 1, 2, 4, 6, 8 and 10 mm/s. The pressure varied from 40 to 100 kPa, with increments of 20 kPa. The needle gauges (D_G) were 25G, 27G and 30G. The methodology of this work, including the experimental procedure and the mathematical analysis, was based on the study of Webb et al. (Webb & Doyle, 2017). However, since the hydrogel used in this study was different, the parameters changed accordingly. The pressure (p) range was obtained via trial and error, and the print speed was increased until 10 mm/s, since it has been the highest print speed used for breast tissue scaffolds' production (Tytgat *et al.*, 2019).

To obtain the optimum parameters, an alternating line with 40 mm long and 10 mm wide was designed using SolidWorks software (Dassault Systems). Then, the STL file was imported into the Repetier Host software (Hot-World GmbH&Co. KG, Germany) and the G-code was obtained using Slic3r. The printed lines were scanned and its thickness (t_{line}) was measured in nine different points, using ImageJ (NIH, USA). With those results, the parameter optimization index (POI) was calculated using equation 1 and normalized by the maximum value.

$$POI = \frac{1}{t_{line} \times D_G \times p} \tag{1}$$

3 RESULTS

In Figure 1, it is represented an example of the printed alternating lines, which, in this case corresponds to a print speed of 6 mm/s, using a 25G needle, at different printing pressures. The average thickness and the standard deviation for all the combinations of printing parameters are represented in Table 1. With these values, the POI was calculated using Equation 1 and the results were normalized by the maximum value. In Figure 2, the normalized values are represented, being the best set of parameters, the one that corresponds to the POI equals to 1.

4 DISCUSSION

The results of Table 1 show that the thickness decreases with the increase of the nozzle gauge, as expected. Moreover, for the same nozzle gauge, the thickness increases with the increase of pressure and the decrease of print speed. In Figure 1, it is also visible the effect of pressure in the line thickness, which increases with the pressure.

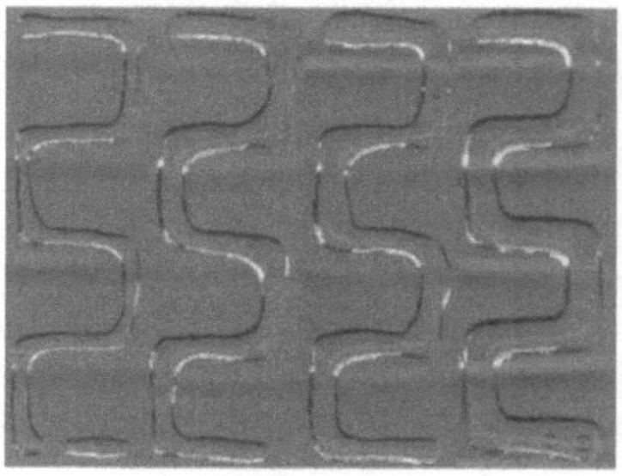

Figure 1. Scan of examples of printed alternating lines. Printing conditions: print speed of 6mm/s and needle gauge of 25G, at 40kPa, 60 kPa, 80 kPa and 100 kPa of pressure (from right to left).

Table 1. Alternating line thickness, in mm, [mean ± standard deviation (range)]. +++ and --- denote configurations that could not create adequate prints by excess or lack of material, respectively. The best configuration is shown in bold.

	1mm/s	2mm/s	4mm/s	6mm/s	8mm/s	10mm/s
40kPa						
25G	6.494 ± 0.959 (5.404-8.851)	5.106 ± 0.506 (4.298-5.872)	4.204 ± 0.329 (3.758-4.65)	3.209 ± 0.500 (2.421-4.183)	2.254 ± 0.339 (1.696-3)	**1.450 ± 0.276 (1.174-2.087)**
27G	3.359 ± 0.446 (2.234-4.149)	2.814 ± 0.390 (2.042-3.510)	2.103 ± 0.237 (1.526-2.544)	1.392 ± 0.171 (1.014-1.781)	---	---
30G	---	---	---	---	---	---
60kPa						
25G	+++	7.372 ± 0.476 (6.708-8.171)	5.958 ± 0.633 (5.143-6.686)	4.739 ± 0.435 (4.243-5.271)	3.188 ± 0.347 (2.667-3.583)	2.719 ± 0.223 (2.250-3.001)
27G	4.443 ± 0.424 (3.347-5.494)	3.588 ± 0.588 (2.400-4.793)	2.361 ± 0.291 (1.075-4.140)	1.987 ± 0.239 (1.120-3.633)	1.573 ± 0.158 (1.114-2.166)	1.468 ± 0.182 (0.739-1.906)
30G	2.360 ± 0.305 (1.891-3.125)	1.989 ± 0.375 (1.458-2.915)	---	---	---	---
80kPa						
25G	+++	+++	6.871 ± 0.772 (4.627-8.682)	5.655 ± 0.793 (3.866-7.035)	4.512 ± 0.655 (3.782-6.260)	3.701 ± 0.504 (3.065-4.826)
27G	7.292 ± 0.912 (5.716-8.825)	5.27 ± 0.808 (4.034-5.723)	3.117 ± 0.396 (1.570-4.532)	2.509 ± 0.467 (1.528-4.595)	1.676 ± 0.240 (1.157-2.598)	1.616 ± 0.261 (1.028-2.614)
30G	3.509 ± 0.438 (2.611-4.042)	2.848 ± 0.516 (2.192-3.74)	1.574 ± 0.262 (1.263-2.021)	---	---	---
100kPa						
25G	+++	+++	8.091 ± 0.489 (7.118-8.818)	7.205 ± 0.639 (6.047-8.189)	6.083 ± 0.836 (4.778-6.943)	5.45 ± 0.440 (4.905-5.988)
27G	+++	6.835 ± 1.233 (4.608-8.206)	3.917 ± 0.449 (2.232-6.781)	3.587 ± 0.557 (1.786-6.507)	2.888 ± 0.460 (2.054-4.299)	2.279 ± 0.374 (1.541-3.541)
30G	4.117 ± 0.586 (3.383-5.234)	3.678 ± 0.467 (3.128-4.532)	2.052 ± 0.190 (1.723-2.426)	1.537 ± 0.184 (1.149-1.787)	1.295 ± 0.143 (0.968-1.81)	1.102 ± 0.155 (0.884-1.326)

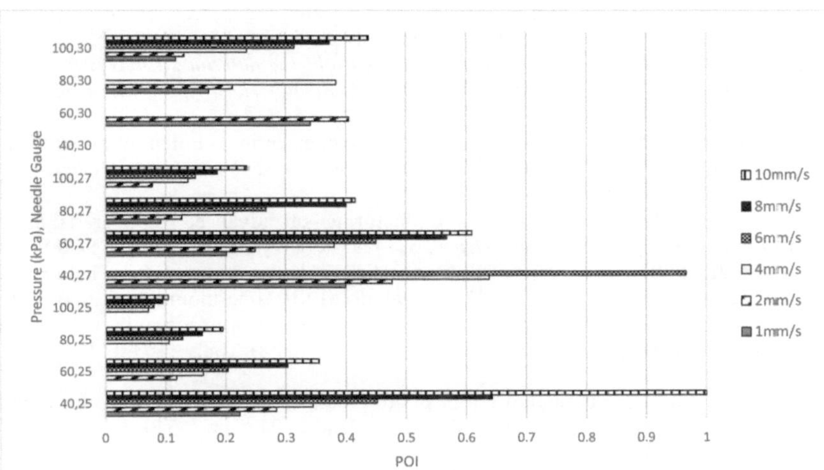

Figure 2. POI for all the combinations of printing parameters analyzed.

According to Webb et al. (Webb and Doyle, 2017), the shear stress and accuracy are directly influenced by the print speed, pressure and nozzle gauge: the shear stress increases with the nozzle gauge and pressure, while the accuracy increases with the print speed and nozzle gauge and decreases with the pressure. Thus, the accuracy can be also correlated to the line thickness, being maximum when the line thickness is minimum. The goal of any printing process is to maximize accuracy and minimize shear stress, which was represented by Webb et al. (Webb and Doyle, 2017) with the equation of POI.

According to Figure 2, we can conclude what is the optimum combination of parameters to print agarose at a concentration of 1.5% (w/w), which corresponds to the set that has a POI equal to 1. Therefore, the optimum printing conditions are 25G of needle gauge, 40kPa of pressure and 10mm/s of print speed.

5 CONCLUSIONS

This work is a step forward in the fields of 3D bioprinting and biomaterials for tissue engineering. The printability of agarose was evaluated which is important for the development of 3D printed scaffolds for soft tissue applications, for example. The optimum printing parameters of agarose were successfully obtained for the concentration tested (i.e. 1.5%(w/w)). This approach can be further extended to more concentrations, hence covering an arbitrarily extended range of printing conditions for agarose.

ACKNOWLEDGEMENTS

The authors gratefully acknowledge funding from FCT, Portugal, under grant 2020.08718. BD, and from project MImBI PTDC/EME APL/29875/2017 financed through FEDER and from research unit LAETA.

REFERENCES

Chae, M. P. *et al.* (2018) *3D bioprinting adipose tissue for breast reconstruction, 3D Bioprinting for Reconstructive Surgery: Techniques and Applications*. Elsevier Ltd. doi: 10.1016/B978-0-08-101103-4.00028-4.

Mandrycky, C. *et al.* (2016) '3D bioprinting for engineering complex tissues', *Biotechnology Advances*. Elsevier Inc., 34(4), pp. 422–434. doi: 10.1016/j.biotechadv.2015.12.011.

Mohseni, M. *et al.* (2018) *Adipose tissue regeneration: Scaffold-Biomaterial strategies and translational perspectives, Biomaterials in Translational Medicine: A Biomaterials Approach*. Elsevier Inc. doi: 10.1016/B978-0-12-813477-1.00013-X.

O'Halloran, N. A. *et al.* (2018) 'Hydrogels in adipose tissue engineering—Potential application in post-mastectomy breast regeneration', *Journal of Tissue Engineering and Regenerative Medicine*, 12(12), pp. 2234–2247. doi: 10.1002/term.2753.

Pereira, R. F. and Bartolo, P. J. (2015) 'Manufacturing of advanced biodegradable polymeric components', *Journal of Applied Polymer Science*, pp. 42458 (1–15). doi: 10.1002/app.42889.

Ratheesh, G. *et al.* (2017) '3D Fabrication of Polymeric Scaffolds for Regenerative Therapy', *ACS Biomaterials Science and Engineering*, 3(7), pp. 1175–1194. doi: 10.1021/acsbiomaterials.6b00370.

Teixeira, A. M. and Martins, P. (2020) 'Mechanical characterisation of an organic phantom candidate for breast tissue', *Journal of Biomaterials Applications*, 34(8), pp. 1163–1170. doi: 10.1177/0885328219895738.

Tytgat, L. *et al.* (2019) 'Additive manufacturing of photo-crosslinked gelatin scaffolds for adipose tissue engineering', *Acta Biomaterialia*. Acta Materialia Inc., 94, pp. 340–350. doi: 10.1016/j.actbio.2019.05.062.

Webb, B. and Doyle, B. J. (2017) 'Parameter optimization for 3D bioprinting of hydrogels', *Bioprinting*. Elsevier B.V., 8, pp. 8–12. doi: 10.1016/j.bprint.2017.09.001.

Zarrintaj, P. *et al.* (2018) 'Agarose-based biomaterials for tissue engineering', *Carbohydrate Polymers*. Elsevier, 187, pp. 66–84. doi: 10.1016/j.carbpol.2018.01.060.

Advances and Current Trends in Biomechanics – Belinha et al. (Eds)
© 2022 Copyright the Author(s), ISBN: 978-1-032-10806-3

Biomechanical study of a Maylard scar during a vaginal birth after caesarean

D.S. Fidalgo, M.V. Pouca, D.A. Oliveira, M.L. Parente & R.N. Jorge
INEGI-LAETA, FEUP, Porto, Portugal

E. Malanowska
Pomeranian Medical University, Szczecin, Poland

K.M. Myers
Columbia University in the City of New York, New York, USA

ABSTRACT: Caesarean section is one of the most common surgeries worldwide. However, the mechanical behaviour of a caesarean scar during a vaginal birth after caesarean (VBAC) is not well understood, due to many questions concerning the uterine wound healing process. The aim of this work is to study the biomechanical Maylard scar stiffness influence during a VBAC through computational methods. A biomechanical model comprising simplified geometries of a foetus and a uterus is therefore used to achieve that objective. The results revealed that an increase of the matrix stiffness of a Maylard scar has an impact on its mechanical behaviour during a VBAC, especially on the maximum principal stress, compared to the uninjured uterine tissue.

1 INTRODUCTION

The caesarean section (CS) surgery rates have been increasing worldwide, both in developed and undeveloped countries (Rusavy *et al.*, 2019). In fact, CS is the most common surgical technique in the United States, where approximately 33% of women who gave birth in 2016 did so by caesarean delivery (Rusavy *et al.*, 2019).

The Maylard incision, which consists of an extended transverse cut characterized by bilateral ligation of the deep inferior epigastric vessels and transection of the anterior rectus abdominis muscle, is an example of CS applied when the baby is at an abnormal position to born (Ortiz Molina *et al.*, 2020).

The professionals have encouraged women with prior caesarean delivery to give birth vaginally on future pregnancies, since 80% of vaginal birth after caesarean (VBAC) happens successfully. However, some authors are concerned with the association between this practice and the risk of uterine rupture (Buhimschi *et al.*, 2005). In fact, obstetricians have realized the importance of the uterine wound healing regarding the dehiscence of uterine scars and their rupture (Roeder *et al.* 2012). Different soft tissues may exhibit a different healing process, leading to different collagen fibre reorientations (Davidson *et al.*, 1997). However, there is not consistent information regarding the fibre orientation in the uterine scars, which represent a great gap towards the understanding of the uterine rupture process during VBAC.

To the best of the authors' knowledge, this will be the first work to perform a numerical simulation of a VBAC. The aim of this study is to develop a biomechanical

DOI: 10.1201/9781003217152-11

approach to analyse the impact of a matrix stiffness increase of a Maylard scar in a VBAC from a mechanical point of view.

2 MATERIAL AND METHODS

A macroscale chemical-mechanical constitutive model will be used and adjusted for this work to replicate a set of uterine contractions and rests (Sharifimajd *et al.* 2016).

The activation of smooth muscle relies on the concentration of the myosin phosphorylation. Myosin can be in one of the following four states: unbound unphosphorylated (α_A), unbound phosphorylated (α_B), bound phosphorylated (α_C) and bound unphosphorylated (α_D). $k_1 - k_7$ represent the transitions rates from one state to another.

The kinetic state model is given by a coupled system of first-order differential equation:

$$\begin{bmatrix} \dot{\alpha}_A \\ \dot{\alpha}_B \\ \dot{\alpha}_C \\ \dot{\alpha}_D \end{bmatrix} = \begin{bmatrix} -k_1 & k_2 & 0 & k_7 \\ k_1 & -(k_2+k_3) & k_4 & 0 \\ 0 & k_3 & -(k_4+k_5) & k_6 \\ 0 & 0 & k_5 & -(k_6+k_7) \end{bmatrix} \times \begin{bmatrix} \alpha_A \\ \alpha_B \\ \alpha_C \\ \alpha_D \end{bmatrix} \tag{1}$$

The contractile element may be represented by a friction clutch, a brake-like mechanism in series with the clutch, and a spring attached to the clutch and the brake. The dynamics of the clutch is characterized by an evolution law:

$$\eta \frac{\dot{}}{\varepsilon_{a\iota}} = -F_{C,i} - \frac{\partial \overline{\psi}_f}{\partial \varepsilon_{a\iota}} + \frac{\partial \overline{\psi}_f}{\partial \overline{\varepsilon_{e\iota}}} \tag{2}$$

where $\frac{\dot{}}{\varepsilon_{a\iota}}$ represents the active strain rate, η is a constant, $F_{C,i}$ is the friction force generated by the clutch in the ith fasciculi family, $\overline{\psi}_f$ the fasciculi isochoric contribution, and $\overline{\varepsilon}_e$ the elastic strain.

A finite element model comprising simplified geometries of a foetus and pregnant uterus was developed. The simplified foetal model was meshed with 10595 nodes and 53732 C3D4H elements. The foetal rotation in the vertical axis was constrained, allowing a stable descent of the foetus throughout the simulation. A hyperelastic behaviour was considered, being applied the Neo-Hookean model with the parameter $C_{10} = 0.13$ MPa and D_1 was set to a low value (0.05 MPa^{-1}), ensuring foetal incompressibility. The uterine model undergone a refinement in the Maylard region, to further introduce the caesarean scar. The refined uterine mesh has 11891 nodes and 8374 C3D8H elements. The longitudinal and circumferential fibres were maintained outside and inside the Maylard scar. One of the the external nodes from the inferior surface was completely fixed, while the remaining nodes from the same border were constrained in the vertical direction (Y axis). The constitutive parameters were established according to Table 1.

A surface-to-surface interaction was applied, with a penalty contact algorithm, between the inner uterine wall and the foetus, including friction, which was set to a value of 0.2 *(Lindgren et al.* 1969). To guide the foetus head throughout the simulations, a contact condition was established between that structure and a discrete rigid surface resembling a spine. Two numerical simulations were conducted, represented by the following initials: (i) Control Model (CM); (ii) VBAC with an increase of the scar matrix stiffness C_{10} *Maylard* (IM). The uterus was divided in 5 layers, L0 to L4, aiming to simulate the diffusion process associated with the electrical stimulus. L0 is the upper uterine region, followed by the L1 zone downwards and so on. Since the electrical part of the model is not considered, it was assumed that the pacemaker area corresponds to the L0 set. Afterwards, the L1 region is stimulated with a fast increase of calcium concentration, triggering another contraction and so on, until the L4 region is reached. An average contraction time of 90 seconds and resting time of 60 seconds were considered (Buttin *et al.*, 2013).

Table 1. Chemical and mechanical parameters applied to the uterus (Sharifimajd et al. 2016; Pasquo et al , 2020).

Mechanical Model		Chemical Model	
Parameter	Value (Unit)	Parameter	Value (Unit)
D_1	1E-5 (MPa^{-1})	k_2	0.6193 (s^{-1})
C_{10}	0.030 (MPa)	k_3	0.1419 (s^{-1})
C_{10} *Maylard*	0.0564 (MPa)	k_7	0.0378 (s^{-1})
ζ_1	0.4 (-)	$\left[Ca^{2+}_{1/2MLCK}\right]_i$	256.98 (nM)
$\mu_{i1}(i = 1, 2)$	0.006 (MPa)	nk_1	8.7613 (-)
$\mu_{i2}(i = 1, 2)$	1.0 (MPa)	α_A	1 (-)
$\bar{\varepsilon}_{opt}$	0.4950 (-)	α_B	0 (-)
γ	1 (-)	α_C	0 (-)
ν_C	0.11 (s^{-1})	α_D	0 (-)
η	0.18 (MPa s)	$\bar{\varepsilon}_{ai}(i = 1, 2)$	0 (-)
k_{xb}	8 (MPa)		
k_C	5.5 (MPa s)		

3 RESULTS

For each simulation we analysed the maximum principal stress evolution in the Maylard scar region (Figure 1), and the vertical displacement of the foetus (Figure 2). Analysing Figure 1, it is possible to observe greater maximum principal stress values for the IM compared to the control model, namely during the third uterine contraction. Concerning Figure 2, the fetal descent mean velocity for the CM is 0.17 mm/s and for the IM is 0.16 mm/s.

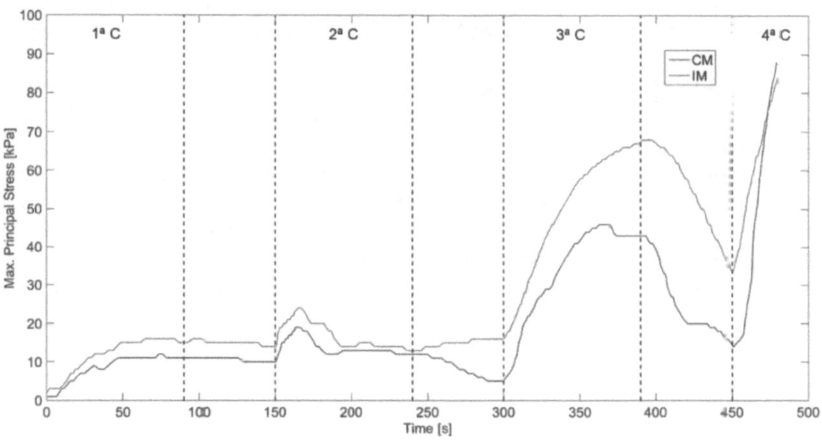

Figure 1. Maximum principal stress evolution for all numerical simulations; "C" stands for contraction.

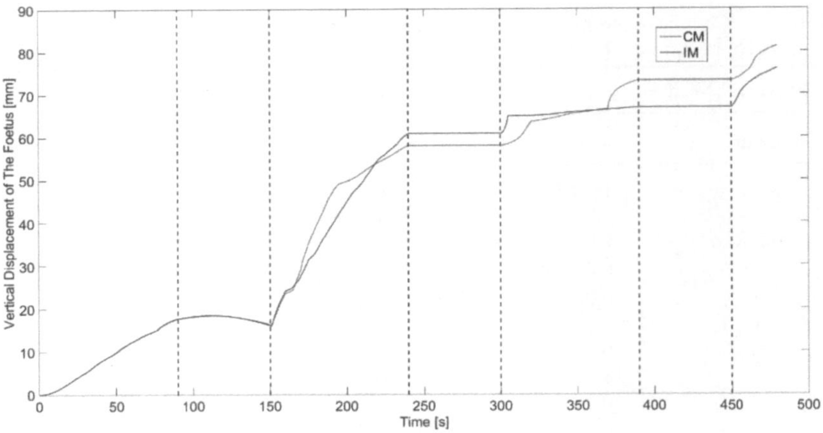

Figure 2. Vertical displacement of the fetal head.

4 DISCUSSION

According to Figure 1, an increase of the Maylard scar stiffness exposes the injured uterine tissue to a greater mechanical stress compared to the control model, potentiating uterine rupture during vaginal delivery. Regarding the CM, the maximum principal stress values obtained for this model are within the physiological range of the pregnant human uterus (Bisplinghoff *et al.*, 2012).

According to the illustrated results in Figure 3, the mean velocity of the foetal descent does not suffer significant alterations (0.16-0.17 mm/s). In fact, according to the statistical results, no significant differences were detected in the duration of the second stage of labour between women undergoing a second vaginal birth (PVB) and women with a first vaginal delivery after a prior caesarean section. The PVB group registered a mean duration of 23.8 minutes, while the VBAC registered 20.8 minutes (Rusavy *et al.*, 2019). These findings lead to the conclusion that a prior Maylard caesarean section will not affect the mean duration of the second stage of labour.

All biomechanical models require some simplifications to be functional. In this work, some simplifications were performed in the uterine-foetal biomechanical model: (i) the constitutive model does not cover the fact that the uterus reaches an almost non-pregnant stage right after the end of the second stage of labour, (ii) the collagen fibre distribution in the uterus was based on the non-pregnant state, (iii) the calcium activation curve was projected to allow a realistic foetal, (iv) it was neglected the viscoelastic passive behaviour on the uterine tissue, the pushing force exerted by the mother, the contraction of the abdominal muscles, the baseline intrauterine pressure, the residual strain in the reference configuration, and the deformability of the foetus head.

5 CONCLUSIONS

The results revealed that an increase of the stiffness of the Maylard scar matrix has an impact on the mechanical behaviour of the caesarean scar during VBAC compared to the control model. This is highlighted by the greater maximum principal stress values registered during the simulations. Further, the mean velocity of the foetal descent was not altered by the stiffness increase, which is in accordance with the literature.

The biomechanical models presented in this work may result in a discussion area for future research. To establish the biomechanical pattern for each patient after caesarean section may allow the development of individual tailored delivery programs.

ACKNOWLEDGMENTS

This research was supported by FCT, which finances the Portuguese Grant 2020.05400.BD, and by LAETA, which granted the project UIDB/50022/2020, and by the project NORTE-01-0145-FEDER-030062 (SIM4SafeBirth).

REFRENCES

Bisplinghoff, J. A., Kemper, A. R. and Duma, S. M. (2012) 'Dynamic material properties of the pregnant human uterus', Journal of Biomechanics, 45(9), pp. 1724–1727. doi: 10.1016/j.jbiomech.2012.04.001.

Buhimschi, C. S. et al. (2005) 'Rupture of the uterine scar during term labour Contractility or biochemistry?', BJOG: An International Journal of Obstetrics and Gynaecology, 112(1), pp. 38–42. doi: 10.1111/j.1471-0528.2004.00300.x.

Buttin, R. et al. (2013) 'Biomechanical simulation of the fetal descent without imposed theoretical trajectory', Computer Methods and Programs in Biomedicine, 111(2), pp. 389–401. doi: 10.1016/j.cmpb.2013.04.005.

Davidson, C. J. et al. (1997) 'Rat tendon morphologic and functional changes resulting from soft tissue mobilization', Medicine and Science in Sports and Exercise, 29(3), pp. 313–319. doi: 10.1097/00005768-199703000-00005.

Lindgren, L. and Holmlund, D. (1969) 'Friction between the fetal head and uterine wall during normal labor and lower uterine spasm.', American journal of obstetrics and gynecology, 103(7), pp. 939–941. doi: 10.1016/S0002-9378(16)34442-8.

Ortiz Molina, E. et al. (2020) 'Maylard's incision: How to make an easy incision for complex pelvic abdominal surgery', International Journal of Gynecological Cancer, 30(1), pp. 154–155. doi: 10.1136/ijgc-2019-000876.

di Pasquo, E. et al. (2020) 'Evaluation of the uterine scar stiffness in women with previous Cesarean section by ultrasound elastography: A cohort study', Clinical Imaging, 64, pp. 53–56. doi: 10.1016/j.clinimag.2020.03.006.

Roeder, H. A., Cramer, S. F. and Leppert, P. C. (2012) 'A look at uterine wound healing through a histopathological study of uterine scars', Reproductive Sciences, 19(5), pp. 463–473. doi: 10.1177/1933719111426603.

Rusavy, Z. et al. (2019) 'Timing of cesarean and its impact on labor duration and genital tract trauma at the first subsequent vaginal birth: A retrospective cohort study', BMC Pregnancy and Childbirth, 19(1), pp. 1–7. doi: 10.1186/s12884-019-2359-7.

Sharifimajd, B., Thore, C. J. and Stålhand, J. (2016) 'Simulating uterine contraction by using an electro-chemo-mechanical model', Biomechanics and Modeling in Mechanobiology, 15(3), pp. 497–510. doi: 10.1007/s10237-015-0703-z.

Advances and Current Trends in Biomechanics – Belinha et al. (Eds)
 ISBN: 978-1-032-10806-3

Optimizing long bone metastases treatment: Thermal necrosis effect with two different cement types and titanium or CF/PEEK intramedullary nail

V.C.C. Oliveira
Centro Hospitalar Universitário do Porto, Instituto de Ciências Biomédicas Abel Salazar, University of Porto (CHUP-ICBAS), Portugal

E.M.M. Fonseca & J. Belinha
LAETA, INEGI, School of Engineering, Mechanical Engineering Department, Polytechnic of Porto (ISEP), Portugal

C.C. Rua & P.A.G. Piloto
Instituto Politécnico de Bragança (IPB), Portugal

R.N. Jorge
LAETA, INEGI, Mechanical Engineering Department, Faculty of Engineering of the University of Porto (FEUP), Portugal

ABSTRACT: Multiple bone metastases management requires pain relief and mechanical stability to achieve functional improvement. Inducing tumour cell necrosis is important for local tumour control. A two dimensional thermal and transient numerical model based on finite element method was built to predict the temperature distribution produced by cement polymerization when mimicking a femoral lytic lesion. The bone cement filled the metastatic lesion and titanium, or CF/PEEK intramedullary nail was used. Additionally, two different cement types were compared. Cement A has low viscosity and cement B high viscosity. The titanium intramedullary nail has shown to be the best heat conductor. Significant thermal necrosis occurs with a thin or thick cement amount, and either with low or high viscosity cement types. When choosing to optimize this synergistic effect, a titanium intramedullary nail and cement A induce the largest necrotic area. The results play a role in the multidisciplinary management of long bone metastases.

1 INTRODUCTION

Metastatic disease reflects an advanced tumour stage and has a high impact on patients' quality of life and survival. As oncological patients' survival has been extended even for multiple bone metastases, it is essential to invest in palliative treatments for improved quality of live. Bone metastases are associated to pain and pathological or imminent fractures that compromise patients' autonomy even for self-care and basic daily activities. The management of bone metastases treatment requires pain relief and mechanical stability to achieve a functional improvement. Inducing tumour cells necrosis to stop or minimize their local growth is important for local tumour control. The use of cementoplasty for bone metastases has been recently increasing, especially for spinal lesions. Cementoplasty is a palliative treatment for bone metastases that could be performed alone or in association to other therapies (Deschamps, 2012). Bone cement filling aims

DOI: 10.1201/9781003217152-12

to treat or prevent pathological fractures and relieves pain in patients with bone metastases. Bone cement has an exothermic reaction where volumetric dimension changes during the polymerization process with heat generation (Pérez *et al.*, 2009). The amount of heat generated depends on bone cement mantle volume and type of cement (Janssen *et al.*, 2012). The low viscosity cement filling leads to a relatively deep cement penetration and a large volume especially when higher bone porosity. The use of high viscosity cement leads to a homogeneous and thinner cement mantle (Janssen *et al.*, 2012). Different authors studied the exothermic reaction of cement polymerization and reported in different publications results regarding the temperature rise using time-dependent polymerization process (Pérez *et al.*, 2009), (Huiskes, 1980). Others proposed empirical models for the prediction of heat generated using experimental and numerical tests (Janssen *et al.*, 2012), (Santos *et al.*, 2006), (Fernandes *et al.*, 2017). In literature, studies reporting experimental measured temperatures indicate thermal necrosis in cortical bone when it reaches a temperature of 45°C (Huiskes, 1980), (Santos *et al.*, 2006). A two-dimensional transient heat conduction model is numerically solved using finite elements in space and finite differences in time. The problem here presented is an approach by coupling a kinetic equation for the polymerization fraction, as referred in literature (Radev *et al.*, 2009). The material with high thermal diffusivity allows the heat to rapidly flow and it quickly conducts heat (Baron *et al.*, 2015). This study aims to evaluate the thermal necrosis effect of bone cement filling to achieve a better local control and to optimize the palliative treatment of multiple bone metastases. A two dimensional thermal and transient numerical model based on finite element method was built to predict the temperature field produced by cement polymerization when mimicking a femoral lytic lesion. Intramedullary nails in Carbon-Fiber Reinforced Polyetheretherketone (CF/PEEK) are gaining attention capable of overcome weaknesses of metal nails, due to their biomechanical properties, and for oncological advantages in imaging follow-up and lower radiation doses. The bone cement polymethylmethacrylate (PMMA) was introduced to fill the metastatic lytic lesion area and titanium or CF/PEEK intramedullary nail were used. Additionally, it was compared two different cement types (A and B). Cement A has low viscosity and cement B has high viscosity, accordingly to literature (Janssen *et al.*, 2012), (Santos *et al.*, 2006). The main objective is to understand the role of such procedure in bone tumour necrosis due to thermal effect. For each numerical model, the temperature distribution produced by the cement polymerization is represented and the thermal necrosis effect through bone-tumour cells penetration was analysed.

2 MATERIALS AND METHODS

Table 1 represents all thermal properties for cortical bone, spongy bone, bone cement, titanium (Ti), and CF/PEEK, assumed constant and isotropic, collected from the literature (Pérez *et al.*, 2009), (Baron *et al.*, 2015), (Ikekwem *et al.*, 2018), (Mike *et al.*, 2019).

In the models without the intramedullary nail, the affected thermal necrosis occurs in a limited region width of 10mm. For an increase amount of 5mm in the metastatic lytic lesion filled with PMMA bone cement, the thermal effect produces always more 10mm of necrotic area in the surrounding spongy bone. Meanwhile, with the intramedullary nail all adjacent bone tissue is affected due to the high thermal conductivity. Figure 1 shows A-P and lateral view radiographs of two patients submitted to internal fixation using a femoral intramedullary

Table 1. Material properties.

Material	Density kg/m^3	Conductivity W/K.m	Specific heat J/kgK	Thermal diffusivity m^2/s
Cortical bone	2100	0.38	1260	1.44E-07
Spongy bone	620	0.39	4926	1.28E-07
Cement A or B	1100	0.20	2000	9.09E-08
Titanium	8900	29.00	377	8.64E-06
CF/PEEK	1540	0.75	1320	6.03E-07

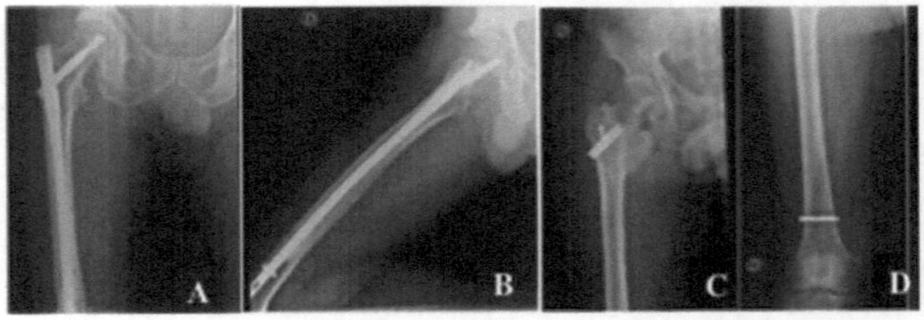

Figure 1. Internal fixation of femur using titanium intramedullary nail (A-B) or CF/PEEK nail (C-D).

nail (Gamma 3 nail, Stryker®). Both patients presented an imminent or pathological fracture due to bone metastases of different primary tumours. The bone lytic lesions were filled with PMMA.

3 RESULTS AND DISCUSSION

In this study, numerical models were produced representing two-dimensional bone geometries with an external diameter equal to 31.2mm and cortical thickness of 7.35mm. In the middle of the model, bone cement was introduced to fill a metastatic lytic lesion with different dimensions equal to 20mm in depth and variable width of 15 and 25mm. A typical mesh with an element length equal to 2.5mm was chosen (Figure 2). The numerical simulations were performed using the finite element method with ANSYS®. The geometrical model was meshed with a 2D thermal solid element (PLANE 77) with 8 nodes and a single degree of freedom, temperature, at each node.

The obtained temperature distribution was compared between models and represented near the filled lytic lesion (as represented on long bone diameter in Figure 2) and selected where the heat flux was greater (Figure 3).

Models representing two different metastatic lesion sizes on long bone diameter compared cement types when using two distinct intramedullary nails, as shown in Figure 4. Each cement type presents the temperature field produced by its polymerization effect in the time instant equal to 384s or 600s, corresponding to the peak temperature of cement A and B, respectively.

The temperature in cement zone reaches the maximum value of 83°C when using cement type A and 68°C when using cement B, and the heat is spread through cortical and spongy bone.

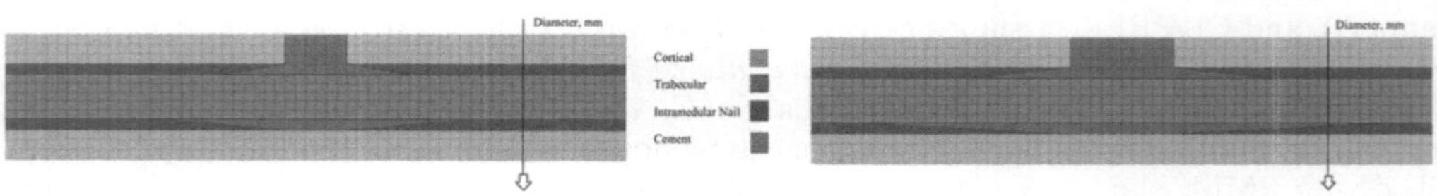

Figure 2. Meshes with cement filling and intramedullary nail: thin cement mantle in the left side (lesion of 15x20mm), and thick mantle in the right side (lesion of 25x20mm).

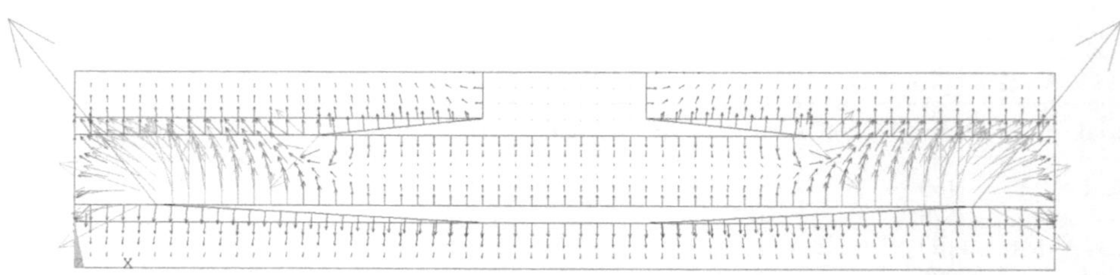

Figure 3. Heat flux resulting from the cement polymerization process.

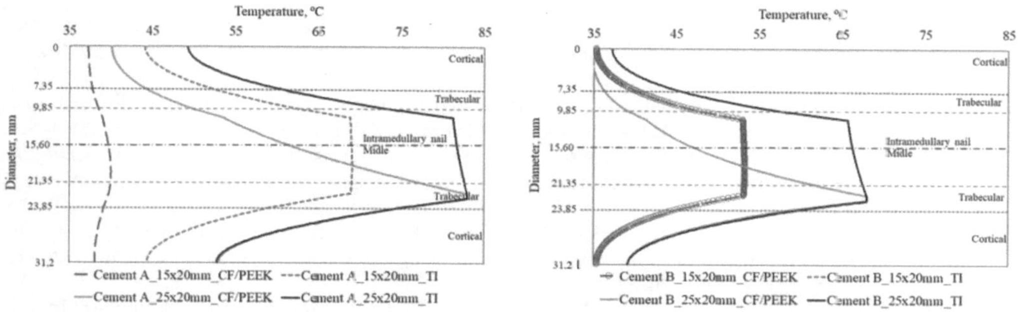

Figure 4. Models analysing cement-bone tumour interface comparing cementoplasty of two different metastatic lesion dimensions using cement type A and B and a titanium or CF/PEEK intramedullary nail.

Comparing both cement types, thermal necrosis occurs in a smaller area with cement B. Furthermore, the greatest difference on the affected area between the two materials occurs when using cement A.

When comparing the temperature field between the specific intramedullary systems, the titanium presents the highest affected area, extending along the nail. This effect is justified by its higher thermal diffusivity and conductivity.

Cement A shows the highest temperature on cement-bone tumour interface, with a thick mantle, reaching 83°C. Using cement B the maximum temperature reaches 68°C, and the heat effect is less widespread. Comparing to the titanium nail, the thermal necrosis decreases in all spongy area due to the lower thermal conductivity of the intramedullary CF/PEEK nail. In both CF/PEEK nail models, bone necrosis was found to affect the spongy area, more effectively in the proximity of the applied cement.

The interaction between all materials was considered as perfect contact and the heat transfer between materials was exclusively by heat conduction. The pressure during the introduction of the intramedullary nail of 11mm diameter compacts bone cement all around the nail through the spongy bone and until contact to both internal and external cortical bone, confirmed on patients' imaging. Furthermore, in this process there is no blood fluid increment.

4 CONCLUSIONS

PMMA is an effective treatment for long bone lytic lesions, verified in all models. Even thin cement mantle produces significant area of thermal necrosis. The low viscosity cement associated to the structural stabilization of a titanium intramedullary nail appears to have the greatest synergistic effect on thermal necrosis extension. This treatment optimizes the local treatment of long bone metastases in clinical application.

REFERENCES

Baron, S., Ahearne, E., Connolly, P., Keaveney, S., Byrne, G. (2015). An Assessment of Medical Grade Cobalt Chromium Alloy ASTM F1537 as a Difficult-to-Cut (DTC) Material. *The Proceedings of the Machine Tool Technologies Research Foundation 2015 Annual Meeting*. San Fransisco, California, USA.

Deschamps, F and Baere, T. (2012). Cementoplasty of bone metastases. *Diagnostic and Interventional Imaging*, 93(9): 685–689.

Fernandes, M.G., Fonseca, E.M.M., Jorge, R.N., Vaz, M., Dias, M.I. (2017). Thermal analysis in drilling of ex vivo bovine bones. *Journal of Mechanics in Medicine and Biology*, 17(5): 16pages.

Huiskes, R. (1980). Some fundamental aspects of human joint replacement: analyses of stresses and heat conduction in bone-prosthesis structures. *Acta Orthopaedica Scandinavica*, 51(supl185): 43–108.

Ikekwem, J.U., Chukwuneke, J.J., Omenyi, S.N. (2018). Thermal Behavior of Bone cement in Hip Replacement. *Physical Science International Journal*, 17(4): 1–14.

Janssen, D., Srinivasan, P., Scheerlinck, T., Verdonschot, N. (2012). Effect of cementing technique and cement type on thermal necrosis in hip resurfacing arthroplasty – a numerical study. *Journal of Orthopaedic Research*, 30(3): 364–70.

Mike, C., Eric D., Philippe, O., Nathalie, G., Colette, L. (2019). Thermal conductivity and diffusivity of carbon-reinforced polyetherketoneketone composites, *Journal of Applied Polymer Science*, 136(38): 1–8.

Pérez, M.A., Nuño, N., Madrala, A., García-Aznar, J.M., Doblaré, M. (2009). Computational modelling of bone cement polymerization: temperature and residual stresses. *Computers in Biology and Medicine*, 39(9): 751–9.

Radev, B.R., Kase, J.A., Askew, M.J., Weiner, S.D. (2009). Potential for thermal damage to articular cartilage by PMMA reconstruction of a bone cavity following tumour excision: a finite element study. *Journal of Biomechanics*, 42(8): 1120–1125.

Santos, J.G., Peixoto, L.S., Nele, M., Melo, P.A., Pinto, J.C. (2006). Theoretical and Experimental Investigation of the Production of PMMA-Based Bone Cement. *Macromolecular Symposia*, 243(1): 1–12.

 ISBN: 978-1-032-10806-3

Protocol to condilar position evaluation in CBCT

B.F. Santos
Faculdade de Medicina Dentária, Universidade Católica Portuguesa, Viseu, Portugal

T.M. Marques, A. Correia & P. Fonseca
Faculdade de Medicina Dentária, Centro de Investigação Interdisciplinar em Saúde, Universidade Católica Portuguesa, Viseu, Portugal

ABSTRACT: The diagnosis of a temporomandibular disorder (TMD) is based on a careful anamnesis, a correct clinical examination and often supported by complementary diagnostic exams, such as Cone-Beam Computed Tomography (CBCT). CBCT is useful in the evaluation of anatomical and anthropometric factors that predispose, trigger, and perpetuate these TMD. Among these factors is the location of the mandibular condyle in the glenoid cavity. Thus, the main objective of this study is to create an imaging protocol for the evaluation of condylar position in CBCT, given the lack of guidelines in this area. The protocol created and presented here allows, besides the objectivity of the imaging evaluation, to make studies on this topic comparable.

1 INTRODUCTION

The study of the temporomandibular joint assumes great importance in Dental Medicine. The most stable musculoskeletal position of the mandible is when the condyle is in its most antero-superior position and centered in the glenoid cavity. According to Gelb (1994) the ideal position of the mandibular condyle is verified when it is in 4 and 7 positions, represented on Figure 1. These positions are defined by drawing a horizontal line tangent to the deepest point of the glenoid cavity (I) and another tangent to the roof of the articular eminence (II), both parallel to the Frankfurt plane. A third line is drawn halfway between these two lines (III). Next, two vertical lines are drawn perpendicular to those drawn previously: the first is tangent to the deepest point of the glenoid cavity (IV) and the second is drawn at the point where line III intersects the posterior wall of the articular eminence (V).

Variations in this position may be related to unbalanced occlusion, parafunctional habits (such as bruxism) and to emotional and physical stress, which, by reducing the adaptive capacity of the stomatognathic system, can lead to the development of a temporomandibular joint (TMJ) disorder. This TMD is manifested essentially by pain (muscular or articular), limited mouth opening, dysfunction and/or articular sounds that increase during mastication or mandibular function and can lead to resorption/erosion of the articular bone surfaces.

Along this line of research, Paknahad & Shahidi (2015) associated the severity of TMD with condylar position and concluded that the more posterior this was, the more severe was TMD condition, perpetuating an anterior displacement of the articular disc. Cone Beam Computed Tomography (CBCT) has been used to analyze condylar position but without criteria. For these reasons, the main aim of this study is to create a protocol for condylar position evaluation in CBCT, based on Gelb's theory.

DOI: 10.1201/9781003217152-13

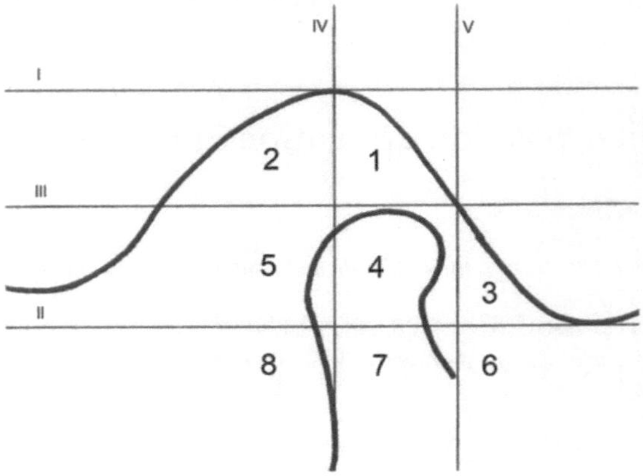

Figure 1. Ideal condilar position (Gelb's theory).

2 MATERIAL AND METHODS

CBCT images were obtained through the Planmeca ProMax® 3D device (Planmeca, Helsinki, Finlândia). To acquire a CBCT, it is necessary to select the field of view (FOV), complying with all beam marker guidelines, as well as all patient positioning instructions determined by the manufacturer. In this study the large FOV (20cm*17cm) was used, with a total filtering of 2.5mm Al + 0.5mm Cu/9kV and 1-16mA. Then, the software divided this zone into small cubes or voxels (400μm), and automatically sectioned it into 3 simultaneous planes: sagittal, axial and coronal.

The effective radiation dose of craniofacial CBCT varies between 30-1,100μSV, depending on the exposure time and volume, which vary according to the patient. As this dose cannot be measured routinely, a dose index for patient exposure (dose area product - DAP) is used to avoid applying high and unnecessary radiation doses, taking into consideration the ALADAIP principle (As Low As Diagnostically Achievable being Indication-oriented and Patient-specific).

Finally, the scans were processed and analyzed using Planmeca Romexis® software, version 5.1.0.R. The entire protocol developed and presented below was obtained based on the tools of that software.

3 RESULTS

For the analysis of the information be accurate and standardized for all participants and studies, it is imperative that the following sequence be followed:

1st step: Skull orientation according to the anatomic planes: Coronal plane was aligned with in-fraorbital foramens; Sagittal plane was aligned according to the Frankfurt plane; and Axial plane was aligned with oval foramens. (Figure 2)

2nd step: "Temporomandibular Joint" tool was selected in the software and axial section is presented. This axial section is positioned so that it has the largest latero-medial diameter of the condyles. The lines corresponding to each condyle are drawn and then, 4 equidistant sagittal cuts perpendicular and 2 coronal cuts parallel to the long axis of the condyle are predefined by the software. The sagittal section 1 is positioned to coincide with the medial pole and the section analyzed is number 2, because the functional area of the joint is closer to the medial pole of the condyle. The thickness of the sections is 0.4mm and the distance between all the sections shown is 5mm. (Figure 3)

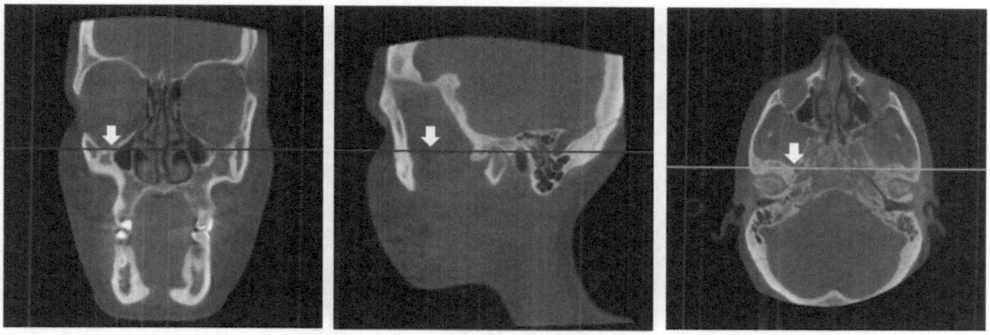

Figure 2. Anatomic orientation of the skull.

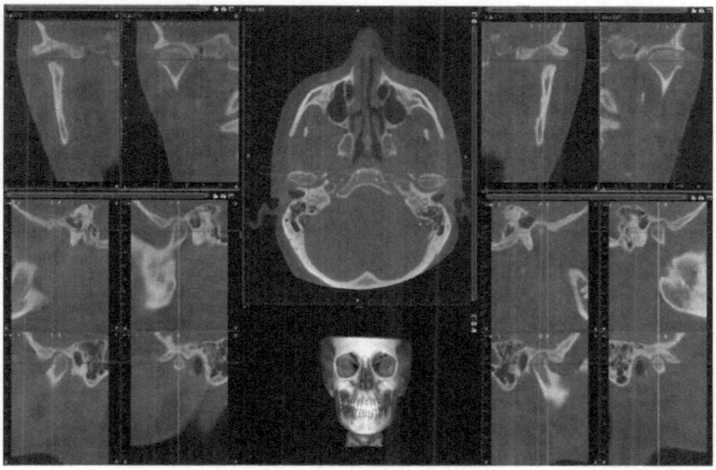

Figure 3. Representation of the sections obtained in the software: The blue line present in the twelve sections represents the Frankfurt plane; the green cuts correspond to coronal sections; and the red cuts correspond to sagittal sections.

4 DISCUSSION

By applying these steps, the condylar position can be bilaterally assessed according to Gelb's theory (Figures 1, 4). And the condyle can be classified according to the areas it mostly occupies: normal (area 4 and 7), posterior (areas 7 and 8; 5 and 8; 4 and 5; 5, 7 and 8; 4, 7 and 8), postero-inferior (area 8), anterior (area 3 and 6), antero-inferior (areas 6; 6 and 7), superior (area 1 and 4), postero-superior (area 2 and 5) and inferior (area 7).

This evaluation is important to prevent or reduce the consequences of an osteoarticular imbalance caused by different growth patterns, functional deviations, skeletal discrepancies or by adaptive changes to muscular or occlusal loads.

It should be noted that due to the high dose of radiation, CBCT should not be performed unnecessarily, and it is essential to comply with the indications of this examination based on the presence of possible complaints, signs and symptoms.

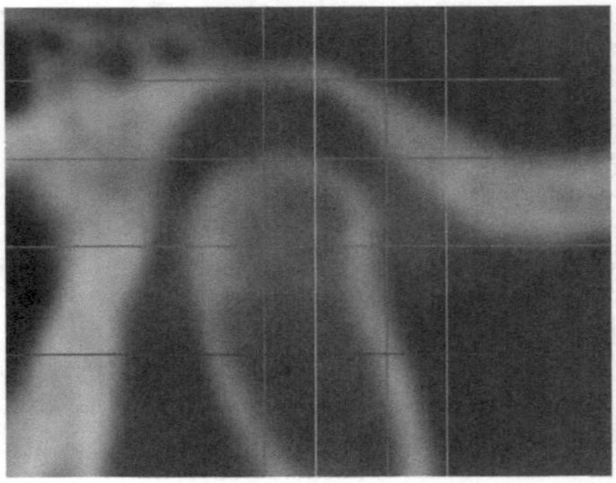

Figure 4. Condylar position analysis (red lines).

5 CONCLUSION

The proposed protocol to CBCT articular analysis proved to be effective. Thus, authors were able to objectify the evaluation of the condylar position and this protocol can be used as a tool to make possible the comparation between results from different studies.

REFERENCES

Gelb, H. 1994. *New concepts in craniomandibular and chronic pain management* (1st ed.), St. Louis: Mosby-Wolfe London.

Leeuw, R. & Klasser, G.D. 2018. *Orofacial pain: guidelines for assessment, diagnosis and management* (6 ed.), Hanover Park, IL: Quintessence.

Okeson, J.P. 2014. *Management of temporomandibular disorder and occlusion* (7st ed.), St. Louis, MO: Mosby Elsevier.

Paknahad, M. & Shahidi, S. 2015. Association between mandibular condylar position and clinical dysfunction index. *Journal of cranio-maxillo-facial surgery*, 43(4): 432–6.

Paknahad, M. et al. 2015. Cone-Beam Computed Tomographic Assessment of Mandibular Condylar Position in Patients with Temporomandibular Joint Dysfunction and in Healthy Subjects. *International journal of dentistry*, 2015, 301796.

Robinson de Senna, B. et al. 2009. Condyle-disk-fossa position and relationship to clinical signs and symptoms of temporomandibular disorders in women. *Oral surgery, oral medicine, oral pathology, oral radiology, and endodontics*, 108(3): e117–e124.

Serra, M.D. & Gavião, M.B.D. 2006. Evaluation of condylar position from transcranial projections in primary dentition. *Dento maxillo facial radiology*, 35(2): 110–6.1

Tabrizi, R. et al. 2016. Evaluation of Condylar Position after Orthognathic Surgery for Treatment of Class II Vertical Maxillary Excess and Mandibular Deficiency by Using Cone-Beam Computed Tomography. *Journal of dentistry, Shiraz, Iran*, 17(4),318–25.

Tamimi, D. & Hatcher, D.C. 2017. *Specialty Imaging: Temporomandibular Joint* (1st ed.), Amyrsis: Elsevier.

Whaites, E. & Drage, N. 2013. *Essentials of Dental Radiography and Radiology* (5th ed.), Churchill Livingstone: Elsevier.

Advances and Current Trends in Biomechanics – Belinha et al. (Eds)
© 2022 Copyright the Author(s), ISBN: 978-1-032-10806-3

IT platform to monitor and control bone-implant loosening in instrumented implants

Inês Simões Peres
Department of Mechanical Engineering, University of Aveiro, Portugal

Jorge A.F. Ferreira & Marco P. Soares dos Santos
Department of Mechanical Engineering, TEMA-Centre for Mechanical Engineering and Automation, University of Aveiro, Portugal

ABSTRACT: Arthroplasties are one of the most common surgical procedures performed worldwide. Although this surgery has excellent outcomes, revision surgeries are still required in 6.45% of the cases after five years and 12.9% after ten years. Different methodologies for monitoring the bone-implant interface have been developed over the past few years. Although these methodologies have demonstrated better results when compared with imaging methodologies, continuous monitoring throughout the patient's daily life is still not possible. This work reports, for the first time, the development a web application that allows monitoring the bone-implant fixation. Through the simulation of the sensing technology, it was possible to visualize and analyze data in different types of plots. This webpage has demonstrated a high potential for continuous monitoring of an instrumented implant.

1 INTRODUCTION

Arthroplasties, which consist of a reconstructive procedure that alters the structure of a joint, is one of the most common surgical procedures performed worldwide, accounted for more than a million replacement surgeries every year (Pivec et al. 2012). Although this surgery has excellent outcomes, revision surgeries are still required in 6.45% of the cases after five years and 12.9% after ten years (Labek et al. 2011). Over the past decade, the number of young patients undergoing this procedure has increased, leading to an increase in the number of revision surgeries in the coming years (Kurtz et al. 2009).

The monitoring of loosening states of the bone-implant interface can be carried out by different methodologies. Currently, this diagnosis is performed by using imaging methods. These are not efficient in detecting micro displacements and do not allow continuous monitoring throughout patients' daily life (Zhang et al. 2002; Mulcahy et al. 1996).

To overcome the imaging limitations new methods have been developed. Vibrational, acoustic, electrical bioimpedance, magnetic inductance, and mechanical deformation methods are examples of new methodologies. Although they demonstrated improvements over imaging methods, they do not allow effective monitoring of target regions, nor continuous monitoring of patients (Cachão et al. 2020).

In this work, the development of a web application that enables the monitoring of a co-surface capacitive sensor network has shown promising results for its use in multifunctional instrumented implants, such that they can be extracorporeally controlled by clinicians.

DOI: 10.1201/9781003217152-14

2 MATERIALS AND METHODS

To enable the monitoring of bone-implant fixation states, an architecture that allows the acquisition of data from a network of planar capacitors, printed in the inner regions near the surface of instrumented bone implants, was developed. Figure 1 presents the overall system view of the proposed architecture.

The system consists of a data acquisition part and a monitoring/control interface. The data acquisition system is composed of a Raspberry Pi, a capacitor to digital converter, and the sensing technology comprising the capacitors sensors network. The monitoring/control part is a webpage that allows the clinician to configure and control the data acquisition part.

The monitoring/control part was developed with a simplified version of the overall architecture. This simplified version comprises only a raspberry pi to simulate the data acquisition part.

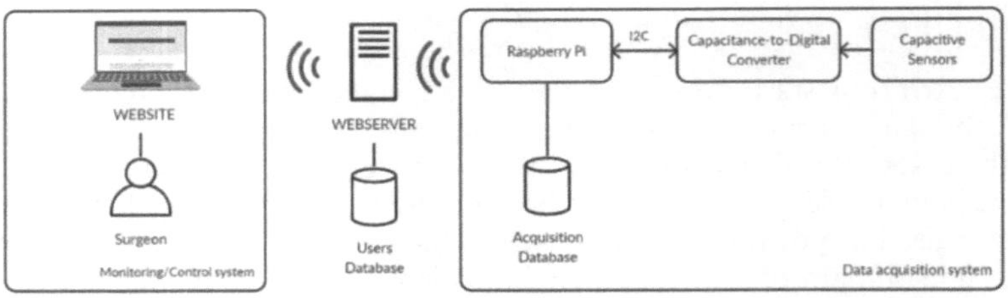

Figure 1. Overall system view.

The functioning of the data acquisition system is highly dependent on the sensing technology architecture. To keep the simulation accurate as the behavior of overall prototype, a printed circuit board was designed, as shown in Figure 2. This board contains a network of 12 capacitors (according to an interdigitated architecture) arranged in a 4x3 matrix design with the following characteristics: 2.5 mm spacing between rows, 4 mm spacing between columns, 0.5 mm gap between electrodes, and a total footprint dimension of each capacitor of 7.5x7.5 mm.

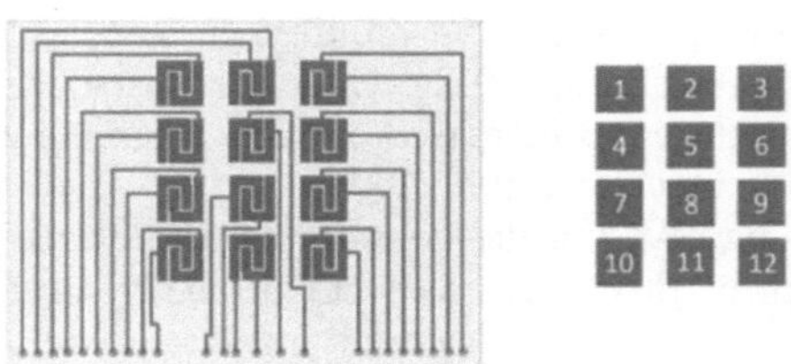

Figure 2. Schematics of the sensing technology and numbering of each capacitor.

The developed webpage has different pages and sections. The homepage describes the purpose of using the website, and indications are provided for the user to know when to create an account and how to fill in the various fields. A section to clarify doubts is available. After the user sends a message, the research group can help the user only if the introduced email is valid. To control/monitor the data acquisition system, the clinician must create an account by filling eight fields (Figure 3) about the clinician, the patient, and the characteristics of the implant.

The *Username* and the *Password* identify the clinician and are the credentials to make the login. The *No. Acquis* represents the number of acquisitions that the sensing technology must perform before converting the mean value into a quantification of the implant loosening (in

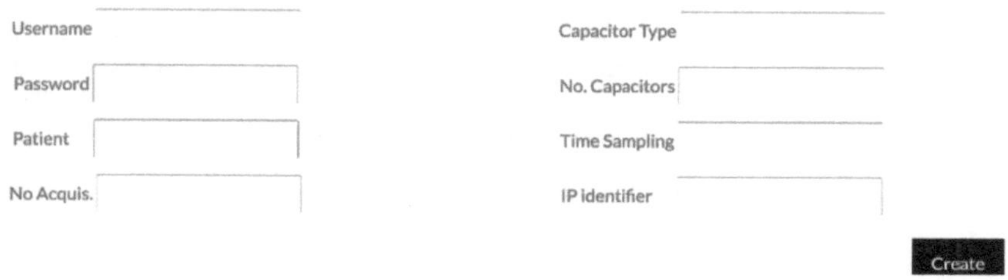

Figure 3. Fields to user creates an account.

terms of bone-implant distance). *Capacitor type* field must be filled with 'a','b' or 'c', depending on the type of capacitor that the implant incorporates. The *No. capacitor* indicates the number of capacitors that the PCB includes. *Time sampling* represents the interval between acquisition measured in days. The *IP identifier* is the IP of the Raspberry Pi connected to the data acquisition system.

After the user creates an account and signs in, the control page is displayed. It is possible to analyze the data in different manners: a three-dimensional representation of the bone-implant distance on a specific day (indicated by the user); a two-dimensional plot that allows the clinician to analyze the distance evolution, of a specific region (capacitor), between two days.

It is also possible to analyze the data in a three-dimensional plot. This feature is a helpful feature during a medical appointment, once it helps to analyze the impact that a specific change in the bone-implant distance has on the joint. Besides, the user can download the collected data to an excel file by pressing a *download* button. Time sampling can be changed, by introducing a new value in the field.

The website was validated through the simulation of an approximation test. In this test, the bone sample is placed above the capacitor and moved towards the capacitor. The values used to perform this test were obtained through a six-order polynomial (Equation 1), which describes this the change on the bone-implant distance, correlating the capacitance and the displacement. It is important to highlight that: (i) the $x = 0$ corresponds to the capacitor-sample contact; and (ii) the positive values corresponds to a compression test.

$$c(x) = \sum_{i=1}^{6} p_i x^i; -4 \leq x \leq 0.25 \ [mm] \tag{1}$$

$p_0 = 2.588$, $p_1 = 0.2925$, $p_2 = 0.4217$, $p_3 = 0.2894$, $p_4 = 0.09628$, $p_5 = 0.01495$, $p_6 = 0.0008502$

The abovementioned polynomial was used to create a table with all values between -4 mm to 0.25 mm with steps of 0.05 mm. This table was saved in the raspberry pi, once this device emulates the data acquisition system. Having this table, it was possible to know the capacitance/displacement values with a linear regression between the two proximal values.

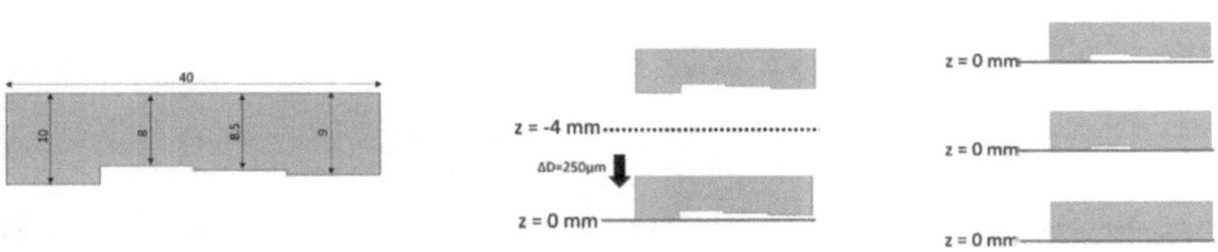

Figure 4. Schematic of the bone sample and scheme illustrating the approximation/regeneration tests.

A bone sample with different heights was defined; 4 mm was considered the initial position of the bone sample to the capacitor. Subsequently, displacements of 0.25 mm were defined until the contact between the capacitor and the higher height region (Figure 4). Afterward, bone regeneration until all the sample regions contacts the capacitor, was also simulated.

3 RESULTS

After the simulation, the web page was used to visualize the data in different manners. The three-dimensional plots were used to analyze the bone loosening on various days (Figure 5). The gap between the bone and the capacitor diminishes until all the capacitor-sample contact is achieved. Note that the 0 mm in the z-axis is represented at the bottom to simulate the capacitor position. The negative values present the distance of the bone relative to the capacitor.

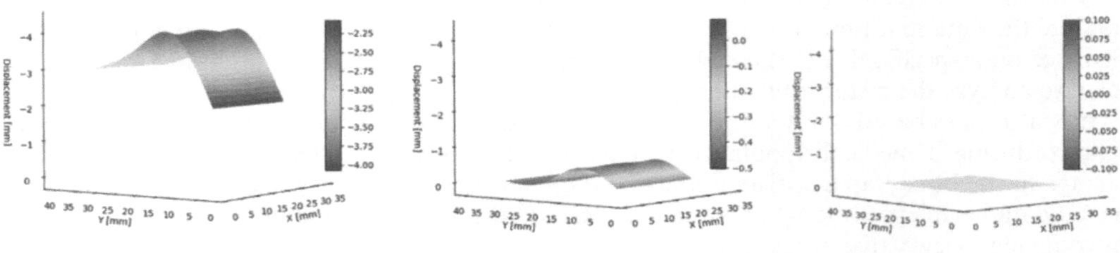

(a) First acquisition (b) Fifteenth acquisition (c) Seventeenth acquisition

Figure 5. Three-dimensional plot for different days of the approximation validation test.

Due to the capacitors network arrangement (matrix 4x3), the acquired values in capacitors of each row are the same. Taken this specification into consideration, the plots of each row are the same. So, it is only demonstrated the first plot of 3 out 4 rows, once the others will be the same. In each plot, the diminishing distance per region is visible as illustrated in Figure 6.

The download operation allows to acquire a file with five columns, corresponding to the capacitor position (x and y columns), the bone-capacitor distance, and the value of capacitance. As expected, the acquired values in the same capacitors' row were the same.

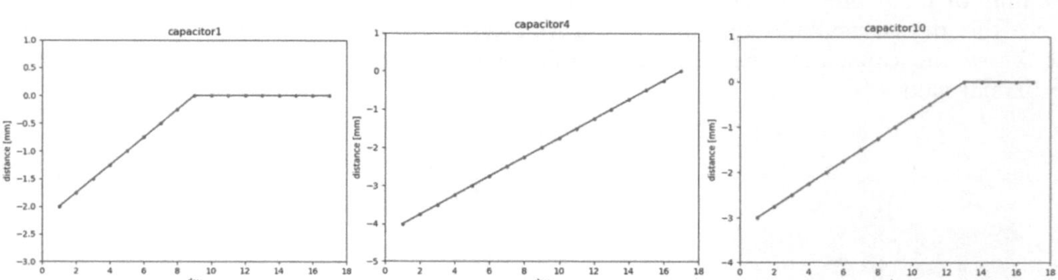

Figure 6. Data acquired from each capacitor of the first row, during the approximation and regeneration test.

4 DISCUSSION AND CONCLUSIONS

In the last decade, different methodologies to improve the monitoring of the bone-implant infixation were proposed. Although these methodologies achieve better results compared to the actual imaging diagnosis, they also present significant limitations, namely: they do not allow the monitoring of target regions; they cannot be incorporated inside the implant; the diagnosis is restricted to a laboratory environment, inhibiting the continuous monitoring of the patients.

The capacitive methodology has important advantages when compared to the others. This technology is able to monitor target regions, it can be integrated within the implant, and it allow the patients continuous monitoring. The website enables data download, which is an important feature, since it allows the clinician to perform further clinical analyses that are not available on the webpage. The possibility of visualizing the data in real-time by clinician, is also important to analyze the impact of loosening in a specific body movement.

The developed application proved to be promising to operate during in-vitro tests within the co-surface capacitive sensor network.

REFERENCES

Cachão, João Henrique, Marco P. Soares Dos Santos, Rodrigo Bernardo, António Ramos, Rainer Bader, Jorge A.F. Ferreira, António Torres Marques, and José A.O. Simões. 2020. "Altering the Course of Technologies to Monitor Loosening States of Endoprosthetic Implants." *Sensors (Switzerland)* 20 (1): 7–10. https://doi.org/10.3390/s20010104.

Kurtz, Steven M., Edmund Lau, Kevin Ong, Ke Zhao, Michael Kelly, and Kevin J. Bozic. 2009. "Future Young Patient Demand for Primary and Revision Joint Replacement: National Projections from 2010 to 2030." In *Clinical Orthopaedics and Related Research*, 467:2606–12. Springer New York. https://doi.org/10.1007/s11999-009-0834-6.

Labek, G, M Thaler, W Janda, M Agreiter, B Stöckl, and Orthopaedic Surgeon. 2011. "Revision Rates after Total Joint Replacement Cumulative Results from Worldwide Joint Register Datasets." *J Bone Joint Surg [Br]*, 93–293. https://doi.org/10.1302/0301-620X.93B3.

Mulcahy, D.M., G.C.C. Fenelon, and D.P. McInerney. 1996. "Aspiration Arthrography of the Hip Joint." *The Journal of Arthroplasty* 11 (1): 64–68. https://doi.org/10.1016/s0883-5403(96)80162-8.

Pivec, Robert, Aaron J. Johnson, Simon C. Mears, and Michael A. Mont. 2012. "Hip Arthroplasty." *The Lancet* 380 (9855): 1768–77. https://doi.org/10.1016/S0140-6736(12)60607-2.

Zhang, Yongde, Aaron W. Putnam, Anneliese D. Heiner, John J. Callaghan, and Thomas D. Brown. 2002. "Reliability of Detecting Prosthesis/Cement Interface Radiolucencies in Total Hip Arthroplasty." *Journal of Orthopaedic Research* 20 (4): 683–87. https://doi.org/10.1016/S0736-0266(02)00005-0.

 ISBN: 978-1-032-10806-3

Biomechanical study of different birthing positions to facilitate the process of labor

M. Borges & T. Mascarenhas
CHUSJ, Faculty of Medicine, University of Porto, Porto, Portugal

D. Oliveira, M. Parente & R.N. Jorge
INEGI-LAETA, Faculty of Engineering, University of Porto, Porto, Portugal

ABSTRACT: During vaginal delivery, several maternal positions can be adopted to facilitate the labor process, although there is limited evidence to support the most ideal position. This work aims at contributing to a better knowledge associated the biomechanics of different birthing positions and their consequences. A biomechanical model composed by the pelvic floor muscles attached to the bones, and a fetus head was used to simulate vaginal deliveries. Two birthing positions were mimicked: non-flexible and flexible sacrum positions. In the former, a 6 mm pubic symphysis widening and a coccyx rotation of 3.6° occurred. In contrast, a lower pubic symphysis widening occurred (3 mm) in the flexible sacrum model, since a higher coccyx rotation was observed. Globally, it can be concluded that positions with the weight off the sacrum appear to be more beneficial for the pelvic girdle bones of the mother.

1 INTRODUCTION

Childbirth can be psychologically and mechanically traumatic for the mother. Extensive physiologic changes occur in the mother to allow the passage of the fetus through the birth canal and, unfortunately, maternal morbidity and mortality remain too high.

To facilitate the process of labor, several maternal postures can be adopted during the second stage of labor. Birthing positions can be divided into positions that take the mother's body weight on or off the sacrum. In positions in which the weight is off the sacrum, such as squatting and sitting positions, the coccyx movement is allowed, being known as flexible coccyx positions. In positions in which the weight is on the sacrum, such as lithotomy and supine positions, the coccyx movement is restricted due to the presence of a surface under it, being known as non-flexible coccyx positions (Zang *et al.* 2020).

As *in vivo* studies in pregnant women are difficult and involve ethical constraints, biomechanical models have been developed to estimate mechanical changes on biological tissues. To the best of our knowledge, only few studies analyzed the musculoskeletal system changes that occur during childbirth using computational models and the changes in the pubic symphysis (PS) were not considered. Thus, it becomes crucial to clarify the adaptation that the maternal pelvis may undergo by allowing the PS widening in order to prevent mother and/or fetus complications.

The purpose of this work is to contribute to a better understanding of the biomechanics of different birthing positions that can be adopted during the second stage of labor and their consequences. A biomechanical model of the mother and fetus was modified to mimic two positions: one in which the coccyx movement is restricted and other that allows its movement.

DOI: 10.1201/9781003217152-15

The movements of the fetal head in vertex presentation and occipitanterior position were simulated.

2 MATERIALS AND METHODS

A finite element model of the mother and the fetus head (Parente *et al.* 2008) was used (Figure 1). The mother's model includes the pelvic girdle bones, the pelvic floor muscles (PFM) and supportive structures. To simulate the real PFM behavior, a quasi-incompressible transversely isotropic hyperelastic constitutive model already used in Parente *et al.* (2009) was and the following constitutive parameters were considered (Parente *et al.* 2009): $c = 0.00185$ MPa, $b = 1.173$, $A = 0.0280$ MPa, $a = 0.6215$, $D_1 = 1.0\times10^{-4}$ MPa^{-1} and $T_0^M = 0.682$ MPa.

In this work, two birthing positions were mimicked: non-flexible and flexible sacrum positions. The sacrococcygeal joint was considered in the latter. Regarding the pelvic girdle bones, the initial PS gap was 4.05 mm, and these bones were considered rigid structures to facilitate the convergence of the simulation. However, in sacrum and coccyx, the cortical and trabecular bone tissues were modeled (Wu *et al.* 2018). Additionally, the main pelvic ligaments were modeled as tension-only elements (Lei *et al.*, 2015). Concerning the sacrospinous and sacrotuberous ligaments, linear beam elements, using a low Young modulus, were superimposed to the truss elements mesh to give numerical stability to the simulation. In sacroiliac joints and in the PS, 3D solid elements were added to mimic the synovial part of these joints and the interpubic disc, respectively, ensuring a realistic simulation of the motion of the pelvic joints and providing some stability to the simulation A Neo-Hookean constitutive model was applied to these elements and to the sacrococcygeal joint elements. Since there is little experimental data on their mechanical properties, material properties of a similar biological tissue were applied (Peña *et al.* 2005): $c_{10}= 6.43$ MPa and $D_1= 1.0\times10^{-4}$ MPa^{-1}. Note that the overall stiffness of the joints is mainly provided by the pelvic ligaments, modeled with the truss elements.

For the boundary conditions, a tie constraint was applied between the lateral supporting structures of the PFM and the pubic bones, and between the sacrospinous and sacrotuberous ligaments and the fetal head. The nodes of the supporting structures corresponding to the insertion of the muscles in the sacrum were considered fixed, as well as the sacrum nodes in the articular faces.

Concerning the finite element model of the fetus, a group of elements inside the head was considered to be rigid and a reference point was assigned to it. During numerical simulations performed with Abaqus software 2018, the vertical movement of the fetal head and its flexion/extension were controlled by the reference point.

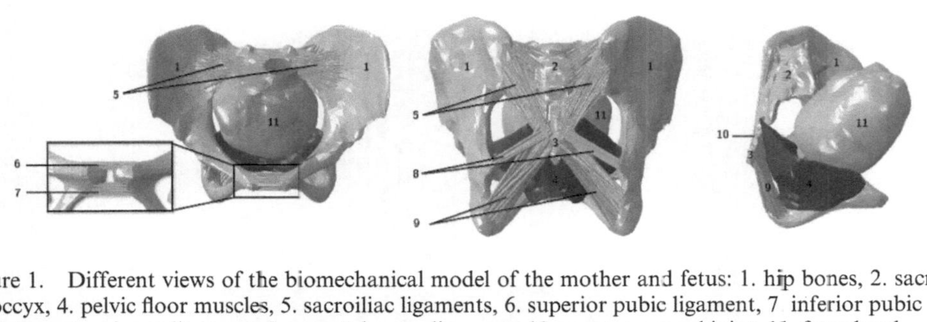

Figure 1. Different views of the biomechanical model of the mother and fetus: 1. hip bones, 2. sacrum, 3. coccyx, 4. pelvic floor muscles, 5. sacroiliac ligaments, 6. superior pubic ligament, 7 inferior pubic ligament, 8. sacrospinous ligament, 9. sacrotuberous ligament, 10. sacrococcygeal joint, 11. fetus head.

3 RESULTS

In this work, numerical simulations of the second stage of labor were performed. Even though the fetus was in the vertex presentation and occipitanterior position, the bones and PFM imposed some constraints on its movements and the predominant position was the left occipitanterior.

During labor, the PS can be more flexible and wider, aiming to facilitate the passage of the fetus through the birth canal. On the other hand, concerning sacrum and coccyx, a small amount of flexion and extension can occur at the sacrococcygeal joint. Thus, the widening of the PS, which corresponds to the distance between the narrowest points of the symphyseal gap, and the rotation of the coccyx were analyzed for each birthing position simulated (Figure 2).

Regarding the flexible sacrum model, the maximum value of the PS widening is 3.2 mm and the maximum coccyx rotation is 15.7°. In the non-flexible sacrum model, a maximum widening of the PS of 6.3 mm and a maximum coccyx rotation of 3.6° occur. These values occur at, approximately, a vertical displacement of the fetus head of 65 mm.

In the PFM, a path was defined in the most inferior portion of this structure and the maximum principal stresses were measured when its maximum occurs (Figure 3). As there are some differences in the space available in the mother's pelvis due to small variations in the cardinal movements, the instant analyzed in the non-flexible sacrum model corresponds to a vertical descent of 70 mm, and 77 mm in the flexible sacrum model.

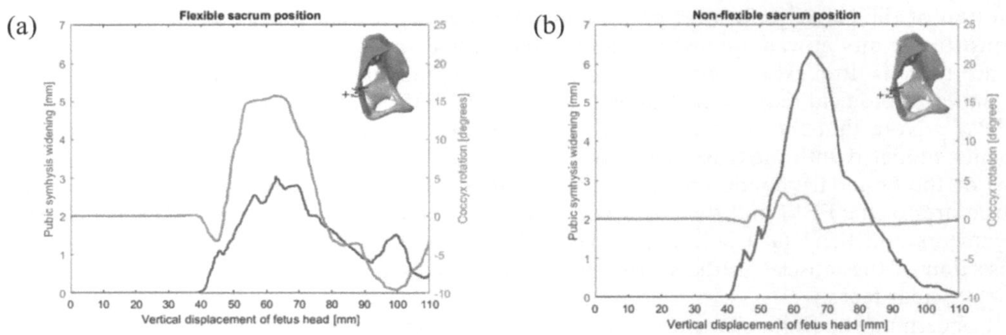

Figure 2. Pubic symphysis widening and coccyx rotation during the vertical displacement of fetus head, assuming two birthing positions: (a) flexible sacrum position and (b) non-flexible sacrum position.

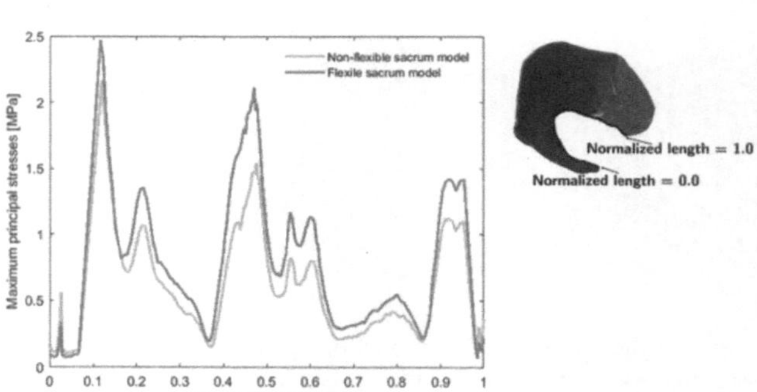

Figure 3. Maximum principal stresses measured at the peak stresses instant along the normalized path in the most inferior portion of the PFM.

It can be observed that higher stress values were obtained in the left area of the levator ani muscle in the flexible sacrum model. Moreover, in the defined path in the PFM, a stretch ratio was calculated, being defined as the ratio between the current and the original tissue length. The maximum value obtained is 1.7 in both models, considering an initial length of 17.1 cm.

4 DISCUSSION

Even though vaginal delivery is a natural process, several injuries can occur. Regarding the PS, a widening of 2-3 mm is expected, and PS diastasis can occur when it is higher than 10 (Segal and Fitzgerald 2015). In this study, there is no rupture of this joint in both positions, although the flexible sacrum model has a lower PS widening than the non-flexible sacrum model. Thus, a higher coccyx rotation of the former is obtained to increase the pelvic space available for the fetus head passage, being withing the physiological limits of the sacrococcygeal joint (Bø *et al.* 2001).

After a 90 mm descent of the fetus head, a flexion of the coccyx is observed in the flexible sacrum model that decreases the pelvic space available. Consequently, an increase of the PS gap is obtained. This movement can be explained by the fact that, in this joint, not all existing ligaments were modeled since the focus of this work is not the study of the ligaments themselves.

Regarding the PFM stresses, the values obtained can be compared with other study (Parente *et al.* 2009), in which the peak of stresses occurred only in the most posterior area of the levator ani muscles. Regarding the stretch ratio, it was also verified that that the maximum stretch ratio occurs in different instances when compared with (Parente *et al.* 2008). These differences are due to the restrictions imposed by the maternal pelvis and the different fetal positions simulated.

The present study demonstrates that, in flexible sacrum positions, the coccyx can move easily, and a lower widening of the PS occurs, appearing to be more beneficial for the mother. These results are consistent with (Reitter *et al.* 2014) that reported that positions where the body's weight is off the sacrum have higher diameters compared to positions in which the coccyx movement is restricted due to the bed under it. Nevertheless, the problem studied is very complex and some limitations should be considered, such as the material properties considered for the ligaments and bones that do not consider hormonal changes that occur during pregnancy.

5 CONCLUSIONS

Overall, the present study shows that different birthing positions cause changes in the pelvic space, so some can be adopted to reduce the risk of obstructed labor and the development of several injuries. Positions where the weight is off the sacrum avoid a high effort of the PS and a high impact on the sacrum and coccyx. Thus, flexible sacrum positions, such as kneeling, standing, squatting and sitting positions should be adopted since, from a biomechanical point of view, they are more beneficial for the bone structure of her pelvis.

ACKNOWLEDGMENTS

This research was supported by Portuguese FCT under research project UIDB/50022/2020 and by the project NORTE-01-0145-FEDER-030062 (SIM4SafeBirth) cofinanced by NORTE2020, through FEDER.

REFERENCES

Bø, K., Lilleås, F., Talseth, T., and Hedland, H., 2001. Dynamic MRI of the pelvic floor muscles in an upright sitting position. *Neurourology and Urodynamics*, 20 (2), 167–174.

Lei, J., Zhang, Y., Wu, G., Wang, Z., and Cai, X., 2015. The Influence of Pelvic Ramus Fracture on the Stability of Fixed Pelvic Complex Fracture. *Computational and Mathematical Methods in Medicine*, 2015.

Parente, M.P.L., Natal Jorge, R.M., Mascarenhas, T., Fernandes, A.A., and Martins, J.A.C., 2008. Deformation of the pelvic floor muscles during a vaginal delivery. *International Urogynecology Journal*, 19 (1), 65–71.

Parente, M.P.L., Natal Jorge, R.M., Mascarenhas, T., Fernandes, A.A., and Martins, J.A.C., 2009. The influence of the material properties on the biomechanical behavior of the pelvic floor muscles during vaginal delivery. *Journal of Biomechanics*, 42 (9), 1301–1306.

Peña, E., Calvo, B., Martínez, M.A., Palanca, D., and Doblaré, M., 2005. Finite element analysis of the effect of meniscal tears and meniscectomies on human knee biomechanics. *Clinical Biomechanics*, 20 (5), 498–507.

Reitter, A., Daviss, B.A., Bisits, A., Schollenberger, A., Vogl, T., Herrmann, E., Louwen, F., and Zangos, S., 2014. Does pregnancy and/or shifting positions create more room in a woman's pelvis? *American Journal of Obstetrics and Gynecology*, 211 (6), 662.e1–662.e9.

Segal, N.A. and Fitzgerald, C.M., 2015. *Musculoskeletal Anatomic, Gait, and Balance Changes in Pregnancy and Risk for Falls*. Musculoskeletal Health in Pregnancy and Postpartum.

Wu, T., Ren, X., Cui, Y., Cheng, X., Peng, S., Hou, Z., and Han, Y., 2018. Biomechanical study of three kinds of internal fixation for the treatment of sacroiliac joint disruption using biomechanical test and finite element analysis. *Journal of Orthopaedic Surgery and Research*, 13 (1), 1–8.

Zang, Y., Lu, H., Zhao, Y., Huang, J., Ren, L., and Li, X., 2020. Effects of flexible sacrum positions during the second stage of labour on maternal and neonatal outcomes: A systematic review and meta-analysis. *Journal of Clinical Nursing*, 29 (17–18), 3154–3169.

Advances and Current Trends in Biomechanics – Belinha et al. (Eds)
 ISBN: 978-1-032-10806-3

Biomechanical analysis of fetal head molding during vaginal delivery

R. Moura & T. Mascarenhas
CHUSJ, Faculty of Medicine of the University of Porto, Porto, Portugal

M.C.P. Vila Pouca, D.A. Oliveira, M.P.L. Parente & R.N. Jorge
INEGI-LAETA, Faculty of Engineering of the University of Porto, Porto, Portugal

ABSTRACT: During childbirth, the fetal head is subjected to tremendous pressure leading to a certain degree of molding. This phenomenon can be harmful to the fetus and plays an essential role in its delivery. This work aims to perform a biomechanical study on fetal head molding by measuring specific diameters that vary during this process. Simulations of the second stage of labor were performed with a finite element model of the fetal head and pelvic floor muscles. A visco-hyperelastic constitutive model was used to characterize the sutures and fontanelles of the fetal head, allowing to obtain a time-dependent response. The main diameters were monitored to calculate the molding index, and it was concluded that prolonged labor reaches higher deformations with a maximum molding of 1.88. The stretch, stresses, and strain measured on the pelvic floor muscles indicate that a fetal head with higher molding leads to fewer injuries on the muscles.

1 INTRODUCTION

Childbirth is a natural physiological process that can lead to adverse events. More than 80% of all vaginal deliveries result in some degree of tissue laceration, which can result in pelvic floor disorders. Levator ani trauma is an often-undiagnosed form of maternal injury that can be harmful and lead to injuries that highly condition women's quality of life (Shek and Dietz 2019).

An extremely important phenomenon during childbirth is the ability of the fetal head to mold, a process designated as fetal head molding. The head adopts an elongated shape due to the pressure of the birth canal and surrounding structures (Bamberg *et al.* 2017). This might result in excessive molding which can be harmful to the fetus. The sutures and fontanelles of the fetal head are the structures that enable the molding process to occur and exhibit viscoelastic behavior (Romanyk *et al.* 2016).

Computational models are essential resources to quantitatively analyze the physiology of labor. Many studies have evaluated the impact that vaginal delivery has on pelvic floor muscles (PFM), namely the effect of different head positions, the influence of pelvic floor activation, the inclusion of obstetric procedures, and the addition of rate-dependent response. However, most studies focus only on the PFM, and none have used a deformable fetal head considering its viscoelastic behavior to analyze the molding progress during the second stage of labor. This work aims to study the molding process quantitatively to deeply understand the biomechanics of childbirth.

DOI: 10.1201/9781003217152-16

2 MATERIAL AND METHODS

A finite element model of the fetal head and pelvic floor muscles, shown in Figure 1, was used (Parente *et al.* 2008). The fetal head is composed of the skin, skull, brain, sutures, and fontanelles. The mentioned structures were modeled with solid tetrahedral elements (C3D4), except from the sutures and fontanelles, that were modeled in this work with membrane elements (M3D3). The pelvic floor muscles are composed of solid hexahedral hybrid elements (C3D8H). A surface defined as a rigid body was used to impose limits of the pelvic bones and improve the convergence of the numerical simulations.

To properly characterize the mentioned tissues, the material parameters summarized in Table 1 were retrieved from literature (Coats and Margulies 2006, Parente *et al.* 2009, Silva *et al.* 2015, Vila Pouca *et al.* 2017). For the skull, skin and brain, a linear elastic constitutive model was used. The sutures and fontanelles were characterized by a visco-hyperelastic constitutive model adapted in this work to plane stress. The Neo-Hookean was the hyperelastic model used and the material constant was obtained through a stress-strain experimental curve of a 2-month-old donor's suture. Regarding the pelvic floor muscles, the transversely isotropic hyperelastic constitutive model proposed by Martins *et al.* (1998) was used.

The numerical simulations were performed in Abaqus®. The second stage of labor was replicated in the vertex presentation and occipito-anterior position. Two simulations were performed considering two different durations for the second stage of labor (20 and 200 minutes). Regarding boundary conditions, the superior nodes of the pelvic floor muscles were fixed, as well as the bone surface. The Augmented Lagrange method was used to impose the contact between the fetal head, muscles, and bone surface. The fetal head molding was evaluated through the molding index (MI), considering the main diameters of the fetal head: maxillo-vertical (MaVD), biparietal (BPD) and suboccipitobregmatic (SOBD).

$$MI = \frac{MaVD^2}{BPD \times SOBD} \tag{1}$$

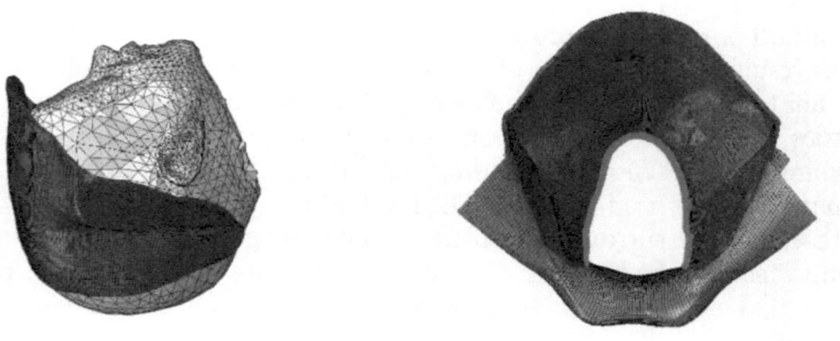

Figure 1. Finite element model of the fetal head (beige), pelvic floor muscles (red), supporting structures (dark gray) and bone surface (gray). A path created to analyze the muscles is represented in red.

Table 1. Material properties used to characterize the model structures.

Sutures and fontanelles			**Pelvic floor muscles**		**Fetal head**	
Neo-Hooke	Viscoelastic parameters		c = 0.020 MPa		E [MPa]	ν (-)
c = 0.027 MPa	$\beta_1 = 1.5$	$\tau_1 = 40$ s	b = 1.750 (-)	**Skull**	250	0.22
	$\beta_2 = 0.7$	$\tau_2 = 800$ s	A = 0.042 MPa	**Brain**	0.0246	0.49
	$\beta_3 = 0.5$	$\tau_3 = 10\,000$ s	a = 0.6215 (-)	**Skin**	0.3	0.25

To analyze the impact of fetal head molding on the pelvic floor muscles, the reaction forces provoked by the passage of the fetal head were measured at the pelvic floor's fixed nodes. Also, the stretch, stresses and strain were evaluated on an inferior path of the pelvic floor (represented in red in Figure 1).

3 RESULTS

In this work, computational simulations of the second stage of vaginal delivery with the fetus in a vertex presentation and occipito-anterior position were performed.

Figure 2 demonstrates the molding index variation during the vertical displacement of the fetus head. One can see that, for the prolonged second stage of labor, a higher molding index was obtained, reaching a value of 1.88. The molding index achieved for the precipitated birth was equal to 1.85. The maximum molding was obtained at approximately 57 mm of descent in both cases.

The MaVD, SOBD and BPD varied +2.13, -4.19 and -4.13 mm, respectively, at the instant of maximum molding for the prolonged labor.

The stretch values were measured in the inferior path of the pelvic floor muscles represented in Figure 1 during the vertical displacement of the fetal head. The maximum stretch achieved was 1.47 corresponding to a fetal descent of 70 mm. The maximum principal stresses and maximum principal logarithmic strain were measured at the instant of maximum stretch along the normalized path defined and are presented in Figure 3.

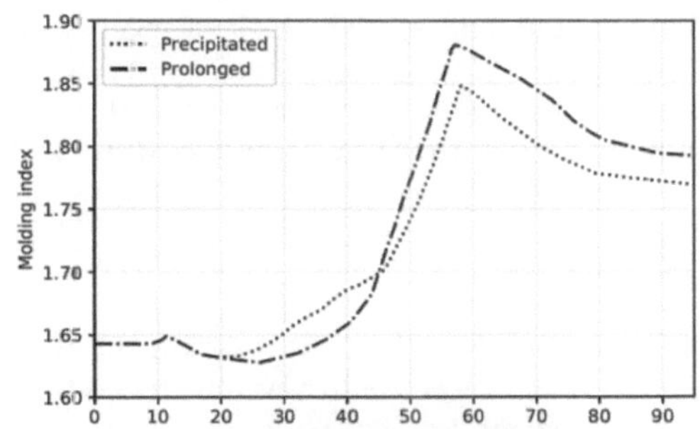

Figure 2. Variation of molding index with vertical displacement of the fetal head.

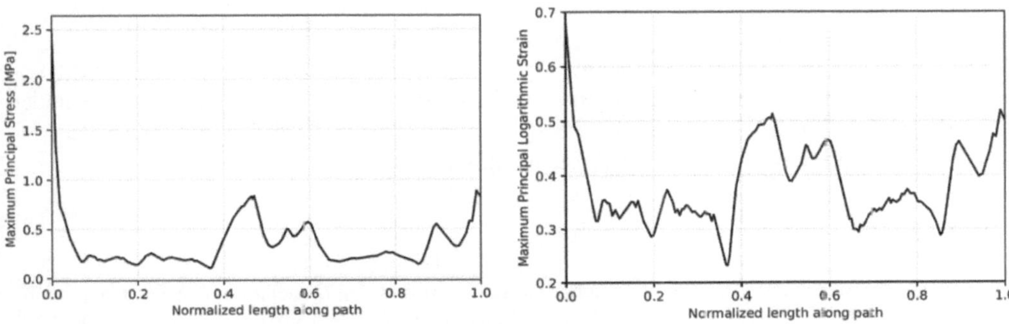

Figure 3. Stress and strain obtained on the pelvic floor muscles.

4 DISCUSSION

Childbirth is a complex biomechanical process that can be harmful to the fetus and mother. Analyzing the fetal head molding during vaginal delivery can help to understand the importance of this phenomenon and to quantitively evaluate the possibility of injuries.

Few numerical studies have analyzed the variation of diameters of the fetal head during the molding process. Literature states that there is an increase in the maxillo-vertical diameter, noticed by the elongation of the fetal head, and a decrease in the suboccipitobregmatic and biparietal diameters. The results obtained in this study are consistent with those of the literature since the aforementioned variations were achieved. Kriewall *et al.* (1977) reported a correlation between a prolonged delivery and an increase in the molding index, which corroborates the results presented.

The stretch of the pelvic floor muscles during the vertical displacement of the fetal head was determined. Parente *et al.* (2009) and Silva *et al.* (2015) have also analyzed the pelvic floor stretching. The former used a rigid fetal head to simulate vaginal delivery and obtained a stretch superior to 1.6. The later recurred to a deformable head considering some deformations of the fetal head and a maximum stretch of 1.504 was registered. In this study, a value of 1.46 was obtained, under the limit indicated in the literature as maximum non-injurious stretch (corresponding to 1.5). The stresses and strain obtained were also inferior to those presented by Silva *et al.* (2015). The presence of a deformable head considering its viscoelastic leads to less stretch, stress and strain of the PFM, indicating that higher molding can lead to less injuries.

The current work has several limitations. Firstly, the maternal finite element model is only composed of the pelvic floor muscles, insufficient to properly evaluate the deformations of the fetal head. Other structures, as the birth canal, also play an important role in the molding process and should be considered in the analysis. The lack of mechanical properties to describe the viscoelastic behavior of sutures and fontanelles can also influence the obtained results.

Despite these limitations, the present study demonstrates that non-invasive methods can be used to assess the degree of molding that the fetus undergoes during childbirth.

5 CONCLUSIONS

This work aimed to quantitatively analyze the fetal head during the second stage of vaginal delivery, considering a time-dependent response. It was concluded that prolonged labor results in higher molding, which can be harmful to the fetus, and that a deformed head causes less muscle injury, demonstrating the importance of the molding process to a successful delivery. The numerical analysis and quantification of physiological processes are very important to deeply understand the mechanics of childbirth.

ACKNOWLEDGMENTS

The authors gratefully acknowledge the support from Portuguese Foundation of Science under research project UIDB/50022/2020 and by the project NORTE-01-0145-FEDER-030062 (SIM4-SafeBirth) cofinanced by NORTE2020, through FEDER.

REFERENCES

Bamberg, C., Deprest, J., Sindhwani, N., Teichgräberg, U., Güttler, F., Dudenhausen, J.W., Kalache, K. D., and Henrich, W., 2017. Evaluating fetal head dimension changes during labor using open magnetic resonance imaging. *Journal of Perinatal Medicine*, 45 (3).

Coats, B. and Margulies, S.S., 2006. Material Properties of Human Infant Skull and Suture at High Rates. *Journal of Neurotrauma*, 23 (8), 1222–1232.

Kriewall, T.J., Stys, S.J., and McPherson, G.K., 1977. Neonatal head shape after delivery: An index of molding. *Journal of Perinatal Medicine*, 5 (6), 260–7.

Martins, J.A.C., Pires, E.B., Salvado, R., and Dinis, P.B., 1998. A numerical model of passive and active behavior of skeletal muscles. *Computer Methods in Applied Mechanics and Engineering*, 151 (3–4), 419–433.

Parente, M.P.L., Natal Jorge, R.M., Mascarenhas, T., Fernandes, A.A., and Martins, J.A.C., 2008. Deformation of the pelvic floor muscles during a vaginal delivery. *International Urogynecology Journal*, 19 (1), 65–71.

Parente, M.P.L., Natal Jorge, R.M., Mascarenhas, T., Fernandes, A.A., and Martins, J.A.C., 2009. The influence of the material properties on the biomechanical behavior of the pelvic floor muscles during vaginal delivery. *Journal of Biomechanics*, 42 (9), 1301–1306.

Romanyk, D.L., Shim, C., Liu, S.S., Lagravere, M.O., Major, P.W., and Carey, J.P., 2016. Viscoelastic response of the midpalatal suture during maxillary expansion treatment. *Orthodontics & Craniofacial Research*, 19 (1), 28–35.

Shek, K.L. and Dietz, H.P., 2019. Vaginal Birth and Pelvic Floor Trauma. *Current Obstetrics and Gynecology Reports*, 8 (2), 15–25.

Silva, M.E.T., Oliveira, D.A., Roza, T.H., Brandão, S., Parente, M.P.L., Mascarenhas, T., and Natal Jorge, R.M., 2015. Study on the influence of the fetus head molding on the biomechanical behavior of the pelvic floor muscles, during vaginal delivery. *Journal of Biomechanics*, 48 (9), 1600–1605.

Vila Pouca, M.C.P., Ferreira, J.P.S., Oliveira, D.A., Parente, M.P.L., and Natal Jorge, R.M., 2017. Viscous effects in pelvic floor muscles during childbirth: A numerical study. *International Journal for Numerical Methods in Biomedical Engineering*, 34 (3), e2927.

Advances and Current Trends in Biomechanics – Belinha et al. (Eds)
© 2022 Copyright the Author(s), ISBN: 978-1-032-10806-3

Muscle activity of gastrocnemius in running with and without ball in basketball

C.M. Amaro
University of Coimbra, Portugal

M.A. Castro
ESTeSC, Polytechnic of Coimbra, Coimbra, Portugal

S. Nolasco
University of Coimbra, Portugal

A.M. Amaro
CEMMPRE, Department of Mechanical Engineering, University of Coimbra, Portugal

L. Roseiro
ISEC, Polytechnic of Coimbra, Coimbra, Portugal

ABSTRACT: In basketball practice, accelerations and decelerations are commonly performed by the athlete, involving the activity of the lower limbs muscles, with special relevance for gastrocnemius. The main goal of the present study was to evaluate the gastrocnemius activity response in two basketball movements: running with and without the ball, comparing two moments of the sports season. The study was approved by a group of nine female basketball athletes, with ages between 18 and 23 years (mean: 19.9 years), and a mean basketball practice of 11 years. All of the volunteer's performed 7 valid repetitions of each movement, which means a total of 126 valid data. No statistically significant changes were detected between the two acquisitions moments, neither for running with the ball nor for running without the ball.

1 INTRODUCTION

Sports activity is extremely important in the context of health, with a clear motivation for its practice. Regular activity helps in the improvement of self-confidence and cognitive function as well as improves skeletal muscle function and cardiovascular health, and reduces depression and anxiety (Haskell et al., 2007; Mickevicius et al., 2018).

Basketball is one of the sports with a higher number of athletes, where the human body undergoes rapid requests and reactions, at high-intensity levels, such as sprints with and without the ball (Castagna et al., 2008). These short and intense actions, carried out in any direction, can induce injuries (Abdelkrim, El Fazaa and El Alti, 2007; Piedra et al., 2020).

The muscle activity can be considered a relevant study parameter in basketball players, related to the fact that gastrocnemius became one of three primary muscles spanning the ankle and the knee, the body parts with a major risk of injuries in basketball (Klyne et al., 2012; Piedra et al., 2020).

DOI: 10.1201/9781003217152-17

The present study aimed to evaluate the gastrocnemius muscle recruitment in two movements: running with (RB) and without the ball (RNB), which involve acceleration and deceleration.

Surface electromyography (sEMG) is a non-invasive technique that allows continuous measurement of activity (Subbu, Weiler and Whyte, 2015). A variety of parameters can be used to assess changes in muscle electrical activity in amplitude and frequency. To normalize the signal amplitude, maximum voluntary contraction (MVC), maximum signal peak, average signal value, or fixed signal value can be used. In this study, the highest value found for the movement appears as a reference for its normalization (Taha et al., 2016), analyzed the performance during training using sEMG, to identify the muscles with a probability of occurrence of injuries. According to these authors, through the correlation of muscle activity with performance, fatigue, and impulse, it is possible to adapt the training to athletes, increasing their performance.

Concerning basketball, some studies have been developed to assess muscle activation in athletes (Żukowska et al., 2016; Matthew and Gretchen, 2018), however, the majority of the analyses focus on the medium and long jump shot, and free throw.

2 MATERIALS AND METHODS

Nine senior female athletes federated in the Portuguese Basketball Federation volunteered to participate in this study, having been evaluated in two season moments: at the beginning, in November (T1), and at the peak of competitive activity, in March (T2). Between these two season moments, the athletes played a total of 20-25 official games. All athletes performed at least seven repetitions, in both T1 and T2, totalizing 126 valid data for each movement.

Table 1 shows the individual and mean relevant characteristics for the sample characterization.

The local ethics committee approved this study, which was performed according to the Declaration of Helsinki. All the participants' data were stored anonymous and confidentially, by the individual registration with a code. All subjects were fully informed of the nature of the study and gave written informed consent.

Surface electromyography data were collected via the BioSignalPlux system after skin preparation to decrease the impedance between the skin and the electrode. This telemetric system contains a data acquisition box and Bluetooth communication with the computer. The sensors used to record sEMG amplified signals with a bandwidth of 25-500 Hz, a common-mode rejection ratio of 100 dB, and an input impedance greater than 100 GOhm. The surface electrodes (Al/AgCl, AMBU BlueSensor N) were placed aligned with the muscle fiber orientation (center-to-center distance of 22 mm) at the most prominent part of the muscle bellies following SENIAM recommendations, Figure 1.

Table 1. Volunteers' relevant characteristics.

Athlete ID	Age [years]	Height [m]	Body mass index [kg.m^{-2}]	Years of practice [years]
1	19	1.79	22.8	7
2	19	1.66	24.3	12
3	19	1.71	18.8	11
4	22	1.64	26.2	17
5	23	1.69	25.3	17
6	21	1.58	28.0	7
7	18	1.62	19.8	11
8	19	1.60	24.0	6
9	19	1.66	19.2	11
Mean ± SD	19.9 ± 1.7	1.66 ± 0.06	23.2 ± 3.3	11 ± 4

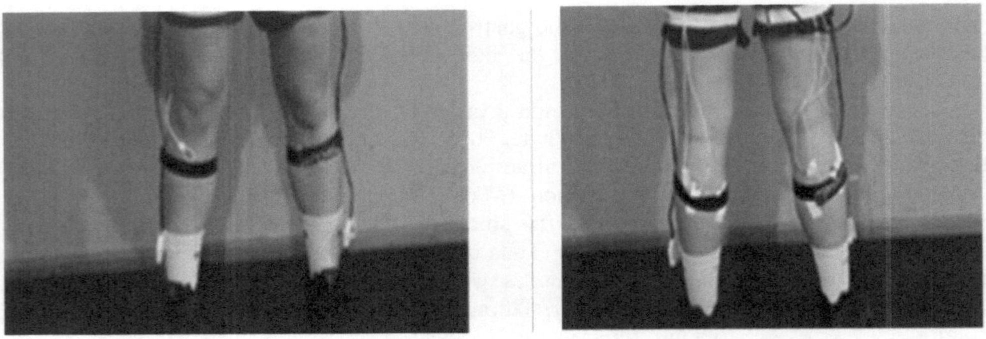

Figure 1. Athletes' sEMG instrumentation.

(a) (b)

Figure 2. Movements: a) running with the ball, b) running without the ball.

Two different movements have been considered in the study, namely running with the ball (RB) and running without the ball (RNB), as shown in Figure 2. The order of the movement performed was randomized, to avoid predisposition in the assessment.

The results were obtained statistically using the IBM SPSS Statistics program. The assumptions of normality were tested with the Shapiro-Wilk test. Due to the non-compliance with the assumption of normality, a Wilcoxon test was used to compare T1 and T2, and the Friedman test was used for multiple non-parametric comparisons. The level of significance was set at $p < 0.05$.

3 RESULTS

The average results ($\overline{X}$) and standard deviation (SD) relative to the percentage of maximum activation during the movement (% MA), are shown in Table 2.

Figure 3 shows an example of the sEMG raw signal, of both legs gastrocnemius lateral and medial, obtained in T1 and T2, for running with the ball (RB).

Table 2. Percentage values of muscle activation compared to MA.

Movement	Muscle	T1 ($\overline{X} \pm$ DP) [kPa]	T2 ($\overline{X} \pm$ DP) [kPa]	p
Running with ball (RB)	LLG	83±15	87±7	0,441
	LMG	85±12	80±22	0,401
	RLG	103±56	87±11	0,051
	RMG	78±18	88±10	0,214
Running without the ball (RNB)	LLG	80±18	86±9	0,374
	LMG	82±11	88±11	**0,025***
	RLG	77±18	86±10	**0,025***
	RMG	78±14	88±9	0,173

* $p < 0,05$; LLG – left lateral gastrocnemius, LMG – left medial gastrocnemius, RLG – right lateral gastrocnemius, RMG – right medial gastrocnemius.

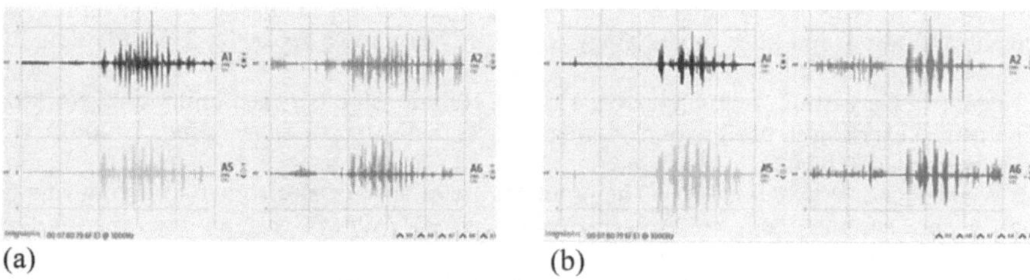

Figure 3. sEMG raw signal for one of the athletes in RB in the two-season moments: (a) T1, (b) T2.

4 DISCUSSION

Through table 3, it is possible to observe that, when comparing RB with RNB, the second presents lower % MA values for T1, whereas in T2 the same does not happen, with the higher similarity between values. Looking at all the values in the table, these are relatively similar to each other, with only a greater discrepancy in the RLG from RB to T1.

Analyzing the significance values, it is observed that we only have a statistically significant difference in two cases, both in the RNB, for the LMG and RLG.

5 CONCLUSIONS

In global terms, the results don't show statistically significant changes between the two times of the season, with only the exception in two cases (LMG and RLG) corresponding to the RNB. However, there is a slight increase over the sporting season in muscle activation values. This trend of increasing values leads to the conclusion that athletes perform the movements most appropriately, thus minimizing the risk of contracting injuries.

It's also possible to conclude that in the case of T2 the results are more homogeneous than in the case of T1, for both RB and RNB.

ACKNOWLEDGEMENTS

This work is sponsored by ERDF funds from the COMPETE program - Operational Program for Competitiveness Factors - and national funds through FCT - Foundation for Science and Technology - within the scope of the UIDB/00285/2020 project.

REFERENCES

Abdelkrim, N. B., El Fazaa, S. & El Alti, J. 2007. Time-motion analysis and physiological data of elite under-19-year-old basketball players during competition. *Br. J. Sports Med.* 41(2): 69–75. doi: 10.1136/bjsm.2006.032318.

Castagna, C., Abt, G., Manzi, V., Annino, G., Padua, E. & D'Ottavio, S. 2008. Effect of recovery mode on repeated sprint ability in young basketball players. *J Strenght Cond. Res.* 22(3): 923–929. doi: 10.1519/JSC.0b013e31816a4281.

Haskell, W.L., Lee, I.M, Pate, R.R., Powell, K.E., Blair, S.N., Franklin, B.A., Macera, C.A., Heath, G. W., Thompson, P.D. & Bauman, A. 2007. Physical activity and public health: updated recommendation for adults from the American College of Sports Medicine and the American Heart Association. *Med. Sci. Sports Exerc.* 39(8): 1423–1434. doi: 10.1249/mss.0b013e3180616b27.

Klyne, D. M., Keays, S.L., Bullock-saxton, J.E. & Newcombe, P.A. 2012. The effect of anterior cruciate ligament rupture on the timing and amplitude of gastrocnemius muscle activation: A study of alterations in EMG measures and their relationship to knee joint stability. *Journal of Electromyography and Kinesiology. Elsevier Ltd* 22(3): 446–455. doi: 10.1016/j.jelekin.2012.01.013.

Matthew, H. & Gretchen, O. 2018. Muscle activation patterns in wheelchair basketball athletes with and without physical disability. *International Journal of Physiatry* 4(1). doi: 10.23937/2572-4215.1510013.

Mickevicius, M. et al. 2018. Early anterior knee pain in male adolescent basketball players is related to body height and abnormal knee morphology. *Physical Therapy in Sport* 32: 273–281. doi: 10.1016/j.ptsp.2018.04.003.

Piedra, A., Peña, J., Ciavattini, V. & Caparrós, T. 2020. Relationship between injury risk, workload, and rate of perceived exertion in professional women's basketball. *Apunts Sports Med.* 55(206): 71–79. doi: 10.1016/j.apunsm.2020.02.004.

Subbu, R., Weiler, R. & Whyte, G. 2015. The practical use of surface electromyography during running: does the evidence support the hype? A narrative review. *BMJ Open Sport and Exercise Medicine* 1(1): 10–12. doi: 10.1136/bmjsem-2015-000026.

Taha, Z., Lee, C.M., Ahamed, N.U., Joseph, S. & Omar, S.F.S. 2016. Performance analysis in strength training: an innovative instrumentation. *Proc. Eng.* 147: 455–460. doi: 10.1016/j.proeng.2016.06.340.

Żukowska, H., Krygowski, D., Szarck-Eckardt, M & Zajac, M. 2016. Flexibility program among lower secondary school students and physical fitness indicators assessed in the convention of health-related fitness (H-RF). *Cent. Eur. J. Sport Sci. Med.* 15(3): 65–72. doi: 10.18276/cej.2016.3-07.

Advances and Current Trends in Biomechanics – Belinha et al. (Eds)
© 2022 Copyright the Author(s), ISBN: 978-1-032-10806-3

Hand-arm vibration in the pass and reception on basketball

A. Cavaleiro
University of Coimbra, Coimbra, Portugal

A.M. Amaro & M.A. Neto
CEMMPRE, Department of Mechanical Engineering, University of Coimbra, Coimbra, Portugal

L. Roseiro
ISEC, Polytechnic of Coimbra, Coimbra, Portugal
CEMMPRE, Department of Mechanical Engineering, University of Coimbra, Coimbra, Portugal

M.A. Castro
ESTeSC, Polytechnic of Coimbra, Coimbra, Portugal
CEMMPRE, Department of Mechanical Engineering, University of Coimbra, Coimbra, Portugal

ABSTRACT: The human being can benefit in many areas from practicing sports. However, it can also be harmful and cause injuries when the body limits are not respected. In particular, vibrations that affect the upper limb above the established limit, can be detrimental to the normal behavior of the body. In a sport like basketball, where the pass and reception play an essential role, studying the vibration induced by the ball, can help to understand the effects of the sport. The main objective of this study is to analyze and compare the values of acceleration, during the pass and reception of the ball, from different experienced groups. Hand-arm vibration of 15 athletes was collected. The group with higher values of acceleration was the intermediate one. The difference, when comparing opposite hands were higher in the most experienced group.

1 INTRODUCTION

Nowadays, it is consensual in the scientific community that maintaining a good level of physical activity, i.e., reaching the objective defined by World Health Organization (WHO), contributes to reducing the risk for many diseases (WHO, 2020). Besides that, at young ages, the profits can be physical and cognitive. Basketball, as a sport played by millions of people around the world (*FIBA*, 2021), gathering players from all age groups, plays a significant role in leading people to a healthier lifestyle. However, basketball is a physical game with a lot of repetitive movements. Although, as in other sports, when the limits are exceeded, adverse effects like injuries, can occur (Newman and Newberg, 2010). When analyzing the sport, one can identify the importance of the upper limb in all the activities that require the ball. Some studies refer that basketball and some movements present in the game, induce hand-arm vibration (HAV). The reception with two hands of a pass thrown to the chest zone appear to induce the highest value registered (Amaro *et al.*, 2016). When people are exposed to HAV, surpassing the established limits, and exposed for long periods, some injuries can be present, with damage to the circulatory and nervous system (Fridén, 2001; Rui *et al.*, 2008). However, the vast majority of these studies are related to work HAV and their health-related problems. Since there are no guidelines for exposure values in sports, for this study, we considered the standard values from ISO 5349 (ISO 5349-1:2001 Mechanical vibration — Measurement and evaluation of human exposure to

DOI: 10.1201/9781003217152-18

hand-transmitted vibration — Part 1: General requirements, 2001). Some studies (Roseiro *et al.*, 2016) have already used this method with valid results. On the other hand, the lack of studies regarding the younger population leaves a gap to be completed in understanding how kids and adolescents react to daily exposure to HAV. Besides that, it is important to know how the experience and adaptation through years of practice can influence the results.

The present study aims to evaluate and compare the values of vibrations induced to the upper limb, on the pass and reception of a basketball ball with two hands, regarding different basketball federated athletes experienced groups.

2 MATERIALS AND METHODS

For the study, 15 male athletes voluntarily participated, all players from the Portuguese Basketball Federation (FPB), in under-12, under-16, and senior class. Informed consent was delivered and signed by the senior players or guardians (in the case of minor volunteers). All the data obtained and used for this analysis was anonymous and confidentially saved. Table 1 shows the anthropometric data from the volunteers. Athletes were divided into 3 groups: the first group (G1) composed of 6 athletes with 0 to 4 (included) years of experience, the second group (G2) composed of 5 athletes with 5 to 9 (included) years of experience, and the third group (G3) by 4 athletes with more than 10 years of experience.

The movement considered in this study was the pass and reception of the ball with both hands, in the chest zone. The two-handed pass is a basic movement, taught since early ages by the coaches, consisting of the extension of the arms to make the ball reach the teammate. Consequently, the other player has to receive the ball, with one or two hands, facing the palms in the direction of the ball and flexing the arms to absorb the impact. In our study, the two-handed pass, the less advanced was chosen, since beginner athletes were included.

To measure the vibration-induced into the hand-arm system a triaxial piezoelectric accelerometer was used. The accelerometer was placed in the dorsal side of the hand, between the 2nd and the 3rd metacarpal bone. To reduce the perturbation from the materials, the athlete used a sports glove with a sewed support. This support was specially designed in AutoCAD® and posteriorly materialized with a 3D printer. Figure 1 shows the athlete with the equipment necessary to do the experimentation.

Each athlete performed 7 repetitions with the accelerometer on each hand, resulting in 210 valid data for the movement. To preserve similar force production during the repetitions of each player the pass was done against a rigid wall and the pressure of the ball was kept always

Table 1. Anthropometric data from the athletes.

Athlete	Age	Height	Weight	Wingspan	Dominant Hand	Practice
	Years	m	Kg	m		Years
1	11	1,42	30	1,45	Left	3
2	11	1,45	37	1,44	Right	2
3	11	1,40	35	1,38	Right	2
4	11	1,40	33	1,45	Right	4
5	11	1,52	40	1,5	Right	1
6	15	1,84	61	1,85	Right	6
7	15	1,73	70	1,73	Right	5
8	17	1,70	72	1,72	Left	8
9	15	1,76	55	1,74	Left	6
10	15	1,77	60	1,78	Right	7
11	29	1,90	90	1,96	Right	20
12	22	1,96	91	1,95	Right	2
13	22	1,85	82	1,94	Right	13
14	20	1,73	74	1,80	Right	14
15	19	1,84	66	1,84	Right	10

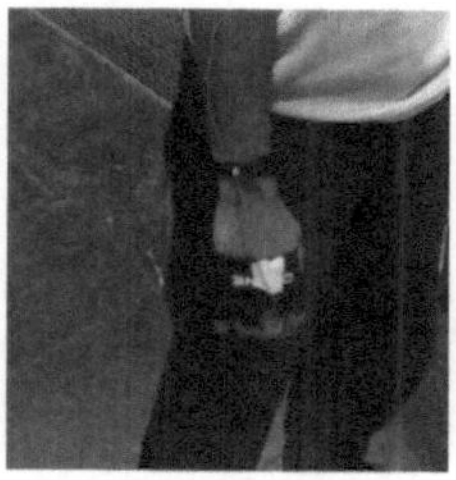

Figure 1. Sport glove and accelerometer in the athlete's hand.

the same. Every repetition was performed at a distance of 3 meters from the wall, using a marker on the ground.

For each hand, the root mean square (RMS) values regarding the three axes of both hands, RMS X D, RMS X E, RMS Y D, RMD Y E, RMS Z D, RMS Z E, were calculated. The total acceleration (AT) value from the right (D) and left hand (E) was also calculated. A program in LabVIEW® software language was used to store and analyze the signal, with a sampling frequency of 1000 Hz, since ISO5349 recommended frequency is 6.3Hz – 1000Hz.

The results were statistically analyzed using IBM SPSS Statistics, Kruskal-Wallis test to verify differences between acceleration values of different groups, accompanied by multiple comparisons of Dunn to compare the opposite hands in the same group.

3 RESULTS

Table 2 displays the average value ($\bar{X}$) and the standard deviation (SD) for the RMS values regarding the X, Y, and Z-axis and total acceleration from the right and left hand considering different experienced groups. The letter p indicates the statistics analysis regarding different groups of the same axis and hand and regarding opposite hands from the same axis and group.

Table 2. RMS values in each axis and the total value of acceleration.

HAV	G1	G2	G3	p
	$\bar{X}$±SD [m/s^2]	$\bar{X}$±SD [m/s^2]	$\bar{X}$±SD [m/s^2]	
RMS X-D	33,3±12,1	41,535±18,7	38,4±12,9	0,252
RMS X-E	30,3±21,7	40,9±42,4	22,4±9,9	0,141
P	0,249	0,686	0,068	
RMS Y-D	27,1±12,2	34,3±12,2	33,9±5,6	0,104
RMS Y-E	25,7±19,5	33,6±24,9	18,3±3,2	0,056
p	0,345	0,893	0,068	
RMS Z-D	33,2±12,3	41,3±26,6	26,5±6,1	0,065
RMS Z-E	26,4±11,0	38,7±31,5	20,1±3,7	0,032
p	0,075	0,345	0,068	
AT-D	56,6±14,1	71,2±37,3	59,0±6,9	0,277
AT-E	49,6±29,1	67,9±58,6	35,9±8,7	0,024
p	0,173	0,893	0,680	

Statistical differences were found in the left hand for the total acceleration and RMS Z E ($p<0,05$). Figure 2 illustrates the different values of RMS and total acceleration considering the each group.

4 DISCUSSION

This study's main goal was to evaluate the acceleration values, both RMS and total, in the right and left hand, considering the three axes, in different experienced basketball groups.

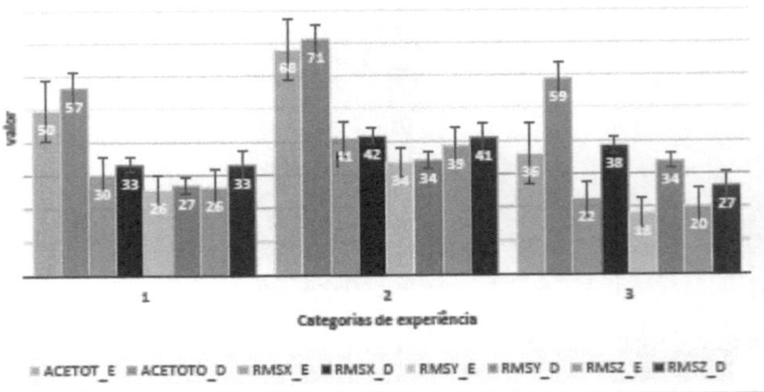

Figure 2. RMS and total acceleration values regarding different groups.

The second group that included athletes with 5 to 9 years of basketball experience, obtained the highest values of acceleration in every variable of the study, although no statistical differences between groups were found. The value for the RMS X-D, the highest value for RMS in every group, was 33,3±12,1 m/s^2 in G1, 41,535±18,7 m/s^2 in G2 and 38,4±12,9 m/s^2 in G3. When we analyze Figure 2, the lowest total acceleration value is recorded in the left hand of the most experienced group, G3,with 35,9±8,7 m/s^2. In the most experienced group, we can see, in Figure 2, a greater discrepancy comparing opposite hands. Every variable, when comparing opposite hands, recorded a higher value for the right hand which could be related with the main tendency of right hand dominance in the studied sample.

5 CONCLUSIONS

In this study, AT-E and RMS Z-E show statistically significant differences ($p<0.05$) between groups. The group that recorded the highest value of total acceleration was the G2, the intermediate one. The differences between opposite hands are more noticeable in the G3, the most experienced group. When comparing opposite hands, the right one recorded the highest value for every variable, despite being a two-handed movement.

REFERENCES

Amaro, A. M. *et al.* (2016) 'Avaliação das vibrações induzidas ao sistema mão-braço numa equipa feminina sub-16.', 27(2002), pp. 67–73.

FIBA - Quick facts (2021). Available at: http://www.fiba.basketball/zh/Page/f439097c-85aa-4996-97d6-65f4b4b193d8/0fe4dd12-c1d0-4fb5-9e36-0f4559c28993 (Accessed: 26 February 2021).

Fridén, J. (2001) 'Vibration damage to the hand: Clinical presentation, prognosis and length and severity of vibration required', *Journal of Hand Surgery*, 26 B(5), pp. 471–474. doi: 10.1054/jhsb.2001.0633.

'ISO 5349–1:2001 Mechanical vibration — Measurement and evaluation of human exposure to hand-transmitted vibration — Part 1: General requirements' (2001), p. 24. doi: doi.org/10.3403/02534966.

Newman, J. S. and Newberg, A. H. (2010) 'Basketball Injuries', *Radiologic Clinics of North America*, 48 (6), pp. 1095–1111. doi: 10.1016/j.rcl.2010.07.007.

Roseiro, L. M. *et al.* (2016) 'Hand-arm and whole-body vibrations induced in cross motorcycle and bicycle drivers', *International Journal of Industrial Ergonomics*, 56(October), pp. 150–160. doi: 10.1016/j.ergon.2016.10.008.

Rui, F. *et al.* (2008) 'A prospective cohort study of manipulative dexterity in vibration-exposed workers', *International Archives of Occupational and Environmental Health*, 81(5), pp. 545–551. doi: 10.1007/s00420-007-0256-2.

WHO (2020) *WHO Guidelines on physical activity and sedentary behaviour, World Health Organization.* Available at: https://apps.who.int/iris/bitstream/handle/10665/325147/WHO-NMH-PND-2019.4-eng.pdf?sequence=1&isAllowed=y%0Ahttp://www.who.int/iris/handle/10665/311664%0Ahttps://apps.who.int/iris/handle/10665/325147.

Advances and Current Trends in Biomechanics – Belinha et al. (Eds)

Muscle activity of the biceps and triceps brachii in basketball

A. Cavaleiro & C.M. Amaro
University of Coimbra, Coimbra, Portugal

A.M. Amaro
CEMMPRE, Department of Mechanical Engineering, University of Coimbra, Coimbra, Portugal

L. Roseiro
ISEC, Polytechnic of Coimbra, Coimbra, Portugal
CEMMPRE, Department of Mechanical Engineering, University of Coimbra, Coimbra, Portugal

M.A. Castro
ESTeSC, Polytechnic of Coimbra, Coimbra, Portugal
CEMMPRE, Department of Mechanical Engineering, University of Coimbra, Coimbra, Portugal

ABSTRACT: Practicing sports is recognized as a beneficial habit for human beings. To know, in detail, every movement it is extremely important to study the muscles that make it possible to happen. In a sport like basketball, where the upper limb plays a crucial role in any action with the ball, the knowledge of the muscle activity from this segment of the body is crucial to evaluate eventual constraints and to define and improve train strategies. The main objective of this study is to analyze and compare the muscle activity in the biceps and triceps brachii, regarding different groups of federate athletes with varied experience levels. sEMG of 15 athletes was collected. The most recruited muscle is the lateral triceps. The most experienced group showed higher levels of muscle activation when compared with the less experienced groups.

1 INTRODUCTION

In the last years, the popularity of basketball has increased among the general population, resulting in millions of participants covering all ages. In this sport, present in the Olympic games since 1936, handling and throwing a ball with a higher weight when compared to other hand dominant sports require strength in the upper limb. Especially when it comes to competition, where the need for precision and effectiveness depends upon, among other things like decision making and technique, on the muscular strength of the muscles involved. Being basketball a team sport it is important that the ball is shared with all the players on the court. To do that, the players use the pass and, consequently, there is someone who receives the ball. A movement that is repeated a lot of times during a game or a practice. Since this movement is so important in the game, it important to know the muscular activity present.

After analyzing the movement, the main muscles that contribute to the movement are the flexors and the extensors of the arm, like the biceps brachii and triceps brachii. Aligned with all the other sports, training is required to refine the skill. Predictably, a movement done by a much-experienced player will be different in some aspects comparing to a less experienced player. According to (Pakosz, 2011), EMG muscle activation is lower in a more experienced player when comparing to a less experienced player, as well as activation time.

DOI: 10.1201/9781003217152-19

A great number of players start playing the sport at a young age, to stack the hours needed to polish the technique required to have success. (Pakosz, 2011) refer that a more experienced player is more accurate in the movement of the pass. Nonetheless, the authors couldn't find any works analyzing the movement of the reception of the pass, very present in the game and with a lot of unsuccess along the run of the game. This unsuccess, appears to be, even more flagrant, when it comes to less experienced and younger players. Since the technique is highly dependent on the moving pattern of the player, and this pattern is influenced by the muscle activity present in the movement (Gribble *et al.*, 2003), we can link the success of the technique with the muscles involved, such as the biceps brachii and triceps brachii. With all that being said, it is decisive to know the differences existent in the movement considering distinct experience levels. However, after reviewing the bibliography that is a lack of studies, in this sport, regarding the early ages and knowing what are the changes that occur as we devote more time to the sport. (Abe, Nozawa and Kondo, 2009) says that electromyography (EMG) can be a system used to measure the motor skill acquired when learning a skill, especially in the early stages of training. Some studies (Uzun *et al.*, 2012; Matthew and Gretchen, 2018) use surface electromyography (sEMG) to measure the muscle activation pattern in wheelchair basketball. In our study, the percentage of muscle activation was normalized with the value of maximum voluntary contraction.

Based on the listed information, and to enrich the literature about this theme and to give data to eventual training protocols leading to improved performance, the present study intends to evaluate and compare the levels of muscular activity on the biceps brachii and triceps brachii in different groups of federate athletes with diverse levels of experience, during the pass and reception of the pass.

2 MATERIALS AND METHODS

In the present study participated a group of volunteers, all masculine amateur players registered in Portuguese Basketball Federation, inscribed in under 12, under 16 classes and senior level. The informed consent was signed by all the legally aged and the education guardians of the minor volunteers. The data used in this study was kept confidential and anonymous. The 15 athletes, 5 of each class, were divided into three distinct groups according to the number of years federated: group 1 - from 0 to 4 years, including; group 2 - from 5 to 9 years, including; group 3- more than 10 years. Table 1 illustrates the characteristics of the volunteers.

The first group (G1) included 6 athletes, the second one (G2) 5 athletes, and the more experienced group (G3) 4. The movements analyzed in this study were the pass and the

Table 1. Anthropometric data of the athletes.

Athlete	Age	Height	Weight	Wingspan	Dominant Hand	Practice
	years	m	Kg	m		years
1	11	1,42	30	1,45	Left	3
2	11	1,45	37	1,44	Right	2
3	11	1,40	35	1,38	Right	2
4	11	1,40	33	1,45	Right	4
5	11	1,52	40	1,5	Right	1
6	15	1,84	61	1,85	Right	6
7	15	1,73	70	1,73	Right	5
8	17	1,70	72	1,72	Left	8
9	15	1,76	55	1,74	Left	6
10	15	1,77	60	1,78	Right	7
11	29	1,90	90	1,96	Right	20
12	22	1,96	91	1,95	Right	2
13	22	1,85	82	1,94	Right	13
14	20	1,73	74	1,80	Right	14
15	19	1,84	66	1,84	Right	10

reception of the pass. For the execution of the first movement, the pass, the athlete has to grab the ball with both hands and extending the arms in the direction that he wants to pass. In the reception the athlete with the palms of the hand facing the ball will absorb the impact flexing the arms, retracting the ball close to the chest. Individually, the athletes realized 7 repetitions of both movements, which gathered 210 results for each movement. The pass was performed against a rigid wall, with the athlete at a distance of 3 meters, from where he did all the repetitions. The pressure of the ball used was controlled. All these restrictions were imposed to maintain relatively similar force production in different subjects.

To collect the muscle activity and activation time, we used surface electromyography, an evidence-certified technique (CJ, 1997), from BioSignalPlus, recording at 1000 Hz, linked to a wireless data acquisitor attached to the back part of the player's shorts. These sensors amplified the signal with a common-mode rejection rate of 100dB, bigger than the recommended 95dB. The electrodes were located in the same direction as the fibers with a 22 mm distance from center-to-center. These electrodes were placed in the muscle belly of right biceps brachii (BD), left biceps brachii (BE), right lateral triceps brachii (TLAD), left lateral triceps brachii (TLAE), right long triceps brachii (TLOD), left long triceps brachii (TLOE), with the reference electrode placed on the clavicular bone. The skin was shaved and cleaned with alcohol on the zone of contact with the electrodes. All the wires were tied with a band for minimal vibration and maximal comfort during the movement, so it was as close to the game situation.

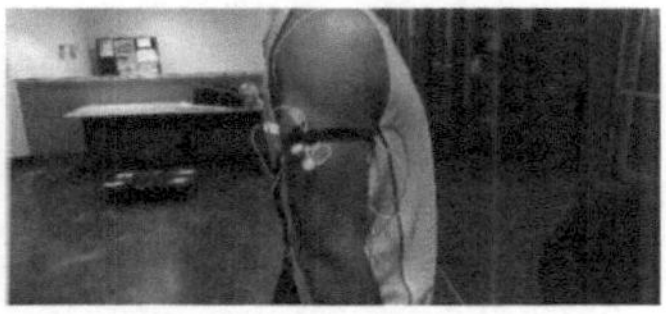

Figure 1. EMG placing on the athlete.

For each muscle, a maximum voluntary contraction test was performed to normalize the values of the maximal percentage of activation (%MMA) of every muscle. The software used for processing the signal of the electrodes was OpenSignals, with a band filter of 20-400 Hz. The results were analyzed using IBM SPSS Statistics, via the Kruskal-Wallis test, non-parametric, to compare the same muscles in different groups followed with multiple comparisons of Dunn to compare opposite side muscles in the same group.

3 RESULTS

Table 2 displays the average value ($\bar{X}$) and the standard deviation (SD) for the %MA normalized with MVC of each muscle considering different experienced groups. The statistics analysis

Table 2. Mean value of the maximal percentage of activation of the different groups.

Muscle	G1	G2	G3	p
	$\bar{X}$±SD [%]	$\bar{X}$±SD [%]	$\bar{X}$±SD [%]	
BD	52,99±39,43	48,79±33,75	79,19±37,95	0,081
BE	49,38±36,35	54,70±41,96	76,22±25,04	0,158
p	0,463	0,345	0,715	
TLAD	76,62±31,12	59,54±65,75	81,68±16,00	0,305
TLAE	69,70±30,23	46,95±32,18	83,29±25,28	0,044
p	0,463	0,500	1,000	
TLOD	55,75±32,05	56,35±23,60	83,39±19,26	0,046
TLOE	49,68±19,65	62,62±30,50	73,36±51,06	0,133
p	0,600	0,345	0,465	

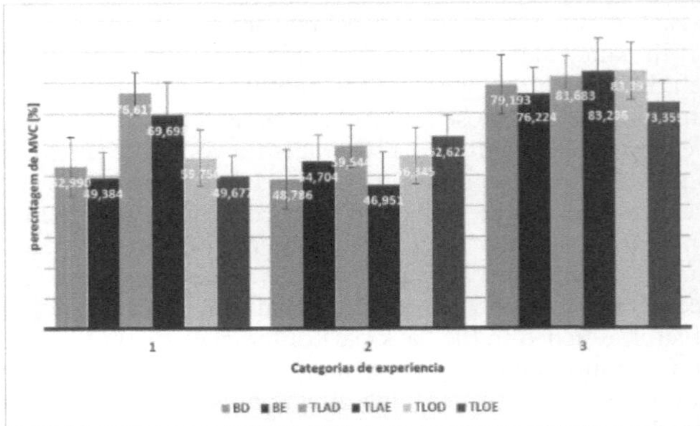

Figure 2. %MA of the muscles among the different groups.

(p) regarding the same muscle in opposite members from the same group and the same muscle in different groups are present in Table 2.

The TLAE and the TLOD showed differences among the groups (p<0,05), with no other variables presenting those differences. Figure 2 shows the different levels of %MA of all the muscles studied.

4 DISCUSSION

With this study, we wanted to study the muscular activity (%MA), normalized by MVC, during the pass and reception of a basketball ball in different experienced groups. With the resource of sEMG we can see in Table 2, and the help of Figure 2, that the most recruited muscle during the movements is the lateral triceps brachii with values around 80 %. For the most experienced group, the %MA of the BD was 79,19±37,95 while the G1 recorded values of 52,99±39,43. In the TLAD, when comparing the G2 with G3, the values of %MA were 46,95±32,18 and 83,29±25,28, respectively. When we analyze Figure 2, the values of the biceps brachii are lower in the less experienced groups when compared with the more experienced ones. The values of TLA and BE decrease from the G1 to G2 aligned with the literature revised. However, all the other variables are in contradiction with other studies affirming that muscle activation levels are lower in more experienced athletes, once they are more efficient in their movement. With a look at the figure _ we can say that the levels of %MA are more homogenous in G3, the groups with more than 10 years of experience.

5 CONCLUSIONS

Through this present study, and with the analysis of the data, several conclusions were obtained. The only muscles with differences (p<0.05) among the groups were TLAE and TLOD. The most recruited muscle is the lateral triceps brachii. The G3, the most experienced, obtain a much higher %MA on biceps brachii when comparing with the less experienced groups. The authors presume that this big gap can be justified by a higher maturation level of the third group, the one with 10 or more years of federate experience.

REFERENCES

Abe, S., Nozawa, T. and Kondo, T. (2009) 'A proposal of EMG-based training support system for basketball dribbling', *Lecture Notes in Computer Science (including subseries Lecture Notes in Artificial*

Intelligence and Lecture Notes in Bioinformatics), 5617 LNCS\(PART 1), pp. 459–465. doi: 10.1007/978-3-642-02556-3_52.

CJ, D. L. (1997) 'The use of Surface Electromyography in Biomechanics', *Journal of Applied Biomechanics*, 13(2), pp. 135–163. doi: 10.1016/j.aqpro.2013.07.003.

Gribble, P. L. *et al.* (2003) 'Role of cocontraction in arm movement accuracy', *Journal of Neurophysiology*, 89(5), pp. 2396–2405. doi: 10.1152/jn.01020.2002.

Matthew, H. and Gretchen, O. (2018) 'Muscle Activation Patterns in Wheelchair Basketball Athletes with and without Physical Disability', *International Journal of Physiatry*, 4(1). doi: 10.23937/2572-4215.1510013.

Pakosz, P. (2011) 'Emg Signal Analysis of Selected Muscles During Shots and Passes in Basketball', *Journal of Health Promotion and Recreation Rzeszow*, (2), pp. 8–14.

Uzun, S. *et al.* (2012) 'Evaluation of muscle fatigue of wheelchair basketball players with spinal cord injury using recurrence quantification analysis of surface EMG', *European Journal of Applied Physiology*, 112(11), pp. 3847–3857. doi: 10.1007/s00421-012-2358-0.

Advances and Current Trends in Biomechanics – Belinha et al. (Eds)
© 2022 Copyright the Author(s), ISBN: 978-1-032-10806-3

Analysis of induced vibrations to the hand-arm system during badminton practice

S. Nolasco
University of Coimbra, Portugal

A.M. Amaro & M.A. Neto
CEMMPRE, Department of Mechanical Engineering, University of Coimbra, Coimbra, Portugal

L. Roseiro
ISEC, Polytechnic of Coimbra, Coimbra, Portugal

M.A. Castro
ESTeSC, Polytechnic of Coimbra, Coimbra, Portugal

ABSTRACT: Badminton is the fastest racket sport in the world and practiced by millions of people, and its potentially harmful effect on the vibrations induced in the hand-arm system of the practitioners justifies detailed analyses. The study aims to evaluate the vibrations transmitted to the Hand-Arm System (HAS), comparing two rackets with different stiffness, during a smash movement, common in badminton sport. A group of volunteers performs a set of smash movements with the rackets instrumented with an accelerometer, and the vibrations induced to the HAS were quantified. The results obtained are present and discussed, showing that higher levels of vibration were obtained with the less rigid racket.

1 INTRODUCTION

Badminton is practiced by more than 200 million practicians (Phomsoupha and Laffaye, 2015). The practice of badminton at a high level requires great physical capacity, as the game is characterized by several direction changes, jumps, accelerations, and sudden decelerations. According to Jorgensen and Winge (1987), for every 1000 hours of practice, 2.9 injuries can occur.

The badminton smash is characterized by being an offensive stroke, usually executed from the bottom of the field and with a downward trajectory (Phomsoupha *et al.*, 2015). The smash can be executed with a forehand (FH) or backhand (BH) grip (Grice, 2008). FH is when the hand palm, at the moment of impact with the shuttle, points to the direction of the smash, Figure 1, with BH being the opposite.

The human body is subject to vibrations in its day-to-day that act on it gradually and according to the intensity of exposure. For low levels of vibration there will be discomfort and reduced efficiency, while for high levels and long periods of exposure, there may be injuries and diseases that can affect blood vessels and circulation (Braga, 2007). In Hand-Arm System (HAS), exposed to vibrations, can affect vascular, neurological, osteoarticular, and muscular systems (Soeiro, 2011). There are no specific standards for the assessment of vibrations in sports practice, thus the standard ISO 5349 - Measurement and evaluation of human exposure to hand-transmitted vibrations, which focuses exclusively on HAS, was used as a guide for the study presented, as it defines principles for the measurement and evaluation of human exposure to vibration transmitted to the HAS. For the

DOI: 10.1201/9781003217152-20

Figure 1. Smash steps with the FH grip.

study carried out, the definition of the HAS, refers to the hand, forearm, and arm, constituents of the upper limbs. This work aims to identify and quantify the level of vibrations induced to HAS during the badminton practice, particularly when performing the smash movement.

2 MATERIALS AND METHODS

Six senior Badminton athletes volunteer to participate in this study, aged 21.83±4.49 years and with an average time of practice of 10.67±5.34 years, signing a consent for participation and data collection. The local ethics committee approved this study, which was performed according to the Declaration of Helsinki., guaranteeing the total confidentiality of the data.

Two rackets with similar geometry was used, differing only in their stiffness. Both are produced by Yonex, belong to the same shaft series, were strung with the same strings, Yonex BG3 at a 125 Newton tension and were used with its factory grip. The racket specifications are shown in Figure 2. Duora 33 is a low-range racket, that is aimed at less experienced athletes and, therefore, more flexible. The Duora 88 is a mid-range racket, aimed at athletes with a higher level of play as claimed by Phomsoupha *et al.* (2015), that more rigid rackets are preferred by more experienced athletes.

The athletes performed 10 shots with maximum intensity, divided into groups of 5 and spaced for 30 seconds each, with the two different rackets.

For the acquisition of the data related to the vibrations, a triaxial piezoelectric accelerometer develop by PCB® with IEPE (*Integrated Eletronic Piezoeletric*) technology was used. The data acquisition was obtained via an acquisition board, NI 9234 of the C series, produced by National Instruments®. This board has four independent channels with a resolution of 24 bits and a dynamic range of 102 dB. The NI cDAQ-9171 chassis is a connection module to the NI 9234 acquisition board, via USB or Wireless, to establish the bridge between the acquisition board and the collection software, installed on a computer. The program for the acquisition, treatment, and analysis of the data related to vibrations used was TREMULA, developed in the LabView software specifically for this type of study (Pérez *et al.*, 2015). Both rackets were instrumented in the same position, 11 cm from the bottom, with the same accelerometer, to decrease the variables in the study in comparative terms (Figure 3). A removable fixing system

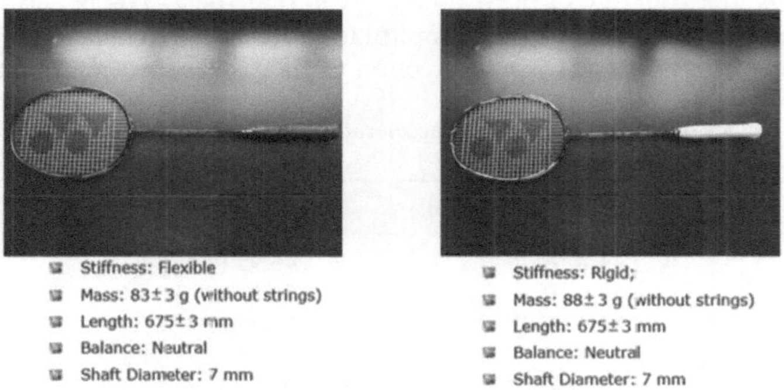

Figure 2. Rackets used for data acquisition (a) Yonex Duora 33 and (b) Yonex Duora 88.

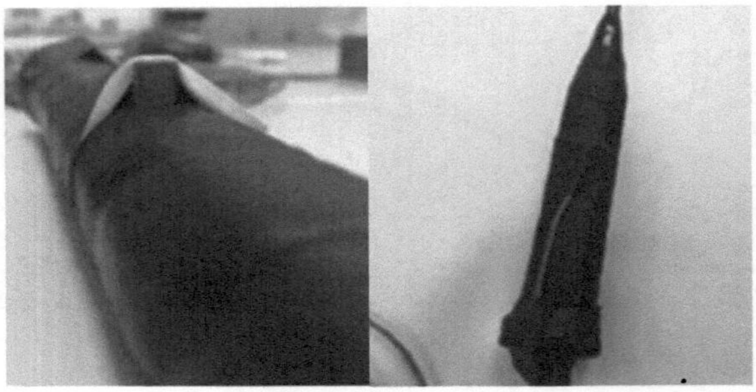

Figure 3. Racket instrumentation with the triaxial piezoelectric accelerometer.

was used to facilitate changes and be little invasive as possible. The barycentric system in the ISO 5349 standard was followed due to the origin of the referential be considered in the interface between the hand and the vibration transmission surface.

A time interval of 0.3 seconds allows visualizing the moment of the impact between the shuttle and the strings, as well as the propagation phase of the generated vibration. This signal was analysed via a Hand-Arm ponderation and the RMS.

The non-parametric test of Wilcoxon for comparison between 2 paired samples with IBM® SPSS Statistics 24.0 was used.

3 RESULTS AND DISCUSSION

The mean results ($\bar{X}$) and the standard deviation (SD) of the values related to weighted accelerations in frequency for the HAS obtained in the tests between rackets with different stiffness (vertical), as well as the level of significance between the variables in study (horizontal) are shown in Table 1. It should be noted that the level of significance chosen was 0.05 and whenever the value is checked, the value is marked with an asterisk (*). RMS represents effective acceleration along each axis and *ahv* the total acceleration.

Two significant differences were found. The first among the 3-axis RMS for the Yonex Duora 33 and the second among the total accelerations between the two rackets. Thus, peer comparison was used to understand where the differences lie. In the first case, the difference is identified between the RMS of the X and the Y-axis, as well as between the X and Z-axis. However, after calculating the adjusted significance, none of these differences proved to be significant. In the second case, the difference between the total acceleration for the two rackets is significant. Considering all movements, when comparing the total acceleration between the two rackets, the difference identified is due to the different properties of each racket.

Regarding the RMS vibrations (Figure 4), the values are maximum in the Z direction, followed by the Y direction with closed values and the X direction is the one with the lowest values. The results are in line with the awaited, given that the Z-axis is perpendicular to the hand palm, being the one in which higher amplitudes were expected. Also, the values obtained on the X-axis are coherent, as this is the direction of the racket's greatest rigidity. Regarding

Table 1. Values of frequency weighted accelerations and level of significance between variables.

	Duora 33 ($\bar{X}$ ±DP)	Duora 88 ($\bar{X}$±DP)	p
RMS X [m/s^2]	48.7±14.6	46.9±14.1	0.173
RMS Y [m/s^2]	58.6±18.2	53.65±21.7	0.249
RMS Z [m/s^2]	63.7±18.9	60.5±17.3	0.600
P	0.042*	0.311	
a_{hv} [m/s^2]	101.1±27.5	95.3±27.1	0.028*

* p < 0,05

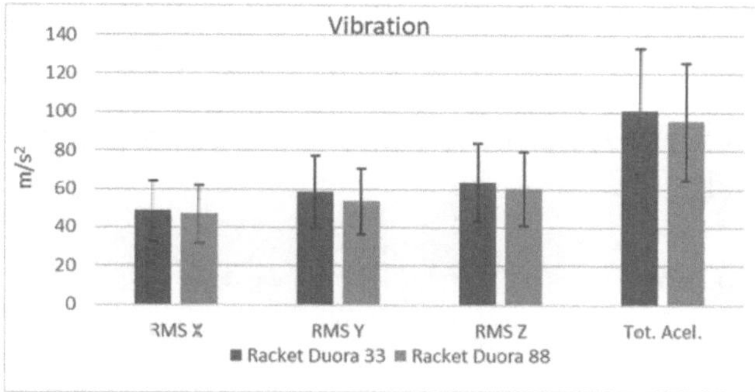

Figure 4. Axis acceleration values and the respective total acceleration as well as their standard deviation.

the Y-axis, the explanation for values so close to the Z-axis may be due to the method of picking up the racket. The existence of hand rotation during the movement can be reflected in a repositioning of the Y-axis in a direction in which the registered accelerations are higher.

The mean values show that Duora 33 racket transmitted vibrations with highest values than Duora 88 (total acceleration). It is, therefore, in the less rigid racket that the values of the vibrations induced to the HAS are higher.

4 CONCLUSIONS

As a result of the analysis of induced vibrations to the hand-arm system, the main conclusion to be drawn is that a racket with greater flexibility, less rigid, induces higher vibrations, due to the amplitude of vibration being greater than that of the more rigid racket. It should be noted that the most flexible rackets are usually more affordable at an economic level, so they are, therefore, the rackets most chosen by athletes who are starting in the sport, but, according to this study, the most harmful for the athlete, especially at a time when the technique is very rudimentary.

ACKNOWLEDGEMENTS

This work is sponsored by ERDF funds from the COMPETE program – Operational Program for Competitiveness Factors – and national funds through FCT – Foundation for Science and Technology – within the scope of the UIDB/00285/2020 project.

REFERENCES

Braga, C. A. (2007) 'Exposição ocupacional a vibrações no sistema mão-braço no sector da construção', p. 112. Available at: http://repositorio-aberto.up.pt/bitstream/10216/12734/2/Textointegral.pdf.

Grice, T. (2008) *Badminton: Steps to Success*. Human Kinetics (Sports instruction series). Available at: https://books.google.pt/books?id=Es5JngEACAAJ.

Jorgensen, U. and Winge, S. (1987) 'Epidemiology of badminton injuries', *International Journal of Sports Medicine*, 8(6), pp. 379–382. doi: 10.1055/s-2008-1025689.

Pérez, M. *et al.* (2015) 'Desenvolvimento de um sistema experimental para a determinação de frequências próprias de vibração', *Mecânica Experimental*, 24, pp. 21–28.

Phomsoupha, M. *et al.* (2015) 'How to use the elasticity of a badminton racket to increase its speed by 80%?', *Computer Methods in Biomechanics and Biomedical Engineering*, 18(sup1), pp. 2028–2029. doi: 10.1080/10255842.2015.1069607.

Phomsoupha, M. and Laffaye, G. (2015) 'The Science of Badminton: Game Characteristics, Anthropometry, Physiology, Visual Fitness and Biomechanics', *Sports Medicine*, 45(4), pp. 473–495. doi: 10.1007/s40279-014-0287-2.

Soeiro, N. (2011) 'Vibrações e o Corpo Humano: uma avaliação', *I Workshop de Vibrações e Acústica*, pp. 1–10.

Advances and Current Trends in Biomechanics – Belinha et al. (Eds)
© 2022 Copyright the Author(s), ISBN: 978-1-032-10806-3

Impact of the racket type on the upper limb muscle activity during badminton smash

S. Nolasco & A. Cavaleiro
University of Coimbra, Portugal

M.A. Castro
ESTeSC, Polytechnic of Coimbra, Coimbra, Portugal

L. Roseiro
ISEC, Polytechnic of Coimbra, Coimbra, Portugal

A.M. Amaro
CEMMPRE, Department of Mechanical Engineering, University of Coimbra, Coimbra, Portugal

ABSTRACT: The practice of a physical activity is fundamental in the context of health and well-being. Badminton is one of the most practiced racquet sports, belonging to the group with great speed in handling the racquet. The analysis of muscle activity contribution to the performance of the movement associated with this sport becomes essential for a better understanding. The present study aimed to evaluate the muscle activity of the arm, bicep, and triceps muscles associated with the smash movement, considering two types of rackets with different stiffness. A group of 6 volunteers, with a mean age of 21.8 (± 4.5) years, and an average badminton practice time of 10.7 (± 5.3) years, performs 10 valid repetitions of the movement and the muscle activity was collected using surface electromyography. The analysis of the results shows that the most rigid racket shows a greater muscular activity of the bicep and the lateral head of the triceps.

1 INTRODUCTION

Badminton, considered the fastest racquet sport, is practiced by millions of persons around the world (Phomsoupha and Laffaye, 2015). To achieve the desired speed during the shuttle, the muscles of the upper limb must produce the appropriate strength for such an undertaking (Lucas-Osma and Collazos-Castro, 2009). The type of racket used can influence the need for muscle strength production.

The badminton smash is characterized by being an offensive stroke, usually executed from the bottom of the field and with a downward trajectory (Phomsoupha *et al.*, 2015). The smash can be executed with a forehand (FH) or backhand (BH) grip (Grice, 2008). FH is when the hand palm, at the moment of impact with the shuttle, points to the direction of the shot, Figure 1, with BH being the opposite.

Surface Electromyography (sEMG) allows the evaluation of muscle activity and is largely used in sports (Belo *et al.*, 2016). In the present study, the muscles studied were the brachial biceps (Bicp), the lateral head of the triceps (TLat), and the long head of the triceps (TLong). Bicp is responsible for flexing the elbow and twisting the forearm. TLat and TLong, together with the medial triceps, are responsible for elbow extension, forearm adduction, and shoulder stabilization. More precisely, TLat is usually recruited for movements that require occasional

DOI: 10.1201/9781003217152-21

Figure 1. Smash steps with the FH grip.

spikes in strength, while TLong is used when sustained force is required or when there is a need for synergistic control of the shoulder and elbow (Lucas-Osma and Collazos-Castro, 2009).

The objective of the present study was to evaluate the muscular activity of the muscles of the arm, brachial biceps, lateral head, and long head of the triceps using surface electromyography, in the smash movement, comparing two rackets of different stiffness.

2 MATERIALS AND METHODS

Ten valid repetitions from 6 right-handed volunteers, with an average age of 21.8 (± 4.5) years and an average badminton practice time of 10.7 (± 5.3) years were collected. The local ethics committee approved this study and was carried out following the Declaration of Helsinki. The smash was made with the forehand grip (FH), using two types of racket (Yonex Duora 33 and Yonex Duora 88). The rackets used are very similar and only differ substantially in their stiffness. They are produced by the same manufacturer, from the same shaft series, were strung with the same 0.74 millimetres Yonex BG3 strings at a 125 Newton tension, and no overgrip was used during the acquisition. The racket specifications are shown in Figure 2. The first is a low-range racket, that is aimed at less experienced athletes and, therefore, more flexible. The second is a mid-range racket, aimed at athletes with a higher level of play as claimed by (Phomsoupha *et al.*, 2015) that more experienced athletes prefer more rigid rackets.

For the acquisition, sEMG BioSignalPlux equipment was used (Figure 3), with 1000 Hz of frequency. The electrodes used are a compound of Ag-AgCl, pre-gelled, disposable, and adherent to the skin and have the following characteristics (BioSignalsPlux, 2019): Gain (ratio between input and output signal): 1000, Range: ±1.5 mV, Bandwidth: 25-500 Hz, Input impedance >100 GOhm.

Muscle activation was assessed using electrodes placed on the muscular bellies of the bicep (Bicp), lateral head (TLat), and long head (TLong) of the triceps, and the reference electrode

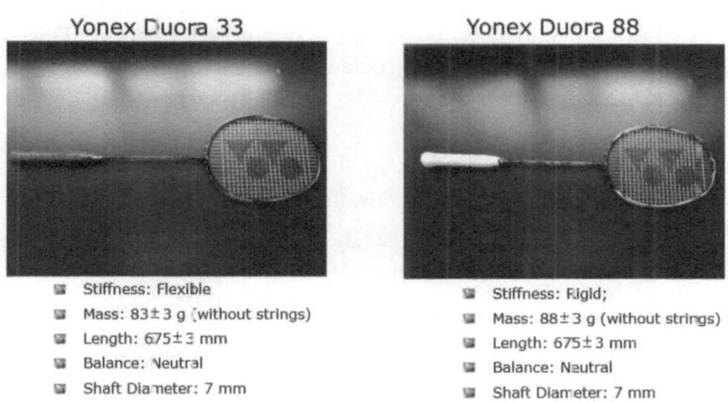

Figure 2. Rackets used for data acquisition.

Figure 3. Electromyographic instrumentation of the athletes.

placed on the lateral third of the right clavicle (Figure 3) after cleaning the skin to decrease the impedance of the interface between the skin and the electrode. A static maximum voluntary contraction (MVC) was measured and used for normalization in the post-processing phase.

The EMG data were digitally filtered (25–400 Hz), full-wave rectified and smoothed through a low-pass filter (12 Hz, fourth-order Butterworth digital filter)

The results were analysed statistically using the IBM SPSS Statistics program, applying Wilcoxon's non-parametric test for comparison between 2 paired samples.

3 RESULTS AND DISCUSSION

The average results (μ) and standard deviation (σ) relative to the maximum voluntary contraction (MVC), in percentage, as well as the comparisons according to the use of different rackets, are shown in Table 1.

Concerning % MVC muscle activation, statistically, significant differences occur between Bicp and TLat for Duora 33 and between TLong and TLat for Duora 88, with $p_{adjusted}=0.012$ for both. This result is an indicator that the most requested muscle in the smash, regardless of the grip, is the TLat which is the agonist muscle, responsible for elbow extension through a concentric or eccentric contraction, depending on the grip, during the smash.

Intramuscular differences occur for normalized amplitude (% MVC) in Bicp and TLat, which are the most requested muscles. Among these, the difference between Duora 88 and Duora 33 is positive. This can be an indication that the vibration transmitted through the rackets influences the muscular activity of the referred muscles which is higher in the less rigid racket, the Duora 33. This is a coherent result since, it's necessary to dissipate lesser energy, the muscle does not need to be so permeable to vibrations, and may have a more intense contraction. This is an interesting result given that, the fastest smash ever, was executed with a racket of high rigidity, which corroborates this result.

Table 1. Percentage of MVC muscle activation for each of the rackets and respective comparisons.

	Músculo	**Duora 33** $\mu\pm\sigma$	**Duora 88** $\mu\pm\sigma$	**p**
% MVC	**Bicp**	58.5±14.1	65.2±16.3	0.028*
	TLat	91.8±17.0	98.2±24.5	0.046*
	TLong	67.4±13.5	58.2±12.3	0.249

* p < 0,05

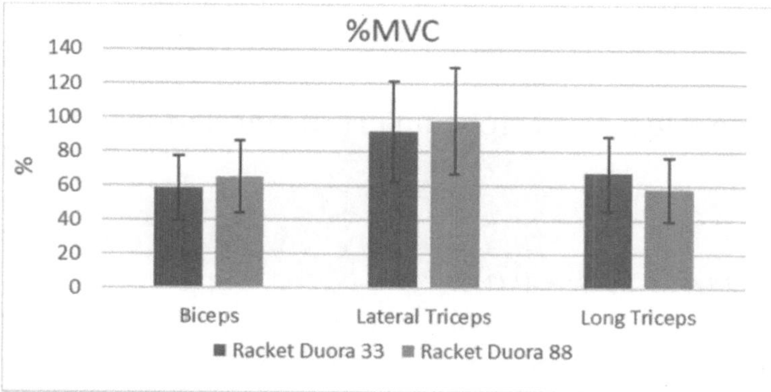

Figure 4. Average of the normalized intermuscular amplitude and their standard deviation.

Figure 4 shows that the most requested muscle is TLat, with Bicp and TLong being requested in a similar, but inferior way. The triceps are responsible for elbow extension, the predominant movement in the smash, regardless of the grip. TLat is responsible for the moments of higher intensity when compared to the other two triceps portions, which can justify such a higher activation of this muscle to the others. It is also clear that Bicp acts as an antagonist and TLong as a stabilizer in the whole smashing process.

4 CONCLUSIONS

The analysis of the results reveals that the lateral head of the triceps is the most requested in terms of muscle intensity and that the stiffer Duora 88 racket requires, for the biceps brachial and lateral head of the triceps, greater intensities of muscle activation during the smash, being responsible not only for the acceleration but also for the deceleration of the arm.

ACKNOWLEDGEMENTS

This work is sponsored by ERDF funds from the COMPETE program – Operational Program for Competitiveness Factors – and national funds through FCT – Foundation for Science and Technology – within the scope of the UIDB/00285/2020 project and Mais Centro Program, Center Region Coordination Committee of EU through European Regional Development Fund..

REFERENCES

Belo, L. R. *et al.* (2016) 'Contribuições da eletromiografia de agulha para o estudo da deglutição em seres humanos', *Revista CEFAC*, 18(5), pp. 1238–1250. doi: 10.1590/1982-021620161852015.

BioSignalsPlux (2019) *Eletromiography*. Available at: https://www.biosignalsplux.com/en/.

Grice, T. (2008) *Badminton: Steps to Success*. Human Kinetics (Sports instruction series). Available at: https://books.google.pt/books?id=Es5JngEACAAJ.

Lucas-Osma, A.M and Collazos-Castro, J.E (2009) *Compartmentalization in the triceps brachii motoneuron nucleus and its relation to muscle architecture, The Journal of comparative neurology*. doi: 10.1002/cne.22123.

Phomsoupha, M. *et al.* (2015) 'How to use the elasticity of a badminton racket to increase its speed by 80%?', *Computer Methods in Biomechanics and Biomedical Engineering*, 18(sup1), pp. 2028–2029. doi: 10.1080/10255842.2015.1069607.

Phomsoupha, M. and Laffaye, G. (2015) 'The Science of Badminton: Game Characteristics, Anthropometry, Physiology, Visual Fitness and Biomechanics', *Sports Medicine*, 45(4), pp. 473–495. doi: 10.1007/s40279-014-0287-2.

Advances and Current Trends in Biomechanics – Belinha et al. (Eds)
© 2022 Copyright the Author(s), ISBN: 978-1-032-10806-3

Preliminary considerations for numerical simulations of assisted delivery

A.B. Estêvão
Faculdade de Engenharia, Universidade do Porto, Portugal

D.A. Oliveira
INEGI, Universidade do Porto, Portugal

M.P.L. Parente
INEGI, Faculdade de Engenharia, Universidade do Porto, Portugal

ABSTRACT: Vaginal delivery is considered to be the leading cause of pelvic floor dysfunction. When in need, the instrument to assist the mother in carrying out vaginal delivery is a vacuum extractor. In this study, initial considerations for numerical simulations of assisted vaginal delivery were analyzed. Therefore, numerical simulations were done with a Finite Element Method of the pelvic floor and the fetus in order to evaluate stresses and strains in maternal pelvic floor muscles. The analysis of the graphics obtained show that for a vertical displacement of the fetal head of approximately 60 mm, the maximum stress in the pelvic floor muscles were 0.82 and 0.29 MPa for a normalized distance of 0 and 0.47 respectively. As for strains, the maximum values were 0.70 and 0.49 for the same values of normalized distances of stresses. These maximum values correspond to muscles that are associated with injuries that occur during childbirth.

1 INTRODUCTION

Any kind of problem that may arise during natural labor that causes the process to be interrupted or prolonged excessively, can lead to a situation of danger not only for the baby but also for the mother. This requires specific measures to speed up birth, one of which is assisted vaginal delivery, which is a common obstetrical procedure and, according to WHO, accounts for 2-23% of deliveries worldwide (Odon, 2011). In the last years, although the total rate of assisted vaginal delivery has been declining, the use of the suction cups has been increasing and it now accounts for nearly 4 times the rate of forceps-assisted vaginal births (Ali and Norwitz, 2009).

During vaginal delivery, when the fetus passes from the womb through the cervix and the vagina, the risk of having pelvic floor muscle injuries is increased (Oliveira *et al.*, 2017), so vaginal delivery is considered to be the leading cause of pelvic floor dysfunction (PFD) (Oliveira *et al.*, 2020). This represents a major public health problem due to its negative impact on the quality of life of various women and due to its high prevalence (Alperin *et al.*, 2016). Several studies have shown that PFD increases with age and affects 9.7% of women aged 20-39 and 49.7% of women aged 80 and over (Alperin *et al.*, 2016; Oliveira *et al.*, 2017). In 2015, Nygaard et al. verified that PFD also has an impact with each pregnancy which can affect 18.4% of women with only one child and 32.4% of women with three or more children.

The main goals of this study are to familiarize with computer modeling and numerical analysis, and with the different *softwares* (Abaqus and SolidWorks). Hence, normal vaginal delivery was simulated, and its impact on pelvic floor muscles was assessed. These models are very important

DOI: 10.1201/9781003217152-22

for this kind of studies, not only to estimate the mechanical changes during pregnancy and vaginal delivery as well as to evaluate failures and sequelae of vaginal birth on pelvic floor muscles (Oliveira *et al.*, 2018). Preliminary considerations were analyzed for the simulation of assisted vaginal delivery. The computational simulations aim to demonstrate the whole labor development with or without vacuum devices, and they are also helpful as an auxiliary tool in clinical environment for some cases that could result in complicated labor.

2 MATERIALS AND METHODS

The biomechanical process of childbirth and the mechanism of labor are still not fully understood. Therefore, in this study, numerical simulations of childbirth were done to evaluate the impact on pelvic floor muscles. Strategies for introducing different suction cups into the delivery process in order to mimic the position of the cup on the surface of the fetal head were also analyzed.

The changes that the pelvic floor muscles suffer during vaginal delivery cannot be quantified in vivo because of ethical and clinical reasons (Parente *et al.*, 2009). Thus, as most biomechanical studies, the finite element method (FEM) was used in this work. The finite element model used for the pelvic floor and for the fetus were developed in the study of Parente et al. (Parente, 2008) and are represented in Figure 1.

Two types of suction cups were modeled during this study: the Kiwi Omnicup vacuum device and the Malmström vacuum extractor. In the literature it was found that the Kiwi Omnicup vacuum device has a diameter of 50 mm (Ali and Norwitz, 2009). This is also one of the diameters available in the Malmström vacuum extractor (González-Diaz *et al.*, 2020). Therefore, for the simulation of assisted vaginal delivery, the 50 mm suction cup of both models indicated will be analyzed in a first stage.

After deciding the different types of dimensions to use in both vacuum extractors' suction cups, it was necessary to create a finite element mesh of the two cups to use in the numerical simulations. The finite element mesh of the Kiwi Omnicup suction cup was already created by Roriz (Roriz, 2015) in his master thesis. To create the Malmström suction cup in 3D, the software SolidWorks was used, and the model obtained can be observed in Figure 2a. Then, this model was imported to the ABAQUS *software* and there, a mesh of the object was created using tetrahedra elements, which can be seen in Figure 2b. Although hexahedral elements are more accurate, the suction cup is not a structure where results will be read, hence the choice for the tetrahedral elements.

In assisted vaginal delivery, the suction cup is placed after the start of delivery. To mimic this moment, it was necessary to study different approaches such as the MODEL CHANGE command, which allows to deactivate and reactivate elements to simulate the removal or insertion of part of the model. It is also necessary to establish the union between the suction cup and the fetus head. The TIE command allows to connect the two surfaces together for the duration of a simulation.

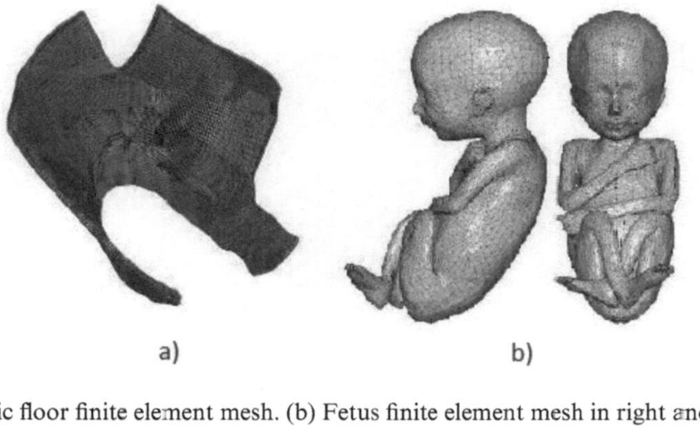

Figure 1. (a) Pelvic floor finite element mesh. (b) Fetus finite element mesh in right and frontal view.

a) b)

Figure 2. (a) Model of the Malmström suction cup in 3D using the SolidWorks software. (b) Finite element mesh of the Malmström suction cup.

3 RESULTS AND DISCUSSION

In addition to the impact on the fetal head, the use of suction cups can also have an impact on the mother, at the level of the pelvic floor muscles. As such, and to further compare this effect, numerical simulations of normal vaginal delivery have been made. The evolution of stress and strain on the pelvic floor during delivery simulations are represented in this section. To obtain these, a curve along the pelvic floor muscles was defined. This curve, named Level 1 can be seen in Figure 3 and according to Parente et al. (Parente *et al.*, 2009) its initial length is 18.1 cm.

Figure 4a shows the distribution of the maximum principal stresses that were obtained on the pelvic floor, in curve Level 1, for a vertical displacement of the fetal head of approximately 60 mm. According to Parente et al (Parente *et al.*, 2009), during fetal descent this position is where stresses are predictably bigger. In Figure 4b, the maximum strains on pelvic floor, also in curve Level 1 for a vertical displacement of the fetal head of approximately 60 mm, can be observed. Both these figures were obtained without suction cups.

As it can be seen in Figure 4a, maximum stresses were obtained when the normalized distance (where values correspond to each point's distance along the Level 1 curve as a fraction of the total length of the curve) were in 0 and 0.47 and the corresponding stress values are 0.82 and 0.29 MPa. If these two points are represented on the Level 1 curve, they coincide with the attachment of the pelvic floor to the bone structure and to the intermediate point of the levator ani muscle, respectively. They are also usually associated with injuries that occur during childbirth (Roriz, 2015). Figure 4b also has two points that stand out where the strain is maximum. These points, equally to the maximum stress graphic, are when normalized distance is at 0 and 0.47, which like said above, are places associated with injuries during vaginal delivery. The respective maximum strains are 0.70 and 0.49.

Figure 3. Level 1 curve along pelvic floor muscles.

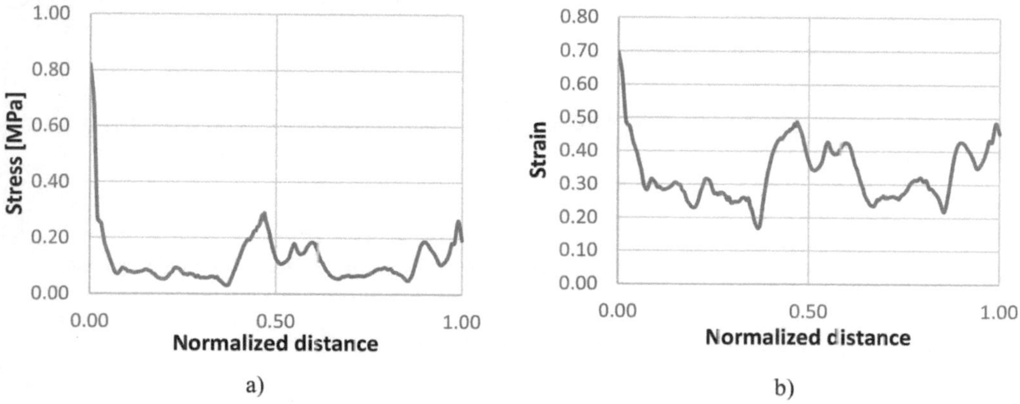

Figure 4. (a) Maximum stresses and (b) maximum strains on pelvic floor along curve Level 1 for a vertical displacement of the fetal head of approximately 60 mm, without suction cup.

4 CONCLUSION

The risk of having pelvic floor muscle injuries increases during vaginal delivery. Pelvic floor muscles, in particular levator ani muscle, are subject to high values of stresses and strains when the fetus passes through the birth canal. For a vertical displacement of the fetal head of approximately 60 mm, the maximum stress obtained in the pelvic floor muscles were 0.82 and 0.29 MPa for a normalized distance of 0 and 0.47 respectively along curve Level 1. As for strains, the maximum values obtained were 0.70 and 0.49 for the same values of normalized distances of stresses. Both these maximum values correspond to muscles that are associated with injuries during vaginal delivery.

REFERENCES

Ali, U. A. and Norwitz, E. R. (2009) 'Vacuum-Assisted Vaginal Delivery', *Reviews In Obstetrics & Gynecology*, 2(1), pp. 5–17.

Alperin, M. *et al.* (2016) 'Impact of vaginal parity and aging on the architectural design of pelvic floor muscles', *American Journal of Obstetrics and Gynecology*. Elsevier Inc., 215(3), pp. 312.e1–312.e9. doi: 10.1016/j.ajog.2016.02.033.

González-Diaz, E. *et al.* (2020) 'Are there differences in the damage to the pelvic floor between malmstrom's and kiwi omnicup vacuums? A multicenter study', *Neurourology and Urodynamics*. John Wiley & Sons, Ltd, 39(1), pp. 190–196. doi: 10.1002/nau.24167.

Odon, J. (2011) 'Assisted vaginal delivery instrument', *World Health Organization*, (mm), p. 2. Available at: http://www.who.int/medical_devices/en/index.html.

Oliveira, D. *et al.* (2018) 'Biomechanical analysis of the damage in the pelvic floor muscles during childbirth', *Lecture Notes in Computational Vision and Biomechanics*, 29, pp. 133–142. doi: 10.1007/978-3-319-71574-2_11.

Oliveira, D. *et al.* (2020) 'Biomechanical Simulation of Vaginal Childbirth: The Colors of the Pelvic Floor Muscles', in Springer (ed.) *Computational Biomechanics for Medicine: Personalisation, Validation and Therapy*, pp. 1–17. doi: 10.1007/978-3-030-15923-8.

Oliveira, D. A. *et al.* (2017) 'The management of episiotomy technique and its effect on pelvic floor muscles during a malposition childbirth', *Computer Methods in Biomechanics and Biomedical Engineering*. Taylor & Francis, 20(11), pp. 1249–1259. doi: 10.1080/10255842.2017.1349762.

Parente, M. P. L. (2008) *Biomechanics of the pelvic floor during vaginal delivery*. Ph.D. thesis. FEUP, UP.

Parente, M. P. L. *et al.* (2009) 'The influence of an occipito-posterior malposition on the biomechanical behavior of the pelvic floor', *European Journal of Obstetrics and Gynecology and Reproductive Biology*. Elsevier, 144S, pp. S166–S169. doi: 10.1016/j.ejogrb.2009.02.033.

Roriz, G. (2015) *Parto computacional assistido*. M.S. thesis. FEUP, UP.

Advances and Current Trends in Biomechanics – Belinha et al. (Eds)
© 2022 Copyright the Author(s), ISBN: 978-1-032-10806-3

On the RANS modelling of the patient-specific thoracic aortic aneurysm

A. Mourato, M. Brito, J. Xavier & L. Gil
UNIDEMI, NOVA School of Science and Technology, NOVA University Lisbon, Caparica, Portugal

A. Tomás
Cardiothoracic Surgery Department, Santa Marta Hospital, Lisbon, Portugal

ABSTRACT: This paper presents a Reynolds-averaged Navier-Stokes (RANS) simulation of patient-specific hemodynamics in thoracic aorta aneurysms. For this purpose, a geometric model using data from a Computed Tomography (CT) performed on a pilot patient was constructed. The numerical model included both the non-Newtonian behavior of the blood and turbulence. The rheological properties of blood were modeled using the Yasuda-Carreau model and the turbulence using the standard k - ε model. The validation of the numerical model was performed by comparing the numerical results with the Magnetic Resonance Imaging (MRI) data. The numerical results showed that the blood ejected from the aortic valve flows at high velocity towards the external curvature of the ascending aorta being deflected in this region to the aortic arch. Therefore, the numerical model can aid the predictions of the hemodynamics inside thoracic aorta aneurysms.

1 INTRODUCTION

During systole the diameter of the ascending aorta increases, accommodating part of the ejected blood. Throughout the diastole, while returning to its original size it redirects the previously accommodated blood to the aortic arch. Thus, the aorta ensures a continuous flow downstream of the aortic arch and decreases the differential between systolic and diastolic pressures (Hall 2011, Belz 1995).

Aneurysms are defined as an anomalous and permanent enlargement of the arteries. These are consequence of changes in hemodynamics that lead to excessive Wall Shear Stress (WSS) values in the aortic wall. Hence, it promotes the degradation of elastin fibers, leading to necrosis of the middle layer of the aortic wall with consequent loss of its elasticity. Despite recent clinical advances, the assessment of risk of an aneurysm rupture was typically assisted by the geometric criterion of maximum diameter, which currently supports surgical indication.

However, there are controversies regarding the pathogenesis and the true risk of acute complications, since more than half of the patients with aortic dissection has a normal aortic diameter (between 40 and 50 mm) at the time of the event (Krüger et al. 2016).

It was currently accepted that the formation of aneurysms has multifactorial etiology. In addition, the stresses in the arterial wall are dependent not only on the geometry of the aneurysm, but also on the structure of the aortic tissue, mechanical properties of the wall and hemodynamic physiological boundary conditions (Soudah et al. 2013). Nonetheless, there are two factors considered to be predominant: genetic and hemodynamic (Decampli, 2017). The integration of biomechanical patient-specific computational models, which also considers the results of imaging tests, can be developed as a tool to support clinical decision regarding the identification of patients who despite having a normal aortic diameter, may benefit from earlier treatment.

DOI: 10.1201/9781003217152-23

Throughout this paper the need for investigation on the development of this type of tools was addressed through the implementation of a numerical model that comprises the non-Newtonian behavior of blood and turbulence using the OpenFOAM software.

2 MATERIALS AND METHODS

The computational domain was defined using CT data. For the geometrical reconstruction the 3D Slicer software was used. The *Segment Editor* function was used to define the portion corresponding to the thoracic aorta in parallel planes to the transversal plane. This selection generated a set of slices which after being interpolated and smoothed (*Volumes* and *Transforms* functions), created the final geometric model. After a correction phase in Blender, the model was imported into the mesh generator software ICEM CFD. Due to the complex geometry, the construction of an unstructured mesh was preferred. This mesh had approximately 10^6 elements and verified all quality criteria, namely the maximum and minimum dimensions of the elements, the height of the first layer of elements and the growth rate close to walls.

In the limits of the model, hemodynamic physiological boundary conditions were defined. In the region close to the aortic valve, an inlet condition was imposed to simulate the presence of the blood jet. A time dependent parabolic velocity profile was defined and the vertical component of the velocity profile was calculated based on the evolution of volumetric blood flow. The other components were calculated based on the slope of the jet. At the limits of the descending aorta, brachiocephalic trunk, left carotid and left subclavian, a time dependent pressure outlet condition was imposed. Both volumetric flow rate and pressure time-series were obtained using data from the MRI perfor med on the pilot patient (Figure 1).

Regarding to the numerical modeling, the open-source CFD toolkit OpenFOAM (Version 2006) was used. OpenFOAM applies a Finite Volume Method (FVM) to numerically solve the RANS equations. The pimpleFOAM solver was used since it solves the RANS equations (Eq. 1 and 2) considering the flow as isothermal and transient and the fluid as incompressible. This solver also allowed the use of rheological and turbulence models and utilized the PIMPLE algorithm to overcome the pressure-velocity coupling problem.

The $k - \varepsilon$ and Yasuda-Carreau were selected for the turbulence and rheological modelling, respectively. The $k - \varepsilon$ model calculates for each time step the value of turbulent viscosity, υ_t (Eq. 3), which has no physical meaning, measures the influence of flow's turbulence and was calculated using known flow characteristics, namely the turbulent kinetic energy, k, and the turbulent energy dissipation rate, ε. The modeling of blood rheology was achieved by calculating the kinematic viscosity of blood, υ_l, for each time step, considering the shear rate applied to the flow (Eq. 4).

$$\frac{dV}{dt} = \nabla \cdot \hat{p} + \nabla \cdot \left(\upsilon_{eff} \nabla \cdot V\right), \text{with} \cdot \upsilon_{eff} = \upsilon_t + \upsilon_l \qquad (1)$$

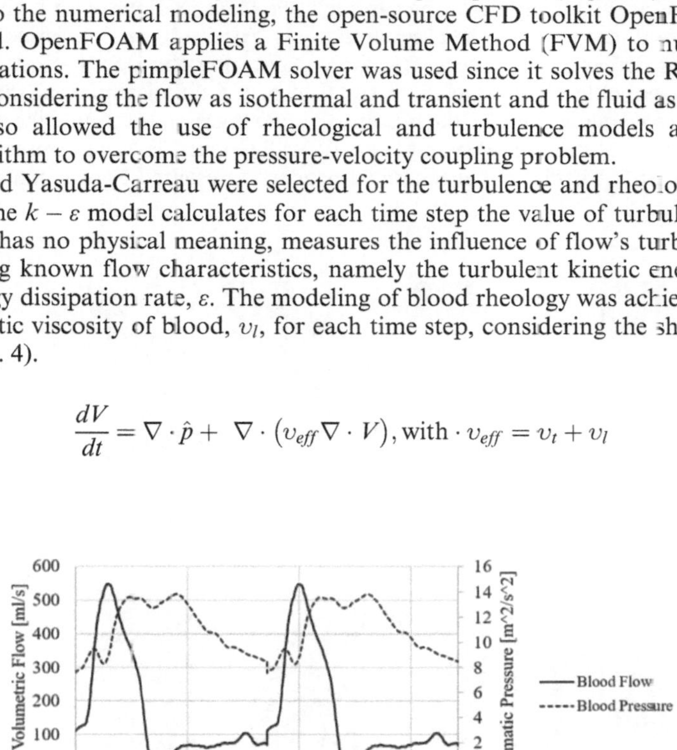

Figure 1. Evolution of volumetric blood flow and pressure over two cardiac cycles.

$$\nabla \cdot \bar{V} = 0 \quad (2)$$

$$\upsilon_t = 0.09\frac{k^2}{\varepsilon} \quad (3)$$

$$\upsilon_l = 1.64 \times 10^{-10}\left(1 + (1,9\,\dot{\gamma})^{0.22}\right)^{1,13} \quad (4)$$

where V = velocity; t = time; $\hat{p}$ = kinematic pressure; υ_{eff}= effective kinematic viscosity; $\bar{V}$ = mean velocity and $\dot{\gamma}$ = shear rate.

3 RESULTS

The modelling was performed in two steps. Firstly, a sensitivity analysis of the mesh was carried out in which it was concluded that the balance between the computational cost and the accuracy of the results occurs for meshes with about 10^6 elements. Secondly, the numerical model was validated by comparing it with the MRI data.

The numerical results for the flow evolution and WSS were calculated throughout a total of 1,71 s (the equivalent to 2 full cardiac cycles) considering a variable time step that was adjusted in a way that the Courant number was less than 5. The initial time step was defined as 10^{-4} s and this value was extremely important for the convergence of the solution that was reached when the numerical residuals of pressure and velocity presented values below 10^{-6} and 10^{-5}, respectively.

The evolution of the flow during a cardiac cycle is shown in Figure 2. It was observed a blood jet formed at the beginning of the cardiac cycle. This jet flows with high kinetic energy towards the external curvature of the ascending aorta. After 120 ms, the jet collided with this region and it was deflected into the aortic arch. Analyzing the flow closest to the internal curvature, it was possible to observe the formation of a recirculation zone characterized by considerably lower velocities and by the presence of two counter-rotating helices.

Regarding to the evolution of WSS in the aortic wall (Figure 3), it was observed that, at the beginning of the cardiac cycle, this mechanical property presented approximately null values.

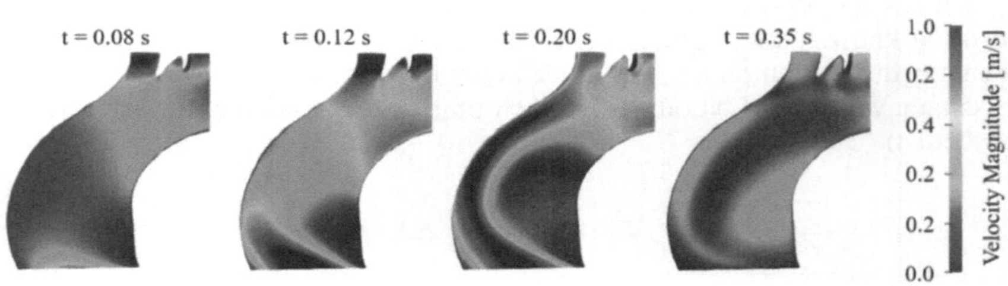

Figure 2. Evolution of blood flow in ascending aorta, throughout 1 cardiac cycle.

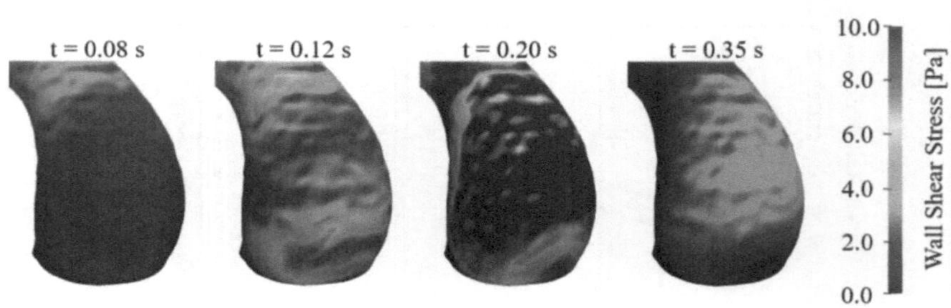

Figure 3. Evolution of WSS in ascending aorta, throughout 1 cardiac cycle.

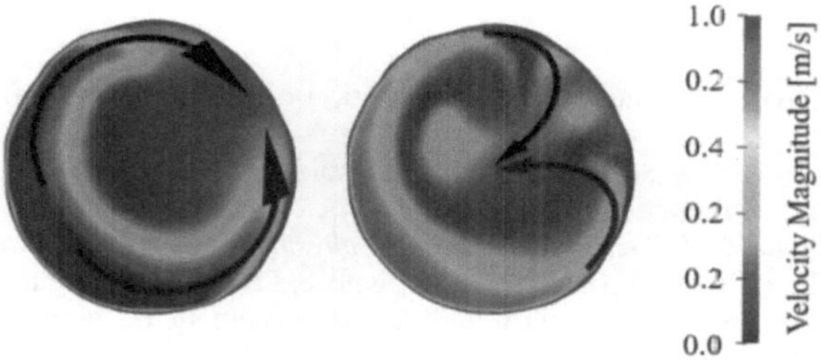

Figure 4. Schematic representation of the formation of the two-counter rotating helixes.

After about 120 ms, the WSS values increased significantly in the region where the blood jet collided with the aortic wall. This increase in WSS spreads along the external curvature as the blood flows in this same region. After the systolic peak, with the decrease in the flow velocity, the WSS also tended to decrease.

4 DISCUSSION

The numerical model was capable to simulate the hemodynamics inside ascending aneurysmatic aortas. This model represents an innovative approach to the process of diagnosing certain pathologies, such as aneurysmal diseases.

Analyzing the results presented in Figure 2, it can be seen that at the beginning of the cardiac cycle the flow presents high velocity values due to the high volume of ejected blood. After the systolic peak ($t > 120$ ms), the blood ejection decreased and the flow inside the ascending aorta slowed down.

Regarding to the formation of the two counter-rotating helixes, this phenomenon was also observed by Heiberg et al. (2007) in 1 out of the 10 analyzed healthy humans. A possible explanation for this event is the fact that part of the blood flows tangentially alongside the aortic wall in both directions, colliding in the opposite region with the consequent deflection of the blood towards the center (Figure 4). In addition to this aspect, it is relevant to explain that the increasing in WSS was due to the impact of the blood jet with the wall. As the jet flows at high velocity, the WSS applied in this region are also high, which promotes the wear and consequent deformation of the aortic wall.

5 CONCLUSION

A numerical model to predict the patient-specific hemodynamics in thoracic aorta aneurysms was developed and validated with regard to MRI data. The numerical results demonstrated the formation of a blood jet, with origin at the aortic valve, which flows at high velocities towards the external curvature of the aorta. The impact of this jet with the wall was transferred as a high WSS, thus promoting its wear.

To improve this model, it is intended in future works to include models that characterize the elastic properties of the aortic wall, such as the Saint Venant-Kirchhoff model, and the fluid-structure interaction. In this way aiming at providing a full integration with clinical practices that are able of classifying the risk index of acute events, such as aortic rupture or aortic dissection.

ACKNOWLEDGMENT

This project was supported by the UNIDEMI and the FCT.IP - UIDB/00667/2020 scholarship.

REFERENCES

Belz, G.G. 1995. Elastic properties and Windkessel function of the human aorta. *Cardiovascular Drugs and Therapy* 9(1): 73–83.

DeCampli, W.M. 2017. Ascending aortopathy with bicuspid aortic valve: More, but not enough, evidence for the hemodynamic theory. *The Journal of Thoracic and Cardiovascular Surgery* 153(1): 6–7.

Hall, J.E. (12th) 2011. *Guyton & Hall textbook of medical physiology*. Rio de Janeiro: Elsevier.

Kilner, P., Yang, G., Mohiaddin, R., Firmin, D. & Longmore, D. 1993. Helical and retrograde secondary flow patterns in the aortic arch studied by three-directional magnetic resonance velocity mapping. *Circulation* 88(5): 2235–2247.

Krüger, T., Forkavets, O., Veseli, K., Lausberg, H., Vöhringer, L., Schneider, W., Bamberg, F. & Schlensak, C. 2016. Ascending aortic elongation and the risk of dissection. *European Journal of Cardio-Thoracic Surgery* 50(2): 241–247.

Soudah, E., Ng, E., Loong, T., Bordone, M., Pua, U. & Narayanan, S. 2013. CFD modelling of abdominal aortic aneurysm on hemodynamic loads using a realistic geometry with CT. *Computational and mathematical methods in medicine* 2013: 1–9.

Advances and Current Trends in Biomechanics – Belinha et al. (Eds)

Electromechanical system for assessing bone consolidation in external fixation systems—a simplified approach

M.F. Paulino, M.A. Neto & A.M. Amaro
CEMMPRE, DEM, University of Coimbra, Portugal

I. Balacó
CHUC, Pediatric Orthopedics Service of the Pediatric Hospital of Coimbra, Coimbra, Portugal

F. Moita & L. Roseiro
ISEC, Polythecnic of Coimbra, Coimbra, Portugal

ABSTRACT: External fixators are widely used in orthopedics, particularly in surgical procedures involving long bone osteotomies. These systems establish the connection between the fragmented bone, guaranteeing the transfer of load until the consolidation of the osteotomy area. System rigidity must ensure the stability, allowing the existence of micro movements, crucial to promote bone callus growth, usually guaranteed by the patient's physical activity. Following the regeneration process, it is important to identify the degree of consolidation, usually performed using bone densitometry. Thus, the development of methodologies that allow the identification of consolidation in bone formation is an open research topic. This work presents an electromechanical system coupled to a unilateral external fixator, aiming to identify the consolidation phase of the regenerated bone. The concept under study uses the behavior of the electric motor that drives the electromechanical system, in particular its consumption, as an indicator of stiffness in the callus area.

1 INTRODUCTION

The external fixation systems are used in various surgical procedures in the context of orthopaedic, particularly in long bone osteotomies, to guarantee the connection between the fragmented bone, allowing the transfer of load between the parts. It is important to guarantee structural rigidity and to allow inter-fragmentary movements, as they are beneficial for rapid bone healing (Cunningham, 2001). Also, the occurrence of micromovements in the focus of the fracture is crucial to promote the growth of the bone callus and control its regeneration (Gardner, 1996; Duda *et al.*, 2001; Hente *et al.*, 2004; Miramini *et al.*, 2016). These micromovements usually occur naturally, through the patient's locomotion.

In the case of surgical procedures, which involve the subsequent regeneration of the bone, the identification of the degree of consolidation in the bone formation area is important, as it allows to define the moment of removal of the fixation systems. Usually, this identification is only possible using images of bone densitometry, which involves the emission of radiation. In this way, the development of methodologies that allow the identification of the consolidation phase without the use of equipment with radiation emission assumes high interest, being able to play a relevant role in the context of medical decision and patient recovery.

DOI: 10.1201/9781003217152-24

This work shows the concept of an electromechanical system, to act with a unilateral external fixator, aiming to identify the consolidation phase of the regenerated bone. The first experimental tests of the concept are present and discussed.

2 MATERIALS AND METHODS

This study uses a simplified tibia-fixator model (Figure 1) composed of two polyamide rods (PA6), (Gardner, 1996; Padovec *et al.*, 2017), and a unilateral *Orthofix*® fixator. This assembly is intended to simulate a 10 mm osteotomy in the central area perpendicular to the mechanical axis of the tibia. The connection between the fixator and the polyamide rods is implemented using four Schanz pins commonly used in this type of surgical procedure, considering the configuration based on the study by (Amaro *et al.*, 2019). The simulation of several phases of the regenerated bone in the osteotomy is carried out applying Sawbones® materials with different stiffness.

The type of fastener considered incorporates a screw elongation method (extender), where the developed electromechanical system is coupled, with a motor and electronic control unit. In this way, it is possible to control the displacement that occurs between the upper pins (fixed clamp) and the lower ones (moving clamp), i.e., between the two rods, also evaluating the displacement that occurs in the area corresponding to the osteotomy. The experimental procedure involves the imposition of a 2 mm displacement in compression, evaluating the behaviour of the motor, namely its electrical consumption, comparing the various stiffness considered to simulate the callus (osteotomy).

To carry out the experimental tests, a platform was created, intended to simulate the situation-type of a bedridden patient. On this platform, fixed supports guarantee the positioning of the system and allows the movement to be carried out in the simplified model. When the movement is induced by the motor, fixing the advance speed at 2 mm/min (Hente *et al.*, 2004), the evaluation of the displacement in the osteotomy was made through a comparator (Mitutoyo®, measurement precision of 1 μm), positioned on one of the faces of the Sawbones® block, as represented in Figure 2.

This type of assembly allows evaluating the influence of the application of micromovements with imposed displacements, for the different materials, related to osteotomy. During the tests, the motor consumption was registered, to obtain a relationship between the current intensity and the stiffness of the material considered. For each material, five tests were carried out, to guarantee their repeatability and reproducibility.

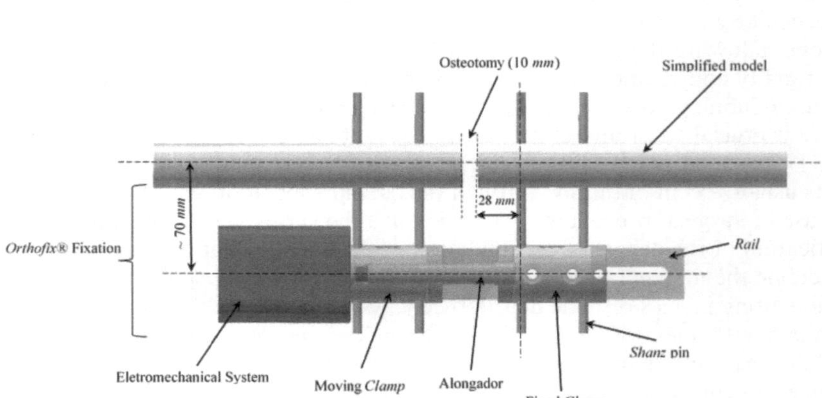

Figure 1. Simplified tibia-external fixator model.

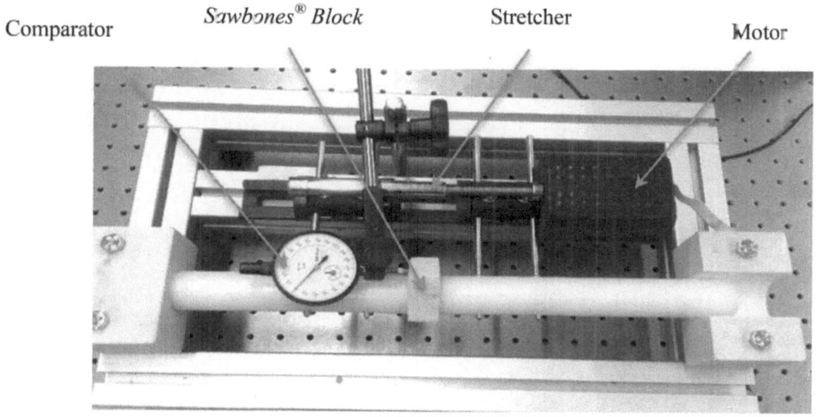

Figure 2. Setup of test with simplified model in horizontal platform.

3 RESULTS AND DISCUSSION

The experimental setup allows the identification of the displacement that occurs in the osteotomy with the variation in stiffness, as seen in Figure 3. With an osteotomy stiffness less than 100MPa the displacement in the callus is about 18% of the imposed value (in stretcher), however, when the callus stiffness increases to values above 600 MPa the displacement is only 6% of the imposed value (in stretcher). As expected, highest stiffness material in the osteotomy decreases the ability to perform movements with the motor. These results show that is possible to estimate the value to impose in the motor to make a 1mm of micromovement in an osteotomy, as suggested in the literature(Kenwright *et al.*, 1991).

The Figure 4 shows the average energy consumption of the motor as a function of the stiffness of the material applied in the osteotomy. These results indicate that the energy consumption of the motor coupled to this type of fastener increases with stiffness in the osteotomy area.

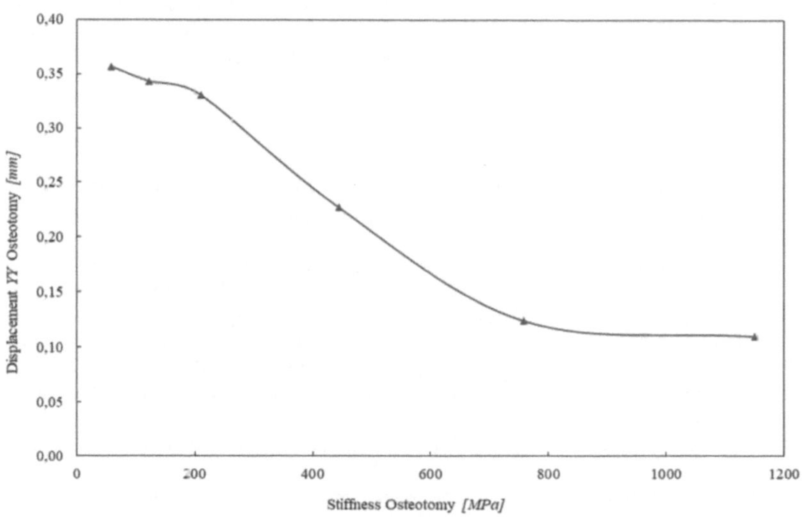

Figure 3. Relationship between stiffness and displacement in osteotomy.

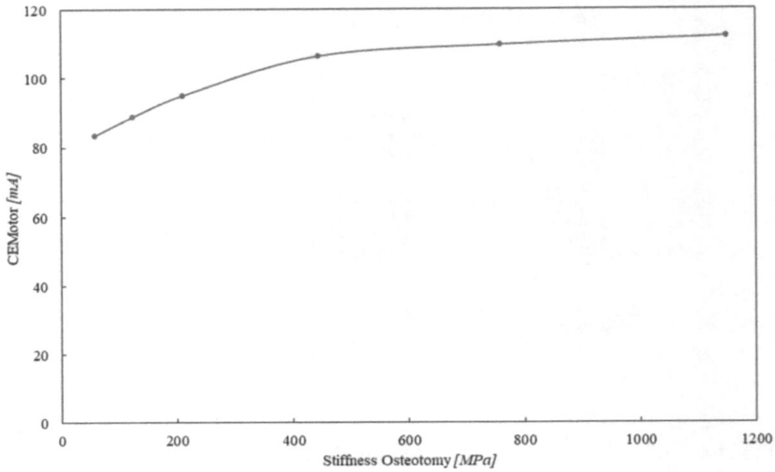

Figure 4. Variation of the energy consumption of the motor with the material stiffness in osteotomy.

4 CONCLUSIONS

The proposed methodology uses information collected from the consumption of a motorized system installed on an external fixator. The experimental results obtained suggest that knowing the CEMotor during bone consolidation process can be a possible way to estimate the state of consolidation of the regenerated bone. This procedure can be considered a promising step in the development of tools to allow the monitoring of the bone consolidation in osteotomies, allowing clinicians to obtain an indicator of the evolution of the recovery and to support the decision of the moment to remove the external fixation system.

ACKNOWLEDGMENT

This work is sponsored by FEDER funds from the COMPETE program - Operational Program for Competitiveness Factors - and national funds through FCT - Foundation for Science and Technology - within the scope of the UIDB/00285/2020 project.

REFERENCES

Amaro, A. *et al.* (2019) 'The Effect of External Fixator Configurations on the Dynamic Compression Load : An Experimental and Numerical Study', *applied sciences*, 10(3), pp. 1–20. doi: 10.3390/app10010003.

Cunningham, J. (2001) 'The biomechanics of frature fixation', *Current Orthopaedics*, 15, pp. 457–464.

Duda, G. *et al.* (2001) 'Mechanical boundary conditions of fracture healing: borderline indications in the treatment of unreamed tibial nailing', *Journal of Biomechanics*, 34, pp. 639–650.

Gardner, T. et al. (1996) 'The influence of external fixators on fracture motion during simulated walking', *Medical Engineering and Physics*, 18, pp. 305–313. doi: 10.1016/1350-4533(95)00056-9.

Hente, R. *et al.* (2004) 'The influence of cyclic compression and distraction on the healing of experimental tibia l fractures', *Journal of Orthopaedic Research*, 22, pp. 709–715.

Kenwright, J. *et al.* (1991) 'Axial Movement and Tibial Fractures', *The journal of bone and Joint Surgery*, 73-B, pp. 654–659.

Miramini, S. *et al.* (2016) 'Influence of fracture geometry on bone healing under locking plate fixations: A comparison between oblique and transverse tibial fractures', *Medical Engineering and Physics*. Elsevier Ltd, 38(10), pp. 1100–1108. doi: 10.1016/j.medengphy.2016.07.007.

Padovec, Z. *et al.* (2017) 'Design, analysis and testing of an external fixation device manufactured from composite materials', *Applied and Computational Mechanics*, 11, pp. 145–154.

Advances and Current Trends in Biomechanics – Belinha et al. (Eds)

Kinematic analysis of the upper limb in reaching—a study in preterm and term young adults

Soraia Pereira, Bruna Almeida, Rubim Santos & Cláudia Silva
Center for Rehabilitation Research (CIR), School of Health of P.PORTO, Porto, Portugal

ABSTRACT: Even in the absence of apparent neurological injury, preterm children, defined by the World Health Organization as babies born before the 37th week of pregnancy, may present movement disorders that are related to atypical neuromotor and behavioral development, noting that these initial deficits have a ripple effect on neurodevelopment. However, despite the evidence that changes in postural control (CP) remain throughout the child's life, no evidence was founded regarding the maintenance of these changes until adulthood, specifically between 18 and 25 years of age, and whether these have implications for the function of the upper limb. Thus, the aim of this study was to evaluate the kinematics of the upper limb and trunk and the behavior of the center of pressure (CoP) during the performance of the reach movement in standing, comparing the dominant upper limb (DL) with the non-dominant upper limb (NDL), in preterm and term young adults.

1 INTRODUCTION

A significant percentage of preterm children, without apparent neurological damage, have disruptions in neuromotor behavior, exhibiting minor or moderate movement disorders that become progressively more evident throughout development, being partly characterized by a dysfunction in the organization of the mechanisms of the postural control. Bearing in mind that postural control is the fundamental basis for motor development (Chen et al. 2015), preterm children, even in the absence of structural damage to the Central Nervous System, may have repercussions on their development and maturation process with interference in the quality of the performance of functional tasks, such as the reach movement (RM), a precursor to most of the daily life tasks involving the upper limb. In fact, studies that evaluated the RM in these children founded that not only the quality of movement is lower compared to term children, but they also had a lower average velocity of movement, a superior number of movement units, a higher frequency of recourse to strategies with bimanual reach, a more irregular trajectory hand movement and a greater number of variations in the acceleration of the hand movement (Guimarães et al. 2013).

Despite the evidence that these changes in postural control remain throughout the child's life (Petersen et al. 2015), it is unknown whether these changes continue to be reflected in deviations from the kinematics of the reach movement in adulthood. On the other hand, several neuroimaging studies carried out in young adults with a premature birth, demonstrated the presence of changes in several cortical and subcortical structures, such as the thalamus and striatum (Meng et al. 2016), as well as the cerebellum and the pre-motor cortex (Lawrence et al. 2014). In association, adults born preterm also have reduced cortico-subcortical connectivity in all important motor areas (Meng et al. 2016, Wheelock et al. 2018) for motor control, namely those who are important for performing the reach movement. However, no evidence was founded regarding the changes in the postural control function and the upper limb

DOI: 10.1201/9781003217152-25

kinematics during the performance of the RM by preterm healthy young adults, specifically between 18 and 25 years of age. So, the aim of the present study was to evaluate the kinematics of the upper limb and trunk and the behavior of the center of pressure (CoP) during the performance of the RM in standing, comparing the dominant upper limb (DL) with the non-dominant upper limb (NDL), in preterm and full term born young adults.

2 MATERIALS AND METHODS

Cross-sectional observational study with 36 young adults aged between 18 and 25 years, divided into two groups: the preterm group (PTG) (gestational age less than 37 weeks) and the full-term group (FTG) (gestational age equal or greater than 37 weeks). Prior to collection, the dominance of the UL was assessed using the Edinburgh Handedness Inventory questionnaire. For the kinematic analysis of the upper limb and trunk, the Qualisys motion capture system and its respective analysis software were used. Two force platforms were used to assess CoP behavior A protocol was developed to evaluate the RM including the camera`s configurations, defined based on the anthropometric characteristics of each participant, and the positions of the reflective markers that were placed on the following bone prominences: distal phalanges of the first and second fingers, distal head of the second metacarpal, radial styloid process, lateral epicondyle, manubrium, acronym and antero-superior iliac spines. A reflective marker was also placed on the target object.

The task consisted in reaching a 0.5L bottle of water, placed on the scapula plane at a height corresponding to the midpoint of the sternum. The movement was performed at a distance corresponding to the functional length of the UL, measured from the acromion to the phalangeal metacarpal joint of the index finger (Pereira et al. 2014). Initially, participants remained in the standing position for 30 secs, with bare feet and with one foot on each of the force platforms and the upper limbs along the body. Subsequently, the participants were instructed to perform the RM at a self-selected speed after a beep, returning to the initial position without the bottle.

For the analysis, was used the average of the six valid repetitions performed by both upper limbs.

The beginning of the movement (t0) was defined when the speed was higher than 2% of the maximum hand speed and the end of the movement (t1) was defined based on the first minimum value of the hand speed. Hand speed was analyzed based on the marker located on the distal phalanx of the first finger. The movement was divided into two phases: the transport phase (phase 1), from t0 to the maximum hand speed, and the anticipatory hand configuration phase (phase 2), from the end of the transport phase to t1, which corresponded to the contact of the hand with the object. The study variables consisted of movement times, speeds, trunk displacement, CoP areas and speeds, maximum hand opening and joint angles for the shoulder and elbow.

3 RESULTS

The final study sample consisted of 19 participants in the PTG (11 females) and 17 in the FTG (10 females), with a median age (and interquartile range) of 20 (3) and 22 (1) years, respectively. The GPT participants had a median gestational age of 34 (5.3) weeks and the FTG of 39 (3) weeks and a birth weight of 2.08 (1.03) kg in the PTG and 3.22 (0, 65) kg in the FTG.

Statistically significant differences were founded in the shoulder initial angle of abduction ($p=0.041$) and final flexion ($p=0.041$) and in the variation of the shoulder abduction ($p = 0.014$) and flexion ($p = 0.014$) in the movement performed with the DL. The FTG presented an initial superior angle of shoulder abduction when the movement was performed by the DL and a superior final angle of shoulder flexion as well as a greater variation of the same angle, being this observed not only when the movement was performed by the DL but also in the movement performed by the NDL. As for the movement of the NDL, differences were observed in the shoulder final flexion (p=0.017) and abduction angles (p=0.002) and in the variation of the shoulder flexion (p=0.024) and abduction (p=0.001) angles. Similar to the observed in the movement of the DL, in the movement performed by the NDL, the FTG presented higher values in both final shoulder flexion angle and in the respective variation. However, with regard to shoulder abduction, in the gesture performed by the NDL, the PTG presented a greater variation in the angle and a superior final angle. Differences were also observed in the variation of the elbow flexion angle in the movement performed by the NDL (p = 0.044), verifying that the PTG had a higher value compared to the FTG (Table 1).

Regarding the spatiotemporal variables of the movement, trunk displacement and CoP behavior, there were no statistically significant differences between the groups. In the comparison between DL and NDL, there were statistically significant differences in the FTG in the total duration of the movement (p = 0.016) and in the duration of the transport phase (p=0.006), which were higher when the gesture was performed by the DL. In addition, the maximum speed of the hand movement was higher, when the movement was performed by the NDL (p=0.000). This latter variable was also statistically different in the PTG. Regarding the kinematic analysis, there were only differences with statistical significance between the DL and the NDL in the initial shoulder flexion angle in the FTG (p = 0.034), which was higher in the movement performed with the NDL.

Table 1. Angular kinematics of the wrist, elbow and shoulder joint of the upper limb.

			Preterm group (n=19)		Full-term group (n=17)		Intergroup inferential analysis	
Variable			Average	Median	Average	Median	Test value	*p* (α=0,05)
Wrist extension (°)	Initial	DL	155,059	-	155,199	-	t=-0,009	0,993
	Final		135,965	-	134,084	-	t=0,745	0,461
	Variation		|-18,627|	-	|-20,831|	-	t=1,061	0,296
	Initial	NDL	153,304	-	154,546	-	t=-0,723	0,475
	Final		135,270	-	134,528	-	t=0,783	0,439
	Variation		|-18,004|	-	|-20,018|	-	t=0,893	0,379
Elbow extension (°)	Initial	DL	143,632	-	143,122	-	t=0,214	0,832
	Final		126,252	-	128,018	-	t=-0,724	0,474
	Variation		|-17,312|	-	|-15,092|	-	t=-0,779	0,442
	Initial	NDL	144,619	-	142,646	-	t=0,943	0,352
	Final		-	126,138	-	130,542	u= 211	0,121
	Variation		-	|-17,689|	-	|-13,672|	u= 213	0,044
Shoulder flexion (°)	Initial	DL	-	2,622	-	1,777	u=112	0,121
	Final		43,352	-	46,141	-	t=-2,123	0,041
	Variation		40,188	-	43,884	-	t=-2,599	0,014
	Initial	NDL	-	2,927	-	2,939	u=138	0,659
	Final		43,106	-	47,565	-	t=-2,504	0,017
	Variation		40,0169	-	44,348	-	t=-2,362	0,024
Shoulder abdution (°)	Initial	DL	11,715	-	13,372	-	t=-2,122	0,041
	Final		30,098	-	27,827	-	t=1,508	0,141
	Variation		18,384	-	14,454	-	t=2,577	0,014
	Initial	NDL	11,346	-	12,537	-	t=-1,041	0,306
	Final		30,998	-	26,754	-	t=3,431	0,002
	Variation		19,655	-	14,217	-	t=3,787	0,001

DL - dominant upper limb; NDL - non-dominant upper limb; p – proof value; t – T-student test; u – Mann-Whitney test; α – significance level

4 DISCUSSION

In this study it was founded that the PTG used a lower angle of flexion of the shoulder, which in biomechanical terms has been associated with an abnormal contribution of the trunk during the reaching movement, in children and adults with neuromotor disorders, muscle weakness, and changes in movement timing and coordination, especially during the transport phase (Mayfield et al. 2017). On the other hand, a greater range of shoulder abduction was observed in the PTG, which may have arisen as a compensatory strategy for the lower range of flexion to decrease the distance between the hand and the object and facilitate its orientation towards the grip. The variation in the elbow extension angle, in the movement performed by the NDL, proved to be superior in the PTG, which may also be a compensatory strategy for the lower flexion angle of the GU, given the coordination between these two joints (Tettamanti et al. 2013).

The results of our study point to a smaller difference in the spatio-temporal variables of the movement between upper limbs in the PTG, in contrast to the FTG, in which there were differences in the total duration of the movement, transport phase and the maximum hand speed and in the kinematics, more specifically at the initial flexion angle of the shoulder. Despite the scientific evidence pointing to the existence of differences between upper limbs in the movement of reaching and grasping an object, evidencing a better performance of the DL in motor tasks of greater precision, kinematic analysis studies in premature children demonstrated greater variability, less consistency in laterality and less differences in kinematics between upper limbs (Assi et al. 2016), which are in agreement with the findings of the present study.

5 CONCLUSION

The study results point to a movement dysfunction in young preterm adults, characterized by a movement pattern of the upper limb in the reaching movement different from the pattern used by the term pairs, however, having no influence on the temporo-spatial variables of the movement.

REFERENCES

Assi, A., Bakouny, Z., Karam, M., Massaad, A., Skalli, W., & Ghanem, I. (10 de September de 2016). Three-dimensional kinematics of upper limb anatomical movements in asymptomatic adults: Dominant vs. non-dominant. Human Movement Science (50), 10–18.

Chen, H.L., Yeh, C.F., Howe, T.H. (2015). Postural control during standing reach in children with Down syndrome. Res Dev Disabil, 38, 345–51.

Guimarães, E.L., Cunha, A.B., Soares, D.A., Tudella, E. (2013). Reaching behavior in preterm infants during the first year of life: a systematic review. Motor Control, 17(4), 340–54.

Lawrence, E.J., Froudist-Walsh, S., Neilan, R., Nam, K.W., Giampietro, V., McGuire, P. (2014). Motor fMRI and cortical grey matter volume in adults born very preterm. Dev Cogn Neurosci, 10, 1–9.

Mayfield, C., Kukke, S., Brochard, S., Stanley, C., Alter, K., & Damiano, D. (2017). Inter-joint coordination analysis of reach-to-grasp kinematics in children and adolescents with obstetrical brachial plexus palsy. Clinical Biomechanics, 46, pp. 15–22.

Meng, C., Bäuml, J.G., Daamen, M., Jaekel, J., Neitzel, J., Scheef, L., et al. (2016) Extensive and interrelated subcortical white and gray matter alterations in preterm-born adults. Brain Struct Funct, 221(4), 2109–21.

Pereira, S., Silva, C.C., Ferreira. S., Silva, C., Oliveira, N., Santos, R. (2014). Anticipatory postural adjustments during sitting reach movement in post-stroke subjects. J Electromyogr Kinesiol, 24(1), 165–71.

Petersen, H., Tulinius, A.T., Georgsdóttir, I., Einarsson, E.J., Patel, M., Haraldsson, Á., et al. (2015). Decreased postural control in adolescents born with extremely low birth weight. Exp Brain Res, 233 (5),1651–62.

Tettamanti, A., Giordano, M., & Gatti, R. (April de 2013). Effects of coupled upper limbs movements on postural stabilisation. Journal of Electromyography and Kinesiology (23), 1222–1228.

Wheelock, M.D., Austin, N.C., Bora, S., Eggebrecht, A.T, Melzer, T.R., Woodward, L.J., et al. (2018). Altered functional network connectivity relates to motor development in children born very preterm. Neuroimage, 183, 574–83.

Advances and Current Trends in Biomechanics – Belinha et al. (Eds)
 ISBN: 978-1-032-10806-3

Using meshless methods to predict angiogenesis due to chemotaxis

A. Guerra
Institute of Science and Innovation in Mechanical and Industrial Engineering (INEGI), Porto, Portugal

J. Belinha
Mechanical Engineering Department, School of Engineering, Polytechnic of Porto (ISEP), Porto, Portugal

R. Natal Jorge
Mechanical Engineering Department, Faculty of Engineering of the University of Porto (FEUP), Porto, Portugal

ABSTRACT: Angiogenesis is the process that ensures the formation of a functional vascular network that sustains tissue homeostasis. The success of this process is dependent on endothelial cell migration, main regulated by chemotaxis. Several growth factors regulate angiogenesis; however, the vascular endothelial growth factor (VEGF) is the main one. In this study, the relevance of using a meshless method to study angiogenesis was addressed. Accordingly, the endothelial cells migrate according to a diffusion-reaction equation, following the VEGF gradient concentration. With this approach, it was possible to simulate angiogenesis and to obtain a capillary network comparable to the one existing *in vivo*, regarding its structure and morphology.

1 INTRODUCTION

Angiogenesis means the formation of new blood vessels from the pre-existing vasculature. This process is essential during wound healing since it provides the reestablishment of the normal blood flow and consequently the sufficient exchange of oxygen and nutrients and the removal of metabolic wastes (Carmeliet and Jain, 2011). The endothelial cell migration is a crucial step during angiogenesis, being this process main regulated by chemotaxis. The vascular endothelial growth factor (VEGF) is the principal regulator of angiogenesis. This soluble factor stimulates endothelial cells to proliferate, to migrate, to differentiate and to survive, allowing new blood vessel formation (Koch and Claesson-Welsh, 2012). Moreover, cutaneous wounds present high levels of VEGF that is produced in response to injury by multiple cells types, such as keratinocytes, macrophages and fibroblasts (Willenborg et al., 2012).

Computational simulation is a very useful tool to fully detail and understand angiogenesis. Some authors have already developed models to describe the effect of chemo-attractant growth factors in the endothelial cell behavior and in the capillary network formation during angiogenesis. The most relevant computational models in these field are discussed with detail in the literature (Guerra et al., 2018). In computational modelling the selection of the numerical method to be employed is an important topic. In the last years, meshless methods started to be used in biology given its usefulness to simulate biologic processes. These methods present some advantages comparatively to other numerical methods (Belinha, 2014). For example, in meshless methods, the field functions are determined through an approximation within a flexible influence-domain. Therefore, meshless methods are capable to analyse structures with complex geometries; to use

DOI: 10.1201/9781003217152-26

high-order continuous weight functions to construct the test functions, allowing to obtain smoother variables fields; to study large deformations problems and moving discontinuities; and to insert locally more nodes where mesh refinement is required to obtain accurate local solutions, without any extra computational cost. However, the use of meshless methods to model angiogenesis in the context of wound healing have been less explored. Since meshless methods allow to analyse any phenomenon that is governed by differential equation (Belinha, 2014), their application in the field of angiogenesis will be explored in this work.

2 MATERIALS AND METHODS

In this work, the angiogenic process was analyzed using a meshless method approach. Accordingly, the discretization technique used was the Radial Point Interpolation Method (RPIM). In the literature, RPIM formulation is described with detail (Belinha, 2014). Our model combines the RPIM with VEGF diffusion, endothelial cell migration, branching process and capillary network formation.

2.1 *Discrete equation system*

In this work, the VEGF diffusion in a homogeneous medium was assumed. To obtain the final numerical solution of the algorithm implemented in the RPIM software, the VEGF diffusion formation needs to be presented as a discrete equation system. Thus, the VEGF diffusion can be numerically simulated as a field problem (Liu and Quek, 2003). The general form of the equation system of 2D linear steady state field problem can be given by the general form of the Helmholtz,

$$D_x \frac{\partial^2 \phi}{\partial x^2} + D_y \frac{\partial^2 \phi}{\partial y^2} - g\phi + Q = 0 \quad (1)$$

where ϕ is the field variable, D_x and D_y represents, respectively, the VEGF diffusion coefficient along dimension x and y, g represents the matrix of chemical infusibility that will be neglected ($g = 0$) and Q is the VEGF concentration. Using the weighted residual approach, it is possible to formulate the meshless system equations:

$$\left[\boldsymbol{K_D} + \boldsymbol{K}_g\right]\boldsymbol{\Phi} - \mathbf{f}_q = 0 \quad (2)$$

where $\boldsymbol{K_D}$ is the diffusion matrix, $\boldsymbol{K}_g$ is the infusibility matrix, $\boldsymbol{\Phi}$ is the VEGF concentration in each node and $\boldsymbol{f}_q$ is VEGF flux vector. Solving the above equation, it is possible to obtain the final VEGF concentration in each node, $\boldsymbol{\Phi}$.

2.2 *Modelling angiogenesis*

In this work, angiogenesis was modeled using a regular nodal mesh with 51 x 51 nodes. The domain geometry was defined in order to fit the biological situation, and four distinct zones/subdomains were identified: the parent vessel where the angiogenesis sprouting is initiated, the extracellular matrix where the capillary network is formed and the wound that releases VEGF. The location of the parent vessel and wound change in the performed simulations. The VEGF concentrations used and its diffusion gradient were obtained from availability literature (Guerra et al., 2020).

The endothelial cell migration is done according to a diffusion-reaction equation for VEGF. After obtaining the VEGF gradient concentration, the position of the new endothelial cell is defined. The direction of the endothelial cell movement is calculated in each iteration by the vector $\boldsymbol{u} = \{u_x, u_y\}$, that follows the VEGF diffusion gradient. Therefore, the RPIM software enters into an iterative loop and in each step the endothelial cells are marked one after the other, allowing to obtain the capillary network. The iteration process stops when the endothelial cell arrives to the domain boundary, to the wound or it is near to another endothelial cell.

The branching process, that means the formation of new capillaries sprouts from pre-existing sprouts, permits the widespread of the vascular network. Accordingly, this process was implemented in the model. Therefore, a new branching location occurs if the endothelial cell was at a distance to the last branching point location greater than the predefined threshold.

3 RESULTS

In this work, the sprouting angiogenesis for different hypothetic biological simulations and using a meshless method approach was simulated (Figure 1). Figure 1a represents the VEGF diffusion gradient, which goes from red to yellow. The red zone is located near the parent vessel and has the lowest VEGF concentration of around 2.5 x 10^{-13} g mm^{-3}. The

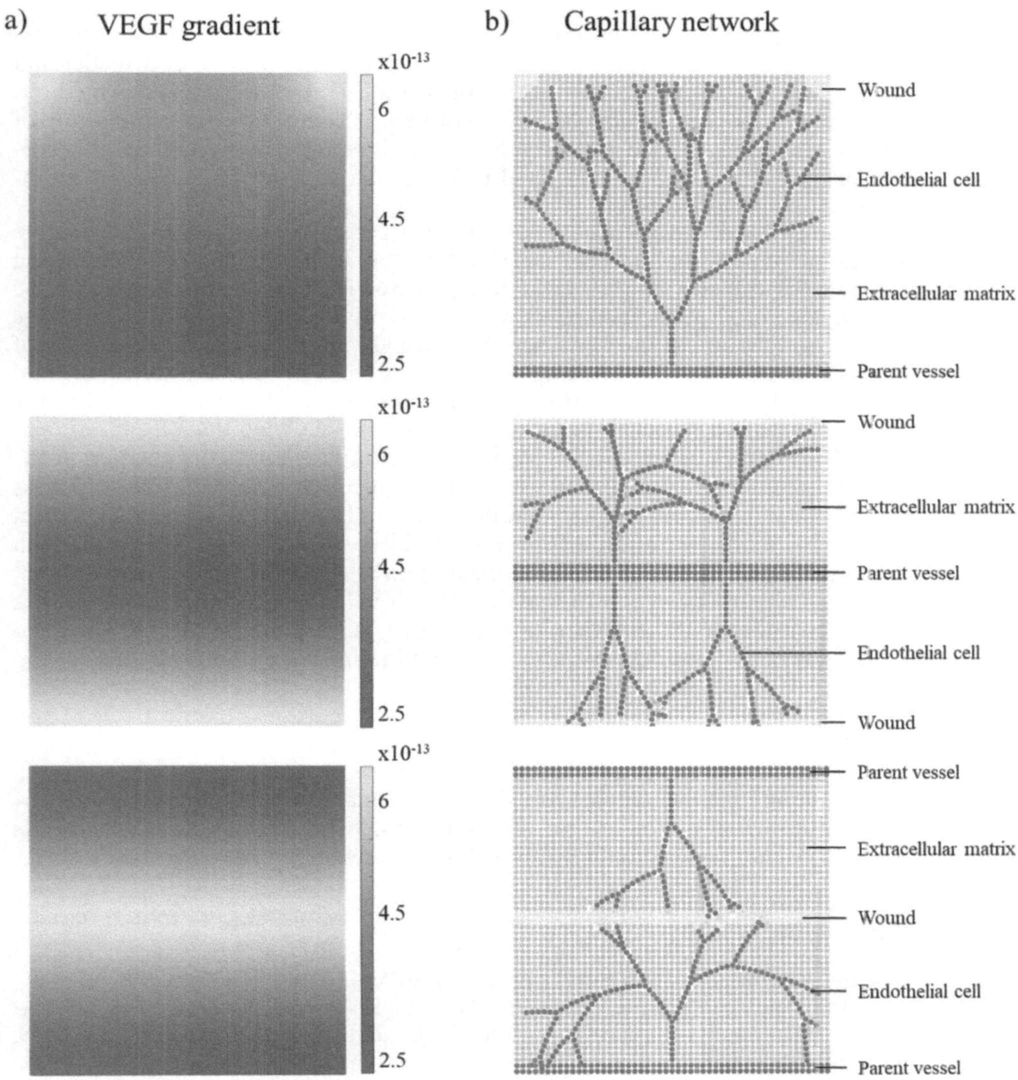

Figure 1. Angiogenesis modelling. a) VEGF diffusion gradient (VEGF concentration in g mm^{-3}); b) Final simulation results from endothelial cell migration using a regular nodal discretization mesh.

yellow zone is located near the wound and has the highest VEGF concentration of around 6.0×10^{-13} g mm^{-3} (Karayiannakis et al., 2003). Figure 1b illustrates the final capillary network obtained with the RPIM software.

Analyzing the VEGF gradient and the capillary network pattern obtained it is possible to verify that endothelial cells migrate from the parent vessel to the wounds or to the regions with higher VEGF concentration, following the VEGF gradient. Moreover, the branching occurrence is presented. In the simulations, the direction of the branching to right or left is random and some fluctuation in the branching angle is allowed. Because of it, the capillary network is not symmetric in the performed simulations. Our numerical results showed that the endothelial cell migration and consequently the capillary network obtained feature the expected pattern.

4 DISCUSSION

Understand angiogenesis is essential to improve wound healing. Several *in vivo* and *in vitro* studies have been developed in order to describe with detail this process (Carmeliet and Jain, 2011). However, given their complexity there is still a lack of experimental methodologies able to fully describe angiogenesis. Accordingly, computational simulation of wound healing angiogenesis started to be implemented since it permits to model this biological process using different approaches and to better understand the regulatory mechanisms and biological species involved.

Most of the studies already developed describe the effect of oxygen and growth factors in endothelial cell proliferation and migration and in the capillary network profile using mesh-depend methods (Guerra et al., 2018). Nevertheless, meshless methods are suitable tolls to model angiogenesis, given its flexibility to represent complex geometries and large deformations and its robustness and accuracy. Accordingly, in this work the pertinence of using RPIM to simulate wound healing angiogenesis was analysed. In the three numerical examples performed (Figure 1), the endothelial cells migrate from the parent vessel to the wounds or to the regions with higher VEGF concentration, following the VEGF gradient concentration. The model here presented, inserts new nodes, or changes the position of the existent nodes, to simulate endothelial cell migration during angiogenesis. This process induces mesh distortion and, consequently, meshless methods are valuable in order to deal successfully with this mesh distortion. Moreover, meshless methods converge faster than mesh-dependent methods, such as the finite element method (Belinha, 2014). Furthermore, in the literature has already been shown that RPIM is a valid tool to model angiogenesis due to chemical diffusion of VEGF (Guerra et al., 2020). Consequently, RPIM formulation can be used to increase angiogenesis knowledge and to improve wound healing.

5 CONCLUSION

In this work, the use of a meshless method to simulate wound healing angiogenesis was performed. The numerical results showed that the capillary network profile obtained presents accurate structure and morphology and resembles the one accessible in biological studies.

ACKNOWLEDGMENT

The authors truly acknowledge the funding provided by Ministério da Ciência, Tecnologia e Ensino Superior - Fundação para a Ciência e a Tecnologia (Portugal), under grant SFRH/BD/133894/2017. Additionally, the authors acknowledge the funding provided by LAETA, under project UIDB/50022/2020.

REFERENCES

Belinha, J. 2014. *Meshless Methods in Biomechanics - Bone Tissue Remodelling Analysis*, Springer International Publishing.

Carmeliet, P. & Jain, R. K. 2011. Molecular mechanisms and clinical applications of angiogenesis. *Nature*, 473, 298–307.

Guerra, A., Belinha, J. & Jorge, R. N. 2018. Modelling skin wound healing angiogenesis: A review. *Journal of Theoretical Biology*, 459, 1–17.

Guerra, A., Belinha, J. & Natal Jorge, R. 2020. A preliminary study of endothelial cell migration during angiogenesis using a meshless method approach. *International Journal for Numerical Methods in Biomedical Engineering*, 36, e3393.

Karayiannakis, A. J., Zbar, A., Polychronidis, A. & Simopoulos, C. 2003. Serum and drainage fluid vascular endothelial growth factor levels in early surgical wounds. *Eur Surg Res*, 35, 492–6.

Koch, S. & Claesson-Welsh, L. 2012. Signal transduction by vascular endothelial growth factor receptors. *Cold Spring Harb Perspect Med*, 2, a006502.

Liu, G. R. & Quek, S. S. 2003. *The finite element method: a practical course*, Elsevier Science - Butterworth-Heinemann.

Willenborg, S., Lucas, T., Van Loo, G., Knipper, J. A., Krieg, T., Haase, I., Brachvogel, B., Hammerschmidt, M., Nagy, A., Ferrara, N., Pasparakis, M. & Eming, S. A. 2012. CCR2 recruits an inflammatory macrophage subpopulation critical for angiogenesis in tissue repair. *Blood*, 120, 613–25.

Advances and Current Trends in Biomechanics – Belinha et al. (Eds)
© 2022 Copyright the Author(s), ISBN: 978-1-032-10806-3

Angiogenesis—numerical simulation of branching and anastomosis processes

A. Guerra
Institute of Science and Innovation in Mechanical and Industrial Engineering (INEGI), Porto, Portugal

J. Belinha
Mechanical Engineering Department, School of Engineering, Polytechnic of Porto (ISEP), Porto, Portugal

R. Natal Jorge
Mechanical Engineering Department, Faculty of Engineering of the University of Porto (FEUP), Porto, Portugal

ABSTRACT: Angiogenesis, the formation of new blood vessels from pre-existent ones, ensures the viability of all cells in the body. Similar to the most biological processes, angiogenesis is highly complex, being modulated by different biochemical species and different cell types. Accordingly, computational models are an advantage since they allow to mimic this process helping to understand it. In this work, an iterative code capable of predicting angiogenesis was presented. Our model mimics specific steps of this process such as the migration of endothelial cells, the branching of the growing capillaries and the anastomosis. The numerical results allowed to achieve a capillary network with similarities to the one obtained *in vivo* studies.

1 INTRODUCTION

Angiogenesis allows the reestablishment of the normal blood flow, the sufficient supply of oxygen and nutrients and the removal of metabolic wastes, essential for cell proliferation and viability (Carmeliet and Jain, 2011). This process is particularly important during wound healing since its success is dependent on the satisfactory delivery of oxygen and nutrients, the formation of a functional vascular network and diffusion of growth factors. In healthy tissues, mature vessels are in a quiescent state. However, during wound healing, quiescent vessels are exposed to proangiogenic factors and sprouting angiogenesis occurs. Then, a previously quiescent endothelial cell is converted into a tip cell. The tip cell is sensible to growth factors gradients in the environment and it starts to migrate, resulting in sprouting vessel elongation. The blood vessel density increases and the vascular sprout will fuse with another neighbouring vessel, in a process called anastomosis. The main regulator of angiogenesis is the vascular endothelial growth factor (VEGF). This soluble factor stimulates endothelial cells to proliferate, to migrate, to differentiate and to survive, allowing new blood vessel formation (Koch and Claesson-Welsh, 2012).

As discussed above, angiogenesis is a complex process being difficult to establish a relationship between changing one biological parameter and its influence in this process. Accordingly, in last decades, computational modelling of angiogenesis has been under focus. This approach allows to study angiogenesis in different scales, using more economic and less time-consuming methodologies, comparatively to the laboratory ones. Indeed, some authors had already developed models to describe the effect of chemo-attractant species in the endothelial cell behaviour during angiogenesis. Some studies analyse the endothelial cell adhesion

DOI: 10.1201/9781003217152-27

and its morphology in the presence of chemoattractant species as well as in its phonotype (quiescent, stalk or tip). Other authors established models able to simultaneously predict the individual cell behaviour and the evolution of species' concentration along distinct time and space scales. The effect of chemotactic growth factors distribution in capillary network formation and structure was also modelled. The most relevant computational models of wound healing angiogenesis are discussed with detail in the literature (Guerra et al., 2018).

In this work, it is presented a numerical model capable to reproduce the endothelial cell migration, branching and anastomosis processes during wound healing. The numerical results demonstrated that endothelial cells migrate following the VEGF gradient concentration, allowing the formation of a capillary network and the vascularization of the tissue.

2 MATERIALS AND METHODS

The discretization technique used in this work was the Radial Point Interpolation Method (RPIM). The full description of the application of RPIM to model angiogenesis is presented in our previous work (Guerra et al., 2020). In the current work, the model combines the RPIM with growth factor diffusion, endothelial cell migration, branching and anastomosis processes in order to obtain a realistic capillary network.

2.1 *Discrete equation system*

In this work, the chemical diffusion in a homogeneous medium was adopted. This formulation should be represented as a discrete equation system in order to obtain the final numerical solution of the algorithm implemented in the RPIM software.

The chemical diffusion can be numerically simulated as a field problem (Liu and Quek, 2003). The general form of the equation system of 2D linear steady state field problems can be given by the general form of the Helmholtz,

$$D_x \frac{\partial^2 \phi}{\partial x^2} + D_y \frac{\partial^2 \phi}{\partial y^2} - g\phi + Q = 0 \tag{1}$$

where ϕ is the field variable, D_x and D_y represents, respectively, the VEGF diffusion coefficient along dimension x and y, g represents the matrix of chemical infusibility that will be neglected ($g = 0$) and Q is the VEGF concentration. Using the weighted residual approach, it is possible to formulate the meshless system equations:

$$[\boldsymbol{K_D} + \boldsymbol{K_g}]\boldsymbol{\Phi} - \boldsymbol{f_q} = 0 \tag{2}$$

where $\boldsymbol{K_D}$ is the diffusion matrix, $\boldsymbol{K_g}$ is the infusibility matrix, $\boldsymbol{\Phi}$ is the VEGF concentration in each node and $\boldsymbol{f_q}$ is VEGF flux vector. Solving the above equation, it is possible to obtain the final VEGF concentration in each node, $\boldsymbol{\Phi}$.

2.2 *Endothelial cell migration*

In our model, the endothelial cell moves according to a diffusion-reaction equation. After obtaining the VEGF gradient concentration, the position of the new tip cell is defined. The direction of the tip cell movement is calculated by the vector $\boldsymbol{u} = \{u_x, u_y\}$, that follows the VEGF diffusion gradient. The vector $\boldsymbol{u}$ variation angle (θ) is calculated in each iteration, $\theta = arctg(u_x/u_y)$. Then, the RPIM software enters into an iterative loop and in each step the tip cells are marked one after the other, allowing to obtain the capillary network. The iteration process stops when the growing capillary arrives to the domain boundary, to the wound or if the tip cell join another endothelial cell.

2.3 *Branching and anastomosis processes*

The branching process, the formation of new capillaries sprouts from pre-existing sprouts or capillaries, is essential in order to allow the widespread of the vascular network and the vascularization of all the tissue. Moreover, anastomosis that means the fusion of a sprout tip to another sprout tip or with another endothelial cell is essential to allow the reestablishment of the normal blood flow. Given its importance for angiogenesis these two processes were implemented in our model. Accordingly, a new branching location occurs if the tip cell was at a distance to the last branching point location greater than the threshold that we defined. The anastomosis occurs if the tip cell was at a distance from another tip or endothelial cell smaller than half of the distance between two nodes of the mesh.

3 RESULTS

In this work, the sprouting angiogenesis from one parent vessel and in the presence of three wounds, using a regular nodal mesh, were simulated (Figure 1). Figure 1a represents the VEGF diffusion gradient, which goes from red to yellow. The red zone is located near the parent vessel and has the basal VEGF concentration of around 2.5 x 10^{-13} g mm^{-3}. The yellow zone is located near the wounds and has a VEGF concentration around 6.0 x 10^{-13} g mm^{-3} (Karayiannakis et al., 2003). Figure 1b represents the endothelial cell migration pattern in sequential time steps. Analyzing the VEGF gradient and the endothelial cell migration profile it is possible to verify that endothelial cells migrate from the parent vessel to the wounds or to the regions with higher VEGF concentration, following the highest VEGF concentration. Moreover, the occurrence of the processes of branching and anastomosis is also observed. Because the direction of the branching to right or left is random and some fluctuation in the branching angle is allowed the capillary network is not symmetric. In the numerical results, we observed that the capillary network and the branching profile follows the expected configuration.

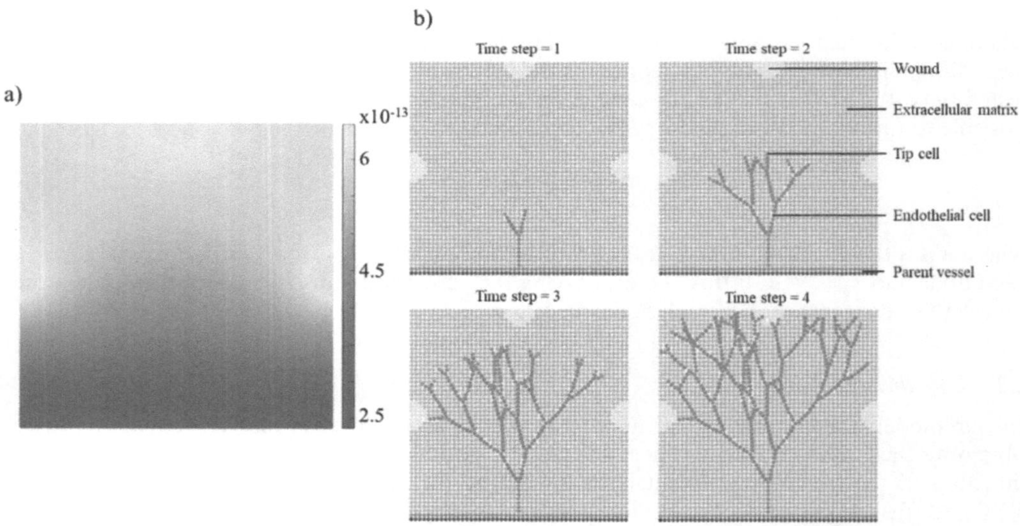

Figure 1. Angiogenesis simulation. a) VEGF diffusion gradient (VEGF concentration in g mm^{-3}); b) Simulation results from tip cell migration in different time steps using a regular nodal discretization mesh.

4 DISCUSSION

The relevance of using computational simulation to model angiogenesis started to be addressed in the last decades considering its potential to try new approaches and new therapies using less time-consuming, cheaper and reproducible methodologies. During wound healing, angiogenesis is essential and its effectiveness is dependent on the migration of endothelial cells, branching and anastomosis. Several cell types that participate in wound healing, such as fibroblasts, endothelial cells, smooth muscle cells, keratinocytes and macrophages, produce VEGF (Bao et al., 2009). Therefore, mimic the endothelial cell migration due to VEGF diffusion gradients is a relevant topic in computational simulation.

In our computational model, the endothelial cells migrate due to chemotaxis. Accordingly, in the performed simulation (Figure 1) the tip cell migrates from the endothelial cell monolayer located in the lower domain boundary to the wounds or to the regions with higher VEGF concentration, following the highest VEGF concentration. These numerical results are in accordance with previous studies of endothelial cells culture *in vitro* that showed that VEGF induces cell migration and chemotactic response (Yancopoulos et al., 2000). This allowed us to conclude that the migration condition is well implemented in the code. In order to simulate with accuracy the angiogenic process during wound healing the branching and anastomosis processes were implemented in our model. Given the configuration of the capillary network obtained and shown in Figure 1, it is possible to verify that that these two processes were correctly implemented. The capillary network obtained resembles the one obtained *in vivo* studies, considering its structure and morphology. Moreover, the achieved capillary network is similar to those obtained by other authors (Sun et al., 2005).

Meshless methods are suitable numerical tools to simulate biomechanical problems due to theirs flexibility to represent complex geometries and large deformations and also their robustness and accuracy. However, the use of these numerical methods to model angiogenesis is less explored. Therefore, in the field of wound healing angiogenesis, RPIM can be used to analyze individual endothelial cells migration accordingly to the VEGF diffusion gradient (Guerra et al., 2020). In the upcoming models, it is intended to include several parameters in the model, such as endothelial cell proliferation rate and VEGF consumption and natural decay. With our approach, in the future, we expect to be able to simulate capillary network with more realistic structure and morphology. Therefore, with the increase in the model complexity it is expected to be able to better understand angiogenesis in order to improve wound healing.

5 CONCLUSION

In this work, a meshless-based model that analyses the endothelial cell migration due to chemotaxis was proposed. Overall, the developed model allowed to simulate angiogenesis and to obtain a capillary network with similarities to the one obtained in *in vivo* assays. The RPIM formulation developed in this work is a valid numerical tool that can be adjusted to biological problems and may also be able to complement biological and medical experiments.

ACKNOWLEDGMENT

The authors truly acknowledge the funding provided by Ministério da Ciência, Tecnologia e Ensino Superior - Fundação para a Ciência e a Tecnologia (Portugal), under grant SFRH/BD/133894/2017. Additionally, the authors acknowledge the funding provided by LAETA, under project UIDB/50022/2020.

REFERENCES

Bao, P., Kodra, A., Tomic-Canic, M., Golinko, M. S., Ehrlich, H. P. & Brem, H. 2009. The Role of Vascular Endothelial Growth Factor in Wound Healing. *The Journal of surgical research*, 153, 347–358.

Carmeliet, P. & Jain, R. K. 2011. Molecular mechanisms and clinical applications of angiogenesis. *Nature*, 473, 298–307.

Guerra, A., Belinha, J. & Jorge, R. N. 2018. Modelling skin wound healing angiogenesis: A review. *Journal of Theoretical Biology*, 459, 1–17.

Guerra, A., Belinha, J. & Natal Jorge, R. 2020. A preliminary study of endothelial cell migration during angiogenesis using a meshless method approach. *International Journal for Numerical Methods in Biomedical Engineering*, 36, e3393.

Karayiannakis, A. J., Zbar, A., Polychronidis, A. & Simopoulos, C. 2003. Serum and drainage fluid vascular endothelial growth factor levels in early surgical wounds. *Eur Surg Res*, 35, 492–6.

Koch, S. & Claesson-Welsh, L. 2012. Signal transduction by vascular endothelial growth factor receptors. *Cold Spring Harb Perspect Med*, 2, a006502.

Liu, G. R. & Quek, S. S. 2003. *The finite element method: a practical course*, Elsevier Science - Butterworth-Heinemann.

Sun, S., Wheeler, M. F., Obeyesekere, M. & Patrick, C. W., JR. 2005. A deterministic model of growth factor-induced angiogenesis. *Bull Math Biol*, 67, 313–37.

Yancopoulos, G. D., Davis, S., Gale, N. W., Rudge, J. S., Wiegand, S. J. & Holash, J. 2000. Vascular-specific growth factors and blood vessel formation. *Nature*, 407, 242–8.

Advances and Current Trends in Biomechanics – Belinha et al. (Eds)

Evaluation of the Semmes-Weinstein Monofilament (SWM) on the diabetic foot assessment

P. Martins & L. Coelho
Instituto Superior de Engenharia do Porto, CIETI - Centro de Inovação em Engenharia e Tecnologia Industrial, Porto, Portugal

ABSTRACT: The loss of protective sensitivity in the diabetic foot requires an early diagnosis due to the imminent possibility of ulceration and amputation of the affected limb. In screening for loss of protective sensitivity, the 10 gf Semmes-Weinstein monofilament (SWM) is the most widely used first-line instrument. However, this decisive instrument lacks calibration and quality control. In this paper we present a study to objectively evaluate the SWM performance using a specific measuring equipment. We conducted a metrological verification of three monofilaments using an automated method under controlled conditions. The obtained results revealed that the SWM compression force depends on the number of repeated uses, the ambient temperature, and the displacement speed and contact with the patient's skin. In fact, all the SWMs showed dependence on external factors and evidenced relative errors between 14 % and 167 %, values that can be considered critical to discriminate the pathology and its evolution.

1 INTRODUCTION

The diabetic foot, a serious complication in diabetic patients and originated by the progressive loss of protective sensitivity, requires an early diagnosis due to the imminent possibility of ulceration and amputation of the affected limb, resulting in evident losses in the patient's ability to move (Mishra *et al.*, 2017; Schaper *et al.*, 2020; van Netten *et al.*, 2020). To assess the loss of protective sensitivity and quickly screen patients who may be considered at risk, the 10 gf Semmes-Weinstein monofilament (SWM) is the most widely used first-line instrument in primary screening and recommended by national and international guidelines (Crawford *et al.*, 2007; Schaper *et al.*, 2020). Basically, this examination consists of promoting a skin touch with the thin SWM filament, in a perpendicular orientation to the skin region to be evaluated, until it has a slight curvature. At this time, a value of equal force will have been applied to what the manufacturer of the instrument declares, checking, by feedback from the patient, if the patient felt the touch or not. This screening should cover several distinct predefined points of the plantar and dorsal regions of the patient's foot, in places that are already well characterized and classified as habitually subject to hyperpressions and with the potential to develop ulcers (Schaper *et al.*, 2020).

The lack of quality control and calibration of the 10 gf SWM, and hence, the unknown associated measurement error, are factors that may have significant influence on the underevaluation or overvaluation of the diagnosis, with impact on the patient's health and on the pressure on healthcare services. On the other hand, the simple "feel"/"don't feel" qualitative assessment, since it is not possible to obtain a numerical value as a result of a particular examination, may be a limiting aspect for a more assertive diagnosis (Dros *et al.*, 2009). Limitations regarding its use in successive applications are also suggested. Its compression force, and the consequent force that is transmitted to the patient, often decay after few successive uses (Chikai & Ino, 2019).

DOI: 10.1201/9781003217152-28

In this paper, we propose to evaluate and analyze the performance of the 10 gf SWM under the effect of three parameters: intensive use; the ambient air temperature; and the displacement speed of the monofilament towards the cutaneous region to be evaluated. To achieve this, the following objectives were defined: (1) to identify the maximum compressive force of each 10 gf SWM, when exposed to a series of successive applications and different ambient temperatures while varying between two displacement velocities; (2) to evaluate compression dynamics through the entire compression phase, since the initial skin contact to the maximum compression of the filament; (3) to determine the relative error of the 10 gf SWM considering the force value that is declared by its manufacturer.

2 MATERIALS AND METHODS

Looking for objective evidence that a given SWM meets the requirements specified by the manufacturer, three 10 gf SWM (Figure 1a) were randomly selected to be tested, in the new condition, from commonly used brands in clinical practice. These monofilaments were named M1, M2 and M3 during our experiments, not to expose their brand or manufacturer.

With a measuring equipment specifically developed for monofilament evaluation (Figure 1b), through an automated method that mimics the international recommended procedures, a metrological characterization of the performance of the SWM was performed under controlled conditions. The equipment is operated by a computer application that allows to fully control position, speed, range, repetitions, and dynamics of the monofilament while reading the applied force/pressure. During the process, the monofilament is attached to a fixation support that moves towards a measuring platform with a foot like shape and a superficial stiffness similar to human skin.

Three distinct testing scenarios were considered: a) *Single measurements*. In this test, single and time isolated measurements of the compressive force of the SWM were performed. For the movement speed towards the force measurement surface we have considered 4 and 8 $mm.s^{-1}$, a range that encompasses the typical approaching speed when in a human made procedure (Chikai & Ino, 2019). For the ambient temperature we have considered the values 20 °C and 25 °C, boundary values for a comfort temperature in a healthcare facility context. The relative humidity (RH) of the air was measured and kept constant at 55 %. Speed tests were carried out with an interval of ten minutes while, for the ambient temperature variation, a twenty-four hours interval period was respected; b) *Series measurements*. In this experience, the intensive use of the SWM was simulated by performing a series of twenty successive applications for each instrument, similar to what occurs in clinical practice, where this instrument can be requested about ten times on each foot (twenty times for both feet). In this case, the ambient temperature parameter varied between 20 °C and 25 °C, maintaining a speed of 8 $mm.s^{-1}$ and air RH at 55 %. Again, testing at different temperatures was twenty-four hours apart; c) *Compression phase*

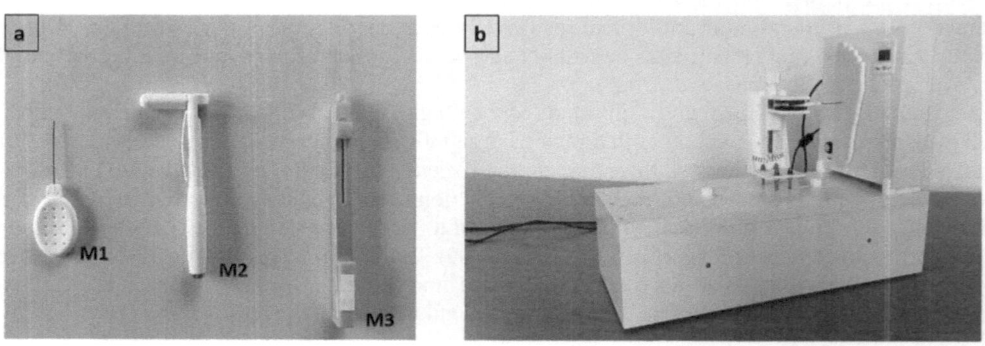

Figure 1. Materials used in the tests: (a) 10 gf SWM selected and (b) automatic measuring equipment.

characterization. The compression phase of the SWM begins when the tip of the monofilament first contacts the skin surface and until it makes the defined buckling while developing the expected 10 gf. Characterizing the entire compression phase implies acquiring multiple measures while the monofilament progresses against the measurement surface. For this, the automated movement of the SWM was controlled to advance 0.08 mm every 0.5 s, with a force measure being acquired at each step. In this scenario, the speed was defined by the length of the advance step, the temperature was varied (20 °C and 25 °C) and air RH maintained at 55 %. Like the previous test, the interval between tests at the two ambient temperatures was maintained.

3 RESULTS

The results obtained for the first scenario (*Single measurement*) are presented in Table 1. We can observe that all SWMs behaved differently and with a high relative error, when compared with the reference 10 gf value. The M1 monofilament was the one that registered the highest force values (20.6 to 26.7 gf), with errors ranging from 106 % to 167 %. On the other hand, M2 monofilament approached more to the 10 gf declared by the manufacturer, however, with values of 11.4 gf to 13.3 gf of compression and relative errors above 10 %, more precisely between 14 % and 33 %. The last instrument, M3, was the only one that placed a compression force below 10 gf, registering values between 7.2 gf and 7.9 gf, still causing relative errors above 20%, specifically from 21 to 28 %. These values can be considered critical to discriminate a possible evolution of the pathology. It has also been observed that increasing the ambient temperature results in a decrease of the buckling force. As for the velocity, it was found that it has a direct action on the performance of the SWM, since the increase in the movement velocity towards the measurement platform, resulted in a greater measured force.

For the second scenario (*Series measurements*), the obtained results are shown in Figure 2a. We can observe that the compression force of the SWM decreases with the number of repeated uses. It also clear that the variation in the ambient temperature influences the behavior of this instrument, since, with the increase in temperature, it considerably decreases its compression force. The M1 instrument is the one with the highest compression force, as well as, in this instrument, the effect of temperature and number of uses is more noticeable, although this trend is reflected in all tested SWM. From the first application to the third or fourth, there is a visible decrease in force, leading to the suspicion that SWMs have a refractory period to recover their viscoelastic properties.

In the third and last scenario (*Characterization of the compression phase*), the analysis of the collected data, depicted in Figure 2b, revealed that all SWMs require approximately 10 mm of displacement to complete the recommended curvature (for the correct evaluation procedure of a patient). The monofilaments showed to have an initial phase, after skin contact occurs, where a progressive increase in force can be observed, until they reach a maximum peak in the first 2 to 3 mm of displacement. In the remaining displacement they maintain approximately

Table 1. Results obtained by single measurements with variation of the parameters of velocity and ambient temperature, with RH at 55 %. The relative error considers the 10 gf reference as declared by the manufacturers.

		Maximum force (gf)		Relative error (%)	
SWM	Speed ($mm.s^{-1}$)	20 °C	25 °C	20 °C	25 °C
M1	4	24.5	20.6	145	106
	8	26.7	22.7	167	127
M2	4	11.7	11.4	17	14
	8	13.3	12.1	33	21
M3	4	7.5	7.2	25	28
	8	7.9	7.5	21	25

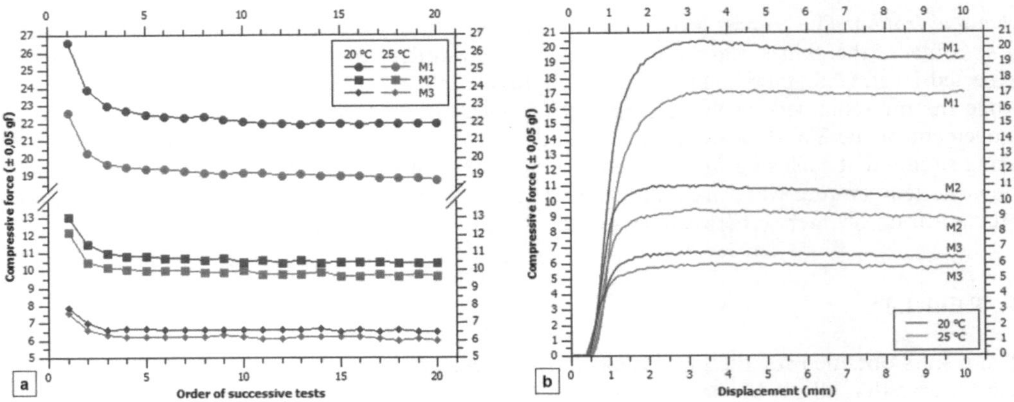

Figure 2. Results of measuring force of 10 gf SWM considering (a) the maximum compression force in a series of successive applications and (b) its performance during the entire compression phase. (Both tests with variation of the ambient temperature; Speed = 8 $mm.s^{-1}$; RH = 55 %).

the compressive force stabilized, but with a slight decreasing tendency, more evident in the instruments M1 and M2. In this test, it was observed that the maximum force is slightly lower than the one recorded in other experiments, because the monofilaments were subjected to a continuous compression effort during the entire displacement and with a very low advance step, which gives enough time for the viscoelastic properties of the material to conform to each measurement that is made, that is, there are no sudden changes. Here, again, the temperature influenced the behavior of all SWMs.

4 DISCUSSION

The devices used in the described tests are similar to those used in clinical practice. The data collected points to the existence of significant variations in the device's characteristics and also to the existence of sensitivity to external parameters such as ambient temperature and humidity. It was also observed that the procedure speed can impact the results. This parameter, in the context of human made monofilament evaluations, is hard to define and it is often neglected. Although the test set is small, not allowing sufficient statistical robustness, the results suggest a reduced confidence about the metrological qualities of this equipment. Additionally, the observed error margins raise questions about the quality of the diabetic foot evaluations that use this instrument and also question the validity of the resulting clinical decisions.

5 CONCLUSIONS

In this study we have described a testing methodology for the 10 gf SWM. It was found that the compression force of the 10 gf SWM is influenced by intensive use, the variation of the ambient conditions, the displacement speed and impact on the skin region. This evidence implies a variation in the force applied to the patient, also causing a greater or lesser relative error associated with the SWM. These results come to question the validity of SWM based evaluations and decisions. The current state of technological evolution allows to replace the SWM with less error prone devices that can easily provide more accurate and objective information. As future work, the authors plan to obtain further confirmation of the obtained results by evaluating a wider range of monofilament devices.

REFERENCES

Chikai, M. & Ino, S. (2019) "Buckling Force Variability of Semmes-Weinstein Monofilaments in Successive Use Determined by Manual and Automated Operation," *Sensors MDPI*, 19(803).

Crawford, F. *et al.* (2007) "Predicting foot ulcers in patients with diabetes: A systematic review and meta-analysis," *Qjm*, 100(2), pp. 65–86.

Dros, J. *et al.* (2009) "Accuracy of Monofilament Testing to Diagnose Peripheral Neuropathy: A Systematic Review," *Annals of Family Medicine*, 7(6), pp. 555–558.

Mishra, S. C. *et al.* (2017) "Diabetic foot," *the BMJ*, 359(SUPP 1), pp. 1–7.

van Netten, J. J. *et al.* (2020) "Prevention of foot ulcers in the at-risk patient with diabetes: a systematic review," *Diabetes/Metabolism Research and Reviews*, 36(S1), pp. 1–22.

Schaper, N. C. *et al.* (2020) "Practical Guidelines on the prevention and management of diabetic foot disease (IWGDF 2019 update)," *Diabetes Metab Res Rev.*, 36(e3266), pp. 1–10.

Advances and Current Trends in Biomechanics – Belinha et al. (Eds)
© 2022 Copyright the Author(s), ISBN: 978-1-032-10806-3

Prediction model for a firefighter skin burn

P.A.G. Piloto
INEGI-LAETA, Instituto Politécnico de Bragança, Bragança, Portugal

A. Rahmani, S. Dahamni & K. Nechab
Hassiba Benbouali University of Chlef, Chlef, Algeria

A. Benarous
Saad Dahlab University of Blida 1, Blida, Algeria

ABSTRACT: Firefighters are usually exposed to extreme conditions, while facing their garments and body to high thermal fluxes, such as flash fires. To prevent burn injuries, it seems important to develop a predictive tool for skin temperature evolution. This paper presents a numerical method based upon experimental measurements of temperature histories within skin and fabric layers. The study is completed with a finite element investigation considering the blood perfusion rate within the skin, moreover, a focus has been made on the time prediction for the skin burn degrees, under several incoming heat fluxes, fabric thicknesses and exposure time. A new proposal is presented to predict the skin burn, depending on the fabric and air gap thickness for flash fires.

1 INTRODUCTION

Firefighters trust their protective garments to protect them during duty, they may be exposed to high heat fluxes, such as flash fires, or exposed to low to moderate heat fluxes, such as urban fires. Flash fires typically release heat fluxes of the order of 80 kW/m^2, while low and moderate urban fires can release up to 40 kW/m^2 (Dahamni et al. 2019). These exposures may cause skin injuries despite short operating times (Ghazy et al. 2010). These injuries range from first-degree burn to third-degree, depending on the fire exposure conditions. One of the criteria used for skin injuries prediction is based on Henriques' integral (Moritz et al. 1947). Several approaches to predict the thermal protective performance of firefighters' garments have been reported in the literature. Torvi (Torvi. 1997) developed a numerical model to validate the ASTM D 4108 bench test, Ghazy and Bergstrom (Ghazy et al. 2010) also developed a numerical model for single-layer protective clothing which considers the combined conduction-radiation heat transfer between fabrics and skin. Following this research, this paper investigates the influence of fabric thickness, incoming heat flux intensity as well as exposure time on skin burn prediction, accounting for blood perfusion rate within the skin.

2 MATERIALS AND METHODS

A typical firefighters' protective garment system is shown in Figure 1 (on the left). This system is modelled based on a nonlinear finite element analysis, accounting for temperature-dependent material properties and suitable boundary conditions based on

DOI: 10.1201/9781003217152-29

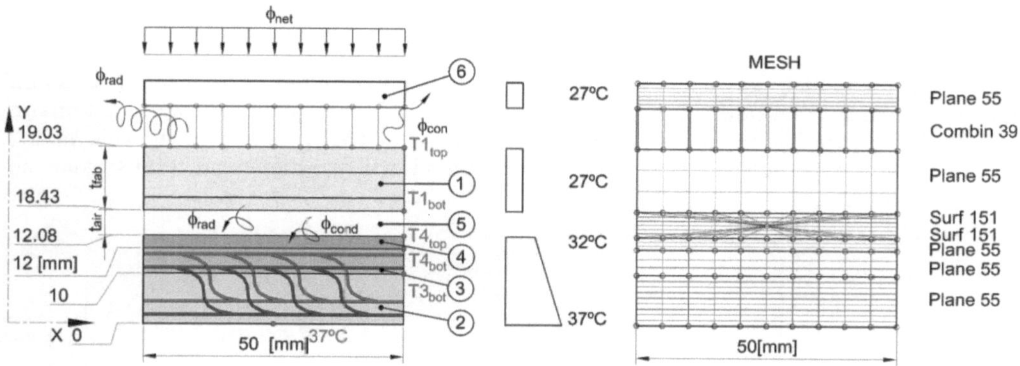

Figure 1. Two-dimensional garment-skin model (left) and finite element mesh (right).

experimental tests. For validation, the adopted model consists of a fire-resistant fabric (1), the human skin (epidermis (4), dermis (3) and subcutaneous (2) layers), and an air gap (5) between the fabric and the skin. The protective garment is submitted to a heat flux of 83 [kW/m^2] during 10 seconds, reflecting 9%. This heat flux is applied on top of the special material (6) (Plane 55) (high conductivity and low heat capacity material) and flows through the material by conduction, and then delivered to the exposed surface of the fabric (1) (Plane 55), using the auxiliary elements (COMBIN39). Meanwhile, energy loss takes place from the exposed surface to the surroundings (27°C) by radiation and convection. At the same time, heat flows through the fabric and delivers heat into the region of the air gap (5) (by radiation and conduction), using special interface surface elements (SURF 151) and normal plane elements (PLANE 55). Conduction is the mode of heat transfer through the skin layers (Plane 55), where blood perfusion (Figure 3) takes place in the dermis and sub-cutaneous layers.

The heat transfer in the skin is governed by the bio-heat equation Eq. 1 (Rai et al. 1999).

$$\frac{\partial}{\partial x}\left(\lambda\frac{\partial T}{\partial x}\right)+\frac{\partial}{\partial y}\left(\lambda\frac{\partial T}{\partial y}\right)+w_b.C_{pb}(T_a-T)=\rho.C_p.\frac{\partial T}{\partial t} \tag{1}$$

The temperature field (T) is determined in each domain and compared with experimental results for validation (Ghazy et al. 2010). The 3rd parcel on the left-hand side applies only to the dermis and subcutaneous regions, assuming the arterial temperature T_a=37 °C. All the other thermal properties depend on the region where the heat flows by conduction, namely λ the thermal conductivity, C_p the specific heat and ρ the specific mass.

The burn damage of the skin may be calculated using Henrique's burn integral equation (Moritz et al.1947), which can be expressed according to Eq. 2. The values for the pre-exponential factor P and the activation energy ΔE may be obtained for the first-degree, second-degree and the third-degree burn injury (Song et al. 2017).

$$\frac{d\Omega}{dt}=\begin{cases}0, & T<44^\circ C\\ Pe^{-(\Delta E/RT)}, & T\geq 44^\circ C\end{cases} \tag{2}$$

Where, Ω is a quantitative measure of burn damage; R is the molar gas constant; T is a time dependant skin temperature. The 1st degree burn occurs when Ω reaches 0.53, the 2nd degree burn occurs when Ω reaches 1 (Song et al. 2017). Using the temperature-history of the basal layer and the 3rd degree, a burn injury also occurs when Ω reaches 1 when the temperature-history of the dermal base layer is considered.

3 RESULTS

To validate the model, the system parameters were tuned according to experimental tests and numerical models from other authors (Ghazy et al. 2010). The results used for validation are presented in Figure 2. The numerical results for $T4_{BOT}$ and $T3_{BOT}$, obtained with this model, agree well with experimental tests while reproducing closely the numerical results from the mentioned authors.

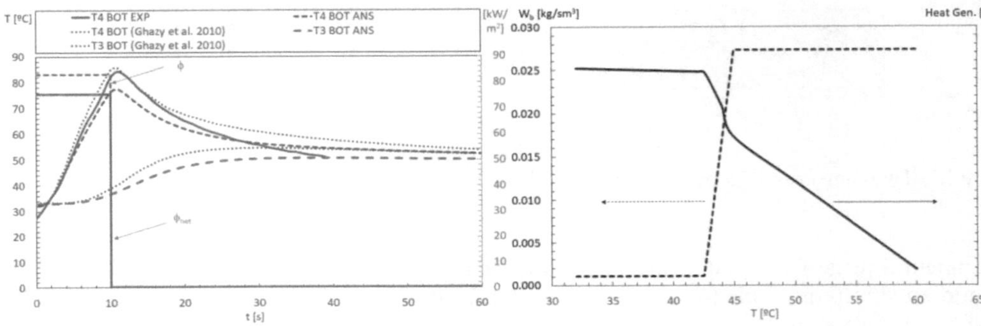

Figure 2. Temperature comparison on basal layer ($T4_{BOT}$) and dermal base ($T3_{BOT}$), with Kevlar/PBI protection (left) and blood perfusion rate depending on skin temperature (right).

A parametric analysis was developed upon a set of simulations for the following parameters: Net heat flux level ϕ_{net} = 15,25,40 and 80 kW/m^2, with a fire exposure time of 10s and 20s, followed by a cool-down period of 60s, protected by different Kevlar/PBI and Nomex thicknesses (t_f = 0.4,0.6,0.8,1.2 and 1.6), assuming several air gap thicknesses (t_{air} = 2,4,6,8 and 10 mm).

Figure 3 Depicts the results for the skin burn, protected with Kevlar/PBI, exposed along 20 s to a heat flux density of 80 KW/m^2. Figure 4 depicts the effect of the heat flux on the skin burn prediction for the first (t1) and second (t2) degree burn.

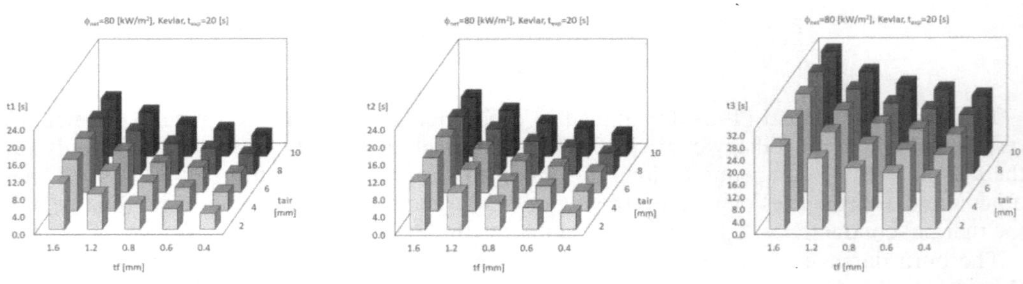

Figure 3. Time to reach the first (t1), second (t2) and third (t3) degree burn, for Kevlar/PBI exposed to 80 KW/m^2, during 20s.

4 DISCUSSION

When exposed to higher heat fluxes, short time is required to achieve the first (t1) and second-degree burn (t2). The time lap t1 and t2 decreases with the decrease of the t_f. The third-degree burn (t3) is reached during the cooling down period for all the simulations performed under ϕ_{net} = 80 kW/m^2, while it only occurs for the ϕ_{net} = 40 kW/m^2 under low thermal resistance (t_{air} ≤ 4mm and t_f ≤ 0.6mm). Similar results were achieved for smaller exposure time when using Kevlar/PBI and Nomex. The time t1 and t2 also decreases with the decrease of the air gap and fabric thickness.

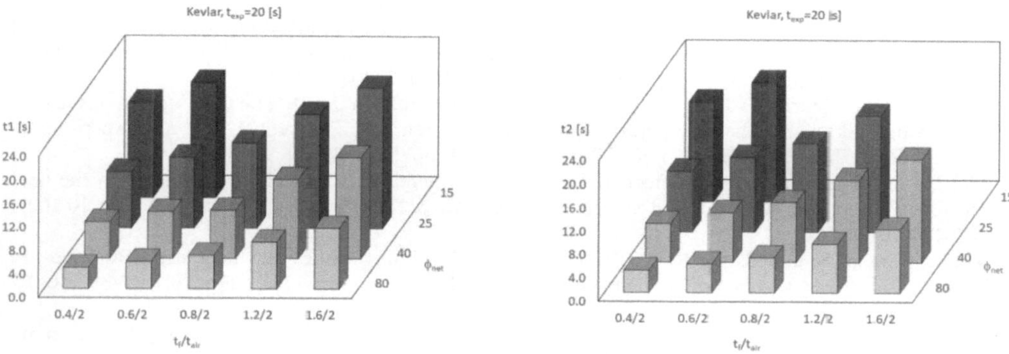

Figure 4. Skin burn prediction (t1 and t2) depending on the net heat flux, for Kevlar/PBI.

Based on the numerical results determined for high heat fluxes (80 KW/m^2) and for high exposure time (20s), a new proposal is presented to determine the time to reach t1, t2 and t3. The proposed model minimizes the sum of the square residuals (SSR) between the numerical results and the approximation. The solution method is based on a nonlinear optimizing solver (Lasdon et al. 1974). The approximation formula is presented in the Eq. 3 and the parameters A,B,C and D are presented in table 1. The sum of the squared residuals (SSR) is also determined.

$$t = A \times t_f + B \times t_{air} + C \times t_f \times t_{air} + D \tag{3}$$

Table 1. The proposed coefficients for the skin burn prediction.

Material	Kevlar/PBI					Nomex				
Time	A	B	C	D	SSR	A	B	C	D	SSR
t1	5.63	0.08	0.14	1.20	1	3.49	0.06	0.14	1.89	1
t2	5.83	0.10	0.13	1.30	1	3.66	0.07	0.14	2.00	1
t3	7.97	0.11	0.42	13.1	11	5.23	0.24	0.15	13.3	4

5 CONCLUSIONS

The validation of a numerical model was obtained from experimental tests developed on a test-bench, using a skin simulant material. After validation, a set of numerical simulations were developed to analyse the effect of several parameters on the skin burn prediction (fabrics thickness, heat flux and exposure time). Three parameters (thickness, the net heat flux and exposure time) were considered to evaluate the thermal behaviour of the skin surface before and during post-fire exposure. Increasing the heat flux will reduce the time to reach the first-degree burn. Thicker garments had monotonically increased the time of the first burn injury and gradually decreasing the maximum temperature for both basal and dermal base layers.

REFERENCES

Dahamni S., Nechab K., Benarous A., Piloto P.A.G. (2019) 'Numerical simulation of incoming thermal fluxes on a firefighter garment', *Revista FLAMMAE - Special Edition 2019*, 5(14). doi: http://dx.doi.org/10.17648/2359-4837/flammae.v5n14.p71-85.

Ghazy, A. and Bergstrom, D. J. (2010) 'Numerical simulation of transient heat transfer in a protective clothing system during a flash fire exposure', *Numerical Heat Transfer; Part A: Applications*, 58(9), pp. 702–724. doi: 10.1080/10407782.2010.516691.

Lasdon, L. S., Fox, R. L. and Ratner, M. W. (1974) 'Nonlinear optimization using the generalized reduced gradient method', Revue française d'automatique, informatique, recherche opérationnelle. Recherche opérationnelle, 8(V3), pp. 73–103. doi: 10.1051/ro/197408V300731.

Moritz, A. R. and Henriques Jr., F. C. (1947) 'Studies of Thermal Injury: I. The Conduction of Heat to and through Skin and the Temperatures Attained Therein. A Theoretical and an Experimental Investigation.', *The American Journal of Pathology*, 23(4), pp. 530–549.

Rai, K. N. and Rai, S. K. (1999) 'Effect of metabolic heat generation and blood perfusion on the heat transfer in the tissues with a blood vessel', *Heat and Mass Transfer*, 35(1), pp. 75–79. doi: 10.1007/s002310050300.

Song, G., Mandal, S. and Rossi, R. (2017) 'Chapter 3: Skin burn injuries and heat stress/fatalities', in *Thermal Protective Clothing for Firefighters*. Woodhead Publishing, pp. 17–26. doi: 10.1016/b978-0-08-101285-7.00003-4.

Torvi, D. A. (1997) *Heat transfer in thin fibrous materials under high heat flux conditions.pdf*. University of Alberta.

Advances and Current Trends in Biomechanics – Belinha et al. (Eds)
© 2022 Copyright the Author(s), ISBN: 978-1-032-10806-3

Experimental comparison of two internal fixation implants in the treatment of femur neck fractures

M. Pinto & V. Maranha
Department of Mechanical Engineering, University of Coimbra, Coimbra, Portugal

A.M. Amaro
CEMMPRE, Department of Mechanical Engineering, University of Coimbra, Coimbra, Portugal

L.M. Roseiro
ISEC, Coimbra Polytechnic, Coimbra, Portugal

M.A. Neto
CEMMPRE, Department of Mechanical Engineering, University of Coimbra, Coimbra, Portugal

ABSTRACT: Fixation of femur head fractures often fails due to lag screw cut-out, especially in osteoporotic bones. This study correlates the experimental results of a new Trochanteric Plate of Contention (TPC), which may improve the resistance to the cut-out failure of internal fixation devices, and the well-known Dynamic Hip Screw (DHS) system. Generally, it is well accepted that the DHS is the first option in the treatment of stable femur fractures (Kokoroghiannis et al., 2012), as well as considered the implant that any new design should be compared with (Kouvidis et al., 2009). Hence, this study compares the strains, stresses, and displacements on these implants under compression tests. The data acquisitions were carried out using electrical and optical sensors.

1 INTRODUCTION

Hip fracture is a main concern of the elderly, most of whom suffer from osteoporosis in addition to various medical and mental diseases that result in their being a high-risk population. Hence, many internal fixation devices have been developed for stabilizing femur fractures. Despite the great number of devices that are available, they can be separated into two main categories: extramedullary (sliding hip screws) and intramedullary (nails) fixation devices (Duymus et al., 2018).

The treatment of stable intertrochanteric fractures is traditionally performed with the dynamic hip screw (DHS) because it presents low failure rates. In contrast, unstable fractures are known to have high complication rates (Ibrahim and Meleppuram, 2017). Duymus et al. (Duymus et al., 2018) in a retrospective study, showed that intramedullar nails are more efficient than extramedullar implants in treating unstable proximal femur fractures. Moreover, the PFLCP implant, which can be also classified as an extramedullary fixation device, should not be the first choice of treatment, but it can be used when there is no option available for other routinely used implants (Shah et al., 2017). Nowadays, as it was at the beginning of this century, the most suitable surgical method for treating unstable fractures is still controversial.

This work deals with extramedullary fixation devices for femur fracture stabilization. It has the primary purpose of comparing the fixation stabilities of the TPC and the Dynamic Hip

DOI: 10.1201/9781003217152-30

Screw (DHS) devices. The TPC medical device comprises a blade plate, a lag screw, and an anti-rotation pin. The innovative blade plate has a geometry that gets inside the femur in the great trochanter region and, after inside, is crossed by the lag screw. The TPC geometry is a compromise between the two main categories of fixation devices. It has a main blade plate that is screwed to the femur diaphysis and extends above the great trochanter, where it is introduced inside the femur and goes until the small trochanter region.

2 MATERIALS AND METHODS

Two synthetic femurs from the Sawbones®, model #343-102, produced using the fourth generation of composite bones, were fractured and, posteriorly, fixed with the two fixation implants, i.e, TPC and DHS. Both devices were instrumented with strain gauges and rosettes, see Figure 1. Each implant has one rosette in the interior face of the blade plate and one gauge in their exterior face. At the same time, spray-painted speckle patterns were experimentally created over the proximal zone of the two femurs, Figure 2.

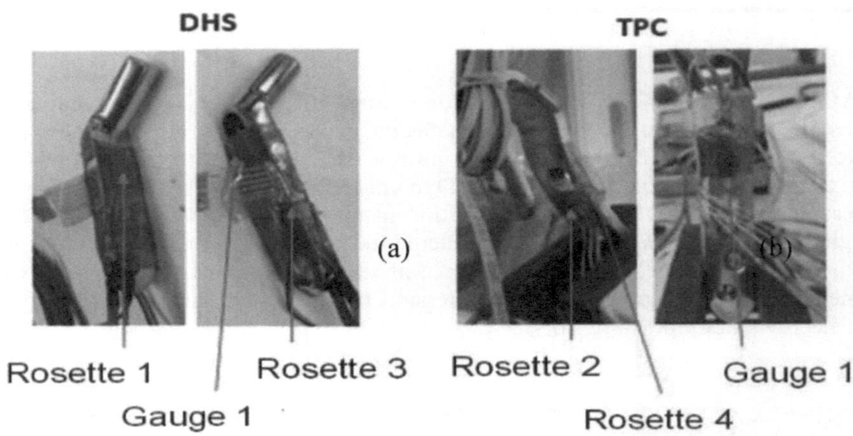

Figure 1. Identification of strain gauges and rosettes on the devices: (a) DHS; (b) TPC.

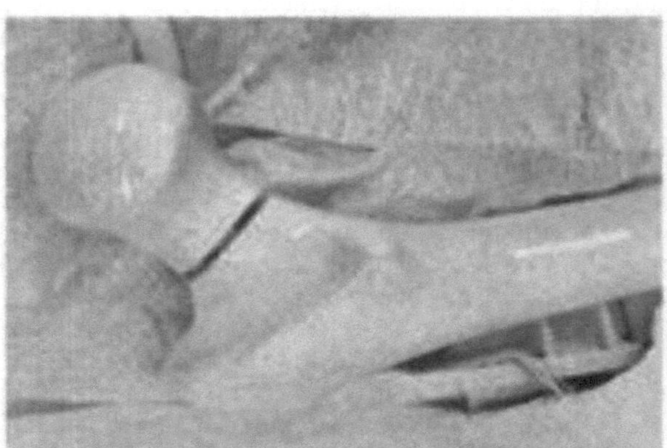

Figure 2. Speckle pattern on the femur (right image).

To ensure reproducibility, both experimental setups were tested under the same conditions: the femurs were fixed to the testing platform and consistently oriented by means of an aluminum block attached to the femoral condyles, holding the femoral shaft at 11° in adduction and 9° in flexion (Lopes *et al.*, 2017). This block was fixed to the platform of a Shimadzu AG-10 universal testing machine (Shimadzu Corporation, Kyoto, Japan), equipped with a 5kN load cell and TRAPEZIUM software, and different levels of force were applied at a displacement rate of 2mm/min. Forces were applied vertically to the femoral head using the machine actuator through an aluminum-Teflon plate. The Teflon material was glued to the aluminum plate, thereby avoiding the transmission of horizontal forces. For all force magnitudes, the maximum force was held for 30s, allowing a constant time for a repeatable amount of creep to take place. Each loading intensity was repeated three times on each specimen, and, between repetitions, the specimen could recover for 10min.

3 RESULTS AND DISCUSSION

The distribution of the vertical displacement in the proximal zone of both femurs is represented in Figure 3. These results were obtained with a compression load of 1000 N and,

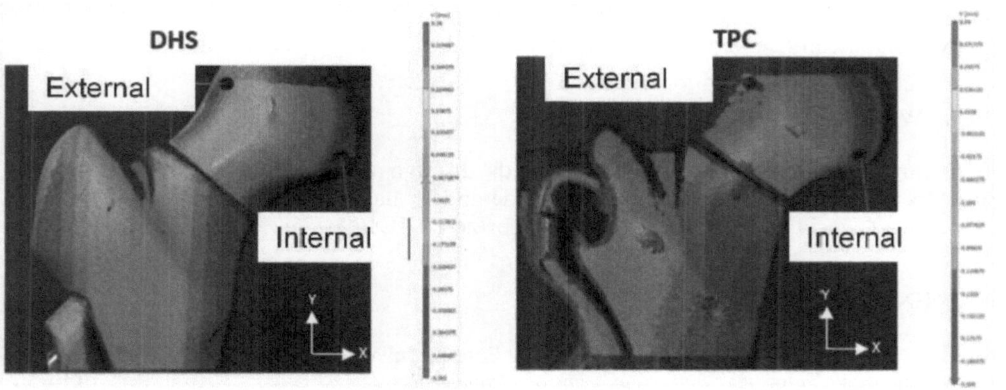

Figure 3. Distribution of the vertical displacement on both experimental setups.

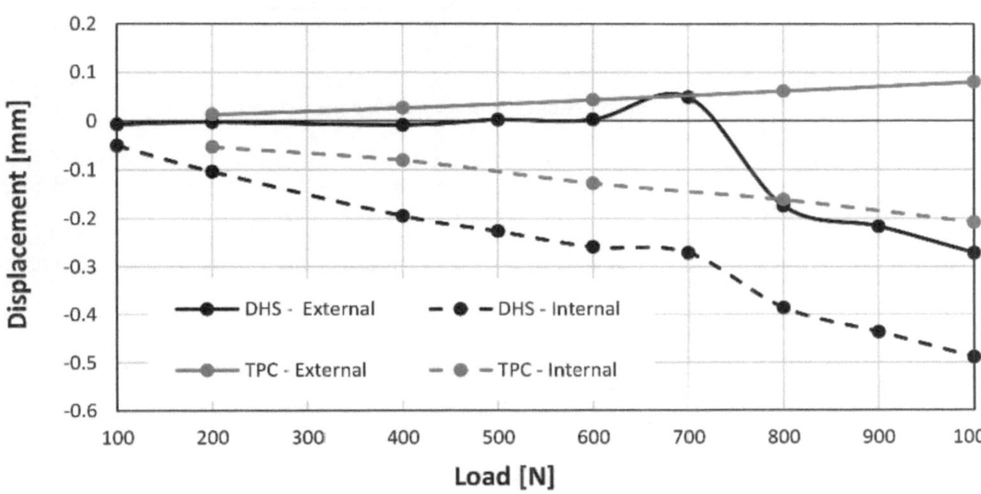

Figure 4. Variation of the vertical displacement for both external and internal points in the femur head for the two fixation implants: DHS and TPC.

simultaneously, using the VIC 3D digital image correlated system. On each one of these images it is possible to identify two points: the first one is positioned on the external face of the femur head whereas the second one is positioned at its internal face. The variation of the vertical displacement of these two points with the applied load is represented in Figure 4.

The femur head's external point fixed with the TPC device always shows positive values, while that fixed with the DHS shows negative values for loads higher than 700N. This last behavior was confirmed visually by an axial rotation of the proximal femur head around the cephalic screw. This rotation shows that the proximal fragment's stability is not ensured by this fixation method and, therefore, early weight-bearing should not be allowed before callus is observed. These results show also that the TPC device has a stiffer behavior than the DHS, even knowing that this type of fracture has especial medical indications to be treated with the DHS.

4 CONCLUSIONS

The experimental results showed that strain values are higher in the DHS than in the TPC device, particularly in the neighborhood of the cephalic screw. The displacement field also showed that the TPC device has a stiffer behavior than the DHS. So, it is expected that this device will significantly enhance the medical device's behavior in osteoporotic femurs.

ACKNOWLEDGMENTS

This research was sponsored by FEDER funds through the programme COMPETE Programa Operacional Factores de Competitividade and by national funds through FCT—Fundação para a Ciência e a Tecnologia, under the project UIDB/00285/2020.

REFERENCES

Duymus, T. M. et al. 2018. Comparison of Intra- and Extramedullary Implants in Treatment of Unstable Intertrochanteric Fractures. *Journal of Clinical Orthopaedics and Trauma*, 10 April. doi: 10.1016/j.jcot.2018.04.003.

Ibrahim, S. and Meleppuram, J. J. 2017. A retrospective analysis of surgically-treated complex proximal femur fractures with proximal femoral locking compression plate. *Revista Brasileira de Ortopedia (English Edition). Georg Thieme Verlag KG* 52(6): 644–650. doi: 10.1016/j.rboe.2016.12.012.

Kokoroghiannis, C. et al. 2012. Evolving concepts of stability and intramedullary fixation of intertrochanteric fractures—A review. *Injury* 43(6): 686–693. doi: 10.1016/J.INJURY.2011.05.031.

Kouvidis, G. K. et al. 2009. Comparison of migration behavior between single and dual lag screw implants for intertrochanteric fracture fixation. *Journal of orthopaedic surgery and research. BioMed Central* 4:16. doi: 10.1186/1749-799X-4-16.

Lopes, V. M. M. et al. 2017. FE and experimental study on how the cortex material properties of synthetic femurs affect strain levels. *Medical Engineering & Physics* 46: 96–109. doi: 10.1016/j.medengphy.2017.06.001

Shah, M. D. et al. 2017. Evaluation of outcome of proximal femur locking compression plate (PFLCP) in unstable proximal femur fractures. *Journal of Clinical Orthopaedics and Trauma*. 8(4): 308–312. doi: 10.1016/j.jcot.2016.11.005.

Advances and Current Trends in Biomechanics – Belinha et al. (Eds)
© 2022 Copyright the Author(s), ISBN: 978-1-032-10806-3

The influence of bone cement mantle thickness on bone-cement interface load transfer—a micro finite element model

A.I. Cabrinha
TEMA, Mechanical Engineering Department, University of Aveiro, Portugal

A.M. Ramos
TEMA, GIB, Mechanical Engineering Department, University of Aveiro, Portugal

ABSTRACT: The extension of cement mantle thickness influence remains unclear. To study the impact of the cement mantle thickness has in the stress and strain states of both bone-cement interface and trabecular bone, a finite element model of the implant-cement-bone system was developed. The trabecular bone was progressively removed, compensating that difference with cement, resulting in cement thicknesses ranging from 2.5 to 7.5 mm. The results show a decrease of the interface overall stresses with the increase of cement mantle thickness. However, in trabeculae situated in the bone middle plane, stress decreased with the increase of cement thickness and, in contrast, a reverse behavior occurred concerning the stress near compact bone. This leads to suggest that a mantle thickness between 5.5 and 7.5 mm provides more balanced results.

1 INTRODUCTION

Cemented hip arthroplasty has become a very successful solution to treat joint diseases, even in geriatric patients or with low bone quality. Although its high success rate, cemented hip arthroplasty continues to lodge limitations and, in some cases, requires revision surgery. Failure can occur in the postoperative period, in short-term, due to thermic necrosis caused by the heat released during polymerization, monomer local toxicity or interface rupture. Failure can also occur at long-term, being associated to factors like the cement mantle thickness. There is few information regarding the influence of cement mantle thickness on the outcome of cemented arthroplasty. When it was initially studied, some authors concluded that thin layers lessen cement mantle fatigue resistance, increasing the failure risk, and that thick layers would increase cement stiffness, increasing the stress in cement-bone interface (Ramaniraka, et al., 2000; Terrier, et al., 2005; Coultrup, et al., 2010). However, when the stems of Charnley-Kerboull and Ceraver Osteal were introduced, they raised questions (Langlais, et al., 2003). The technic, named line-to-line, consists of removing all the trabecular bone from the medullary cavity and inserting the steam that goes almost from an edge to another and fill the rest with cement, which results in a very thin cement layer, sometimes thinner than 1 mm. This phenomenon has proven to give excellent results in the short and long term, and it has become known as "The French Paradox". More recently, Calvert et al. (Clavert, et al., 2019), suggested that the optimal cement thickness depended on the anatomical location of the implant that is analyzed. Other studies concluded that the optimal cement thickness depended on the patient's bone morphology and quality and can vary between cases (Lamvohee, et al., 2014). Others underline that having a uniform, non-deficient layer is more important than the thickness itself (Clauss & Breusch, 2019). Moreover, other authors referred the importance that

DOI: 10.1201/9781003217152-31

cement mantle thickness has in gap formation, determining where failure occurs (Ramos & Simões, 2009). Kumar et al. (Kumar, et al., 2020) also concluded that the cement mantle thickness was important within the failure mechanism. According to the authors, thin layers promote failure in implant-cement interface, because gap formation concentrates on this area. On the other hand, when the layer is thick, gap formation was dependent on implant geometry and bone removal process. Throughout literature, it is clear that the cement mantle thickness influences the outcome of cemented arthroplasty. However, the extension of that influence is yet not completely understood and in some cases it is controversial. Aiming to better understand this topic, a study of the cement mantle thickness influence on the stresses and strains at the cement-bone interface and trabecular bone was performed. To achieve this, an axisymmetric finite element model of the implant-cement-bone, located in the epiphysis of a long bone was developed, in which the cement thickness was progressively increased, resulting in values ranging from 2.5 up to 7.5 mm.

2 MATERIALS AND METHODS

Micro computed tomography (μCT) scan, with 20μm resolution, was used to reconstruct a Sawbones polyurethane foam with 7.5 PCF density. A Simpleware SanIP software was used to reconstruct the bone. The other components were modeled in CAD software Solidworks. The stem was modeled into a 10 mm base cone with an angle of 1.43° in ASTM F75 CoCr. Bone cement was defined to have 2.5 mm thickness, in the base model, as a Zimmer Regular of 2.7 GPa (Lewis, 1997). Trabecular bone was reconstructed in, approximately, 30mm in the exterior. Cortical bone was modeled into a cylindrical geometry in epoxy resin, with a uniform 2 mm thickness. All materials were considered to be homogeneous, isotropic and linear elastic. In order to reduce computational costs, axisymmetry in relation to the stem revolution axis was used. The simulation was carried out by applying a 11.5 N force at the stem extremity and fixating the cortical bone. The stem and cement had a node-to-surface contact, with a 0.3 static friction contact coefficient. Both cement and trabecular bone and cortical bone were considered bonded. A mesh refinement was performed, resulting in convergence of 390 200 DOF mesh.

To study the influence of bone cement mantle thickness, the trabecular bone was progressively removed, and the difference was added in the cement thickness. This resulted in a range of 2.5 to 7.5 mm cement thicknesses. To evaluate the influence of cement thickness on the mechanical behavior of the cement-bone interface, the values of maximum principal stress were calculated in both materials at the interface between them. In the trabecular bone, the values of maximum principal stress were calculated in three areas: at 1mm from the interface, in the middle plane and at 1mm from cortical bone.

Figure 1. Illustration of the axisymmetric model developed to study the influence of cement mantle thickness. Identification of the model elements and constrains.

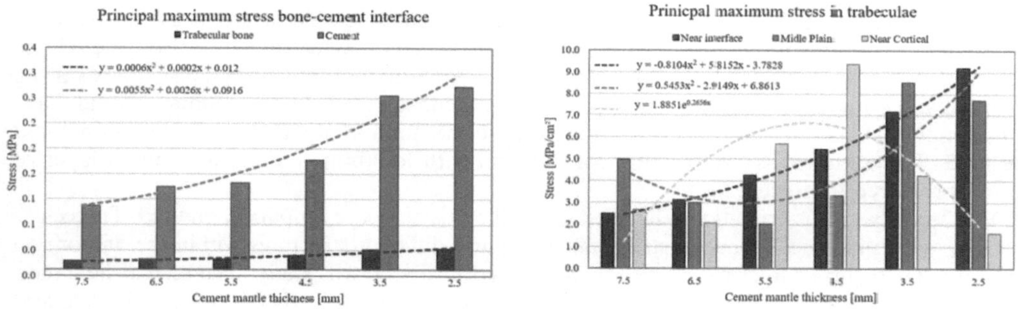

Figure 2. In the left side, principal maximum mean stress in the components with the increase of cement mantle thickness graphic. In the right side, principal maximum mean stress at the trabecular bone, in the three different zones analyzed with the increase of cement thickness.

3 RESULTS

The influence of the cement mantle thickness on the mechanical behavior of the implant is notorious. The results show that an increase in cement mantle thickness decreased the stresses that both cement and trabecular bone were subjected to. At the bone-cement interface (Figure 2a), the stresses of both materials decreased gradually with the increase of cement mantle thickness, having a total decrease of 66% for the cement and 59% for the bone, between the maximum and minimum cement thickness. The same happened for the strains, which decreased a total of 67% and 54%, respectively. However, the observed in the trabecular bone was substantially different (Figure 2b). In the middle plain, the increase of cement mantle thickness decreased bone stresses until a thickness of 5.5mm was obtained, in a total of 74%. From that point on, the stresses increased in about 2.5 times. At the trabeculae situated near the cortical bone, an opposite behavior occurred. The trabeculae stresses increased almost 6 times with the increase of cement thickness. It then tended to decrease with the increase of cement mantle thickness.

4 DISCUSSION

While at the cement-bone interface (and the trabeculae of the support bone near this interface) the increase of cement mantle thickness gradually decreases the stresses in both cement and cancellous bone, the stresses in trabeculae situated in the middle plain of the support bone and near the cortical bone seemed to be dependent on the cement mantle thickness. Albeit thinner thicknesses (2.5-3.5 mm) provide the support bone near cortical bone lower stresses, the trabeculae in the support bone middle plain and bone-cement interface stress levels are very high (2 to 6 times higher). On the other hand, taking a cement thickness of 4.5 mm into account, both the support bone middle plain trabeculae and the cement-bone interface are subjected to moderate stress levels, but, in contrast, the trabeculae near cortical are under very high stresses. Thus, the present work suggests that thicknesses between 5.5 and 7.5 seem to provide both materials, in all areas, with balanced acceptable solicitations, implying that this range of values provide better results. Our findings differ from those of Ramaniraka et al. (Ramaniraka, et al., 2000). Analyzing the femoral component, the authors observed that thicknesses inferior to 2mm induced excessive stress and micromotions on the bone-cement interface, and that thicker mantles also increased the interface stress and micromotion. Terrier et al. (Terrier, et al., 2005) found similar results. The authors studied the load transfer across the glenoidal component and stated that thicknesses between 1.0 and 1.5 would be ideal and that values bellow that range fragilized both the cement mantle and the bone-cement interface, but values above the range stiffened the cement, increasing the stress in the interface. This is in not in line with what was found by Clavert et al. (Clavert, et al., 2019). The author also analyzed the glenoidal component and found that the layer should be as thin as possible between the posterior part of the implant and the bub-chondral bone (not exceeding 1mm) and should be optimized around the keel of the component. The results of the author suggest that the optimal cement thickness varies within the implant.

The optimal cement thickness might be dependent on other factors such as the patient's morphology and their bone quality. Lamvohee et al. (Lamvohee, et al., 2014) studied the effect of cement mantle thickness, acetabular size, bone quality and body mass index on the acetabular component stresses. They found that diameters inferior to 50mm should have a minimal thickness of 4mm, whilst for diameters of 58mm a minimal thickness of 1mm could be used. The authors also stated that for patients with low bone quality a minimal layer of 4mm is required to prevent failure.

This study has some limitations. The first one is the axisymmetric simplification used. Trabecular bone is a morphological complex structure. It acquires different structures within the anatomical position. Thus, this characteristic was not considered in the present study. The second limitation consists of the type of load applied. In this study a static tension load was used, which induced compression and shear to the interface and bone. In reality, the bone is subjected to a combination of tension, compression and shear, with constant alteration in its magnitudes. Finally, the effects of bone remodeling were not reproduced. Despite these limitations, the focus of the work was to observe the influence of the cement mantle thickness on the load transfer of the interface, rather than the values. We hope that they will assist in further understanding this influence.

5 CONCLUSIONS

The following conclusions may be drawn from the present study: (1) cement mantle thickness influences the load transfer of the interface, (2) the increase in cement thickness increases the interfacial stresses; (3) in the present conditions, a cement thickness of 5.5 to 7.5 seems to be optimal. It seems that the influence of cement mantle thickness is a rather complex issue. The optimal thickness is dependent of several factors, such as bone morphology and quality, design of the implant, anatomical position. There is still a lot of uncertainty regarding the optimal cement thickness. Further work is required to better understand the extension of cement mantle influence on the outcome of cemented arthroplasty.

REFERENCES

Clauss, M. & Breusch, S., 2019. The 'French paradox' may not be a paradox after all – but for what reason?. *Bone & Joint Research*, 8(1), pp. 1–2.

Clavert, P., Bouchaïb, J., Kling, A. & Kempf, J., 2019. Does the cement mantle thickness influence the glenoid loosening in anatomic total shoulder arthroplasty? An experimental study. *Journal of Orthopaedic Science*, 24(1), pp. 81–86.

Coultrup, O., Hunt, C., Wroblewski, B. & Taylor, M., 2010. Computational assessment of the effect of polyethylene wear rate, mantle thickness, and porosity on the mechanical failure of the acetabular cement mantle: EFFECT OF POLYETHYLENE WEAR RATE, MANTLE THICKNESS, AND POROSITY ON CEMENT MANTLE FAILURE. *Journal of Orthopaedic Research*, 28(5), pp. 565–570.

Kumar, A. et al., 2020. Influence of interface crack and non-uniform cement thickness on mixed-mode stress intensity factor and prediction of interface failure of cemented acetabular cup. *Theoretical and Applied Fracture Mechanics*, Volume 107.

Lamvohee, J. et al., 2014. Total Hip Replacement: Tensile Stress in Bone Cement is influenced by Cement Mantle Thickness, Acetabular Size, Bone Quality, and Body Mass Index. *Journal of Computer Science & Systems Biology*, 7(3), pp. 72–78.

Langlais, F., Kerboull, M., Sedel, L. & Ling, R., 2003. THE 'FRENCH PARADOX'. *The Journal of Bone and Joint Surgery. British volume*, 85-B(1), pp. 17–20.

Lewis, G., 1997. Properties of acrylic bone cement: State of the art review. *Journal of Biomedical Materials Research*, 38(2), pp. 155–182.

Ramaniraka, N. A., Rakotomanana, L. R. & Leyvraz, P., 2000. The fixation of the cemented femoral component: EFFECTS OF STEM STIFFNESS, CEMENT THICKNESS AND ROUGHNESS OF THE CEMENT-BONE SURFACE. *The Journal of Bone and Joint Surgery. British volume*, 82-B(2), pp. 297–303.

Ramos, A. & Simões, J. A., 2009. The influence of cement mantle thickness and stem geometry on fatigue damage in two different cemented hip femoral prostheses. *Journal of Biomechanics*, 42(15), pp. 2602–2610.

Terrier, T., Büchler, P. & Farron, A., 2005. Bone–cement interface of the glenoid component: Stress analysis for varying cement thickness. *Clinical Biomechanics*, 20(7), pp. 710–717.

 ISBN: 978-1-032-10806-3

Influence of the gender in the rotational shot-put, in Athletics' elite athletes

P. Miranda-Oliveira
Federação Portuguesa de Atletismo, Lisboa, Portugal
ESTG, Politécnico de Leiria, Portugal

D. Correia
Juventude Vidigalense, Leiria, Portugal

M.M. Vieira
ESTG, Politécnico de Leiria, Portugal
SEG-CEMMPRE, Universidade de Coimbra, Portugal

R.B. Ruben
CDRsp-ESTG, Politécnico de Leiria, Portugal

O. Fernandes
Comprehensive Health Research Center, Universidade de Évora, Portugal

ABSTRACT: This study aimed to determine the temporal parameters in the rotational shot-put phases and the shot-put release conditions in elite athletes. To analyze your correlation with the result and to analyze the differences between the male and female athletes. Eight elite athletes were analyzed. The 3D kinematic analysis was applied, through two high-speed cameras and a calibration volume (3.67mx2.05mx3.68m) to use the 3DDLT. The Spearman and Pearson correlation tests and the Independence sample T-test were applied. The results showed that the release velocity is a determinant factor to higher performance. Furthermore, the athletes with high performance performed a final phase with smaller temporal results in the 1st single support and double support. When we compared the male and female athletes, the results showed that the female athletes with small performance need to perform at a proper rhythm in the shot-put throw's final phase.

1 INTRODUCTION

The shot-put is an explosive event, of the Track and Field, that require high power levels (Błażkiewicz et al. 2016, Terzis et al. 2007). Over the years, several techniques were performed in the shot-put, as glide, rotational, leg reverse, and shuffle (Schofield et al. 2019). The rotational technique is a highly complex movement. It consists of performing two rotations in a limited space (Coh et al. 2008), to produce high force levels in a shorter time before throwing the shot (Schofield et al. 2019). The shot-put success is defined by the shot release conditions, release velocity, release angle and release height (Błażkiewicz et al. 2016, Schofield et al. 2019, Coh et al. 2008, Lee et al. 2015). The release velocity is the parameter with a higher association with the competition result (Błażkiewicz et al. 2016, Lee et al. 2015). Furthermore, the inverse relationship between the release velocity and release angle contributes to the athlete's performance (Lee et al. 2015). Another aspect

DOI: 10.1201/9781003217152-32

analyzed by the coach is the rhythm of the throwing, in your several phases. In biomechanics analysis, the rhythm is defined by calculation the temporal parameters in the different phases. To analyze the last published biomechanics report by World Athletics about the 2017 World Championships in Athletics, we can consider the following shot-put phases: 1st single support, flight phase, 2nd single support, double support, and release (Dinsdale et al. 2017). Where, the delivery phase, also, known, as the double support or power phase is determinant in shot-put throw because it produces 90% of the throw's velocity and during 200ms, in mean (Schofield et al. 2019). When analyzed studies about the influences of the rhythm analysis of the rotational shot-putter and the differences between male and female athletes, few studies were found. This study aimed to determine the temporal parameters in the rotational shot-put phases and the shot-put release conditions in elite athletes. (1) To analyze your correlation with the result and (2) to analyze the differences between the male (M) and female (F) athletes.

2 MATERIAL AND METHODS

Twenty-five valid throws performed by the eight first positions of the European Throwing Cup (ETC) 2018, Leiria that carried out the rotational technique and that signed the Inform Consent approved by the Ethical Council of the Politécnico de Leiria were analyzed.

Two high-speed cameras (Panasonic FZ200) to record at 100Hz in the high definition were used and put in the posterior frontal plan and the right-side sagittal plan to apply the tridimensional kinematics analysis. A calibration volume with 3.67m x 2.05m x 3.68m was used. The Kinovea® motion analysis system was used to determine the absolute and relative temporal parameters in the different rotational shot-put phases (Phases: 1st single support, flight phase, 2nd single support, double support, and the total time of the throw), also to digitize the shot-put path to calculate the release conditions (release velocity, release angle and release angle). The relative temporal parameters were determined through Equation 1.

$$(Phase\ time)/(Total\ time) \times 100 \tag{1}$$

The tridimensional direct linear transformation (3DDLT) to convert the digital coordinates in metrics coordinates (Abdel-Aziz & Karara 2015) and the low pass filter to filter the metric coordinates was realized through the Scilab®. A Butterworth low pass filter and the cutting frequency was determined by spectral power analysis (Winter 2005), a cut-off frequency between 10Hz and 20Hz was applied.

The statistical analysis was performed through Jamovi®. The normality was tested through Shapiro-Wilk ($p > 0.05$). The Pearson and Spearman tests were used to analyze the association between the result and the temporal parameters and the release parameters. The association levels, $r \leq 0.3$, small; r between 0.3 and 0.5, moderated; and $r > 0.5$, high were considered (Hopkins et al. 2009). The Independence sample T-test was applied to compare the temporal parameters and release parameters between male athletes and female athletes.

3 RESULTS

The mean and standard deviation, the normality analysis for all analyzed parameters, and the comparison between male athletes and female athletes were shown in Table 1.

The correlation between the result and the temporal parameters and the result and the release parameters were shown in Table 2.

Table 1. Mean and standard deviation of the temporal parameters and release parameters.

	Female Mean	±SD		Male Mean	±SD		p
Throws (N)	17			8			
Result (m)	17.0	±1.13	¥	20.2	±1.09		p<0.001 ψ
1SS (ms)	509	±43.5		440	±22.0		p<0.001 ψ
FP (ms)	68.8	±26.7		123	±54.7	¥	p=0.035 ψ
2SS (ms)	218	±16.7		180	±41.1		p=0.042 ψ
DS (ms)	202	±20.4		184	±21.3		p=0.054
TT (ms)	998	±61.6		926	±21.3		p=0.004 ψ
Release velocity	12.5	±0.66	¥	13.4	±0.43		p=0.003 ψ
Release angle	33.8	±4.24		36.9	±1.62		p=0.054
Release height	1.94	±0.17		2.15	±0.10		p=0.004 ψ
1SS_rel (%)	51.0	±3.25	¥	47.5	±1.52		p=0.029 ψ
FP_rel (%)	6.85	±2.60		13.2	±5.73	¥	p=0.016 ψ
2SS_rel (%)	21.9	±1.53	¥	19.5	±4.71		p=0.793
DS_rel (%)	20.2	±1.62		19.9	±2.57		p=0.687

¥ non-normal - p≤0.05; ψ) significant differences between male athletes and female athletes; 1SS - 1st single support; FP - flight phase; 2SS - 2nd single support; DS - double support; total time of the throw (TT); 1SS_rel – 1st single support relative time; FP_rel - flight phase relative time; 2SS_rel – 2nd single support relative time; DS_rel - double support relative time.

Table 2. Correlation between the result and temporal and release parameters.

	Female Result (m)		Male Result (m)	
1SS (ms)	r=-0.424	p=0.090	r=0.623	p=0.099
FP (ms)	r=-0.588	p=0.013 ¥	r=0.794	p=0.019 ¥
2SS (ms)	r=-0.503	p=0.040 ¥	r=-0.797	p=0.018 ¥
DS (ms)	r=-0.493	p=0.044 ¥	r=-0.808	p=0.015 ¥
TT (ms)	r=-0.828	p<0.001 ψ	r=0.305	p=0.462
Release velocity	r=0.766	p<0.001 ψ	r=0.975	p<0.001 ψ
Release angle	r=-0.432	p=0.083	r=-0.003	p=0.994
Release height	r=-0.167	p=0.522	r=-0.181	p=0.668
1SS_rel (%)	r=0.207	p=0.426	r=0.754	p=0.031 ¥
FP_rel (%)	r=-0.521	p=0.032 ¥	r=0.826	p=0.011 ¥
2SS_rel (%)	r=-0.096	p=0.713	r=-0.778	p=0.023 ¥
DS_rel (%)	r=0.076	p=0.773	r=-0.776	p=0.023 ¥

Significance: ¥ p<0.05; ψ p<0.001; 1SS - 1st single support; FP - flight phase; 2SS - 2nd single support; DS – double support; total time of the throw (TT); 1SS_rel - 1st single support relative time; FP_rel - flight phase relative time; 2SS_rel - 2nd single support relative time; DS_rel – double support relative time.

4 DISCUSSION

The purpose of this study was to determine the temporal parameters in the rotational shot-put phases and the shot-put release conditions in elite athletes. (1) To analyze your correlation with the result and (2) to analyze the differences between the male and female athletes.

25 (17F; 8M) valid throws carried out by the first eight positions of the competition was analyzed.

When associated the result with the release conditions, in both groups, the release velocity was a determinant factor to high performances (M: r = 0.975, p < 0.001; F: r = 0.766, p <

0.001), (Błażkiewicz et al. 2016, Schofield et al. 2019, Coh et al. 2008, Lee et al. 2015). Only female athletes, we verified an inverse relationship between release speed and release angle ($r = -0.755$, $p < 0.001$) indicated a significant effect in the throw distance (Lee et al. 2015). To compare the results between male athletes and female athletes, we can suggest an optimize of the release angle in female athletes.

When analyzed the rhythm of the female athletes, we verified best performance was small time in the flight phase (absolute time: $r = -0.588$, $p = 0.013$; relative time: $r = -0.521$, $p = 0.032$), in the 2nd single support (absolute time: $r = -0.424$, $p = 0.090$) and the double support (absolute time: $r = -0.493$, $p = 0.044$). Also, this study verified athletes with a small total time of the throw (total time: $r = -0.829$, $p < 0.001$) obtained the best performances. So, the coach needs to pay attention to decreasing the time in the flight phase, 2nd single support, and double support, but is important they do not lose amplitude over the throw.

In male athletes, this study suggested that the best athletes obtained a high time in the 1st single support (relative time: $r = 0.754$, $p = 0.031$) and in the flight phase (absolute time: $r = 0.794$, $p = 0.019$; relative time: $r = 0.826$, $p = 0.011$). Indeed, the male athletes with the best result started the throw slower. However, they had a faster 2nd single support (absolute time: $r = -0.797$, $p = 0.018$; relative time: $r = -0.778$, $p = 0.023$) and a double support (absolute time: $r = -0.808$, $p = 0.015$; relative time: $r = -0.776$, $p = 0.023$). These results reinforce that a faster power phase (or double support phase) allows the transfer of high energy to the shot, increasing the release velocity and obtaining a high competition result (Schofield et al. 2019).

When compared the analyzed parameters between male and female athletes we obtained significant statistical differences in all parameters, except in the 2nd single support (relative time: $p = 0.793$), double support (absolute time: $p = 0.054$; relative time: $p = 0.687$), and release angle ($p = 0.054$). These results suggested that all athletes started the throw differently, and this aspect influencing, mainly, the release velocity, and respectively result. The final phase of the throw didn't have significant statistical differences suggesting that booth athletes' groups needed to be fast in the final phase, to increase release velocity to obtain a higher distance throw (Schofield et al. 2019).

5 CONCLUSION

The study verified, for both groups, the release velocity as the determinant factor to the athletes' performance. Also, in release conditions, we analyzed an inverse correlation between the angle release and velocity release to female athletes, a factor to review by the coach.

To analyze the differences between rotational shot-put male athlete and female athlete we verified that all athletes started with different rhythms (1st single support phase, flight phase, and 2nd single support phase) but finished with similar rhythms (double support phase). We concluded they needed to do a fast rhythm in the final phase (double support) of the throw to accelerate the shot and increase the release velocity, an essential factor to improve your performance.

ACKNOWLEDGMENT

To the European Athletics to allow the biomechanical project in the European Throwing Cup 2018, Leiria.

To the Juventude Vidigalense to the support during the implementation of the biomechanical project.

To the Federação Portuguesa de Atletismo to the given support, before, during, and after the implementation of the biomechanical project and to brand this project in your site and social networks.

To the Politécnico de Leiria to give support, namely, through the Ethical Council, to approve of the Inform Consent and to allow the collaboration of the students in the biomechanical project.

To the athletes involved in the study and that participate through the Inform Consent.

To the volunteers to supporting the kinematic collections during the competition.

To the 2018 and 2020 class of the Biomecânica do Movimento Curricular Units, of the Biomechanics degree, of the ESTG from the Politécnico de Leiria to support in the data processing.

REFERENCES

Abdel-Aziz, Y. I., & Karara, H. M. (2015). Direct Linear Transformation from Comparator Coordinates into Object Space Coordinates in Close-Range Photogrammetry*. Photogrammetric Engineering & Remote Sensing, 81(2), 103–107.

Błażkiewicz, M., Łysoń, B., Chmielewski, A., Wit, A. (2016). Transfer of mechanical energy during the shot put. J Hum Kinet, 52, 139–146.

Coh, M., Stuhec, S., Supej, M. (2008). Comparative biomechanical analysis of the rotational shot put technique. Coll Antropol, 32(1), 249–56.

Dinsdale, A., Thomas, A., Bissas, A., & Merlino, S. (2017). Biomechanical Report IAAF World Championships London 2017: Shot Put Throw.

Hopkins, W. G., Marshall, S. W., Batterham, A. M., & Hanin, J. (2009). Progressive statistics for studies in sports medicine and exercise science. Medicine and Science in Sports and Exercise, 41(1), 3–12.

Lee S., Davis R., Judge L.W., Kwon Y.H., Han K., Kim J., Kim J., Kim J. (2015). Gender-Based Correlation Profiles Among the Release Factors and Distance Thrown in Paralympic Seated Shot Put. Adapt Phys Activ Q., 32(4), 318–30.

Schofield, M., Cronin, J. B., Macadam, P., & Hébert-Losier, K. (2019) Rotational shot put: a phase analysis of current kinematic knowledge, Sports Biomechanics.

Terzis, G., Karampatsos, G., Georgiadis, G. (2007) Neuromuscular control and performance in shot-put athletes. J Sports Med Phys Fitness, 47(3), 284–90.

Winter A., D. (2005). Biomechanics and motor control of human movement (3rd ed.).

Advances and Current Trends in Biomechanics – Belinha et al. (Eds)
© 2022 Copyright the Author(s), ISBN: 978-1-032-10806-3

Using a meshless algorithm to numerically simulate cell division

M.I.A. Barbosa
Institute of Science and Innovation in Mechanical and Industrial Engineering (INEGI), University of Porto, Portugal

J. Belinha
Department of Mechanical Engineering, School of Engineering, Polytechnic of Porto, Portugal

R.M. Natal Jorge
Department of Mechanical Engineering, University of Porto, Portugal

A. Carvalho
Institute for Research and Innovation in Health (I3S), University of Porto, Portugal

ABSTRACT: In recent years, numerical methods have proved to be a capable computational tool for complementing experimental studies, namely in the study of cells and associated diseases. Thus, several methods and models have been proposed to solve different scenarios. Meshless methods emerged as an alternative to traditional numerical methods, and promising results have already been obtained. However, regarding the study of cells with the use of these methods, the number of works found in the current literature is quite limited. This way, the aim of the present work is to analyse the viability of a novel numerical approach, combining a meshless method with a tumour cell growth algorithm. In addition, the effect of the oxygen and glucose concentrations in the tumour growth was also studied. With this, it was possible to obtain satisfactory and coherent results suggesting that RPIM is an efficient alternative numerical technique for this type of studies.

1 INTRODUCTION

During the process of cell division, two new cells are formed from one cell, through a complex set of reactions and mechanisms that are mainly dependent on oxygen concentration, glucose concentration, and pH (King and Cidlowski, 1998, Casciari et al., 1992).

However, during this process, different types of mutations can occur and tumour cells tend to appear. These cells present an uncontrolled growth and are responsible for the appearance of cancer, one of the main causes of death worldwide (Luengo-Fernandez et al., 2013, Cristini and Lowengrub, 2010).

To understand the behaviour of cells under normal and pathological conditions, numerical simulations have been proposed as an alternative to conventional methods, since they avoid ethical issues and reduce the experimental time and costs (Bathe and Zhang, 2004, Marom and Linden, 1990). In this field, meshless methods emerged as an appealing substitute to conventional methods to study cells, since these present higher accuracy and, with these methods, it is easier to introduce and/or eliminate nodes from the discretization set (Belinha, 2016, Dehghan and Mohammadi, 2017).

In this work, the meshless method RPIM, is used to simulate the process of tumour cell growth. Since it is a recent method, the number of works presented in the literature

DOI: 10.1201/9781003217152-33

is reduced and, to the best knowledge of the authors, this method was never used in the analysis of tumour cell growth. Thus, in order to understand the viability of the RPIM in this field, it is important to overcome this gap in the literature (Dehghan and Mohammadi, 2017).

Furthermore, the aim of this work is to propose an initial novel algorithm that is capable of describing individual tumour cellular growth, dependent on oxygen and glucose concentration, solved by the RPIM. Then, the influence of the concentration of these two components was verified, taking into account severe conditions of hypoxia and glucose deprivation, as well as, the ability of the RPIM to proficiently simulate the complete division process.

2 MATERIAL AND METHODS

In the present work, the RPIM was used to simulate the process of tumour cell growth. Like other traditional discrete methods, the first step is to define the geometry of the problem, the solid domain, its limits, and its boundary conditions. However, contrary to mesh-dependent discretization techniques, which usually discretize the domain resorting to elements, the RPIM perform this discretization resorting to a nodal set without a known relation between the nodes. Thus, the main difference is the imposition of the nodal connectivity, which is done by overlapping influence-domains, and the construction of the shape functions, which are obtained through the RPI technique. This combines radial basis functions (RBF) with polynomial basis functions which are posteriorly used to obtain the field variables that are under study (Belinha, 2014).

Concerning the used algorithm, it was developed and ran in MATLAB®. In the proposed approach, the concentrations of oxygen and glucose are defined as the variables that affect the cell growth. Briefly, the input data is defined (i.e. the diameter of the cell, the number of maximum iterations, the cell and medium properties, the initial concentration and the diffusion coefficients of oxygen and glucose, etc), based on data presented in the literature (Casciari et al., 1992), and a growth law was established, in order to initiate and impose the growth of the cell over the iterations (Santos et al., 2017). If the concentrations of oxygen and glucose are higher than 0, then the cell continues to grow until it achieves the double of its size, while oxygen and glucose are consumed. When the size of the cell doubles, the process of division occur. The cell is forbidden to grow and forced to divide into two new cells, and the program stops.

3 RESULTS

The algorithm was tested in order to verify its viability under different values of concentration of oxygen and glucose. For this purpose, a normal case was defined, in which the concentrations of oxygen and glucose were defined as $3.5\text{x}10^{-16}$ mol/μm^2 and $2.75\text{x}10^{-14}$ mol/μm^2, respectively, and compared with two different situations: oxygen and glucose depletion. Note that, in both cases, the other variables were not modified.

After running the algorithm in MATLAB®, the colour maps of the evolution of the cell, in terms of volume, were obtained and analysed in order to verify how these concentrations affect the process of growth and division of the cell.

In the figure below (Figure 1), the colour maps of the evolution of the cell under normal concentrations of oxygen and glucose are shown. As a side note, the sequence of frames that is presented in Figure 1 does not correspond to sequential iterations.

Analysing Figure 1, it is possible to verify that the tumour cell grows and achieves the step of division, generating two new cells at the end of the process, since the concentration values of oxygen and glucose were normal, which do not condition the process of cellular growth.

On the other hand, when the concentrations of glucose and oxygen were decreased, the same was not verified, as showed in Figure 2 and 3, respectively.

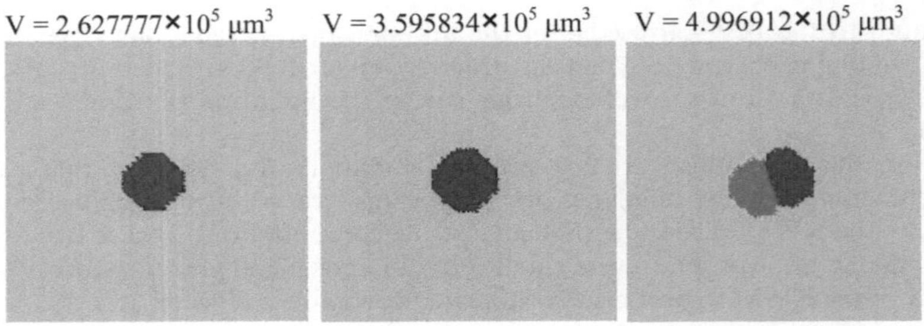

Figure 1. Colour maps of the evolution of the cell in terms of volume (V) under normal concentrations of oxygen and glucose. Green represents the medium, blue the first cell and orange the new cell generated by division.

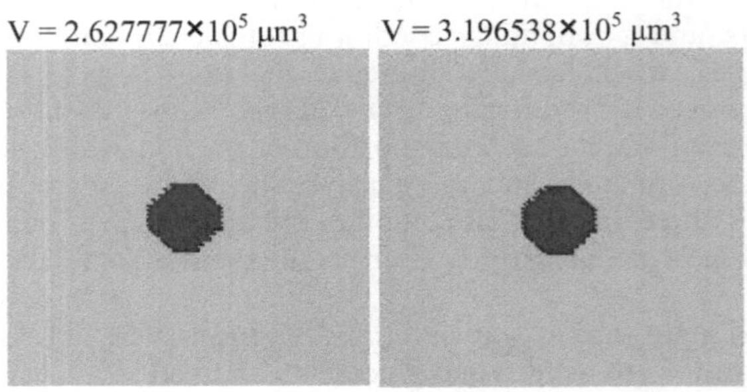

Figure 2. Colour maps of the evolution of the cell in terms of volume (V) under normal concentrations of oxygen and lower levels of glucose. Green represents the medium and blue the cell.

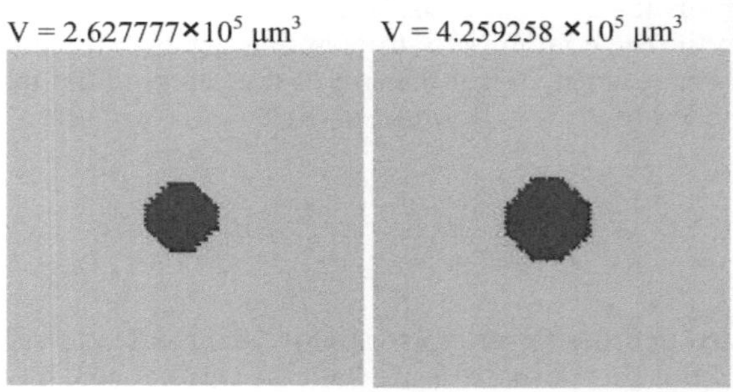

Figure 3. Colour maps of the evolution of the cell in terms of volume (V) under normal concentrations of glucose and lower levels of oxygen. Green represents the medium and blue the cell.

Initially, in both cases, the cell starts to grow, however, as it consumes the oxygen and glucose, the cell stops the growth process and, consequently, the occurrence of the process of division is prevented, since the double of the volume is not achieved.

4 DISCUSSION

Analysing the results, when normal conditions were considered, it was possible to verify that the volume of the cell tends to grow exponentially, over time, as expected and according to the data that was initially defined. Moreover, the cell also achieves the double of its volume, after the period of time that was initially defined, and the division of the cell into two new ones occurs, as defined by the algorithm, when this condition is satisfied.

When a physiological extracellular environment exists, cells are allowed to grow normally since the source of nutrients is sufficient to assure that the process of growth and division occurs adequately. However, it is expected that, when the concentrations of oxygen and glucose are under the normal concentration threshold, the process of growth and division is affected. Usually, cells tend to reduce their proliferation rate or even stop the whole process. In severe conditions, they can also initiate apoptosis and die. Concerning the study of the influence of oxygen and glucose deprivation, it was possible to see that, in both cases, the decrease of the concentration promotes a lower final volume achieved by the cell since the process stopped after some iterations. Such implied that the process of cellular division did not occur. This can be explained by the fact that, depending on the case under study, oxygen or glucose were completely consumed from the medium. These results are in accordance with the expected since the growth of cells is only conditioned by the existence of these two species.

So, according to the obtained results, the algorithm seems to generate coherent and suitable results for simulating the process of cellular growth and division. However, note that this process is quite simplified. Under these conditions, cells tend to initiate the process of angiogenesis to assure their survival, which affects the balance between cellular growth and death. This way, other processes should be considered to understand the real influence of glucose and oxygen during the process of cellular growth.

5 CONCLUSION

Several methods have been used to solve different biomedical models, the same has occurred in the cancer field. Meshless methods, like the RPIM, appeared as an alternative to traditional methods, such as the Finite Element Methods, and although they are more computationally expensive, they generate more accurate results. However, the number of works in the literature using these methods is still reduced when compared with the other ones. In the present work, the RPIM was combined with a new algorithm to simulate tumour cell growth, dependent on the concentration of oxygen and glucose. In the end, it was possible to conclude that this novel method seems to be suitable and adequate for this type of study and analyses, since it led to adequate and satisfactory results.

Moreover, since the growth of the cell is dependent on the concentrations of oxygen and glucose, the study of the influence of lower values of concentration of these two species was performed. In both cases, the growth of the cell was affected by the decrease of the concentration in the extracellular medium, which was coherent with what was defined by the algorithm and with what occurs in the natural cellular environment. In both cases, the cell did not double its volume, and the growth was stopped. This implied that the process of division did not occur.

On the other hand, it is important to highlight that the algorithm is still in a preliminary phase. Thus, several improvements should be included in future work such as, for example, the study of more cells, instead of just one, or the introduction of the process of angiogenesis, in order to achieve more realistic results. Despite that, the obtained results were promising and the algorithm seems to have potential to study this type of problem.

ACKNOWLEDGMENTS

The authors truly acknowledge the funding provided by Ministério da Ciência, Tecnologia e Ensino Superior – Fundação para a Ciência e a Tecnologia (Portugal), under grants: SFRH/BD/146272/2019. Additionally, the authors gratefully acknowledge the funding provided by LAETA, under project UIDB/50022/2020.

REFERENCES

Bathe, K. J. & Zhang, H. 2004. Finite element developments for general fluid flows with structural interactions. *International Journal for Numerical Methods in Engineering*, 60, 213–232.

Belinha, J. 2014. Meshless methods in biomechanics. *Bone tissue remodelling analysis*.

Belinha, J. 2016. Meshless Methods: The Future of Computational Biomechanical Simulation. *Journal of Biometrics & Biostatistics*, 07.

Casciari, J. J., Sotirchos, S. V. & Sutherland, R. M. 1992. Variations in tumor cell growth rates and metabolism with oxygen concentration, glucose concentration, and extracellular pH. *Journal of cellular physiology*, 151, 386–394.

Cristini, V. & Lowengrub, J. 2010. *Multiscale modeling of cancer: an integrated experimental and mathematical modeling approach*, Cambridge University Press.

Dehghan, M. & Mohammadi, V. 2017. Comparison between two meshless methods based on collocation technique for the numerical solution of four-species tumor growth model. *Communications in Nonlinear Science and Numerical Simulation*, 44, 204–219.

King, K. & Cidlowski, J. 1998. Cell cycle regulation and apoptosis. *Annual review of physiology*, 60, 601–617.

Luengo-Fernandez, R., Leal, J., Gray, A. & Sullivan, R. 2013. Economic burden of cancer across the European Union: a population-based cost analysis. *The lancet oncology*, 14, 1165–1174.

Marom, S. A. & Linden, M. J. 1990. Computer aided stress analysis of long bones utilizing computed tomography. *Journal of biomechanics*, 23, 399–404.

Santos, C., Belinha, J., Jorge, R. N. & Vinga, S. 2017. *A New Numerical Methodology to Simulate Tumour Cell Proliferation*. MSc, University of Porto.

Advances and Current Trends in Biomechanics – Belinha et al. (Eds)
 ISBN: 978-1-032-10806-3

Numerical simulation of normal and pathological erythrocytes using advanced discretization numerical techniques

M.I.A. Barbosa
Institute of Science and Innovation in Mechanical and Industrial Engineering (INEGI), University of Porto, Portugal

J. Belinha
Department of Mechanical Engineering, School of Engineering, Polytechnic of Porto, Portugal

R.M. Natal Jorge
Department of Mechanical Engineering, University of Porto, Portugal

ABSTRACT: Red blood cells, also known as erythrocytes, are the main constituent of blood and are responsible for the vital function of transporting oxygen to all tissues of the body. Like other cells, when afflicted by a specific pathology, red blood cells might have their normal mechanical properties changed. These alterations are specific to each pathology and can, for example, change the natural frequency of vibration of the cells. Thus, the objective of the present work was to find the first four modes of vibration of normal and pathological red blood cells, resorting to three numerical methods: the Finite Element Method (FEM), a mesh-based method, the Radial Point Interpolation Method (RPIM), and the Natural Neighbour Radial Point Interpolation Method (NNRPIM). Also, the viability of the methods, when applied to this type of analysis, was verified. In the end, it was possible to conclude that all methods promote similar and suitable results.

1 INTRODUCTION

Red blood cells are characterized by a simple biological structure, with cytoplasm made up of 95% haemoglobin, which allows them to take oxygen to and from all the tissues of the human body (Buys et al., 2013).

Regarding these specific cells, due to some pathologies, such as, for example, sickle cell anaemia, spherocytosis, cancer or malaria, their geometries and mechanical properties can be modified and their functions compromised. Consequently, the whole human body can be affected. This way, these pathologies represent a major problem and a challenge for the medical field nowadays, due to the huge impact in the daily life of patients and the enormous number of new cases and deaths every year (Tomaiuolo, 2014, Mohandas and Gallagher, 2008). So, to study these modifications it is essential to overcome these numbers and to find new methods and treatments that could help with this objective.

All bodies present a specific natural vibration frequency. However, this can be modified when an alteration occurs due to, for example, a pathology (Monteiro et al., 2011, Awojobi, 1971). Regarding red blood cells, this vibration frequency is expected to be altered when pathological red blood cells are compared with normal ones, and such modification can be studied, for example, resorting to biomechanical simulations, which are currently considered as a powerful tool to perform different types of analyses and studies, namely in this particular

DOI: 10.1201/9781003217152-34

field of pathologies (Bodnár and Sequeira, 2008, Ataullakhanov and Panteleev, 2005). These modifications can also be seen as useful for the appearance of new methods of diagnose and treatments based on the induction of a resonance vibration in order to destroy abnormal cells and consequently help to overcome these kind of diseases (Moreno et al., 2010).

Taking this into account, the aim of the present work is to analyse qualitatively the first four modes of natural vibration of normal and pathological red blood cells, using three numerical methods: the FEM, which is a mesh-based method and nowadays the most popular numerical method used, and the RPIM and the NNRPIM, two novel meshless methods that were never used for this type of analyses and studies. Thus, the viability of the application of these methods to this type of analyses was also studied and, for this purpose, a qualitative study of the colour maps of the four modes of vibration was performed, always taking into account a comparison with the results obtained with FEM.

2 MATERIAL AND METHODS

As previously mentioned, in this work, one mesh-based method (FEM) and two meshless methods (RPIM and NNRPIM) were used to analyse health and pathological red blood cells. These last two methods emerged as an alternative to mesh-based methods, since they are capable of promoting more accurate results. In opposition to the FEM, which discretizes the domain in elements that have a pre-established relation between them, meshless methods discretize the problem domain with a set of nodes without a pre-established relationship between them. Besides that, in meshless methods, the nodal connectivity is established using, for example, influence-domains or influence-cells, instead of elements, as used in FEM. Regarding the construction of the shape functions, the two meshless methods used in this work resort to the radial point interpolation technique, which combines a polynomial basis function with a radial basis function (Belinha et al., 2016, Belytschko et al., 1996).

Regarding the models, three different 3D geometries of red blood cells were created. One representing a normal healthy red blood cell (H), with a biconcave shape, another a red blood cell affected by sickle cell anaemia (SCA), with a shape of a sickle, and a third one representing a red blood cell affected by ovalocytosis (O), with an oval shape without biconcave feature which is characteristic of health red blood cells. The models were created in Meshmixer® and Femap® (student version), using data presented in the literature (Tomaiuolo, 2014). Moreover, simple elastic material assumptions were assumed, and the analyses were performed in FEMAS® (Finite Element and Meshless Analysis Software (cmech.webs.com)), which is a freeware academic software. The mechanical properties (Young's modulus and Poisson's coefficient) assumed in this work were also based on literature (Tomaiuolo, 2014).

In the end, the first four modes of vibration of each model were obtained and compared with the three numerical methods, in a qualitative analysis. All models were allowed to vibrate freely, which means that no degree of freedom was constrained.

3 RESULTS

In order to compare the results of the three discretization techniques, the obtained ibration modes colour maps were acquired and depicted. Thus, the results obtained for the second natural mode of vibration are presented in Table 1. Note that, as previously mentioned, this are just qualitative results, since the colour maps of the natural vibration mode fields of each solution are normalized between 0 and 1: $\boldsymbol{u}_{norm} = (\mathbf{max}(\boldsymbol{u}))^{-1}\boldsymbol{u}$.

Analysing Table 1, it is possible to visualize that the results obtained with the RPIM, the NNRPIM and the FEM are quite similar, in terms of the maximum vibration frequency and qualitative natural vibration mode field configuration. Regarding the maximum vibration frequency, for all models, all methods led to similar values. In Table 1, due to the solid uneven spatial rotation, the similarity between the obtained vibration mode fields is possibly not very

Table 1. Colour map of the second free vibration mode ($\hat{\omega}_2$) obtained with the different methods.

		$\hat{\omega}_2(rad/s)$	
Model	FEM	RPIM	NNRPIM
H	$\hat{\omega}_{2max}$ = 1.995473 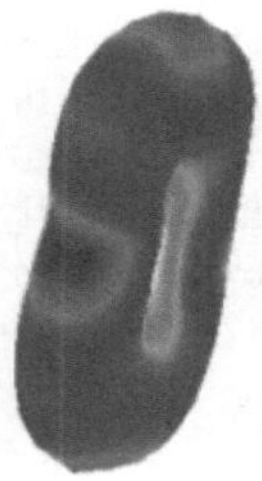	$\hat{\omega}_{2max}$= 1.747961 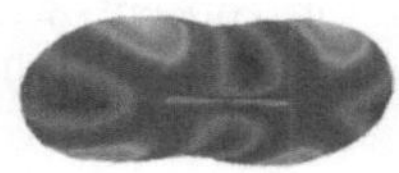	$\hat{\omega}_{2max}$ = 1.706409 
SCA	$\hat{\omega}_{2max}$ = 1.455187 	$\hat{\omega}_{2max}$ = 1.217512 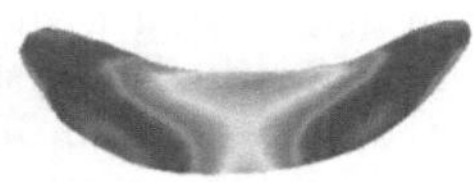	$\hat{\omega}_{2max}$ =1.194898 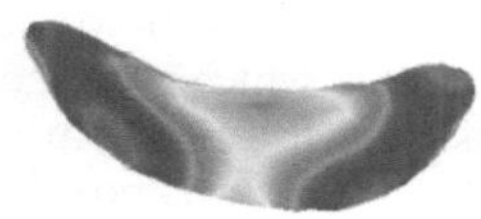
O	$\hat{\omega}_{2max}$ = 3.820723 	$\hat{\omega}_{2max}$ = 3.723813 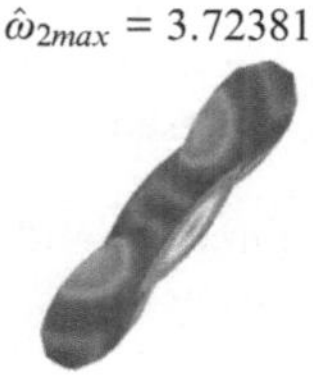	$\hat{\omega}_{2max}$ = 3.768774 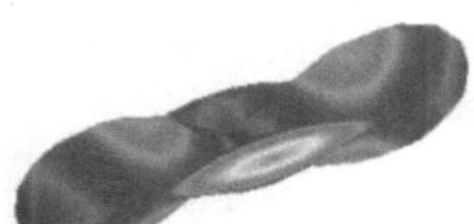

u_{norm}: MAX  MIN

perceptible. Since all discretization methods provide similar solutions, this representation option was adopted to offer additional visual information.

As expected, when this second vibration mode is compared with the first, the value of the maximum vibration frequency increases ($\hat{\omega}_2 > \hat{\omega}_1$) and the corresponding vibration modes are also distinct. Notice that different vibration frequencies promote different vibration modes, leading to different reactions to vibration and, consequently, different final deformations and stresses fields.

4 DISCUSSION

Analysing the results obtained, it is possible to verify that the meshless methods present similar results to the ones obtained with the FEM, in terms of vibration frequencies and vibration modes. For each analysed model (pathology) it was verified that the model geometry influences significantly the vibration frequency and mode.

Such can be explained by the fact that all models present different mechanical properties and initial geometries, which consequently implies different a different dynamic response. Note that the vibration frequency and mode is characteristic of each body.

From the obtained results, it can be concluded that a H-RBC possesses a second vibration frequency between 1.70rad/s and 2.00rad/s. On the other hand, SCA-RBC possesses a second vibration frequency between 1.19rad/s and 1.46rad/s and O-RBC shows a second vibration frequency between 3.72rad/s and 3.82rad/s. These results are a strong indicator that each kind of RBC possess a distinct and identifier second vibration frequency. Similar results were observed for the first, third and fourth vibration frequencies (dissimilar vibration frequencies for distinct RBF shapes). Thus, inducing by ultrasound a vibration frequency between 1.19rad/s and 1.46rad/s will make SCA-RBC vibrate by resonance and a similar resonance effect will be observed for O-RBC if the induced vibration frequency is between 3.72rad/s and 3.82rad/s.

Regarding the method performance it was observed that FEM, the RPIM and the NNRPIM promote identical results in all cases. However, it was observed that NNRPIM (despite its accuracy) is a heavy computational procedure. This extra computational cost is due to the use of the natural neighbour concept, which allows to automatically impose the nodal connectivity and construct the integration mesh but, at the same time, is a very computationally demanding.

5 CONCLUSION

The aim of the present work was to analyse qualitatively the first four modes of vibration of normal and pathological red blood cells, using a mesh-based method, and two meshless methods, which were never used in this kind of analysis, the RPIM and the NNRPIM. For this, two different pathologies were considered and 3D models were created and analysed. After that, it was possible to conclude that the meshless methods are capable to deliver results similar with the FEM, thus, representing a valid alternative to FEM. Meshless methods allow to discretize the problem domain directly from medical images, using the pixel spatial information. Such discretization flexibility represents an asset in the analysis of these microscale structures.

It was also verified that each RBC geometry (and corresponding material property) leads to a distinct vibration frequency and vibration mode. This information can be used to develop new diagnosis and/or treatments based on ultrasound technology.

To conclude, this work helps to qualitatively comprehend the behaviour of red blood cells in healthy and pathological conditions and to verify the viability of the meshless methods when compared to traditional mesh-based methods. Moreover, this type of analyses might be valuable in the discovery of, for example, new treatments, and in the dissemination of these methods, that are much less used than the FEM, even though they are known to have superior accuracy. Thus, future studies should be carried out in order to enlarge the range of knowledge and the literature in this area.

ACKNOWLEDGMENTS

The authors truly acknowledge the funding provided by Ministério da Ciência, Tecnologia e Ensino Superior – Fundação para a Ciência e a Tecnologia (Portugal), under grants: SFRH/BD/146272/2019. Additionally, the authors gratefully acknowledge the funding provided by LAETA, under project UIDB/50022/2020.

REFERENCES

Ataullakhanov, F. I. & Panteleev, M. A. 2005. Mathematical modeling and computer simulation in blood coagulation. *Pathophysiology of haemostasis and thrombosis*, 34, 60–70.

Awojobi, A. 1971. Approximate solution of high-frequency-factor vibrations of rigid bodies on elastic media.

Belinha, J., Araújo, A., Ferreira, A., Dinis, L. & Jorge, R. N. 2016. The analysis of laminated plates using distinct advanced discretization meshless techniques. *Composite Structures*, 143, 165–179.

Belytschko, T., Krongauz, Y., Organ, D., Fleming, M. & Krysl, P. 1996. Meshless methods: an overview and recent developments. *Computer methods in applied mechanics and engineering*, 139, 3–47.

Bodnar, T. & Sequeira, A. 2008. Numerical simulation of the coagulation dynamics of blood. *Computational and Mathematical Methods in Medicine*, 9, 83–104.

Buys, A. V., Van Rooy, M.-J., Ssoma, P., Van Papendorp, D., Lipinski, B. & Pretorius, E. 2013. Changes in red blood cell membrane structure in type 2 diabetes: a scanning electron and atomic force microscopy study. *Cardiovascular diabetology*, 12, 1–7.

Mohandas, N. & Gallagher, P. G. 2008. Red cell membrane: past, present, and future. *Blood*, 112, 3939–3948.

Monteiro, M. O., Pinto, N., Marin, P. J., Santos-Filho, S. D. & Bernardo-Filho, M. 2011. Effect on osmotic fragility of red blood cells of whole blood submitted to vibrations in an oscillating platform. *African Journal of Biotechnology*, 10, 14197–14202.

Moreno, M., Raniero, L., Arisawa, E. Â. L., do Espírito Santo, A. M., dos Santos, E. A. P., Bitar, R. A. & Martin, A. A. 2010. Raman spectroscopy study of breast disease. *Theoretical Chemistry Accounts*, 125, 329–334.

Tomaiuolo, G. 2014. Biomechanical properties of red blood cells in health and disease towards microfluidics. *Biomicrofluidics*, 8, 051501.

Advances and Current Trends in Biomechanics – Belinha et al. (Eds)

Is the Equimetrix stability index capable of differentiating feet positions?

Pedro Fonseca, Ricardo Sebastião, Manoela Sousa, Márcio Goethel, Filipa Sousa, Leandro Machado & João Paulo Vilas-Boas
LABIOMEP, Porto Biomechanics Laboratory, University of Porto, Porto, Portugal
FADEUP, Faculty of Sport of the University of Porto (CIFI2D), Porto, Portugal

Pierre Barralon & Igone Idigoras
Tecnalia, Basque Research and Technology Alliance (BRTA), Spain

ABSTRACT: Balance control is a fundamental ability to regulate the quasi-static human erect standing, and its assessment is fundamental in providing information regarding the risk of falling The Equimetrix was developed as an affordable and easy to use device that, using a webcam for measuring the migration of the center of mass, and a sensitive mat to measure the base of support, can integrate balance information in the form of a real-time stability index (SI). This work aims at verifying if the SI can differentiate feet positions during static stance and if this discriminatory ability is comparable to traditional balance parameters. Results demonstrate that the SI has a higher predictive capability to detect stable and metastable stances than other traditional parameters, such as the center of mass mean path velocity.

1 INTRODUCTION

Postural control consists in the ability to regulate the body in order to keep one's balance. Several methods can be used to evaluate this regulation, but a simple approach would be to consider the ability to maintain the vertical projection of the body's center of mass (CoM) within the base of support (BoS) defined by the lower limbs (Teranishi et al., 2013).

The Equimetrix system was developed to assess static and dynamic balance by allowing the monitoring and estimation of the CoM position and BoS area. It consists of an instrumented mat capable of measuring the BoS area with a resolution of 2.3 cm^2, and a webcam, positioned on the subject's posterior trunk to measure displacement from a reference frame established by an optical pattern fixed to the mat's border, as depicted in Figure 1.

The information from both the mat and the camera is then consolidated in the form of a balance indicator, called the Stability Index (SI), which is a metric of the subject's ability to maintain postural control in real-time. While the SI is not a direct measure of balance, it can provide information regarding how stable a participant is during a given task. This feedback can be useful during rehabilitation or training sessions, where immediate feedback allows the implementation of corrections.

The aim of this study is to evaluate if the SI calculated by Equimetrix is effectively capable of differentiating feet stance positions, and how the SI relates to traditional balance parameters.

DOI: 10.1201/9781003217152-35

Figure 1. Representation of the Equimetrix sensing mat with the reference pattern at its border.

2 MATERIALS AND METHODS

An experimental sample was recruited comprising 8 male (30.8 ± 8.5 years old; 178.4 ± 6.6 cm; 81.8 ± 10.8 kg) and 12 female (33.8 ± 9.4 years old; 161.9 ± 5.5 cm; 63.2 ± 6.8 kg) volunteers. Muscular-skeleton lesions in the past six months, neurological or motor disorders and visual or hearing impairments were established as exclusion criteria. All participants presented right lower limb dominance. This study was approved by the Faculty of Sports of the University of Porto Ethics Committee under protocol CEFAD 26/2020.

In order to calculate the participants' center of mass and its migration, 58 retroreflective markers in a full-body configuration were placed on relevant anatomical points, and their position recorded with a 12-camera motion capture system (Qualisys AB, Sweden) operating at 200 Hz. Simultaneously, the Equimetrix camera was attached to the participant's back, and its field of view adjusted according to the manufacturer instructions so that it would capture the reference pattern. The participant was instructed to stand barefoot on the Equimetrix sensory mat in an erect static position for 60 seconds, with the upper limbs comfortably extended and parallel to the trunk, while gazing a point at eye level. A single balance trial was recorded simultaneously with the kinemetric capture and Equimetrix system at four feet positions: bipodal (BI), feet together (FT), full tandem (fullT) and unipodal (UNI). During BI, feet were kept parallel with a 15 mm distance between malleoli; FT required the participant to keep both feet in contact at the heels and at the first metatarsal medial aspect; during fullT, participants had to place their dominant foot in front of the non-dominant foot, with the heel touching the toes; and in UNI, only the dominant foot was kept in contact with the ground, without knee flexion, while the non-dominant was raised.

Synchronization between the motion capture system and the Equimetrix was not possible by electronic means. Therefore, a mechanical synchronization was devised in which a step was performed by a researcher on the Equimetrix sensory mat, under which a force platform, synchronized with the motion capture system, was located. This step was performed at the beginning and ending of a trial, and it would appear simultaneously as a sudden increase in the BoS and the force, allowing the establishment of a common temporal frame.

Body markers' trajectories were identified and interpolated in Qualisys Track Manager (Qualisys AB, Sweden), and this information was then used to calculate a kinematic center of mass (K_CoM) from a six-degrees of freedom model in Visual 3D (C-Motion, USA). A custom routine was developed in Matlab R2019a (Mathworks, USA) in order to further process the K_CoM. This procedure included the removal of offset in the stabilogram in order to center its variation around zero, the synchronization with the SI signal from Equimetrix, and the calculation of parameters of interest. From the stabilogram, mean path velocity (MPV), sway range and the root-mean-square (RMS) sway were calculated for the anterior-posterior (AP) and medial-lateral (ML) direction, with the overall mean distance (MD) and length being also calculated. From the statokinesiogram, the 95% predictive ellipse area was calculated. The SI signals were retrieved directly from the Equimetrix software, and averaged for the trial's duration, for each feet position.

Table 1. Comparison of SI amd K_CoM variables for each feet position.

	MPV			Range Ellipse					Sway RMS	
	SI	AP	ML	AP	ML	Area	MD	Length	AP	ML
BI/FT	0.50	*	0.16	*	0.74	*	*	*	0.94	*
BI/fullT	*	*	*	*	0.15	*	*	*	0.03	*
BI/UNI	*	*	*	*	*	*	*	*	*	*
FT/fullT	*	*	0.11	0.18	0.31	0.10	0.10	0.07	0.01	0.25
FT/UNI	*	*	*	0.01	*	*	*	*	*	0.05
fullT/UNI	0.98	0.06	*	0.40	*	0.14	0.05	*	0.06	0.18

* : $p < 0.008$

The paired samples Student's *t*-test was used to compare the SI and K_CoM parameters between all feet positions. The significance level was adjusted according to the Bonferroni correction for a series of six comparisons. Two categorical groups of stable and metastable stances were established (Kibele et al., 2015) and a logistic regression (backward, Wald) was performed. Further logistic regressions (enter method) were performed, individually, with parameters that proved statistically significant predictors. All statistical procedures were performed in SPSS 26 (IBM, USA) with a significance level of $\alpha = 0.05$.

3 RESULTS

The multiple paired samples Student's *t*-test had a significance level adjusted according to the Bonferroni correction for six comparison (adjusted $\alpha = 0.008$). The results of each stance comparison can be observed in Table 1.

None of the analyzed parameters were capable of differentiating all feet positions. Considering the SI variable, there were statistically significant differences between BI and fullT, BI and UNI, FT and fullT, and FT and UNI. Therefore, two categorical groups were created based on these differences. The BI and FT positions were merged into the Stable category, while fullT and UNI were merged into the Metastable category.

The backward logistic regression required nine steps to achieve maximum likelihood, with only the predictors SI and MPV in the AP direction (MPV-AP), showing statistical significance to the model ($p < 0.01$), which has an overall prediction capability of 95.0% A logistic regression using only SI as a predictor, provides a model with a 91.3% ability of predicting the stance condition, while a model only with MPV-AP as an overall prediction of 80.0%. These results are presented in Table 2.

4 DISCUSSION

Balance testing is performed with several different feet stance in order to evaluate performance under conditions usually found in daily living (e.g. bipedal), but also in more challenging positions (e.g. full tandem). Although results demonstrate that the SI is not able to differentiate between each of the feet stances assessed, neither can the other analyzed parameters. If feet stances with higher area and similar length or width distribution are grouped, the SI shows a good ability to perform this differentiation. This is particularly interesting, as only MPV-AP, from the selected parameters, can also perform this differentiation.

The fact that MPV-AP was also able to differentiate stances was expected, since unstable positions such as UNI have an increased migration velocity (Lopes & David, 2018) in order to control the CoM position. However, in a position such as fullT, where the BoS was increased in the AP direction, this was not expected. Nevertheless, Kirby et al. (1987) reported similar observation when studying the feet position influence on the CoP displacement.

These results revealed that, although the model with two predictors (SI and MPV-AP) had the best R^2 values, the model using only SI as a predictor had not only higher R^2 values, but also better overall prediction percentages when compared to the model with just MPV-AP

Table 2. Statistical results of the three logistic regressions with the SI and K_CoM parameters as predictors of stance.

Logistic Regression	Predictor	β	Wald's X^2(df) (*p*)	R^2 Cox/R^2 Snell	Overall Class
Full	Constant	15.07	6.52 (1) (0.01)	0.67/0.89	95.0%
	SI	-9.26	8.73 (1) (0.03)		
	MPV-AP	2.25	7.94 (1) (<0.01)		
SI	Constant	15.71	19.07 (1) (<0.01)	0.57/0.76	91.3%
	SI	-6.54	19.74 (1) (<0.01)		
MPV-AP	Constant	-6.12	17.77 (1) (<0.01)	0.49/0.65	80.0%
	MPV-AP	1.89	20.46 (1) (<0.01)		

(91.3% vs. 80.0%). This is certainly related to the fact that MPV-AP can differentiate BI and FT, while the SI cannot, thus reducing the predictive model strength. This indicates that the SI provided by Equimetrix can differentiate between a stable and unstable feet stance, with a superior prediction capacity than only using the MPV-AP.

5 CONCLUSIONS

The Equimetrix device can differentiate between stable and metastable feet positions during quasi-static balance. Of all the traditional balance parameters considered in this study, only the MPV-AP was able to detect the same feet stances as SI.

Being able to detect stable and metastable positions in real-time is an interesting feature for the Equimetrix device, as it will enable immediate corrective interventions in posture. However, this study results indicate that the use of MPV-AP along with SI will provide even more discriminatory ability to the system.

ACKNOWLEDGMENTS

The authors would like to thank the volunteers for their generous and enthusiastic participation. This study was conducted as part of the individual project BABEN under the Centro Internacional sobre el Envejecimiento (CENIE), Fundación General de la Universidad de Salamanca project 0348_CIE_6_E and the Programa de Cooperación Interreg V-A, España–Portugal (POCTEP) 2014–2020 from the Fondo Europeo de Desarrollo Regional (FEDER).

REFERENCES

Kibele, A., Granacher, U., Muehlbauer, T. & Behm, D. G. 2015. Stable, Unstable and Metastable States of Equilibrium: Definitions and Applications to Human Movement. *Journal of sports science & medicine*, 14, 885–887.

Kirby, R. L., Price, N. A. & MacLeod, D. A. 1987. The influence of foot position on standing balance. *Journal of Biomechanics*, 20, 423–7.

Lopes, G. & David, A. C. 2018. Bipedal and unipedal stance in Brazilian football 7-a-side athletes with cerebral palsy. *Fisioterapia e Pesquisa*, 25, 303–308.

Teranishi, T., Sakurai, H., Ohtsuka, K., Yamada, M., Tsuzuki, A., Miyasaka, H., Tanino, G., Abbas, O. & Kondo, I. 2013. Relationship between Feet Position and Anterior-posterior Center of Pressure (COP) location –What are the Determining Factors in the Adjustment of Center of Gravity?&ndash. *Journal of Physical Therapy Science*, 25, 41–43.

Advances and Current Trends in Biomechanics – Belinha et al. (Eds)
© 2022 Copyright the Author(s), ISBN: 978-1-032-10806-3

Postural control in military personnel: Effect of load and footwear

Pedro Fonseca, Ricardo Sebastião, Manoela Sousa, Leandro Machado, Filipa Sousa & João Paulo Vilas-Boas
LABIOMEP, Porto Biomechanics Laboratory, University of Porto, Porto, Portugal
FADEUP, Faculty of Sport of the University of Porto (CIFI2D), Porto, Portugal

Inês M. Sousa
LABIOMEP, Porto Biomechanics Laboratory, University of Porto, Porto, Portugal
Universidade Católica Portuguesa, Porto, Portugal

Mário Vaz
LABIOMEP, Porto Biomechanics Laboratory, University of Porto, Porto, Portugal
INEGI, Institute of Science and Innovation in Mechanical and Industrial Engineering, Porto, Portugal

ABSTRACT: Balance is a fundamental skill to allow the erect standing, either during quasi-static situations or during gait. This activity is regulated by the integration of the visual, vestibular and somatosensory systems, which can be affected by intrinsic and extrinsic factors. Understanding the alterations induced by footwear or load as extrinsic factors with potential to change balance ability, offer an important aspect when designing or choosing military footwear. Therefore, this study aims to analyze how the use of military footwear and load carriage can affect the quasi-static balance of military personnel. An experimental sample of soldiers, wearing neutral shoes and two models of military footwear, was evaluated in terms of loaded and unloaded quasi-static balance on a dedicated balance assessment system. Results demonstrate that the analyzed footwear do not significantly affect the selected balance parameters, but load does induce an effect in balance, which is more pronounced in the medial-lateral direction.

1 INTRODUCTION

Human balance is a complex mechanism involved in the maintenance of static erect positions, but also necessary to enable locomotion and the ability to perform complex movements. Therefore, the human body relies on the adequate function and integration of the visual, vestibular and somatosensory systems to provide the necessary sensory inputs to regulate balance. Any interference with these systems, either from intrinsic of extrinsic factors, will result in a loss of balance ability (Chander et al., 2019b) and an increase in the risk of falling.

Footwear is a commodity intended to provide a degree of protection and comfort to the feet and, in some cases, increased functionality. Different models will provide different advantages compared to barefoot walking, such as temperature maintenance, shock absorption, or plantar surface protection by alteration of the force distribution along its surface (Chander et al., 2019a, Rosenblad-Wallin, 1988). Military footwear is a special case where a boot needs not only to preserver the soldier's feet health and comfort (Rosenblad-Wallin, 1988), but also to provide additional functionality in order to face the demands of the terrain found in an operations theatre. Several materials, designs and features can be used, most of the times with the objective of offering operational reliability (Chander et al., 2019a, Chander et al., 2019b,

DOI: 10.1201/9781003217152-36

Rosenblad-Wallin, 1988), but not necessarily improving biomechanical performance (Chander et al., 2019a).

The sensory feedback at the feet is important in the maintenance of balance (Maurer et al., 2001). However, since in a shod condition the feet will be in contact with the footwear's artificial interface and not with the ground, the insole properties will act as extrinsic factors that can interfere with balance regulation. Therefore, characteristics such as the shaft height, mass, sole geometry, hardness or thickness, and footwear material are relevant for the understanding of how a footwear model can induce chances in balance (Chander et al., 2019b).

Additionally, soldiers may be required to carry loaded backpacks. Although the load may change according to the mission, march type and duration, it can reach 30 kg (US Army, 2017), usually placed on their trunk's back. This overweight situation will increase the stress on the plantar regions of the feet, on the footwear insoles, and increase the trunk's inertia, thus potentiating its oscillation. The leaning of the trunk under the increased load may require the soldier to perform postural adaptations, since the higher mass in their back will shift the center of mass.

Considering that footwear and load carriage are fundamental in the military service, and that both the footwear and additional weight may be extrinsic factors with the potential to alter balance, this study aims to: (i) investigate how the boots used by the Portuguese army change static balance, and up to what ex-tent the use of a loaded backpack can influence it and (ii) to compare the enhanced sole properties to understand if this can be a favorable extrinsic factor.

2 MATERIALS AND METHODS

An experimental sample of 12 soldiers from the Transmissions Regiment of Porto of the Portuguese Army participated in this study, and included 10 male (28.6 ± 5.0 years, 1.78 ± 0.06 m, 81.30 ± 9.08 kg) and 2 female volunteers (27.0 ± 2.0 years, 1.61 ± 0.02 m, 62.69 ± 1.01 kg). Exclusion criteria were established as the existence of musculoskeletal lesions in the past six months, any diagnosed neurological or motor disorder, as well as visual or hearing impairments.

Three models of footwear were used in this study: a standard neutral shoe with flat sole (NF), a standard boot of the Portuguese army (MF) and a military tactical boot with enhanced rubber sole (EMF). While each participant brought their own MF, an appropriate NF and EMF was provided by the authors to each participant for the purpose of the evaluation. Additionally, a military backpack with a weight of 27.0 ± 0.5 kg was used to assess the effect of load in all shod conditions.

Quasi-static balance while wearing each footwear model, in the unloaded and loaded conditions (without and with the backpack, respectively) was assessed on a Biodex Balance System (Biodex, Inc., USA) with the participants standing with the feet parallel at shoulder's width, upper limbs extended and parallel to the trunk, and gazing a point at eye level at 1.5 m of distance.

For each footwear model and load condition, a 60 seconds trial was performed in a randomized order. Trials were repeated in case of movement of the arms or readjustment of the feet or backpack. Participants had the opportunity to perform at least one non-valid trial in order to familiarize themselves with the testing procedures.

The resulting sway angle data collected with the Biodex Balance System were exported and processed with a Matlab R2019a routine (The MathWorks Inc., USA). This procedure included the removal of the stabilograms mean, thus centering the statokinesigram around zero. This routine was also used to calculate the sway angle Path Velocity and Range in the anterior-posterior (AP) and medial-lateral (ML) directions.

Data was assessed for univariate and multivariate outliers, and tested for normality with the Shapiro-Wilk test. A two-way repeated measures MANOVA was conducted to compare the main effects of footwear (NF, MF and EMF), load (without backpack and with backpack) and their interaction on the combined variables of interest. Descriptive statistics are represented as mean and standard deviation. All statistical procedures were performed in SPSS 26 (IBM, USA) with a significance level of $\alpha = 0.05$.

This study was approved by the HSJ Ethics Committee with the number 114/19, and conducted in accordance with the Helsinki Declaration. Before any experimental procedure, each participant read and signed an informed consent.

3 RESULTS

The analysis of the two-way repeated measures MANOVA found no main effect of the footwear in the analyzed balance parameters ($F(8, 4) = 0.417$, $p = 0.86$, Power = 0.09). Only the load factor presented a statistically significant effect ($F(4, 8) = 4.36$, $p < 0.05$, Power = 0.71). The interaction effect (Load x Footwear) was not significant as well ($F(8, 4) = 3.489$, $p = 0.12$, Power = 0.43).

Therefore, all footwear models were pooled in the same group (footwear), and the load effect analyzed. An increase in all the parameters can be observed during the loaded test condition, either in the AP or ML direction. However, the values registered in the ML direction seem to have a higher relative increase than those in the AP direction, as denoted by the relative variation in Table 1. Variations in both directions are associated to a high effect size and observed power.

Table 1. Results of the balance parameters during the unloaded and loaded condition with the neutral and military footwear pooled in the same group.

Variable	Without Backpack Mean (SD)	Without Backpack Mean (SD)	Variation (η_p^2)	$F(1, 11)$ (p)	Power
Path Velocity AP (°/s)	1.00 (0.26)	1.19 (0.36)	19.0% (0.58)	14.87 (<0.01)	0.94
Path Velocity ML (°/s)	0.63 (0.19)	0.82 (0.27)	30.2% (0.60)	16.59 (<0.01)	0.96
Range AP (°)	3.27 (1.06)	4.13 (1.64)	26.3% (0.61)	17.48 (<0.01)	0.97
Range ML (°)	1.48 (0.87)	2.32 (1.14)	56.8% (0.61)	17.30 (<0.01)	0.97

AP: anterior-posterior; ML: medial-lateral; SD: standard-deviation.

4 DISCUSSION

This study compared the effect of using two models of MF when compared with a NF with minimal insole and features, with and without the addition on load by means of a military backpack. Although an empirical understanding of the differences between footwear models would have expected to find an effect of the worn model in balance, that did not happen.

The NF reduced insole thickness and zero-drop height between the heel and forefoot is in great contrast with the characteristics of the MF and EMF. Such height differential has been reported to produce a forward shift of the foot's center of pressure, thus requiring postural and kinematical adaptations that can be measured during balance assessment (Menant et al., 2008). Unfortunately, sole thickness, heel hardness and heel to forefoot height drop were not measured in this study, and therefore cannot be used to better understand the absence of differences between footwear models. The study of Chander et al. (2019b) gives the understanding that differences between military footwear models are difficult to identify, and this may explain why differences between MF and EMF were not found. Similar results were obtained by Chander et al. (2019b) and DeBusk et al. (2018), when analyzing a regular and minimalist military footwear model, where no footwear effect was found during quasi-static erect balance evaluations with eyes open. Still, this does not provide an adequate explanation to why this study cannot find a footwear effect when comparing models with so different characteristics, especially between the NF and EMF.

Nevertheless, the results observed in this study demonstrate that an increased load provided by a backpack is an active factor causing significant changes to the sample's balance. When comparing load conditions, regardless of the footwear model, it is possible to observe an increase in all variables with load, which is also in accordance to the results of Chander et al. (2019b) and DeBusk et al. (2018). Although the increased mass on the trunk may have resulted in a set of postural adaptations, such as a trunk flexion, balance parameters in the AP direction did not increased as much as those in the ML direction.

The variation on AP direction has been reported to be dependent on visual input, with an increase during eyes closed evaluations (Howcroft et al., 2017). Since this study was performed with eyes open, and a reference point was available, an adequate deep perception may have provided the necessary feedback for AP control, although Reynard et al. (2019) reported this may not be an important factor when standing with feet apart.

The ankle's higher freedom of movement in the AP direction may also serve a role in postural control in that direction. However, the lower variation found in this study is more in line with the hypothesis of the participant enacting a better AP control in order to avoid falling, due to the increased trunk inertia. A hip strategy is usually associated with the ML control, and our results seem to show that increase load exacerbated the sway in this direction. One hypothesis for this increase may be related to the establishment of a ML strategy of balance in order to reduce the AP migrations, relying on the reduced amplitude of motion available at the ankle in this direction. While this strategy may be helped by the high boot shaft of the MF and EMF, that does not happen with the NF.

5 CONCLUSIONS

This study has found no difference between a neutral footwear and the Portuguese military footwear or an enhanced military footwear, putting in evidence that load is the only factor influencing balance. The increase in balance parameters in the ML direction are usually related to increased risk of falling in elderly individuals, but insufficient information exists in order to know if that is the case with the military.

These results allow the establishment of new hypothesis, namely the enacting of a better AP control and a ML sway strategy when carrying loads. However, further studies are needed to verify the validity of these observations and hypothesis, especially with the inclusion of additional balance parameters.

This study is yet another evidence that current military footwear does not provide superior balance performance during quasi-static situations. This information may be used as support for the continued development of new footwear solutions.

ACKNOWLEDGMENTS

The authors would like to thank all the soldiers that participated in this study, their ranking officers for allowing their participation, and the Portuguese Army for lending the necessary materials for the study. This study was conducted under the project FAMEST - Footwear, Advanced Materials, Equipment's and Software Technologies - POCI-01-0247-FEDER-024529, co-financed by the Programa Operacional Competitividade e Internacionalização (POCI), of Portugal2020 and the Fundo Europeu de Desenvolvimento Regional (FEDER).

REFERENCES

Chander, H., Knight, A. C. & Carruth, D. 2019a. Does Minimalist Footwear Design Aid in Postural Stability and Fall Prevention in Ergonomics? *Ergonomics in Design*, 27, 22–25.

Chander, H., Knight, A. C., Garner, J. C., Wade, C., Carruth, D., Wilson, S. J., Gdovin, J. R. & Williams, C. C. 2019b. Impact of military type footwear and load carrying workload on postural stability. *Ergonomics*, 62, 103–114.

DeBusk, H., Hill, C. M., Chander, H., Knight, A. C. & Babski-Reeves, K. 2018. Influence of military workload and footwear on static and dynamic balance performance. *International Journal of Industrial Ergonomics*, 64, 51–58.

Howcroft, J., Lemaire, E. D., Kofman, J. & McIlroy, W. E. 2017. Elderly fall risk prediction using static posturography. *PLOS ONE*, 12, e0172398.

Maurer, C., Mergner, T., Bolha, B. & Hlavacka, F. 2001. Human balance control during cutaneous stimulation of the plantar soles. *Neuroscience Letters*, 302, 45–48.

Menant, J. C., Steele, J. R., Menz, H. B., Munro, B. J. & Lord, S. R. 2008. Optimizing footwear for older people at risk of falls. *J Rehabil Res Dev*, 45, 1167–81.

Reynard, F., Christe, D. & Terrier, P. 2019. Postural control in healthy adults: Determinants of trunk sway assessed with a chest-worn accelerometer in 12 quiet standing tasks. *PloS one*, 14, e0211051–e0211051.

Rosenblad-Wallin, E. F. S. 1988. The design and evaluation of military footwear based upon the concept of healthy feet and user requirement studies. *Ergonomics*, 31, 1245–1263.

US Army 2017. Foot Marches: Army Techniques Publication 3-21.18. *In:* ARMY, H. D. O. T. (ed.). Washington D.C.: U.S. Government Printing Office.

Advances and Current Trends in Biomechanics – Belinha et al. (Eds)

The effect of the smartphone use on gait velocity

M. Saraiva
Robocorp Laboratoy, Polytechnic Institute of Coimbra, Portugal
Faculty of Sports, University of Porto, Portugal

M.A. Castro
CEMMPRE, University of Coimbra, Portugal
Robocorp Laboratoy, Polytechnic Institute of Coimbra, Portugal
ESTESC-Coimbra Health School, Portugal

A. Cavaleiro
University of Coimbra, Portugal

J.P. Vilas-Boas
LABIOMEP-UP, Faculty of Sports, CIFI2D, University of Porto, Portugal

ABSTRACT: The use of smartphone has increased amongst all ages and it use during walking causes serious injuries. Understanding the motor control processes of gait while using a smartphone seems to be of high practical relevance. The aim of this study was to analyze the gait velocity while using a smartphone in young adults. Fifteen healthy young adults (age= 21.27±3.03 years, mean±SD) were recruited. The gait velocity was assessed in two conditions: walking at a normal pace and velocity (single motor task) and walking while texting on smartphone (motor dual-task), using the Qualisys® Motion Capture System. Compared to walking without using a smartphone, the subjects walked at slower velocity when using a smartphone ($p<0.05$). In conclusion, the use of smartphone has a negative impact on gait velocity, suggesting the risk of falls and injuries in young adults, also, can be an indicator of more evident gait changes in old age.

1 INTRODUCTION

Many daily activities require the performance of multiple tasks and involve the integration of cognitive and motor skills (Plummer et al., 2013). Thus, efficient postural control is fundamental to the success of most daily tasks.

Changes in the gait patterns during concurrently cognitive or motor tasks performed have been reported and related to an increased risk of falls (Hausdorff et al., 2008).

Gait velocity is associated with balance and functional ability and it has been considered a predictor of health status (Fritz et al., 2009), functional decline, and a clinical indicator of well-being, frailty, and mortality (Studenski et al., 2011). For this reason, the gait velocity is an important study parameter.

The smartphone has become a device most used in our daily lives and amongst all ages. Using smartphones during walking increases pedestrians' risk of accidents, injuries, or death. (Nasar & Troyer, 2013).

Waking while using a smartphone may affect motor control and response capacity, due to the dual-task effect compromising gait efficiency and safety. So, this study aimed to compare gait velocity in young adults when walking while using a smartphone and walking freely.

DOI: 10.1201/9781003217152-37

2 MATERIALS AND METHODS

Subjects were 15 healthy young adults (age= 21.27±3.03 years, height=169±10 cm; body mass index= 23.06±3.39, mean ± SD). The sample characteristics are presented in Table 1.

Table 1. Sample characteristics (mean and standard deviations (SD)).

Parameters	Mean	SD
Age (years)	21.27	3.03
Height (cm)	169	10
Weight (Kg)	68.49	16.19
BMI* (kg/m^2)	23.06	3.39

* Body Mass Index.

The inclusion criteria for this study were: healthy young adults (18 – 35 years). Exclusion criteria included (1) diseases or injuries that affect the musculoskeletal, nervous, and cardio-respiratory system, (2) being on medication, (3) recent surgeries (less than two months) that interfere with gait, (4) abnormalities or not corrected-to-normal vision, (5) vestibular disorders.

After checking the eligibility criteria fifteen young adults were recruited. All procedures were done in compliance with the Declaration of Helsinki and after a detailed description of the objective and research methodology, all participants signed informed consent. This study was approved by the Ethics Committee of the Polytechnic of Coimbra (approval number: 27_CEPC2/2019).

A questionnaire was used to assess the sociodemographic characteristics of the participants and to extract some data about usual smartphone use.

All participants underwent anthropometric measurements (weight and height) before performing motor tasks. Then, they were instructed to walk in two conditions during 60 s: walking at normal pace and velocity (single motor task) and walking while typing on a smartphone (simulate texting on the smartphone) with preferred hand or both hands according to what they were used to (motor dual-task). Each participant repeated twice each task, with one minute rest period between trials. All participants performed 3 gait cycles before data collection, allowing them to stabilize gait velocity.

No priority was given to texting and walking tasks. The participants were instructed to use their personal smartphone and their normal method of texting on the smartphone while natural walking.

The gait velocity analysis was performed using Qualysis® Motion Capture System (Qualysis AB, Kvarnbergsgatan 2, 411 05 Göteborg, Sweden) with 10 Oquos® Optoelectronic cameras high speed and resolution of 1.3 to 12 megapixels with a 200Hz measurement frequency, which simultaneously followed the movements of 69 reflective markers on the full-body model according to the IOR skin marked protocol in a previously calibrated volume with an error less 0.7 mm (Figure 1). The reflective markers were always fixed with a double-side tape onto landmarks on the participants' skin by the same researcher. The gait velocity was measured from the marker on the xiphoid process and data processing was performed using Qualysis Track Manager software (Qualysis AB, Sweden).

The statistical analysis in this study was conducted using IBM-SPSS 25.0 software. Quantitative data are reported as mean ± SD. The Shapiro-Wilks test was used to confirm the normal distribution of all continuous variables. The equality of variances was assessed by computing Levene´s test. The gait velocity differences between motor single and dual-task were assessed by a paired t-test. The statistical significance level was set to $p<0.05$.

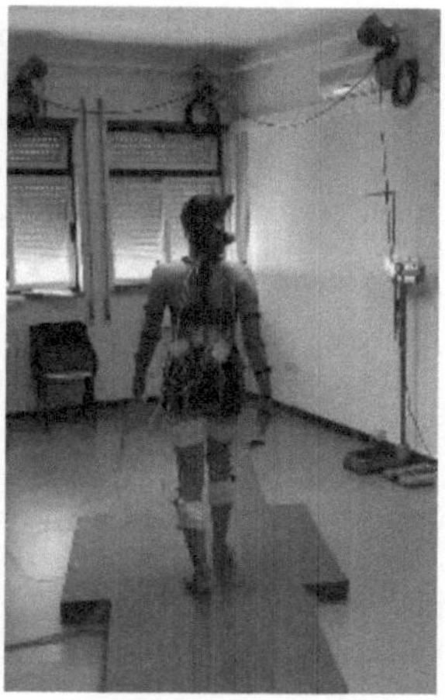

Figure 1. Data collection.

3 RESULTS

Most participants reported regular smartphone use for an average of 2 to 4 hours per day and all participants type on the smartphone with both hands. The gait velocity was 1.39 m/s on single motor task (only walking) and in the motor dual-task (walking while texting on the smartphone) the gait velocity was 1.29 m/s. Walking while texting on the smartphone (motor dual-task) significantly decreased gait velocity ($p<0.05$) comparatively to only walking (single motor task) (Figure 2).

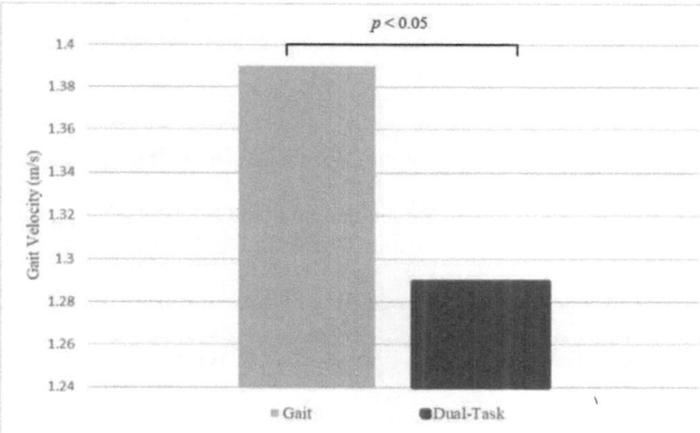

Figure 2. Comparison of gait velocity between single motor task (gait) and motor dual-task.

4 DISCUSSION

The main purpose of this study was to investigate changes in gait velocity caused by smartphone use during motor dual-task. Results demonstrated a gait velocity decreased during walking while texting on a smartphone compared to a simple walking task (single task).

Results are consistent with previous studies that reported decreased gait velocity and changes in other spatiotemporal gait parameters during smartphone use (Jeon et al., 2016; Licence et al., 2015).

Lamberg et al. (2012) found a gait velocity decrease of 16% and 33% when participants walk while talking on a cell phone compared to only walking and when they walk while texting on a cell phone, respectively. These results show a decrease in gait velocity in the dual-task performance compared to a single task.

During walking the arm movement is used as a strategy to increase the dynamic stability of the gait (Punt et al., 2015). When texting the smartphone with both hands the upper limbs are positioned in front of the trunk, which reduces and limits the natural movement of the trunk and arms, contributing to reduce the ability to walk maintaining dynamic balance (Schabrun et al., 2014). For this reason, the pedestrian accidents, the risk of falls, and musculoskeletal injuries are greater.

The smartphone use during the gait performance causes limitations of body movement and decreased visual information from the environment, contributing to changes in gait pattern (decreased gait velocity).

The present research adds valuable information to the knowledge around the effects of smartphone use on gait velocity in young adults: it is essential to prevent changes in gait caused by the use of a smartphone and to pay attention to the impact on the fall risk, pedestrian accidents, and musculoskeletal injuries. The reduction of gait velocity while texting on a smartphone in young adults can be a predictor of more evident gait changes and a marker of neurodegeneration and pedestrian injuries in old age.

5 CONCLUSION

Walking while texting on a smartphone modifies gait performance. The use of smartphones has a negative impact on gait velocity in young adults, suggesting an increased risk of falls, pedestrian accidents, and injuries.

We suggest specific training that included simultaneously motor and cognitive tasks to improve dual-task performance in both young and older people.

REFERENCES

Fritz, S., & Lusardi, M. 2009. White paper: "walking speed: the sixth vital sign". *J Geriatr Phys Ther*, 32, 46–9.

Hausdorff, J.M., Schweiger,A., Herman,T., Yogev-Seligmann, G. & Giladi, N. 2008. Task Decrements in Gait among Healthy Older Adults: Contributing Factors. *J Gerontol A Biol Sci Med Sci.*, 63(12): 1335–1343

Jeon, S., Kim, C., Song, S. & Lee G. 2016. Changes in gait pattern during multitask using smartphones. *Work*; 53: 241–7

Lamberg, E.M. & Muratori, L.M. 2012. Cell phones change the way we walk. *Gait & Posture*. 35: 688–6890

Licence, S., Smith, R., McGuigan, M.P. & Earnest, C.P. 2015. Gait pattern alterations during walking, texting and walking and texting during cognitively distractive tasks while negotiating common pedestrian obstacles. *PLoS One*, 10(7): e0133281

Nasar, J.L. & Troyer, D. 2013. Pedestrian injuries due to mobile phone use in public places. *Accident Analysis and Prevention*, 57: 91–95

Plummer, P., Eskes, G., Wallace, S. et al. 2015. Cognitive-motor interference during functional mobility after stroke: state of the science and implications for future research. *Arch Phys Med Rehabil.*, 94(12)

Punt, M., Bruijn, S.M., Wittink, H., van Dieën, J.H. 2015. Effect of arm swing strategy on local dynamic stability of human gait. *Gait Posture*, 41(2): 504–9

Schabrun, S.M., van den Hoorn, W., Moorcroft, A., Greenland, C. & Hodges, P.W. 2014. Texting and Walking: Strategies for Postural Control and Implications for Safety. *PLoS One*, 9(1): e8312

Studenski, S., Perera, S., Patel, K., Rosano, C., Faulkner, K., Inzitari, M., et al. (2011). Gait speed and survival in older adults. JAMA, 305(1), 50–58.

Advances and Current Trends in Biomechanics – Belinha et al. (Eds)

Plantar pressure analysis of a shoe perceived as biomechanically superior

Ricardo Sebastião, Manoela Sousa, Pedro Fonseca, Leandro Machado, Filipa Sousa & João Paulo Vilas-Boas
LABIOMEP, Porto Biomechanics Laboratory, University of Porto, Porto, Portugal
FADEUP, Faculty of Sport of the University of Porto (CIFI2D), Porto, Portugal

Denise Soares
KinesioLab, Institute of Intercultural and Transdisciplinary Studies of Almada, Piaget Institute of Almada, Almada, Portugal

Mário Vaz
LABIOMEP, Porto Biomechanics Laboratory, University of Porto, Porto, Portugal
INEGI, Institute of Science and Innovation in Mechanical and Industrial Engineering, Porto, Portugal

ABSTRACT: Plantar pressure analysis is important to understand the interactions between the shoe and the plantar surface. The aim of this study was to compare three different types of shoes: a model perceived as biomechanically superior (BS), a subject's regular use shoe (OS) and a neutral footwear model (NS), in terms of peak plantar pressure and maximum contact area. When compared with the NS, the BS and OS models presented lower peak plantar pressure and higher maximum contact area values, in a significant number of foot regions. When compared to each other, lower peak plantar pressures were observed in the OS central forefoot and lateral hindfoot regions. In conclusion, a biomechanically superior model tends to perform better than a neutral shoe, but a shoe adapted to the subject's foot may be as good, or even better, than a biomechanically superior one, in terms of plantar pressure distribution.

1 INTRODUCTION

The analysis of in-shoe plantar pressure provides important information on the forces acting upon the plantar surface and the shoe. The protective function of the shoes on the biological structures may vary depending on the shoe type and the materials used, some of which are associated with increased overload (Clinghan et al., 2008). Better cushioning proprieties, provided by materials with superior resilience and conferring greater contact area between the insole and the foot's plantar surface, are usually found in more sophisticated shoes, such as in running shoes. These have been reported to provide a greater attenuation of the impact force when compared to less developed models (Dickinson et al., 1985). However, there is evidence that less enhanced shoes, with low and medium cost, provide identical or even better plantar pressure dampening proprieties (Clinghan et al., 2008). The purpose of this study is to compare a shoe perceived as biomechanically superior, with a neutral shoe and the subject's daily use footwear, to determine which one can present superior performance in terms of plantar pressure attenuation.

DOI: 10.1201/9781003217152-38

2 MATERIALS AND METHODS

A sample of 10 (7 females and 3 males) physically active subjects (31.00 ± 6.49 years old; 71.07 ± 1.12 kg; 1.65 ± 0.11 m), all right limb dominant, were recruited for this study. Three types of shoes were selected: a shoe perceived as biomechanically superior (BS), the subject's own shoe (OS) and a neutral footwear model (NS). The BS was selected according to its global score and comfort level (93%), as rated by 1265 users of an athletic shoes review website (https://runrepeat.com/) in October 2018. The OS corresponded to the participant's own preferred daily use footwear. The NS consisted of a model with minimum sole thickness and virtually no foot accommodation proprieties. The in-sole plantar pressure was recorded with the use of the Pedar-X system (novel Gmbh, Germany) at a 100 Hz sampling frequency. After a 5-minutes accommodation period, subjects were instructed to walk at a self-selected pace, along a 10-meter walkway. A total of five trials for each footwear were recorded, in a randomized order.

The collected data was processed with a custom Matlab R2019a routine (The MathWorks Inc., USA). The foot was divided in 10 regions (Castro et al., 2013), and peak plantar pressure (PPP) and maximum contact area (MCA) were calculated for each region: hallux, other toes, medial forefoot, central forefoot, lateral forefoot, medial midfoot, lateral midfoot, medial hindfoot, central hindfoot and lateral hindfoot. One step of the dominant lower limb was analyzed per trial.

An *a priori* power analysis, conducted from a pilot study and using G*Power v3.1.9.2 (Faul et al., 2007), revealed that, for differences between two paired means with four variables of interest, a statistical power of 0.8 was achievable with a sample of 10 subjects, for effect sizes between 0.49 and 2.06. All other statistical analysis were performed in SPSS 26 (IBM, USA) with a significance level of $\alpha = 0.05$. Data normality was assessed with the Shapiro-Wilk test. For normality distributed data, paired-samples Student's *t*-test were conducted to compare PPP and MCA in each region of interest and for each shoe type. Non-parametric data were analyzed with Wilcoxon signed ranks tests. The Bonferroni correction was applied for three multiple comparisons (adjusted $\alpha = 0.017$).

3 RESULTS

The results of the comparisons between BS and OS are presented in Table 1. Descriptive statistics are presented as mean ± standard deviation and median (interquartile range) for normally distributed and non-parametric variables, respectively. The effect sizes Cohen's *d* and *r* are presented along paired-samples Student *t*-tests and Wilcoxon signed rank tests, correspondingly. An asterisk denotes a statistically significant difference between shoes.

Table 1. Peak plantar pressure and maximum contact area in each foot region of the BS and OS models.

Foot region	Shoe	Peak pressure (kPa)	*p*	Effect size	Max Area (cm^2)	*p*	Effect size
Halux	BS	241.29 ± 71.04	0.289	0.36	8.75 (1.57)	0.180	0.30
	OS	267.95 ± 83.95			8.75 (1.51)		
Other toes	BS	143.75 (57.55)	0.285	0.24	12.37 (2.33)	0.066	0.41
	OS	126.25 (64.13)			12.49 (1.79)		
Medial forefoot	BS	184.90 ± 42.27	0.033	0.80	10.25 (1.68)	0.600	0.12
	OS	163.95 ± 40.78			10.25 (1.11)		
Central forefoot	BS	228.92 (94.63)	0.007*	0.60	15.40 (2.23)	0.043	0.45
	OS	167.25 (69.13)			15.40 (2.46)		
Lateral forefoot	BS	154.72 ± 29.07	0.259	0.38	10.24 (1.5)	0.500	0.15

(*Continued*)

Table 1. (*Continued*)

Foot region	Shoe	Peak pressure (kPa)	*p*	Effect size	Max Area (cm^2)	*p*	Effect size
	OS	144.90 ± 39.07			10.24 (1.79)		
Medial midfoot	BS OS	62.06 ± 11.56 61.80 ± 16.80	0.924	0.03	13.99 ± 4.87 16.89 ± 2.84	0.076	0.64
Lateral midfoot	BS OS	99.09 ± 22.19 91.25 ± 34.14	0.252	0.39	16.56 (2.91) 16.56 (2.86)	0.180	0.30
Medial hindfoot	BS OS	176.53 ± 20.71 161.90 ± 19.41	0.088	0.60	8.88 (1.08) 8.88 (1.57)	1.000	0.00
Central hindfoot	BS OS	194.50 (36.21) 183.00 (46.75)	0.047	0.44	15.3 (2.6) 15.3 (2.6)	1.000	0.00
Lateral hindfoot	BS OS	178.58 (26.06) 155.00 (35.00)	0.009*	0.58	8.92 (1.08) 8.92 (1.19)	0.655	0.10

Statistically significant differences were found in terms of PPP values, namely in the central forefoot ($Z = -2.701, p = 0.007, r = 0.60$) and lateral hindfoot ($Z = -2.601, p = 0.009, r = 0.58$), with higher PPP values being observed on the BS model. No other statistically significant differences were found.

The results of the comparisons between BS and NS are summarized in Table 2.

Table 2. Peak plantar pressure and maximum contact area in each foot region of the BS and NS models.

Foot region	Shoe	Peak pressure (kPa)	*p*	Effect size	Max Area (cm^2)	*p*	Effect size
Halux	BS NS	241.29 ± 71.04 349.66 ± 98.02	0.008*	1.07	8.75 (1.57) 8.75 (1.67)	0.500	0.15
Other toes	BS NS	143.75 (57.55) 134.00 (119.34)	0.646	0.10	13.04 ± 1.76 12.92 ± 2.10	0.759	0.10
Medial forefoot	BS NS	184.90 ± 42.27 228.66 ± 83.04	0.063	0.67	10.44 ± 1.64 9.65 ± 1.29	0.164	0.48
Central forefoot	BS NS	241.31 ± 49.90 307.94 ± 77.40	0.002*	1.39	16.09 ± 1.62 15.79 ± 1.08	0.612	0.17
Lateral forefoot	BS NS	154.72 ± 29.07 235.51 ± 65.98	0.001*	1.53	10.24 (1.5) 10.80 (1.7)	0.600	0.12
Medial midfoot	BS NS	62.06 ± 11.56 26.24 ± 17.40	<0.001*	2.57	13.99 ± 4.87 2.59 ± 1.95	<0.001*	2.23
Lateral midfoot	BS NS	99.09 ± 22.19 81.40 ± 24.24	0.021	0.88	16.56 (2.91) 12.50 (6.40)	0.005*	0.63
Medial hindfoot	BS NS	176.53 ± 20.71 238.96 ± 47.31	0.002*	1.34	8.88 (1.08) 7.62 (2.13)	0.011*	0.57
Central hindfoot	BS NS	194.50 (36.21) 296.00 (48.69)	0.005*	0.63	15.30 (2.60) 15.30 (2.60)	0.317	0.22
Lateral hindfoot	BS NS	178.58 (26.06) 235.25 (75.50)	0.028	0.49	8.92 (1.08) 8.66 (1.82)	0.161	0.31

Higher PPP values were found, in the NS, in the following regions: hallux, ($t(9) = -3.376$, $p = 0.008$, $d = 1.07$), central forefoot ($t(9) = -4.409$, $p = 0.002$, $d = 1.39$), lateral forefoot ($t(9) = -4.830$, $p = 0.001$, $d = 1.53$),, medial hindfoot ($t(9) = -4.234$, $p = 0.002$, $d = 1.34$) and central hindfoot ($Z = -2.803$, $p = 0.005$, $r = 0.63$). Lower values were found in the medial midfoot ($t(9) = 8.130$, $p < 0.001$, $d = 2.57$). Higher MCA values were reported, in the BS model, in medial midfoot ($t(9) = 7.046$, $p < 0.001$, $d = 2.23$), lateral midfoot ($Z = -2.803$, $p = 0.005$, $r = 0.63$) and medial hindfoot ($Z = -2.533$, $p = 0.011$, $r = 0.57$). No other statistically differences were found.

The results of the comparisons between OS and NS are summarized in Table 3.

Table 3. Peak plantar pressure and maximum contact area in each foot region of the OS and NS models.

Foot region	Shoe	Peak pressure (kPa)	p	Effect size	Max Area (cm^2)	p	Effect size
Halux	OS NS	267.95 ± 83.95 349.66 ± 98.02	0.004*	1.22	8.75 (1.51) 8.75 (1.67)	0.043	0.45
Other toes	OS NS	126.25 (64.13) 134.00 (119.34)	0.333	0.22	12.49 (1.79) 12.49 (2.58)	0.128	0.34
Medial forefoot	OS NS	163.95 ± 40.78 228.66 ± 83.04	0.004*	1.22	10.25 (1.11) 9.96 (2.03)	0.008*	0.60
Central forefoot	OS NS	167.25 (69.13) 287.25 (83.63)	0.005*	0.63	15.40 (2.46) 15.40 (1.99)	0.173	0.30
Lateral forefoot	OS NS	144.90 ± 39.07 235.51 ± 65.98	<0.001*	2.15	10.24 (1.79) 10.80 (1.70)	1.000	<0.001
Medial midfoot	OS NS	61.80 ± 16.80 26.24 ± 17.40	<0.001*	2.57	16.89 ± 2.84 2.59 ± 1.95	<0.001*	6.26
Lateral midfoot	OS NS	91.25 ± 34.14 81.40 ± 24.24	0.268	0.37	16.56 (2.86) 12.50 (6.40)	0.005*	0.63
Medial hindfoot	OS NS	161.90 ± 19.41 238.96 ± 47.31	<0.001*	2.30	8.88 (1.57) 7.62 (2.13)	0.008*	0.60
Central hindfoot	OS NS	182.60 ± 26.57 303.85 ± 28.91	<0.001*	5.04	15.30 (2.60) 15.30 (2.60)	0.317	0.22
Lateral hindfoot	OS NS	156.30 ± 26.26 241.43 ± 44.81	0.001*	1.44	8.92 (1.19) 8.66 (1.82)	0.161	0.31

In terms of PPP, statistically significant differences were found in the following regions: hallux ($t(9) = -3.850$, $p = 0.004$, $d = 1.22$), medial forefoot ($t(9) = -3.867$, $p = 0.004$, $d = 1.22$), central forefoot ($Z = -2.803$, $p = 0.005$, $r = 0.63$), lateral forefoot ($t(9) = -6.787$, $p < 0.001$, $d = 2.15$), medial midfoot ($t(9) = 8.114$, $p < 0.001$, $d = 2.57$), medial hindfoot ($t(9) = -7.257$, $p < 0.001$, $d = 2.30$, central hindfoot ($t(9) = -15.94$, $p < 0.001$, $d = 5.04$) and lateral hindfoot ($t(9) = -4.537$, $p = 0.001$, $d = 1.44$). Higher values were found in the NS, aside from the medial midfoot. Higher MCA were reported, in the OS, in the medial forefoot ($Z = -2.666$, $p = 0.008$, $r = 0.60$), medial midfoot ($t(9) = 19.796$, $p < 0.001$, $d = 6.26$), lateral midfoot ($Z = -2.803$, $p = 0.005$, $r = 0.63$) and medial hindfoot ($Z = - 2.668$, $p = 0.008$, $r = 0.60$), with no other statistical differences observed.

4 DISCUSSION

The BS and OS presented lower PPP and higher MCA values, compared to the NS. It's minimum insole thickness, lead to an increase in PPP and a decrease in MCA values. The BS and OS are built with better materials, providing foot accommodation. This fact should provide a reduction in PPP, when compared to less specialized models (Lullini et al., 2020). The BS presented higher PPP values in the central forefoot and lateral hindfoot, when compared to the OS. It was expected that the BS, a pricier model, built with more sophisticated materials, would enable plantar pressure attenuation. However, performance is not related to cost, with some low and medium cost shoes performing better than more expensive ones, in terms of plantar pressure dampening (Clinghan et al., 2008). The OS, molded to the subject's feet, presented as less hard, mainly on the insole. This lead to lower PPP when compared to new, often harder shoes, as the BS (Lam et al., 2017). This fact prompts the need for more studies in order to determine the characteristics of biomechanically superior shoes and their relationship to performance, cost and comfort.

5 CONCLUSIONS

The results from this study indicate that a shoe perceived as biomechanically superior and an already adapted own shoe are able to perform better that a neutral shoe, in terms of plantar pressure dampening. The BS presented higher PPP values in the central forefoot and lateral hindfoot regions, compared to the OS. In conclusion, a shoe adapted to the subject foot can be as good, if not better, than a biomechanically superior one, in terms of plantar pressure distribution.

ACKNOWLEDGMENTS

This study was conducted under the project FAMEST – POCI-01-0247-FEDER-024529, co-financed by *Programa Operacional Competitividade e Internacionalização* (POCI), through *Portugal 2020* and *Fundo Europeu de Desenvolvimento Regional (FEDER).*

REFERENCES

Castro, M., Abreu, S., Sousa, H., Machado, L., Santos, R. & Vilas-Boas, J. P. 2013. Ground reaction forces and plantar pressure distribution during occasional loaded gait. *Applied Ergonomics*, 44, 503–509.

Clinghan, R., Arnold, G., Drew, T., Cochrane, L. & Abboud, R. 2008. Do you get value for money when you buy an expensive pair of running shoes? *British Journal of Sports Medicine*, 42, 189–193.

Dickinson, J. A., Cook, S. D. & Leinhardt, T. M. 1985. The measurement of shock waves following heel strike while running. *Journal of Biomechanics*, 18, 415–422.

Faul, F., Erdfelder, E., Lang, A. G. & Buchner, A. 2007. G*Power 3: a flexible statistical power analysis program for the social, behavioral, and biomedical sciences. *Behavior Research Methods*, 39, 175–191.

Lam, W. K., Ng, W. X. & Kong, P. W. 2017. Influence of shoe midsole hardness on plantar pressure distribution in four basketball-related movements. *Research in Sports Medicine*, 25, 37–47.

Lullini, G., Giangrande, A., Caravaggi, P., Leardini, A. & Berti, L. 2020. Functional Evaluation of a Shock Absorbing Insole During Military Training in a Group of Soldiers: A Pilot Study. *Military Medicine*, 185, 643–648.

Advances and Current Trends in Biomechanics – Belinha et al. (Eds)
© 2022 Copyright the Author(s), ISBN: 978-1-032-10806-3

Military overloaded gait: Effect on biomechanical variables

M. Sousa, P. Fonseca, R. Sebastião, S. Morais, I. de Sousa, L. Machado, F. Sousa & J.P. Vilas-Boas
Center of Research, Education, Innovation and intervention in Sport, Faculty of Sport, University of Porto, Portugal
Porto Biomechanics Laboratory, University of Porto, Portugal

D. Soares
Kinesiolab, ISEIT, Piaget Institute of Almada, Portugal

M. Vaz
Institute of Science and Innovation in Mechanical Engineering and Industrial, University of Porto, Portugal
Porto Biomechanics Laboratory, University of Porto, Portugal

ABSTRACT: The objective of this study was to analyze the effect of military specific load carriage, when wearing a neutral footwear, on gait linear kinematics and ground reaction forces. Fourteen military personnel were evaluated in the two conditions: non-backpack and backpack. Linear kinematics and ground reaction forces (GRF) were measured and compared between test conditions. Changes were found for stance and swing time, as well as for the initial, final and total double limb support time. For GRF, changes were observed in the anterior-posterior (final stance), medial-lateral and vertical (initial stance) directions. These alterations may be the result of an attempt to keep stability and a forward momentum during gait.

1 INTRODUCTION

Military personnel is exposed to several demanding biomechanical tasks during training and operations, such as long periods standing, walking on unstable surfaces and carrying heavy loads (Attwells et al. 2006). The amount of load carried can influence both training and operations, with historical record indicating that load carriage, shoulder straps and belts result in reduced performance, and unnecessary deaths (Knapik et al. 2004). Hence, understanding the biomechanics aspects related to the backpack carriage can contribute to a better understanding of its effects on performance and decrease or minimize these consequences. The effects of load carriage on linear kinematics are not yet clear. An increase in double limb support (Birrell & Haslam, 2009; Fox et al. 2020), decreased stride length, and increased stride frequency (Fox et al. 2020; Rice et al. 2017) are found in studies with soldiers using backpacks. Although these findings are commonly reported, there is a lack of consistency in the protocols used, leading to non-reproducible results (Birrell & Haslam, 2009). Thus, quantifying and understanding changes in linear kinematics can be a determinant factor in identifying risk factors and preserving a soldier's health.

The ground reaction forces (GRF) is an important measure of the forces acting during gait and load carriage (Birrell et al. 2007). Several studies have identified the increase in vertical and anterior-posterior forces as directly proportional to the added load (Birrell et al. 2007;

DOI: 10.1201/9781003217152-39

Fox et al. 2020; Birrell & Haslam, 2009) and unchanged for medial-lateral direction (Birrell et al. 2007). On the other hand, Simpson et al. (2018), while analyzing differences in load, observed similar values between 30% and 40% of the body's weight, indicating attenuation of the force production. Thus, the proportionality of the increase in GRF is contradictory (Birrell & Haslam, 2007). Given this bias, it is important to accurately assess the impact of force on military personnel while carrying loads, since the forces and other mechanisms are associated with the occurrence of lower limb injuries, such as stress fractures due to excessive, repetitive strain on the bone (Rice et al. 2019).

A better understanding of loaded gait by military personnel can reveal not only substantial information, but also induce a positive impact on this population. Therefore, the aim of this study is to analyze the effect of load carriage in gait of military volunteers, when wearing a neutral footwear, in terms of gait linear kinematics and ground reaction forces.

2 MATERIALS AND METHODS

Fourteen military volunteers from the Portuguese Army were recruited, including 2 females (27.00 ± 2.00 y, 1.61 ± 0.02 m, 62.69 ± 1.01 kg) and 12 males (28.17 ± 4.89 y, 1.78 ± 0.04 m, 83.13 ± 7.15 kg). Only participants who did not have any musculoskeletal injuries or diagnosed foot pathology in the last 6 months were eligible for this study. All participants read and signed an informed consent. The study was conducted in accordance with the Helsinki Declaration and was approved by the HSJ Ethics Committee with the number 114/19.

Two gait conditions were established. In the first condition, the participant's performed gait without backpack (NBP), while in the second condition they had to carry a 27.1 ± 0.6 kg backpack (BP). In both conditions, the participants were assessed while walking along a 10-meter walkway at a normal and comfortable self-selected pace, and wearing a neutral footwear with flat insole. The participants were allowed to performed several non-valid gait trials in order to familiarize themselves with the data collection protocol. Once the participant reported enough familiarization, data collection began with the NBP, followed by the BP. Five trials were collected for each condition. In order to calculate the gait's linear kinematics, retroreflective markers with a full-body setup were placed on participants, and data collected with a 12-camera motion capture system (Qualisys AB, Sweden) operating at 200 Hz. Also, four force platforms (Bertec Inc., USA) were located in the middle of the calibrated volume, operating at 2000 Hz, in a configuration that enabled the measurement of two consecutive steps. Kinematic and kinetic data synchronization was guaranteed by electronically integration of the kinematic and GRF information through a dedicated analog-to-digital converter (National Instruments, USA) and recorded with the Qualisys Track Manager Software (Qualisys AB, Sweden).

Both the kinetic and kinematic data were processed in Visual3D (C Motion Inc., USA). Only data from the right lower limb was assessed. Linear kinematics was time normalized to the gait cycle's duration. The GRF amplitude was normalized to the subject's body weight during the NBP, and to the subject's plus backpack weight in the BP, and time normalized to the stance duration. Analysis of the GRF was performed in terms of their three main descriptors in the anterior-posterior (AP), medial-lateral (ML) and vertical direction (V) (Richards, 2008). Data normality was assessed with the Shapiro Wilk test. For normally distributed variables, results are described as mean and standard deviation (SD). The Paired Student's *t*-test was used to compare the variables between NBP and BP conditions. All tests were performed using SPSS 26 (IBM, USA) and a significance level of $\alpha = 0.05$.

3 RESULTS

When comparing the linear kinematics between conditions, significant differences were observed for stance, $t(13) = -3.05$; $p = 0.009$ and swing time, $t(13) = 3.29$; $p = 0.006$, initial, $t(13) = -8.78$; $p < 0.01$, final, $t(13) = -8.00$; $p < 0.01$ and total double limb support, $t(13) = -8.75$; $p < 0.01$.

Specifically, higher means were found for the BP condition for stance time, initial, final and total double limb support, except for swing time. In what concerns the time normalized parameters, significant differences were observed for stance, $t(13) = -7.72$; $p < 0.01$ and swing time, $t(13) = 9.65$; $p < 0.01$, initial, $t(13) = -11.55$; $p < 0.01$, final, $t(13) = -8.29$; $p < 0.01$ and total double limb support, $t(13) = -10.08$; $p < 0.01$. Descriptive results can be observed in Table 1.

For the GRF descriptors (Table 2), significant differences were observed between NBP and BP for AP3, $t(13) = 4.28$; $p = 0.001$, ML1, $t(13) = -9.44$; $p < 0.01$, ML2, $t(13) = 5.81$; $p < 0.01$, ML3, $t(13) = 4.84$; $p < 0.01$ and V1, $t(13) = 2.39$; $p = 0.033$. For ground contact time significant differences were observed between NBP and BP for AP2, $t(13) = -2.44$; $p = 0.029$, AP3, $t(13) = 3.45$; $p = 0.004$, ML3, $t(13) = 3.62$; $p = 0.003$ and V3, $t(13) = 6.88$; $p < 0.01$.

Table 1. Mean (SD) of the gait linear kinematics without (NBP) and with backpack (BP).

Variables	Absolute NBP	Absolute BP	Normalized (% gait cycle) NBP	Normalized (% gait cycle) BP
Speed (m/s)	1.11 (0.08)	1.09 (0.09)	-	-
Stride Width (m)	0.13 (0.02)	0.13 (0.02)	-	-
Stride Length (m)	1.30 (0.08)	1.28 (0.09)	-	-
Step Length (m)	0.64 (0.05)	0.64 (0.05)	-	-
Cycle Time (s)	1.17 (0.07)	1.17 (0.09)	-	-
Steps per minute	103.22 (6.35)	102.74 (7.40)	-	-
Strides per minute	51.43 (3.17)	51.35 (3.86)	-	-
Step Time (s)	0.58 (0.03)	0.58 (0.04)	49.87 (0.002)	50.02 (0.006)
Stance Time (s)	0.72 (0.04)	0.75 (0.05)*	61.89 (0.007)	64.01 (0.008)*
Swing Time (s)	0.44 (0.02)	0.42 (0.03)*	38.20 (0.007)	36.04 (0.007)*
Initial Double Limb Support (s)	0.14 (0.01)	0.16 (0.01)*	12.05 (0.008)	14.14 (0.009)*
Final Double Limb Support (s)	0.13 (0.01)	0.16 (0.01)*	11.71 (0.007)	13.86 (0.008)*
Double Limb Support (s)	0.27 (0.02)	0.32 (0.02)*	23.77 (0.014)	27.98 (0.017)*

* Significant difference between conditions ($p < 0.05$).

Table 2. Mean (SD) of ground reaction forces without (NBP) and with backpack (BP).

	Ground Reaction Force (% BW) NBP	Ground Reaction Force (% BW) BP	Ground Contact Time (% gait cycle) NBP	Ground Contact Time (% gait cycle) BP
AP1	-16.13 (2.22)	-16.23 (2.40)	15.71 (2.65)	15.16 (1.81)
AP2	0.12 (0.07)	0.18 (0.08)	55.47 (3.50)	57.22 (2.82)*
AP3	17.15 (1.73)	15.76 (1.73)*	86.24 (1.43)	85.35 (1.26)*
ML1	-3.92 (0.85)	-2.65 (0.67)*	3.36 (1.24)	2.95 (0.79)
ML2	6.36 (0.93)	5.35 (0.79)*	23.34 (3.95)	25.56 (4.74)
ML3	7.46 (1.06)	6.63 (0.90)*	77.74 (4.74)	74.40 (2.91)*
V1	101.12 (4.56)	97.66 (5.69)*	23.00 (1.44)	23.93 (3.29)
V2	80.46 (2.54)	79.72 (4.47)	44.30 (3.79)	43.35 (4.34)
V3	109.09 (4.66)	107.67 (4.64)	77.26 (0.85)	79.69 (1.64)*

* Significant difference between conditions ($p < 0.05$).

4 DISCUSSION

This study investigated the effect of military specific load carriage, when wearing a neutral footwear, on gait linear kinematics and ground reaction forces. The main biomechanical changes found in linear kinematics were an increase in the stance time, as well as in the initial, final and total double limb support time and a shorter swing duration. According to Birrell & Haslam (2009) the effects of linear kinematics with the use of the backpack are not yet clear, except for double support time. Our results corroborate this statement, as well as other studies (Attwells et al. 2006; Birrell & Haslam, 2009) that observed an increase in double support. An increase during the stance time and, consequently, a decrease during the swing time were expected, given the gait cycle is divided by these two periods. In addition, to guarantee stability under load, there is an increase in stance time and therefore a greater double support time, due to a longer feet contact with the ground. In general, the results of this study can be explained as an attempt to achieve greater stability to support the load on the back (Birrell & Haslam 2009) and, consequently, trying to keep the same gait pattern when walking (Attwells et al. 2006; Birrell & Haslam, 2009).

In relation to GRF analysis, results for the BP condition are characterized by a reduction in the AP3 force value, this may be an indication of an increased trunk flexion providing enough inertia to keep a forward movement. In the ML variables, there was a reduction in the loaded condition, which may indicate an attempt to increase the stability of the ML due to the body's center of mass being further from its neutral position when load is added (Birrell et al. 2007). On the other hand, some studies demonstrate an increase in the ML force with added load (Birrell et al. 2007; Birrell & Haslam, 2010). Fox et al. (2020) suggests that this may be related to a decrease in stability and, consequently, to an increase in the ML GRF oscillation during load carriage to maintain postural stability. The unusual low value of V1 may be related to a higher knee flexion due to the increased load (Birrell et al. 2007; Birrell & Haslam, 2009). This behavior can also be related to the normalization of the total transported mass (backpack and subject) or to an adaptation to the overload condition to minimize possible adverse effects of the high vertical GRF on the musculoskeletal system (Castro et al. 2015). In addition, the mechanisms behind these changes in the GRF are not yet clear, but it may be caused by the participant's characteristics or as strategy to reduce the impact of the force with the ground and maintain the same performance (e.g. the same gait speed). In relation to ground contact time, Rice et al. (2017) reported that a longer ground contact time could be a protection mechanism. Our result for AP2 and V3 were also superior with load carriage, but not for AP3 and ML3. These results can express a protective mechanism to reduce body instability caused by poor load distribution (i.e. only on the back) during gait.

Although both limbs were evaluated in this study, the results of only the right limb were presented. Angular kinematics has also not been reported, but the biomechanical variables presented can support changes in gait pattern.

5 CONCLUSION

As expected, gait in a loaded condition showed changes in the evaluated parameters. The use of a backpack influenced the magnitude of the GRF and the pattern of load distribution during gait. In particular, a decrease in the GRF in the vertical direction, unlike many other studies. Thus, it is suggested the integration of other biomechanical variables in an attempt to understand these results and thus contribute to other new studies.

ACKNOWLEDGMENTS

The authors would like to thank the Military from the Portuguese Army who participated in the study. This study is part of the Project Mobilizer FAMEST - POCI-01-0247-FEDER

-024529, funded by the *Programa Operacional Competitividade e Internacionalização* (POCI), by Portugal 2020 and *Fundo Europeu de Desenvolvimento Regional* (FEDER).

REFERENCES

Attwells, R.L. Birrell, S.A. Hooper, R.H. Mansfield, N.J. 2006. Influence of carrying heavy loads on soldiers' posture, movements and gait. *Ergonomics*, 49(14):1527–37.

Birrell, A. Hooper, R.H. Haslam, R.A. 2007. The effect of military load carriage on ground reaction forces. *Gait & Posture* 26:611–614.

Birrell, A. & Haslam, R.A. 2009. The effect of military load carriage on 3-D lower limb kinematics and spatiotemporal parameters. *Ergonomics*, 52(10):1298–1304.

Birrell, S. A. & Haslam, R. A. (2010). The effect of load distribution within military load carriage systems on the kinetics of human gait. *Applied Ergonomics*, 41(4): 585–590.

Castro, M. Figueiredo, M. Abreu, S. Sousa, H. Machado, L. Santos, R. & Vilas-Boas, J. P. 2015. The influence of gait cadence on the ground reaction forces and plantar pressures during load carriage of young adults. *Applied Ergonomics*, 49:41–46.

Fox, B. D. Judge, L. W. Dickin, D. C. & Wang, H. 2020. Biomechanics of Military Load Carriage and Resulting Musculoskeletal Injury: A Review. *Journal Orthopedics & Orthopedic Surgery*, 1(1): 6–11.

Knapik, J. J. Reynolds, K. L. & Harman, E. 2004. Soldier Load Carriage: Historical, Physiological, Biomechanical, and Medical Aspects. *Military Medicine*, 169: 45–55.

Rice, H. Fallowfield, J. & Dixon, S. 2019. Altered forefoot function following a military training activity. *Gait & Posture*, 74: 182–186.

Richards, J. 2008. Biomechanics in Clinic and Research: An interactive teaching and learning course. Elsevier Health Sciences, London, United Kingdom.

Simpson, J. D. DeBusk, H. Hill, C. Knight, A. & Chander, H. 2018. The role of military footwear and workload on ground reaction forces during a simulated lateral ankle sprain mechanism. *The Foot*, 34: 53–57.

Advances and Current Trends in Biomechanics – Belinha et al. (Eds)

Cochlear implantation using a straight cochlear implant electrode—a numerical study

B. Areias
Institute of Science and Innovation in Mechanical and Industrial Engineering, Portugal

M.P.L. Parente & R.M. Natal Jorge
Faculty of Engineering of the University of Porto, Portugal

F. Gentil
School of Health, Polytechnic Institute of Porto, Portugal
WIDEX, Portugal

ABSTRACT: Cochlear implants have been developed to assist adults and children who have significant hearing loss. This device converts the sound into electrical signals that are delivered directly to spiral ganglion cells of the cochlear nerve. Atraumatic electrode array and preservation of the residual hearing after cochlear implantation surgery play a critical role in ensuring positive outcomes. In this respect, it is necessary to develop more flexible electrodes array. The platinum/iridium wires within the electrode array are mostly responsible for the stiffness of the electrode. With this work two different wires arrangement will be studied and, the insertion force and the contact pressure between them will be compared, in order to find the least aggressive.

1 INTRODUCTION

Cochlear implants are electrical prostheses that enable to experience auditory sensations via a direct electrical stimulation of the cochlear nerve. These devices are able to replace the function of damaged inner hair cells. However, when a patient with severe to profound sensorineural hearing loss showed some residual hearing, the insertion of the electrode array must be performed carefully, otherwise the patient may lose your residual hearing.

The cochlear implant systems are generally composed by two component systems: the external speech processor that records the sound and the internal component which converts it into a logical sequence of electrical impulses. This information is transmitted through electromagnetic induction to the receiver coil located under the skin. The receiver coil delivers the appropriate electrical signals to the electrode array, which has been inserted into the scala tympani. The contacts along the electrode array stimulate the cochlear nerve, resulting in a release of electric impulses towards the brain where the signal is then interpreted (Lenarz, T. 2017). A cochlear implant electrode array usually consists in platinum contacts and input wires (platinum-iridium 90/10 with insulating Teflon coating) embedded in a silicone carrier material.

The cochlea is a hydromechanical frequency analyzer. Its function is to perform a real time spectral decomposition of the complex acoustic signal in producing a spatial frequency map. The human auditory system has the capability of detecting and analyzing sounds with frequencies from 20 Hz to 20 kHz and an intensity range up to 120 dB (Ni, G. et al. 2014.).

There are some complications that may be encountered during an electrode insertion, such as partial insertion, scalar translocation, and tip fold-over. Electrode migration is a post-

DOI: 10.1201/9781003217152-40

surgical complication associated with an outward movement of the electrode array due to inherent springiness of the electrode array or to pulling forces external to the scala tympani (Ishiyama, A. et al. 2020). Lim et al. (Lim, Y.S. et al. 2005) carried out a study based in six electrodes with different arrangements. Each electrode comprises sixteen metal wires within the silicone matrix. The results showed that low insertion forces and contact pressures at the cochlear implant tip are obtained when the wires are stacked horizontally. The horizontal plane was defined as the plane which contains the basilar membrane. Nowadays, current cochlear implants contain between 12 and 24 electrodes.

However, through a bibliographic research it was concluded that original papers within this scope are very scarce in the literature. Thus, the numerical work developed here intends to increase the knowledge about the wire arrangement within the silicone matrix, simulating two different possible arrangements. The finite element method was used to describe the complete insertion of the electrode array.

2 MATERIALS AND METHODS

3D anatomical model geometry of the human bony labyrinth was obtained from a magnetic resonance imaging scan, of a male patient without cochlear pathologies, 85 years old. The straight electrode array represented in Figure 1 was based on the standard commercially available CI622 cochlear implant manufactured by the Cochlear medical device company (Cochlear 2021). The CI622 electrode has a length of 25mm and consists of 22 platinum electrode contacts. The basal end has a diameter of 0.55mm and the hemispherical tip measures 0.3mm. The mesh definition of electrode array is summarized at the Table 1.

The mechanical properties of the electrode array are summarized in the Table 2. The matrix (silicone) was considered elastic whereas the 22 Platinum/Iridium wires (each wire has a diameter

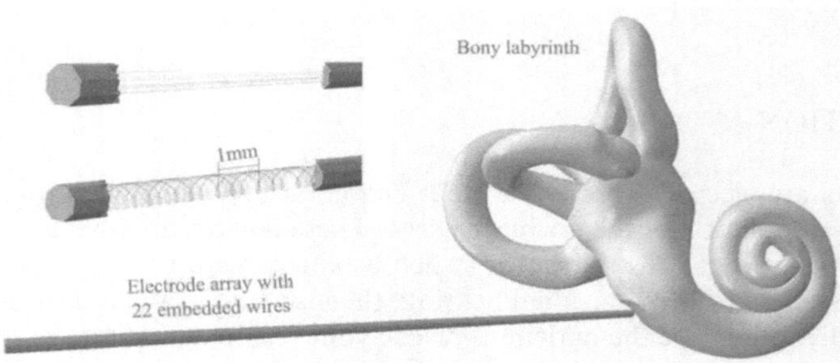

Figure 1. 3D model with two different wires arrangement.

Table 1. Mesh definition of the electrode array.

	Silicone matrix	Straight wires	Spiral wires
N° nodes	20172	2315/2315	6087/3888
N° elements	15942	2293/2293	6065/3866
Element type	C3D8R	B31/B31	B31/B31

Table 2. Mechanical properties of the electrode array.

	E [MPa]	v[−]	ρ [kg.m^{-3}]	σ_{p11}[MPa]	ε_{p11} [-]	σ_{p12}[MPa]	ε_{p12}[-]
Wires	202300	0.378	21600	295	0	567	0.3
Silicone matrix	180	0.45	1120				

of 0.025mm) were considered with elasto-plastic properties. The mechanical properties were based in the CES EduPack software (Granta EduPack 2020), in the work of Merker et al. (Merker, J. et al. 2001) and Kha et al. (Kha, H. N. et al. 2004). The B31 elements were embedded in the C3D8R elements using an embedded element technique available in Abaqus FE software.

The bony labyrinth was considered rigid and fixed. The friction between the lateral wall and the electrode array was not considered in our study. The contact pressure around the cochlea is distributed under the lateral wall of the scala tympani since the electrode array was inserted through the round window.

A displacement of 25mm was imposed at the proximal end of the electrode array, thus considering a full insertion. Some simplifications were implemented in the real cochlea, thus the platinum contacts, the scala media, the scala vestibuli and the organ of Corti were not considered in all numerical simulations. The Abaqus/Explicit FE software package was used for modeling the cochlear implantation.

3 RESULTS

The straight and the spiral wire configurations were studied. Figure 2 shows the insertion force during the insertion of both wire configurations. The insertion process is complete after a displacement of 25mm at the proximal end of the electrode array. In both configurations the insertion force tends to increase with the increasing insertion depth. The sharpest increase of the insertion force was observed during the adaptation phase or after the contact initiation between the electrode array and the lateral wall, and also after completing one turn inside the cochlea, after 21mm. The insertion force obtained in the straight configuration is 2.5 times greater than the spiral configuration with an insertion depth of 25mm.

Figure 3 shows the mean contact pressure in the distal portion of the electrode array during the insertion process for both straight and spiral configurations. Both configurations show an increase in contact pressure during all the insertion process. It was found that the straight configuration has a higher mean contact pressure than the spiral configuration, although the difference is small. The peak of the mean contact pressure was obtained for the straight configuration, thus showing that the straight configuration may be more traumatic. Distal portion refers to half of the electrode array that contains the hemispherical tip. At the proximal portion, the spiral configuration showed a lower mean contact pressure, mainly in the final part of the insertion process.

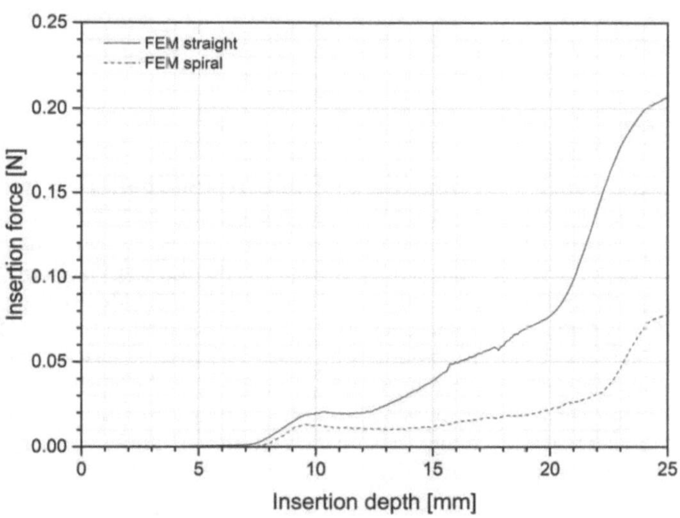

Figure 2. Insertion force vs insertion depth of the two different wires configuration.

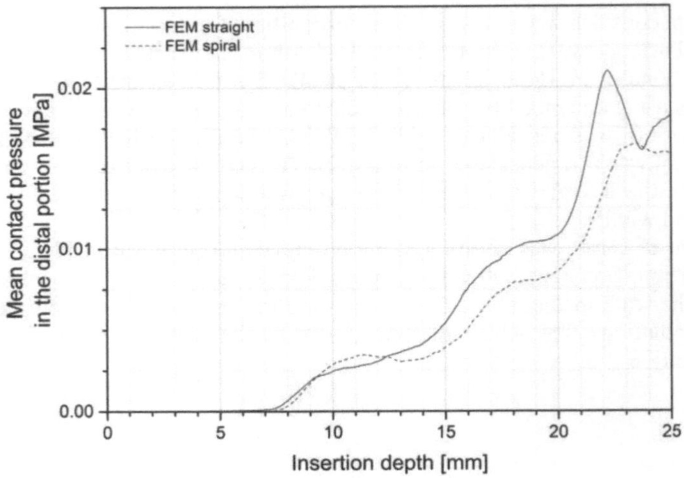

Figure 3. Mean contact pressure in the distal portion of the electrode array among the two different wires configuration.

4 DISCUSSION

The spiral wire configuration was the configuration with the lower insertion force and mean contact pressure. The contact pressure was evaluated in two different areas of the electrode array, thus the proximal and distal portions were separated and then compared with each other. We can see in both proximal and distal portion that for the most part of the insertion process, the mean contact pressure is the higher in the straight configuration.

The insertion force is an important factor during the insertion, since a sudden increase during the process often means more intracochlear damage. Clinical evidence suggests that intracochlear damage during electrode array insertion negatively affects residual low frequency hearing. However, the 3D finite element model developed has some simplifications, among which the removal of the scala vestibuli and scala media. This simplification was due to the fact that the magnetic resonance has no resolution to see such scalae. The model admits a constant insertion speed of the electrode array, that in fact is practically impossible and depends on the surgeon's experience.

5 CONCLUSIONS

A 3D cochlear model can be thought as a tool to carry out numerical analysis, in which researchers can predict the response of the system to different stimulation intensities and frequencies. The electrode array should have a geometry and a mechanical behavior that minimize the trauma during its insertion. The configuration of the platinum/iridium wires embedded in the silicone play a crucial role in the global stiffness of the electrode array.

All studied configurations have a tendency to increase the insertion force and the mean contact pressure with increasing the insertion depth. However, the increase in the insertion force and the mean contact pressure is more pronounced in the straight configuration. The numerical results showed that we can consider the straight configuration, as the configuration with higher risk of the cochlear wall damage. In our view point it is now necessary to develop and complete the 3D numerical model in order to predict other types of complications, such as perforations in the basilar membrane.

ACKNOWLEDGMENT

The authors truly acknowledge the funding provided by Ministério da Ciência, Tecnologia e Ensino Superior - Fundação para a Ciência e a Tecnologia (Portugal), under grant: SFRH/BD/129397/2017. This research was also supported by the Portuguese Foundation of Science and Technology under the research project UIDB/50022/2020.

REFERENCES

Cochlear 2021. Cochlear Nucleus CI622 with Slim Straight [Online]. Available: https://www.cochlear.com/us/en/professionals/products/cochlear-implants/surgery/surgical-guides [Accessed 15th February 2021].

Ishiyama, A. et al. 2020. Potential insertion complications with cochlear implant electrodes. Cochlear Implants International, 21, 206–219.

Kha, H. N. et al. 2004. Stiffness properties for Nucleus standard straight and contour electrode arrays. Medical Engineering & Physics, 26, 677–685.

Lenarz, T. 2017. Cochlear implant - state of the art. GMS Current Topics in Otorhinolaryngology - Head and Neck Surgery, 16, Doc04.

Lim, Y.S. et al. 2005. Three-dimensional analysis of electrode behavior in a human cochlear model. Medical Engineering & Physics, 27, 695–703.

Merker, J. et al. 2001. High temperature mechanical properties of the platinum group metals: Elastic properties of platinum, rhodium and iridium and their alloys at high temperatures. Platinum Met. Rev., 45.

Ni, G. et al. 2014. Modelling cochlear mechanics. Biomed Res Int, 2014, 150637.

Granta EduPack 2020. Granta Design Limited 2020.

Advances and Current Trends in Biomechanics – Belinha et al. (Eds)
© 2022 Copyright the Author(s), ISBN: 978-1-032-10806-3

Bilayered arcade-like scaffolds for articular cartilage repair

Â. Semitela & J. Couto
Centre of Mechanical Technology and Automation (TEMA), Department of Mechanical Engineering, University of Aveiro, Portugal

A. Capitão & A.F. Mendes
Centre for Neuroscience and Cell Biology, University of Coimbra, Portugal

P.A.A.P Marques & A. Completo
Centre of Mechanical Technology and Automation (TEMA), Department of Mechanical Engineering, University of Aveiro, Portugal

ABSTRACT: Articular cartilage is a highly organized tissue adapted to the complex mechanical loading in joints. Given the limited self-healing of this tissue, tissue engineering (TE) strategies have explored the development of 3D anisotropic fibrous scaffolds with the implementation of specific mechanical stimulus. However, this functionally is dependent on the ability to recreate the depth-dependent collagen fibre alignment on the 3D fibrous scaffolds, which has been considerably challenging. In this work, bilayered structures with different fibre orientations were fabricated via polycaprolactone and gelatin electrospinning and polyethylene glycol (PEG) particles electrospraying. After sacrificial PEG particles removal, large inter-fibre spaces were created, that were compatible with chondrocyte migration. The *in vitro* studies confirmed the biocompatibility of the scaffolds and their ability to guarantee cell attachment and migration through the scaffold. Further, the mechanical stimulation applied through unconfined compression substantially improved chondrocyte response. These results confirmed the potential of the developed 3D bilayered scaffolds for articular cartilage TE.

1 INTRODUCTION

Articular cartilage is a highly specialized tissue of semi-rigid consistency with outstanding properties of deformability, resistance to mechanical loading and low-friction gliding, essential to joint function. However, this tissue does not possess vascularization or neural networks which prevents its intrinsic ability to self-regenerate (Correa and Lietman, 2017). Current clinical treatment strategies have varying success rates, but average long-term results are unsatisfactory (Correa 2017). Due to these limitations, tissue engineering (TE) strategies have been explored as an alternative solution to develop tissue constructs capable of mimicking the depth-dependent organization of the fibrous collagen network of the native articular cartilage tissue. Electrospinning has proved to be capable of creating anisotropic scaffolds by sequentially producing different fibre sizes and orientations in a continuous construct to mimic the structural, topographic organization and mechanical properties of the articular cartilage (McCullen 2012; Steele 2014). Due to nearly two-dimensional nature of these scaffolds, post-processing strategies have attracted substantial interest due to their high flexibility in creating the desired final scaffold shape. In fact, our group has recently developed a new methodology to fabricate bilayered electrospun scaffolds with an arcade-like fibrous organization, that was able to induce depth-

DOI: 10.1201/9781003217152-41

dependent chondrocyte attachment and migration *in vitro* (Girão 2020). Here, to improve the bio-inductive properties of the scaffolds, Gelatin (GEL) has been incorporated within the polycaprolactone (PCL) system to improve the bio-inductive properties, as previously demonstrated (Semitela 2020). Additionally, a specific mechanical stimulus was applied through cyclic unconfined compression, which has been known to accelerate ECM production and, consequently, increase the elastic moduli of the tissue formed (McCullen 2012).

2 MATERIALS AND METHODS

2.1 *Fabrication of the bilayered fibrous electrospun scaffolds*

PCL (Sigma-Aldrich) and GEL (Sigma-Aldrich) were dissolved separately in 2,2,2-trifluoroethanol (TFE; TCI) at a concentration of 10% (w/v), stirred vigorously at room temperature (RT) overnight, and mixed at a ratio of 50:50 with 0.2% (v/v) of acetic acid (Sigma-Aldrich). A PEG (Sigma-Aldrich) solution (1.5 g/mL) in chloroform (ACROS Organics) was prepared at 50 °C, cooled at RT, and then poured into a 10 mL plastic syringe to be electrosprayed simultaneously with the electrospinning of the PCL + GEL solution following the specifications described elsewhere (Semitela 2020, 2021), in a NANON-01 A electrospinning setup (MECC). Rectangles with 5 by 40 mm were cut from this mesh, rolled into a spiral – deep layer – and placed in the collector for a new deposition of PCL+GEL fibers and PEG microparticles – superficial layer. The resulting scaffolds were then washed in a graded series of ethanol aqueous solutions and distilled water to dissolve the PEG microparticles and freeze-dried for further structural, mechanical and biological characterization.

2.2 *Characterization of the bilayered fibrous electrospun scaffolds*

Scanning electron microscopy (SEM) was used to visualize the topography of the bilayered scaffolds with and without the PEG microparticles using Hitachi TM4000 plus (5 kV). PEG microparticle diameter and pore size was determined using Image-Pro Plus software ($n > 50$).

The water absorption capacity of the scaffolds was determined by swelling of freeze-dried scaffolds (with known weights) in distilled water for 24 hours at 37 °C.

The compressive properties of the bilayered scaffolds were measured under quasi-static unconfined compression in wet conditions at a rate of 0.5 mm/min, using a Shimadzu MMT-101 N (Shimadzu Scientific Instruments) with a load cell of 100 N. The compressive moduli of the samples were calculated through the tangent modulus of the linear portion of the stress–strain curve at low strain.

An immortalized human chondrocyte cell line C28/I2 (kindly provided by Prof. Mary Goldring, Hospital for Special Surgery, New York and Harvard University) was used. Cells were maintained at 37 °C in a humidified atmosphere of 5% CO_2 in air, in Dulbecco's Modified Eagle Medium (DMEM)/Nutrient Mixture F-12 Ham 1:1 (v/v) (DMEM: Gibco, Life Technologies; F-12: Sigma-Aldrich) supplemented with 10% (v/v) non-heat-inactivated fetal bovine serum (Gibco, Life Technologies), 1% (v/v) penicillin/streptomycin (Grisp). Medium refreshments were performed two times a week. Cells were harvested at pre-confluence using trypsin/EDTA solution (0.05%/0.02%, Sigma-Aldrich). The scaffolds were disinfected in 70% (v/v) ethanol aqueous solution. Then, 0.25×10^6 cells were seeded on the bottom and top of each scaffold followed by incubation at 37 °C for 2 hours. After an initial attachment period of 5 days, 5 cell-laden scaffolds were placed in a patented bioreactor (Bandeiras 2015) and subjected to unconfined cyclic compression for 12 days, following a specific stimulation regime: 2 hours of sinusoidal compression (0.5 Hz, 0 – 10 % strain) and 4 hours of rest; repeated 4 times a day. 5 static controls were considered. Medium was refreshed two times a week.

Cell viability was calculated based on the metabolic activity of the chondrocyte-laden scaffolds after 1, 7, 14 and 17 days of culture measured by a resazurin method. Resazurin solution (0.1 mg/mL in phosphate buffered saline (PBS; Sigma-Aldrich); ACROS Organics) was added to fresh medium at a final concentration of 10 % (v/v). Scaffolds were incubated in this solution at

37°C for 4 hours in the dark, after which 100 μL per well was transferred to a 96-well plate and absorbance (Abs) at 570 and 600nm was measured in a microplate reader (Synergy HTX, BioTek). For each day, final absorbance values for each sample were calculated as the ratio Abs570/Abs600nm minus the Abs570/Abs600 nm ratio of a negative control (scaffold without cells). The absorbance values of cells incubated in the tissue culture polystyrene on the first time point were taken as 100% and the percentage of viable cells was calculated from these control values.

After the final time point, chondrocyte morphology and distribution within the scaffolds was assessed by SEM and haematoxylin and eosin staining. The dynamically stimulated scaffolds and the respective static controls were fixed in 4 % (w/v) paraformaldehyde in PBS, embedded in frozen embedding medium (Cryomatrix, Thermo scientific) and sectioned transversally in a cryostat (MEV, SLEE cryostat). The scaffolds' cross-sections were dehydrated with graded concentrations of ethanol aqueous solutions, treated with hexamethyldisilane (TCI), kept in a fume hood for air drying and visualized by SEM. Several scaffolds' sections were also stained with hematoxylin and eosin solutions to highlight in purple-blue the cells in the bilayered scaffolds.

3 RESULTS AND DISCUSSION

The bilayered scaffolds were successfully fabricated, in a first stage with the PEG particles, with an average diameter of 34.82 ± 8.21 μm (Figure 1a). After their removal large interfibre spaces with a high percentage of sizes ranging from 10 to 20 μm are left (Figure 1b), which are suitable for chondrocyte migration. The scaffolds were able to absorb almost immediately eleven times their weight in water, demonstrating their hydrophilic character, mainly attributed to the presence of GEL in the fibres (Figure 1c) (Semitela 2021). The stress-strain curves of the compression assays (Figure 1d) clearly depicted an elastic behaviour with a compression modulus of nearly 2 kPa. This value was significantly lower than the ones reported for articular cartilage (89.5 ± 48.6 kPa) (McCullen 2012). Nevertheless, there are reports stating that scaffolds with lower compression

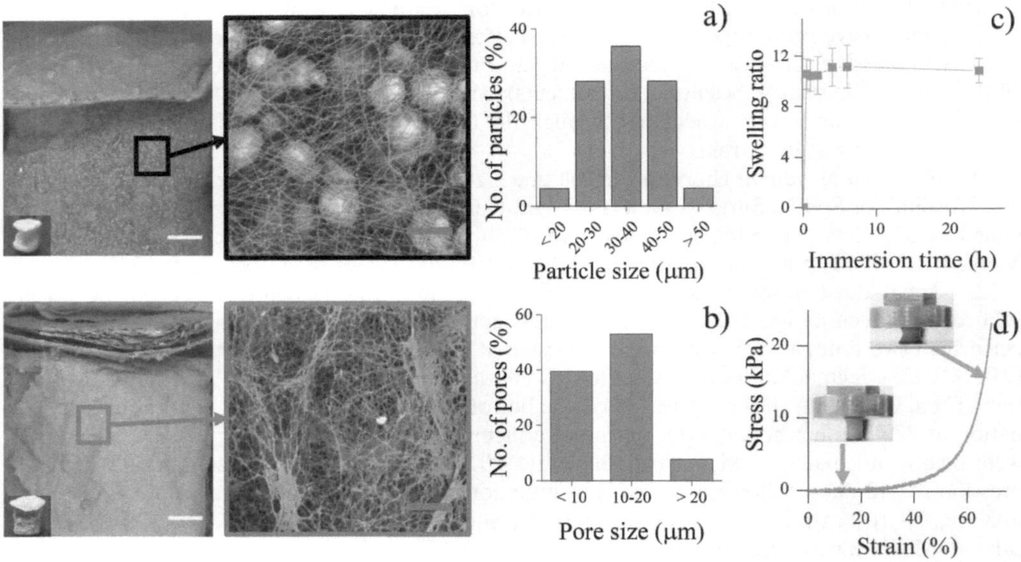

Figure 1. Structural and mechanical characterization of the bilayered scaffolds: Topography of the scaffolds with PEG microparticles and the respective PEG particle size distribution (a); and after PEG removal and the respective pore size distribution (b); swelling ratio up to 24 hours of immersion (c) and mechanical response to compression (d). Scale bars: white: 500 μm, black: 100 μm and green: 50 μm.

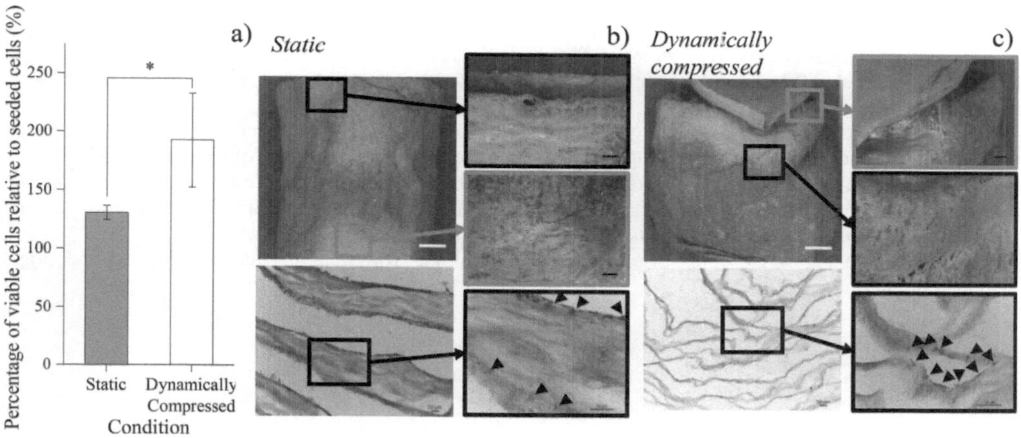

Figure 2. In vitro characterization of the bilayered scaffold: percentage of viable chondrocytes in the scaffolds (a) and the respective SEM and haematoxylin and eosin staining images after 17 days of culture under static (b) and dynamically compressed conditions (c). Statistical analysis by One-way ANOVA followed by post hoc Tukey's test; *$p<0.05$, where * denotes statistically significant differences between culture conditions. Scale bars: white: 500 μm, black: 100 μm and green: 50 μm.

moduli should be favourable to chondrogenesis by avoiding the stress-shielding effect, and consequently, allowing mechanical stimulation of the chondrocytes (Steele 2014).

The *in vitro* studies revealed that these scaffolds allowed substantial chondrocyte attachment and proliferation, not only due to the arcade-like fibrous configuration, but also due to GEL incorporation and enlarged porosity provided by the PEG removal (Semitela 2020). The dynamically compressed environment seemed to be extremely beneficial to chondrocyte proliferation and migration, as the cell-seeded scaffolds subjected to compression displayed higher percentage of viable chondrocytes (Figure 2a) as well as evidence of chondrocyte migration between the layers (Figure 2c), which was not visible under static conditions (Figure 1b).

4 CONCLUSIONS

Thes bilayered scaffolds presented in this work offer suitable chondrocyte attachment and mechanical support and substantially promoted their proliferation and migration, not only under dynamically compressed conditions, but also in static conditions, rendering them as suitable scaffolds for articular cartilage TE applications.

ACKNOWLEDGEMENTS

This work was supported by the Portuguese funding of Program COMPETE-FEDER, Programa Operacional Competitividade e Internacionalização through the projects POCI-01-0145-FEDER-028424 and CENTRO-01-0145-FEDER-022083. Also, by Fundação para a Ciência e Tecnologia I.P. (FCT, IP) through the projects PTDC/EME-SIS/28424/2017, UIDB/00481/2020 and UIDP/00481/2020. The authors thank to FCT for the PhD grant SFRH/BD/133129/2017.

REFERENCES

Bandeiras, C., et al. 2015. Tissue Engineered Cartilage in Unconfined Compression: Biomechanical Analysis. *Materials Today: Proceedings* 2: 355–364.

Correa, D., Lietman, S.A., 2017. Articular cartilage repair: Current needs, methods and research directions. *Seminars in Cell and Developmental Biology* 62: 67–77

Girão, A.F., et al. 2020. Microfabrication of a biomimetic arcade-like electrospun scaffold for cartilage tissue engineering applications. *Journal of Materials Science: Materials in Medicine* 31: 69
McCullen, S.D. et al. 2012. Anisotropic fibrous scaffolds for articular cartilage regeneration. *Tissue Engineering Part A* 18: 2073–2083.
Semitela, Â., et al. 2020. Electrospinning of bioactive polycaprolactone-gelatin nanofibres with increased pore size for cartilage tissue engineering applications. Journal of Biomaterials Applications 35:471–484.
Semitela, Â., et al. 2021. Boosting in vitro cartilage tissue engineering through the fabrication of polycaprolactone-gelatin 3D scaffolds with specific depth-dependent fiber alignments and mechanical stimulation. *Journal of the Mechanical Behavior of Biomedical Materials* 117: 104373.
Steele, J.A.M., et al. 2014. Combinatorial scaffold morphologies for zonal articular cartilage engineering. *Acta Biomaterialia* 10: 2065–2075.

Advances and Current Trends in Biomechanics – Belinha et al. (Eds)
© 2022 Copyright the Author(s), ISBN: 978-1-032-10806-3

Automated fabrication of 3D chondrocyte-laden anisotropic scaffolds for articular cartilage

Â. Semitela & A.L. Pereira
Department of Mechanical Engineering, Centre of Mechanical Technology and Automation (TEMA), University of Aveiro, Portugal

A. Capitão & A.F. Mendes
Centre for Neuroscience and Cell Biology, University of Coimbra, Portugal

P.A.A.P Marques & A. Completo
Department of Mechanical Engineering, Centre of Mechanical Technology and Automation (TEMA), University of Aveiro, Portugal

ABSTRACT: Tissue engineering (TE) strategies for repairing and regenerating articular cartilage face critical challenges to approximate the biochemical and biomechanical microenvironment of native tissue, particularly regarding collagen fibril depth-orientation and chondrocyte distribution. Here, a recently developed electromechanically 3D electrospinning platform was employed to develop three-dimensional (3D) anisotropic electrospun scaffolds in a fully automated manner with simultaneous chondrocyte incorporation. As expected, the 3D scaffolds possessed an arcade-like fibrous configuration with a uniform chondrocyte distribution. Overall, the results suggest that this combined approach has potential for cartilage TE.

1 INTRODUCTION

Articular cartilage self-repair is limited because the intrinsic regenerative ability of the tissue is low and cartilage lesions due to trauma or diseases tend to progressively intensify. Moreover, current treatment options usually lead to long-term unsatisfactory results because the tissue formed lacks the native anisotropic configuration, particularly the collagen fibres alignment, progressing from parallel in the superficial region, to random in the middle zone, and finally perpendicular oriented in the deep zone, as well an appropriate chondrocyte distribution, resulting in inferior biochemical and biomechanical properties (Correa 2017). Given these limitations, there is an increasing demand for tissue engineering (TE) approaches to develop tissue constructs with these characteristics. Several reports have pointed out the relevance of the electrospinning as a scaffold fabrication technique for cartilage TE due to similarities between the resultant polymer-based fibres and the collagen fibres in the extracellular matrix of the native tissue (Correa 2017). Still, despite the extensive research, few studies have assessed the importance of a controlled collagen fibril depth-orientation and chondrocyte distribution on the biochemical and biomechanical properties of engineered cartilage (McCullen 2012; Steele 2014; Munir 2020; Girão 2018, 2020; Semitela 2021). This mainly stems from the difficulty to create three-dimensional (3D) electrospun scaffolds with specific fibre alignment as well as a uniform chondrocyte distribution throughout these structures. Substantial progress has been made recently through the development of 3D electrospun scaffolds by means of post-processing strategies of the electrospun meshes (Girão 2018, 2020; Semitela 2021). However, these strategies are not automated, resulting in high sample variability, and the

DOI: 10.1201/9781003217152-42

respective chondrocyte infiltration is limited. This work is a preliminary effort to develop in a fully automated manner chondrocyte-laden 3D anisotropic tissue constructs using a newly developed electromechanically 3D electrospinning platform.

2 MATERIALS AND METHODS

All experiments were performed in a recently developed electromechanically 3D electrospinning platform (Completo 2020) (Figure 1a and b), thoroughly cleaned with 70 % (v/v) ethanol aqueous solution (ChemLab) beforehand and sterilized with UV radiation. The remaining used instruments were already sterile or autoclaved at 121 °C beforehand. All the electrospinning systems were assembled and cleaned prior to use with a phosphate-buffered saline (PBS; Sigma-Aldrich) supplemented with 1% (v/v) penicillin/streptomycin (P/S; Grisp) and 2.5 μg/mL solution of Amphotericin B (Sigma-Aldrich).

2.1 *Preparation of the polymer solution*

PCL (Sigma-Aldrich) and GEL (Sigma-Aldrich) were dissolved separately in 2,2,2-trifluoroethanol (TFE; TCI) at a concentration of 10% (w/v), stirred vigorously at room temperature (RT) overnight, and mixed at a ratio of 60:40 with 0.2% (v/v) of acetic acid (Sigma-Aldrich). The resulting solution was filtered-sterilized using a 0.45 μm filter and poured into a 5 mL plastic syringe.

2.2 *Preparation of the the chondrocyte suspension*

The immortalized human chondrocyte cell line, C28/I2 (kindly provided by Prof. Mary Goldring, Hospital for Special Surgery, New York and Harvard University), was maintained at 37 °C in a humidified atmosphere of 5% CO_2 in air, in Dulbecco's Modified Eagle Medium (DMEM)/Nutrient Mixture F-12 Ham (F-12) 1:1 v/v (DMEM – Gibco, Life Technologies; F-12, Sigma-Aldrich), supplemented with 10 % (v/v) non-heat inactivated fetal bovine serum (Gibco, Life Technologies), 1 % (v/v) P/S (Grisp) and 0.25 μg/mL Amphotericin B. The medium was refreshed twice a week. Chondrocytes were harvested at pre–confluence using trypsin/EDTA solution (0.05%/0.02%, Sigma-Aldrich) and 70,000 chondrocytes were mixed with 3 wt% alginate (Sigma-Aldrich) solution in culture medium, previously filter-sterilized.

2.3 *Fabrication and characterization of the chondrocyte-laden 3D scaffolds*

PCL+GEL solution, placed in the electromechanically 3D electrospinning platform (Figure 1a and b), was fed at a controlled flow rate of 2.5 mL/h to a blunt needle with a tip diameter of 0.4 mm (21 G), with an applied positive voltage of 15 kV. The polymer jet was ejected onto at the rotating collector at 1500 rpm, with an applied negative voltage of – 8 kV using a needle to collector distance of 13 cm. The fibres were then placed on the support according to a specific programmed alignment (Figure 1c). After each layer of PCL+GEL fibres, a drop of chondrocyte suspension was placed on top of the fibres. This process was repeated eight times. The resulting 3D scaffolds were then removed from the support, placed in 48-well plate and cultured at 37 °C in a humidified atmosphere of 5% CO_2 in air in culture medium for 7 days.

Chondrocyte viability was calculated based on the metabolic activity of the chondrocyte-laden scaffolds after 1 and 7 days of culture measured by a resazurin method. Resazurin solution (0.1 mg/mL in phosphate buffered saline (PBS; Sigma-Aldrich); ACROS Organics) was added to fresh medium at a final concentration of 10 % (v/v). Scaffolds were incubated in this solution at 37°C for 4 hours in the dark, after which 100 μL per well was transferred to a 96-well plate and absorbance (Abs) at 570 and 600nm was measured in a microplate reader

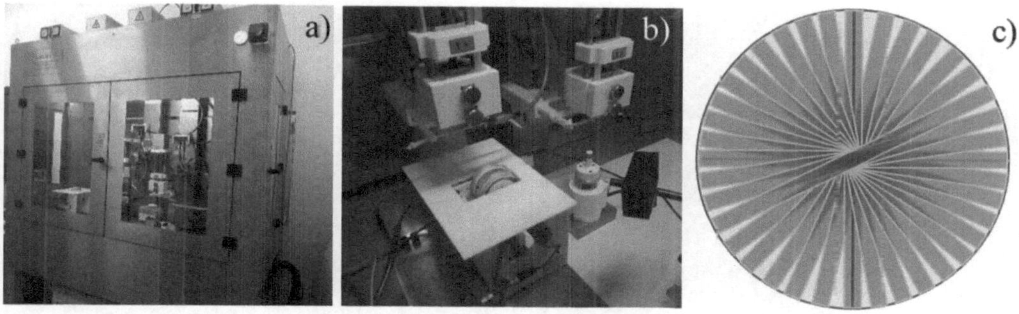

Figure 1. Developed electromechanically 3D electrospinning platform (a and b) and the programmed fibre alignment (b).

(Synergy HTX, BioTek). For each day, final absorbance values for each sample were calculated as the ratio Abs570/Abs600nm minus the Abs570/Abs600 nm ratio of a negative control (scaffold without cells). The absorbance values of cells incubated in the tissue culture polystyrene on the first time point were taken as 100% and the percentage of viable cells was calculated from these control values.

After the seventh day of culture, chondrocyte morphology and distribution within the scaffolds was assessed by SEM and haematoxylin and eosin staining. The 3D scaffolds were fixed in 4 % (w/v) paraformaldehyde in PBS, embedded in frozen embedding medium (Cryomatrix, Thermo scientific) and sectioned transversally in a cryostat (MEV, SLEE cryostat). The scaffolds' cross-sections were dehydrated with graded concentrations of ethanol aqueous solutions, treated with hexamethyldisilane (TCI), kept in a fume hood for air drying and visualized by SEM. Several scaffolds' sections were also stained with haematoxylin and eosin solutions to highlight in purple-blue the cells in the bilayered scaffolds.

3 RESULTS AND DISCUSSION

Chondrocyte-laden 3D scaffolds were successfully fabricated (Figure 2a), electromechanically 3D electrospinning platform.

Despite the fact that a substantial percentage of chondrocytes did not survive the layering process – only around 20 % of the seeded chondrocytes were viable after 1 day of culture –, these were able to proliferate over the 7-day culture period, as a 2-fold increase on the percentage of viable chondrocytes was observed. The low initial chondrocyte survival could be attributed to the detrimental effects of several factors: the toxic solvents used for the polymer electrospinning, chondrocyte dehydration on top of the fibres and the inadequate process environmental conditions (low CO_2 concentration and low temperature) (Canbolat 2011). However, over time in culture, these factors no longer were significant, favouring the increase of the percentage of viable chondrocytes. Chondrocytes were effectively incorporated within the fibrous matrix, which was confirmed on the SEM images (Figure 2b) and haematoxylin-eosin staining (Figure 2c) of the 3D scaffolds' sections of the scaffolds. Additionally, chondrocyte distribution was fairly homogeneous throughout the height of the 3D scaffold.

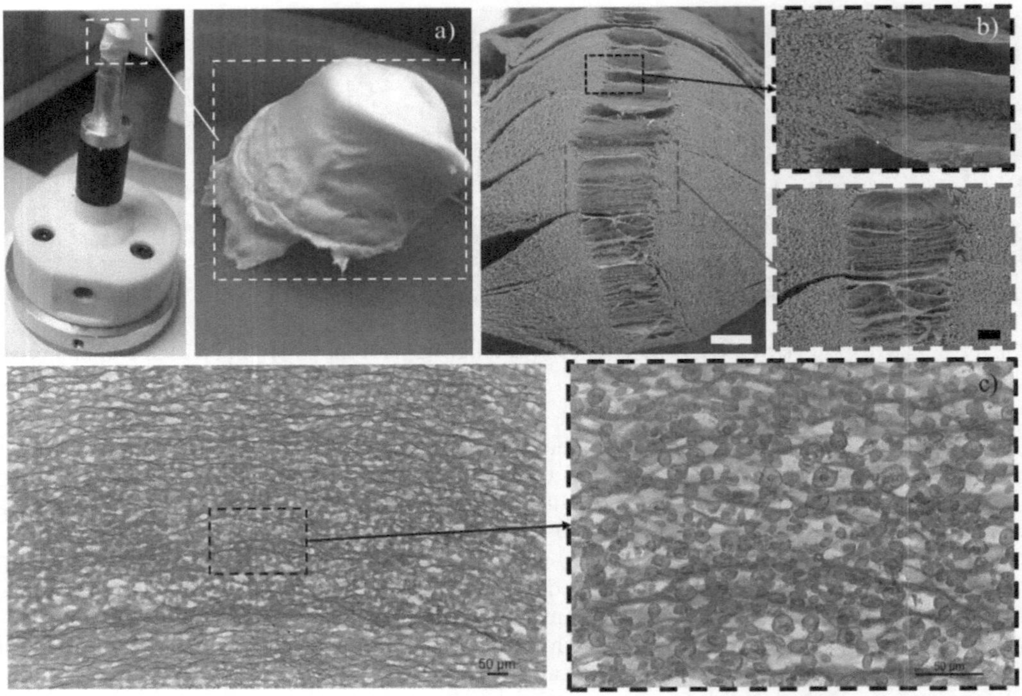

Figure 2. Chondrocyte-laden 3D scaffold (a); SEM images (b) and hematoxylin-eosin staining (c) of the 3D chondrocyte-laden scaffolds' sections after 7 days of culture. Scale bars: white: 500 μm, black: 200 μm and red: 100 μm.

4 CONCLUSIONS

Chondrocyte-laden 3D anisotropic electrospun scaffolds were successfully developed with a controlled fibre depth-orientation and chondrocyte distribution, showing the potential of the automated electrospinning system for cartilage TE.

ACKNOWLEDGEMENTS

This work was supported by the Portuguese funding of Program COMPETE-FEDER, Programa Operacional Competitividade e Internacionalização through the projects POCI-01-0145-FEDER-028424 and CENTRO-01-0145-FEDER-022083. Also, by Fundação para a Ciência e Tecnologia I.P. (FCT, IP) through the projects PTDC/EME-SIS/28424/2017, UIDB/00481/2020 and UIDP/00481/2020. The authors thank to FCT for the PhD grant SFRH/BD/133129/2017.

REFERENCES

Canbolat, et al. 2011. Mammalian cell viability in electrospun composite nanofiber structures. *Macromolecular Bioscience* 11: 1346–1356.

Completo, A., Marques, P.A.A.P. 2020. *Electrospinning system and process for large-scale manufacturing of aligned 3D fiber matrices*. European Patent Office (EPO)-EP3670714 (A1).

Correa, D., Lietman, S.A., 2017. Articular cartilage repair: Current needs, methods and research directions. *Seminars in Cell and Developmental Biology* 62: 67–77

Girão, A.F., et al. 2018. Mimicking nature: Fabrication of 3D anisotropic electrospun polycaprolactone scaffolds for cartilage tissue engineering applications. *Composites Part B: Engineering* 154: 99–107

Girão, A.F., et al. 2020. Microfabrication of a biomimetic arcade-like electrospun scaffold for cartilage tissue engineering applications. *Journal of Materials Science: Materials in Medicine* 31: 69
McCullen, S.D. et al. 2012. Anisotropic fibrous scaffolds for articular cartilage regeneration. *Tissue Engineering Part A* 18: 2073–2083.
Munir, N. et al. 2020. Integrational Technologies for the Development of Three-Dimensional Scaffolds as Platforms in Cartilage Tissue Engineering. *ACS Omega* 5: 12623–12636
Semitela, Â., et al. 2021. Boosting in vitro cartilage tissue engineering through the fabrication of polycaprolactone-gelatin 3D scaffolds with specific depth-dependent fiber alignments and mechanical stimulation. *Journal of the Mechanical Behaviour of Biomedical Materials* 117: 104373.
Steele, J.A.M., et al. 2014. Combinatorial scaffold morphologies for zonal articular cartilage engineering. *Acta Biomaterialia* 10: 2065–2075.

Advances and Current Trends in Biomechanics – Belinha et al. (Eds)
© 2022 Copyright the Author(s), ISBN: 978-1-032-10806-3

Correlation between different tests to diagnose flatfoot condition

J. Marouvo
RoboCorp Laboratory, i2A, Polytechnic Institute of Coimbra, Coimbra, Portugal
Faculty of Sport (FADEUP), University of Porto, Porto, Portugal

M.A. Castro
Department of Physiotherapy, Coimbra Health School (IPC), Coimbra, Portugal
Center for Mechanical Engineering, Materials and Processes, University of Coimbra, Coimbra, Portugal

F. Sousa
Porto Biomechanics Laboratory (LABIOMEP-UP), University of Porto, Porto, Portugal
Faculty of Sport (FADEUP), CIFI2D, University of Porto, Porto, Portugal

ABSTRACT: The Navicular Drop test, FootPrint Arch Angle, and the Resting Calcaneal Stance Position, are different ways to assess foot posture alteration, namely flatfoot condition that can lead to both kinematic and kinetic adaptation along the whole body. This study aims to analyze the relationship between three different methods to diagnose a flatfoot condition. Sixteen participants were included in this observational descriptive study. The Navicular Drop test, FootPrint Arch Angle, and the Resting Calcaneal Stance Position were applied in each subjects. Data did not show statistically significant correlation between the different tests which can impair the flatfoot condition assessment. The diagnose accuracy of flatfoot condition can be compromised depending on the tests and different used combination.

1 INTRODUCTION

Foot problems are related to impaired lower-limb mobility and postural stability, having a detrimental impact on the quality of life, and are reported as a common concern in the community (Sung 2018). Foot alignment, as the most distal part of the lower-limb kinematic chain as well as providing support to maintain postural stability, has an important role in upright standing, walking, and running (Farokhmanesh *et al.* 2014, Tahmasebi *et al.* 2015). Also, foot alignment alterations may affect the spine biomechanics that consequently can lead to spine instability, muscle imbalance, and structural adjustments (Farokhmanesh *et al.* 2014). Thus, foot posture, through performed lower-limb motion pattern can induce several injuries and has been associated with abnormal foot motion during gait pattern (Buldt *et al.* 2013). A flat foot condition can have several causes, namely neurological or muscular restrictions, ligament or joint laxity, and range of motion and muscle activity impairments (Hunt *et al.* 2004). Regarding this type of foot posture condition, subjects presented a higher risk of developing mechanical overloading injuries compared to others' foot posture conditions (Hunt *et al.* 2004, Buldt *et al.* 2013). Besides, among foot posture, namely the flatfoot condition, induces a greater foot sole contact area, altered plantar pressure pattern, and an increased proximal lower-limbs joints motion. The foot sole is considered a sensitive map through its numerous cutaneous mechanoreceptors. Those provide imperative afferent information for balance, postural stability, gait pattern and, muscular activity response (Kim *et al.* 2015). Besides, increase

DOI: 10.1201/9781003217152-43

afferent input from the foot sole can affects negatively postural awareness, increasing instability, and fall risks. This alteration can be present in flatfoot subjects which can affect shock absorption, muscular activity, stability, and therefore gait pattern (Zuil-Escobar *et al.* 2018, 2019). In the literature, authors often refer to different methods to assess subjects' foot posture (Zuil-Escobar *et al.* 2018, 2019, Cho *et al.* 2019). Those can be divides into visual observation, radiographs, FootPrints, and clinical measurements (Khanna *et al.* 2019). The FootPrint Arch Angle (AA), the Navicular Drop Test (NDT), and the Resting Calcaneal Stance Position (RCSP) are three different tests that can be used to assess the flatfoot condition. The first requires the use of ink or digital system recorder, and the last two are simple and can be used quickly in a clinical environment using few instruments (Cho *et al.* 2019), respectively a ruler and a goniometer, and all have no side effects on testes subjects (Zuil-Escobar *et al.* 2018, 2019). Although, all the assessment methods present several limitations depending on the examiner or the chosen test (Zuil-Escobar *et al.* 2018, 2019, Cho *et al.* 2019, Khanna *et al.*, 2019). However, as all methods are considered non-invasive, the ink method presents several biases and it's practitioner-dependent, while digital systems are expensive though being user-friendly (Zuil-Escobar *et al.* 2018, Khanna *et al.*, 2019). Few papers investigated the correlation between FootPrints and the NDT but the correlations between the NDT, AA, and the RCSP were not previously estimated. Then, the purpose of this study was to investigate the relationship between the NDT, the AA, and the RCSP tests in subjects with flatfoot condition.

2 MATERIALS AND METHODS

This observation descriptive study was carried out at the RoboCorp Laboratory – Physiotherapy, at the Coimbra Health School, at the Polytechnic Institute of Coimbra, after the Ethics Committee of Polytechnic Institute of Coimbra (13_CEPC2/2019) approval. The sample size was calculated using the *G*power 3.1.5* software (Franz Fraul, Kiel, Germany) based on the study previously published by Zuil-Escobar & al (2018). A required sample size of 13 was determined by achieving an estimated alpha level of 0.05, and a power of 0.95. Consequently, 16 bilateral flatfoot participants were recruited (24,75 years ± 5,05 SD). Before any assessment, all subjects were informed about the study purpose, procedures, and all completed a consent form. Only the subjects who presented a bilateral NDT value of >9mm were included in this study which characterized those as flatfooted (Kim *et al.* 2015). Thus, subjects who presented the following conditions were excluded: a) ankle sprain in the last 6 months. b) physiotherapy treatment program or history of injury including bilateral ankle injury; c) bone fracture associated with an ankle sprain, such as avulsion fracture or osteochondral ankle injury; d) injury or surgery to the spine, hip, knee, or ankle. Furthermore, the procedures were realized using a plastic ruler to measure the NDT, a rigid plastic goniometer for the RCSP (Enraf-Nonius B. V, Rotterdam, The Netherlands), and a plantar pressure platform with 100Hz frequency (PhysioSensing, Sensing Future Technologies, Coimbra, Portugal) to measure the AA. Flatfoot condition is considered respectively when the AA value is >42° (Tahmasebi *et al.* 2015) and the RSCP value is >4° (Kim *et al.* 2015). All assessments were realized by a single physiotherapist with more than 6 years of experience and the subjects were asked to stand in a bilateral weight-bearing barefoot position. Therefore, for the NDT assessment, the physiotherapist placed the rigid plastic-made ruler perpendicularly to the ground. Then, the ground-navicular bone distance (millimeters) was registered both in a relaxed and induced neutral position where the assessment position difference quantifies the flatfoot severity (Sung 2018). Consequently, the RCSP was used to assess the calcaneal angle formed by the longitudinal bisecting lines of the calcaneus and the distal third of the leg, which were drawn previously in a prone position. For both the NDT and the RCSP assessment, the mean of three measurements for each defines the final score value which was calculated using the *IBM SPSS Statistics 25.0* software (IBM Corporation, New York, USA). Thus, for the AA assessment, subjects stood upon the force platform and were asked to maintain the look horizontally at a reference point for 5 seconds to stabilize the position before recording the data. If any participants failed to maintain their position, the trial was repeated without notifying the participant. For the AA, the data was initially converted to an image format to be further processed with the help of the *Image J* software (National Institute for Health, Bethesda, USA). Finally, the overall data were statistically processed and analyzed with the *IBM SPSS Statistics 25.0* software (IBM Corporation, New York, USA). The descriptive statistics, mean and standard deviation, were calculated for all variables regarding all subjects. Associations between the NDT, the AA, and the RCSP were determined by Pearson's correlation analysis (*95% ICC*).

3 RESULTS

Concerning the mean value and the standard deviation of each test, the NDT presented a value of 11.23 mm (± *1.45*), the RCSP a value of 4.72° (± *1.56*), and finally the AA a value of 61.13° (± *12.21*). Regarding all correlation values, the NDT and the RCSP presented a 0.397 score which corresponds to a positive moderate coefficient. The NDT and the AA correlation based on Pearson's analysis presented a 0.184 score, corresponding to a positive low coefficient, and the RCSP and AA presented a -0.118 score corresponding to a negative low coefficient. However, none of the correlations presented a statistically significant value (*p>.05*). All the Pearson's correlation values are presented in Table 1.

Table 1. Pearson's correlation between the different methods.

		Navicular Drop Test	Arch Angle
Resting Calcaneal Stance Position	*Correlation value*	0.397	-0.118
	p-value	,122	,378
Navicular Drop Test	*Correlation value*		0.184
	p-value		,297

4 DISCUSSION

This study is, to our knowledge, the first to look at these previously mentioned combination test correlations to diagnose flatfoot condition. Previous works regarded just a few tests to assess accuracy, reliability, and correlations, or with the inclusion criteria to restrict only the flatfoot subjects or even without foot posture assessment as inclusion criteria. In flatfoot subjects classified by the NDT score value, we did not find any statistically significant correlations between all methods assessment, namely the NDT, the RCSP, and the AA (*p>.05*). Also, regarding the mean value of each test, all presented values higher than the cut-off score previously established by several authors (Zuil-Escobar *et al.* 2018, 2019). In our scientific research, no previously published papers relating to the relationship between those test combinations were found. Our results concerning the NDT correlations with the FootPrint parameters, namely the AA are controversial compared to two previously published scientific articles. Zuil-Escobar & al (2019) in subjects with flatfoot condition, found a statistically significant correlation value between the NDT and the AA (*p<.01*) which value corresponds to a negative very high score (r = -0.732). Although, in another study, Zuil-Escobar & al (2018) found in volunteers without foot posture differentiation, a statistically significant correlation (*p<.05*) between the NDT and the AA. This result presented a negative moderate score (r = -0.643). Contrarily, no scientific paper was found in our literature research concerning the correlation between the RCSP and the NDT and nor with the AA. Those controversial findings compared to ours can be due to functional alterations present in flatfoot subjects. In our research, the authors stated a greater foot complex mobility in subjects with a flatfoot condition. also, those present a larger range of motion variability (Sung 2018). Those alterations may influence the appearance of kinematics and kinetic impairments as well as an increase postural sway which represents a decreased postural stability (Kim *et al.* 2015, Sung 2018). Contrarily to the FootPrint parameters, the RCSP and the NDT were developed to ensure a novel method to assess foot posture but present several limitations. Those are tests that investigated respectively the hindfoot and the midfoot mobility. Several FootPrint parameters ensure the analysis of the plantar foot area, leading to a more precise diagnose of the different foot posture conditions, as long as investigating the sole load distribution, which varies among subjects. However, the AA, contrarily to the most used FootPrint parameters, is a method that analyzes only the angle form by the contact area of the fore relative to midfoot. Most of the other FootPrint parameters analyze mostly the foot sole contact area The Arch Index is the only method that depends on the toeless foot contact area while the others evaluate different fore-, mid- or hindfoot parameters (Zuil-Escobar *et al.* 2018). Papers that related correlations between few various tests present moderate to low values between tests but strong values for the inter and intra-reliability making them useful and easy to apply in a clinical environment (Zuil-Escobar *et al.* 2018, Cho *et al.* 2019). According to our results, the combination of those tests can mislead to an incorrect foot posture assessment which increases these controversial assessment methods (Sung 2018). Despite this, as this study shares various foot posture tests assessment and correlations among them, several limitations can compromise the outcomes and results. In this study, the NDT assessment served as inclusion criteria. Therefore, only subjects who presented a bilateral flatfoot condition were included in this study, regarding the NDT score. However, like the foot complex lays on several joints and inter-associations, it will be interesting to evaluate the same correlation regarding FootPrint parameters as inclusion criteria instead of the NDT, whose assessment of foot posture is based on mobility. Furthermore, this method is practitioner-dependent and can be biased by a lack of experience. Another study limitation is the non-characterization of subjects' weight, since scientific literature relates increased weight as a risk factor to develop higher foot arch values, and therefore and increase foot sole contact area.

5 CONCLUSION

This study related the correlations score acquired between several methods, namely the AA, the RCSP, and the NDT. Regarding foot posture, the correlations between those tests did not present statistically significant results to identified flatfoot subjects. According to these results, the diagnosis accuracy of the flatfoot condition can be compromised depending on either the chosen test or the NDT combined with the AA or the RCSP. The NDT and the RCSP showed to mislead foot posture condition assessment, unlike Footprints parameters that can be important evaluation tools in a clinical environment.

ACKNOWLEDGMENTS

RoboCorp Laboratory co-funded by QREN under the Programa Mais Centro, of the Coordination Commission of the Central Region and the European Union through the European Regional Development Fund.

Body Keeper Carlos Morgado Santos – Physiotherapy, Osteopathy and Posturology Clinic, Nutrition and Psychology, Coimbra, Portugal

REFERENCES

Buldt, A.K., Murley, G.S., Butterworth, P., Levinger, P., Menz, H.B., and Landorf, K.B., 2013. The relationship between foot posture and lower limb kinematics during walking: A systematic review. *Gait and Posture*, 38 (3), 363–372.

Cho, Y., Park, J.-W., and Nam, K., 2019. The relationship between foot posture index and resting calcaneal stance position in elementary school students. *Gait & Posture*, 74 (August), 142–147.

Farokhmanesh, K., Shirzadian, T., Mahboubi, M., and Shahri, M.N., 2014. Effect of Foot Hyperpronation on Lumbar Lordosis and Thoracic Kyphosis in Standing Position Using 3-Dimensional Ultrasound-Based Motion Analysis System. *Global Journal of Health Science*, 6 (5), 254–260.

Hunt, A.E. and Smith, R.M., 2004. Mechanics and control of the flat versus normal foot during the stance phase of walking. *Clinical Biomechanics*, 19 (4), 391–397.

Khanna, N. and Premavathy, D., 2019. Assessing occurrence of flatfoot condition among student population. *Drug Invention Today*, 11 (10), 2477–2480.

Kim, J. ah, Lim, O. bin, and Yi, C. hwi, 2015. Difference in static and dynamic stability between flexible flatfeet and neutral feet. *Gait and Posture*, 41 (2), 546–550.

Sung, P.S., 2018. The Sensitivity of Thresholds by Ground Reaction Force and Postural Stability in Subjects With and Without Navicular Drop. *The Journal of Foot and Ankle Surgery*, 57 (4), 742–746.

Tahmasebi, R., Karimi, M.T., Satvati, B., and Fatoye, F., 2015. Evaluation of Standing Stability in Individuals With Flatfeet. *Foot & Ankle Specialist*, 8 (3), 168–174.

Zuil-Escobar, J.C., Martínez-Cepa, C.B., Martín-Urrialde, J.A., and Gómez-Conesa, A., 2018. Medial Longitudinal Arch: Accuracy, Reliability, and Correlation Between Navicular Drop Test and Footprint Parameters. *Journal of Manipulative and Physiological Therapeutics*, 41 (8), 672–679.

Zuil-Escobar, J.C., Martínez-Cepa, C.B., Martín-Urrialde, J.A., and Gómez-Conesa, A., 2019. Evaluating the Medial Longitudinal Arch of the Foot: Correlations, Reliability, and Accuracy in People With a Low Arch. *Physical Therapy*, 99 (3), 364–372.

Advances and Current Trends in Biomechanics – Belinha et al. (Eds)
© 2022 Copyright the Author(s), ISBN: 978-1-032-10806-3

Relationship between various tests to diagnose bilateral neutral foot subjects

J. Marouvo
RoboCorp Laboratory, i2A, Polytechnic Institute of Coimbra, Coimbra, Portugal
Faculty of Sport (FADEUP), University of Porto, Porto, Portugal

M.A. Castro
Department of Physiotherapy, Coimbra Health School (IPC), Coimbra, Portugal
Center for Mechanical Engineering, Materials and Processes, University of Coimbra, Coimbra, Portugal

F. Sousa
Porto Biomechanics Laboratory (LABIOMEP-UP), University of Porto, Porto, Portugal
Faculty of Sport (FADEUP), CIFI2D, University of Porto, Porto, Portugal

ABSTRACT: To assess foot posture alteration, various tests and indexes can be used to assess the Medial Longitudinal Arch heigh, to further prevent lower limbs injuries, like the Navicular Drop Test, FootPrint Arch Angle, and the Resting Calcaneal Stance Position. The aim of this study is to assess the relationship between three 3 different tests to categorize neutral-foot subjects. Twenty participants (14 males and 6 females) were included in this study and exposed to the Navicular Drop Test, FootPrint Arch Angle, and the Resting Calcaneal Stance Position. Data showed a moderate significant correlation between tests that can decrease the accuracy to identify neutral-foot subjects.

1 INTRODUCTION

Foot posture is generally characterized by the alignment of the foot skeleton and may vary considerably between individuals (Buldt *et al.* 2018, Sung 2018). Also, it can induce altered plantar pressure patterns and proximal joints motion. Too, neuromotor responses to the altered sensory afferents signals affect muscle function, foot and lower-limb biomechanics, as the Central Nervous System (CNS) uses muscle coactivation as a motor control mechanism to modulate joint stiffness and therefore postural stability (Feldman 2016, Peterka 2018). These can occur globally and locally through postural and functional joint stabilization. Because stability requires cognitive resources to process and integrate somatosensory input, any additional process can reduce stability maintenance and thereby increase instability. This information is processed in the CNS to create neuromotor necessary output commands to maintain postural stability. Besides, proper mobility of the ankle joint has an important role in functional activities and to realize several postural corrections (Feldman 2016, Peterka 2018). Besides, foot sole afferent input affects postural awareness and regarding different foot posture conditions, it can be triggered by neurological, muscular, ligamentous, joint, motion and muscle activity impairment (Buldt *et al.* 2018). In neutral foot subjects, those patterns show higher pressure, force, and contact area values in the medial arch, central forefoot and hallux, while these variables are lower in the lateral and medial forefoot (Buldt *et al.* 2018). Static and dynamic postural controls are required during daily living activities. Since lower-

DOI: 10.1201/9781003217152-44

limbs postural alterations lead to postural variations in the pelvic girdle and increase low back pain risk, foot alignment and therefore foot posture should be pondered as an important and effective factor (Farokhmanesh *et al.* 2014). Thus, foot posture, through altered lower-limb motion patterns can induce injuries and have been associated with abnormal foot motion in posture and gait pattern (Buldt *et al.* 2018). Regarding the existing scientific literature, several tests and indexes can be used to ensure the foot posture type of each individual (Zuil-Escobar *et al.* 2018, Cho *et al.* 2019) like the Medial Longitudinal and Footprint Arch angle (AA), Navicular Drop Test (NDT), Resting Calcaneal Stance Position (RCSP). Some FootPrint parameters analyze the entire toe-less plantar foot area which are considered as precise methods to diagnose foot posture, as those methods explore the plantar load distribution. Though, unlike the others FootPrint parameters, the AA is a method that analyzes the angle form by the contact area of the fore relative to midfoot. According to Zuil-Escobar & al (2018), the NDT can be useful for clinical settings assessment, but this method is practitioner-dependent and can be biased by a lack of experience of the practitioner. Contrarily to the FootPrint parameters, the NDT was developed as a novel method to assess foot posture which investigated the midfoot mobility. Thought, associations between the NDT, the AA, and the RCSP were not previously estimated in neutral foot subjects. This study aimed to investigate the association between the NDT, the AA, and the RCSP tests in subjects with a bilateral neutral foot condition.

2 MATERIALS AND METHODS

An observation descriptive study was realized after the approval of the Ethics Committee of Polytechnic Institute of Coimbra (13_CEPC2/2019), at the RoboCorp Laboratory – Physiotherapy, at the Coimbra Health School, at the Polytechnic Institute of Coimbra. Before any procedure, the *G*power 3.1.5* software (Franz Fraul, Kiel, Germany) was used to calculate the sample size. A size of 13 was required and determined by achieving an estimated power of 0.95 and an alpha level of 0.05, based on the study previously published by Zuil-Escobar & al (2018). According to this criteria, twenty bilateral neutral foot participants were included in this study (21,90 years ± 3,31 SD) comprising a total of forty feet. Initially, information about the study purpose and procedures was given to all the participants as well as a required consent form. Participants who presented an NDT value of >9mm were excluded to restrain the inclusion of only neutral foot subjects (Kim *et al.* 2015). Also, the following several exclusion criteria were applied: a) ankle sprain in the last 6 months; b) physiotherapy treatment program or history of injury including bilateral ankle injury; c) bone fracture associated with an ankle sprain, such as avulsion fracture or osteochondral ankle injury; d) injury or surgery to the spine, hip, knee, or ankle. Moreover, for the assessment, a plantar pressure platform with 100Hz frequency (PhysioSensing, Sensing Future Technologies, Coimbra, Portugal) was utilized to measure the AA, as well as a plastic-made ruler to measure the NDT and finally a rigid plastic-made goniometer for the RCSP (Enraf-Nonius B.V, Rotterdam, The Netherlands). Regarding those methods, a flatfoot is considered when the AA value is <42° (Tahmasebi *et al.* 2015), the RSCP value is <4° (Kim *et al.* 2015) and finally, the NDT value is <9mm (Kim *et al.* 2015). For the procedures, a single physiotherapist with more than 6 years of experience realized all the procedures, with the participants in bilateral weight-bearing barefoot stance position. Subjects stood upon a force platform to assess the AA, and all participants were asked to maintain the look horizontally at a reference point for 5 seconds to stabilize the standing position to reach a minimal body sway before the data recording. If any participants failed to maintain the position, the trial was repeated with given any information to the participant. Consequently, for the NDT assessment, the physiotherapist placed the rigid plastic-made ruler perpendicularly to the ground. Then, the ground-navicular bone distance (millimeters) was registered both in a relaxed and induced foot neutral position by the physiotherapist. The difference between both positions quantifies the navicular drop severity (Sung 2018). Finally, the physiotherapist used the goniometer to assess the calcaneal angle formed by the longitudinal bisecting lines of the calcaneus and the distal third of the leg. All lines were drawn previously in a prone position by the same physiotherapist. The AA data was initially converted to an image format and processed with the *Image J* software (National Institute for Health, Bethesda, USA) to extract the AA values. For the two other tests, the mean of three measurements for each defines the final score value which was calculated using the *IBM SPSS Statistics 25.0* software (IBM Corporation, New York, USA). Then, the overall data were statistically processed and analyzed with the *IBM SPSS Statistics 25.0* software (IBM Corporation, New York, USA). For all variables, the descriptive statistics, mean, and standard deviation were calculated. Correlations between the NDT, the AA, and the RCSP were determined according to Pearson's correlation analysis (*95% ICC*).

3 RESULTS

Regarding the sample characteristics, the NDT presented a mean value of 5.44 mm (± *2.27*), the RCSP a value of 2.12° (± *1.77*), and finally the AA a value of 42.67° (± *4.04*). According to the results of the correlation, NDT and RCSP presented a 0.326 score which corresponds to a positive moderate coefficient with statistical significance ($p = .040$). The NDT and AA correlation presented a -0.179 score corresponding to a negative low coefficient and RSCP and AA presented a -0.328 corresponding to a negative moderate coefficient. Although, those last ones didn't present statistically significant results ($p > .05$). All the Pearson's correlation values are presented in Table 1.

Table 1. Pearson's correlation between the different methods.

		Navicular Drop Test	Arch Angle
Resting Calcaneal Stance Position	*Correlation value*	0.326	-0.328
	p-value	,040	,067
Navicular Drop Test	*Correlation value*		-0.179
	p-value		,328

4 DISCUSSION

Among our scientific research, this is the first study to analyze the association between the RCSP and the two other tests, namely the NDT and AA in neutral foot subjects. Previous papers related the correlation between the NDT and several FootPrint parameters, like the Arch Index, FootPrint Index, Chippaux-Smirak, and Staheli indexes. Along with correlation, in these papers, the authors investigated the accuracy and reliability of those methods with different inclusion criteria among several foot posture conditions. In neutral foot subjects, we only found a statistically significant correlation ($p = .040$) between the NDT and the RCSP, with a coefficient value of 0,326. The other correlation did not present statistically significant results ($p > .05$). In our research, we did not find any paper which related the correlation between the RCSP and the other tests. Our result regarding the NDT correlations with the FootPrint parameters, namely the AA, is controversial compared to a paper realized by Zuil-Escobar & al (2018). In this study, the authors investigated the correlation between different methods to assess foot posture conditions. They found a statistically significant correlation ($p < .05$) between the NDT and the AA with a negative moderate coefficient (r = -0.643). The NDT refers to midfoot mobility to assess foot posture while the RCSP to the hindfoot, while the FootPrint parameters suggest a whole assessment of foot posture by sole load distribution. Also, the NDT and RCSP are two tests that did not assess the foot area contact to the ground, instead of FootPrint parameters. Though, since the AA is characterized by the angle between the medial line and the most medial aspect of the midfoot area, the load distribution influence negatively this variable leading to a misunderstanding of the score and thereby to foot posture assessment (Tahmasebi *et al.* 2015, Zuil-Escobar *et al.* 2018). Also, in neutral foot subjects, research seems to highlight controversial results regarding foot sole distribution. Several authors may suggest an even weight distribution among the fore- and hindfoot, while some refer a higher pressure on the heel than on the forefoot. In the literature, a consensus stated a greater load in the forefoot, the second and third metatarsal heads. According to these, the correct foot posture assessment can be misled due to high variability among neutral foot subjects regarding both plantar pressure distribution and area, and the ankle and foot complex mobility. Thereby, in the results score of the mobility tests, the same individual may present a different plantar pressure pattern that can impair the correlation coefficient and its interpretation (Fernández-Seguín *et al.* 2014, Sung 2018). Finally, analyzing the means value and standard deviation for each used test, some incoherence was found regarding the cut-off values of each foot posture assessment test. Some scores didn't reach the cut-off values to diagnose foot posture conditions. Regarding each test mean value, in our study, the NDT and RCSP presented values within the range previously established to identify neutral foot conditions (Zuil-Escobar *et al.* 2018). Likewise, for the AA, values greater than *42°* correspond to flatfoot condition (Tahmasebi *et al.* 2015, Zuil-Escobar *et al.* 2018). However, the AA showed a higher value than the cut-off score (*mean = 42.67± 4.04*) misleading the NF condition assessment. Finally, several limitations can negatively influence the outcomes and results of the correlation among the different foot posture methods. The NDT was used in this study as inclusion criteria, which is considered a user-friendly method with a low cost and limitation. However, as neutral foot subjects may exhibit several plantar pressure patterns, further research needs to be accomplished with a FootPrint parameters index as inclusion criteria to allow a higher foot posture accuracy and correlation among tests. Since literature ensures several plantar pressure distribution patterns among participants' weight, this parameter needs to be included in further studies to characterize the accuracy of those methods.

5 CONCLUSION

In this study, the correlations were analyzed for several foot posture assessment methods, regarding respectively the AA, the RCSP, and the NDT. Also, in neutral foot subjects, only the correlation between the NDT and the RCSP showed interesting results, which is in accordance with previously published papers. However, according to these results, the diagnosis accuracy of the flatfoot condition can be compromised depending on either the chosen test or the NDT combined with the AA or the RCSP, as the correlation presented a moderate coefficient. Finally, the RCSP and the AA showed to mislead foot posture condition assessment, unlike other Footprints parameters, considered as important evaluation tools in a clinical environment.

ACKNOWLEDGMENTS

RoboCorp Laboratory co-funded by QREN under the Programa Mais Centro, of the Coordination Commission of the Central Region and the European Union through the European Regional Development Fund.

Body Keeper Carlos Morgado Santos – Physiotherapy, Osteopathy and Posturology Clinic, Nutrition and Psychology, Coimbra, Portugal

REFERENCES

Buldt, A.K., Forghany, S., Landorf, K.B., Levinger, P., Murley, G.S., and Menz, H.B., 2018. Foot posture is associated with plantar pressure during gait: A comparison of normal, planus and cavus feet. *Gait and Posture*, 62 (March), 235–240.

Cho, Y., Park, J.-W., and Nam, K., 2019. The relationship between foot posture index and resting calcaneal stance position in elementary school students. *Gait & Posture*, 74 (August), 142–147.

Farokhmanesh, K., Shirzadian, T., Mahboubi, M., and Shahri, M.N., 2014. Effect of Foot Hyperpronation on Lumbar Lordosis and Thoracic Kyphosis in Standing Position Using 3-Dimensional Ultrasound-Based Motion Analysis System. *Global Journal of Health Science*, 6 (5), 254–260.

Feldman, A.G., 2016. The Relationship Between Postural and Movement Stability. In: J. Laczko and M. L. Latash, eds. Cham: Springer International Publishing, 105–120.

Fernández-Seguín, L.M., Diaz Mancha, J.A., Sánchez Rodríguez, R., Escamilla Martínez, E., Gómez Martín, B., and Ramos Ortega, J., 2014. Comparison of plantar pressures and contact area between normal and cavus foot. *Gait and Posture*, 39 (2), 789–792.

Kim, J. ah, Lim, O. bin, and Yi, C. hwi, 2015. Difference in static and dynamic stability between flexible flatfeet and neutral feet. *Gait and Posture*, 41 (2), 546–550.

Peterka, R.J., 2018. Sensory integration for human balance control. *In*: *Handbook of Clinical Neurology*. Elsevier B.V., 27–42.

Sung, P.S., 2018. The Sensitivity of Thresholds by Ground Reaction Force and Postural Stability in Subjects With and Without Navicular Drop. *The Journal of Foot and Ankle Surgery*, 57 (4), 742–746.

Tahmasebi, R., Karimi, M.T., Satvati, B., and Fatoye, F., 2015. Evaluation of Standing Stability in Individuals With Flatfeet. *Foot & Ankle Specialist*, 8 (3), 168–174.

Zuil-Escobar, J.C., Martínez-Cepa, C.B., Martín-Urrialde, J.A., and Gómez-Conesa, A., 2018. Medial Longitudinal Arch: Accuracy, Reliability, and Correlation Between Navicular Drop Test and Footprint Parameters. *Journal of Manipulative and Physiological Therapeutics*, 41 (8), 672–679.

Advances and Current Trends in Biomechanics – Belinha et al. (Eds)

Postural alert system for sitting position

A. Alves, R. Almeida, D. Moreira, S. Silva, A. Marques & P. Guimarães
ISEP-P. Porto, School of Engineering, Polytechnic of Porto, Porto, Portugal

ABSTRACT: We spent most of our days sitting, and, usually, in a not so correct posture, which causes some damage to our health. Detection and classification of the posture is one of the main applications in the medical field because it allows to detect whether the position taken is correct or incorrect, and, if it is incorrect it has the goal to reduce the negative consequences for the user's health by correcting the position. This work presents the development of a chair that aims to alert the user to the type of posture that is being taken and to store this information for future analysis.

1 INTRODUCTION

The rapid technological development in industrialized countries, automation and computerization of jobs, which occurred from the second half of the twentieth century, were the main factors assumed by the increasingly frequent adoption of the sitting position at jobs (Marques et al., 2010). Nowadays, we can see that more and more people spend a large part of their day sitting on an office chair, car seat or even on a couch (Mengjie et al., 2017).

Office workers, for example, spend an average of 75% of their day sitting, and more than half in long periods of a sedentary lifestyle. In general, most people spend 8.07 hours a day in the position in question when they are not working (Muppavram et al., 2018). In this sense, it is easy to understand that this reality has consequences for personal health since bad postures can trigger generalized chronic diseases/spinal disorders, causing back pain and movement dysfunction (Flutur et al., 2019, Shiri et al., 2019).

In order to minimize the health problems arising from this issue, it's important that the development and implementation of ergonomic products that enhance good posture occur. In this application context, an ergonomic chair will have to be adjustable taking into accont that the target audience will be quite diverse and with great variability (Pulzatto et al., 2017).

Thus, and according to Fialho et al. (Fialho et al., 2007), the main factors of ergonomic evaluation are the dimensions (length, width and height), inclination between the seat and the backrest, presence of free space between seat-backrest, type of front edge of the seat, stability and presence of straight edges. However, it is important to alert the user who is in the wrong position, so systems that monitor some postural parameters and consequently create discomfort in order to enhance the change in posture are increasingly adopted (Martins et al., 2013)

In this work, a solution was developed, whose objective is to alert the user of their incorrect posture, through constant monitoring that is recorded and can be subsequently evaluated by a clinician. There is only a limited amount of positions that the prototype can identify. Those positions were proposed by the authors of this article with no scientifical background to support them since that was not the goal of the project at the time.

DOI: 10.1201/9781003217152-45

2 MATERIALS AND METHODS

As it was said previously, there is a general need in modern society to correct the sitting posture to prevent several diseases. To do so, we developed a children-size chair with several sensors incorporated in it to monitor, detect and, if needed, correct the user's posture.

Using *SolidWorks* software, the design of the prototype of the chair and its sensors was elaborated.

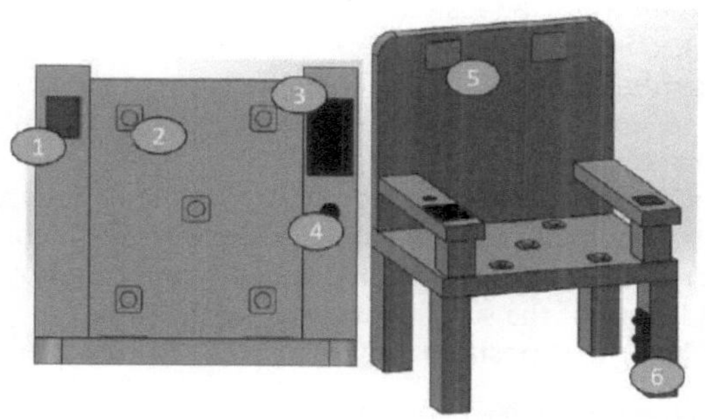

Figure 1. Prototype of the chair. Upper view (left) and front view (right). (1) RFID; (2) Pressure sensor. (3) LCD; (4) Buzzer; (5) FSR; (6) Photoelectric sensor.

Observing the previous figure, it is possible to verify that the chair consists of 6 different types of modules. Identified with number 1 it is a module that intends to identify the user who is using the chair, also called RFID.

Number 2 points to one of the 5 pressure sensors placed on the seat of the chair that has the goal to identify the position taken on the seat.

Regarding number 3, a module is visible that aims to show certain messages, for example, to say what needs to be corrected in the person's posture when it is incorrect. In addition, when a position is not known, it is shown only as "Incorrect" and, when a correct posture is taken, this information is shown on the module in question, which is also called LCD.

Identified by the number 4 it is a sound module that is used to warn when an incorrect posture is taken. This module is also called a buzzer.

With the module identified with number 5, it is possible to check the position of the back of the person using the chair. This module is also called FSR.

Finally, identified by number 6 there is a module to check whether the position of the legs is correct or incorrect, when the person is sitting. It can also be called a photoelectric module.

In addition to the modules visible in previous figure, a module that allows obtaining the date and time, called RTC, was also used. By applying materials given, the design of the prototype came true as shown in the Figure 2.

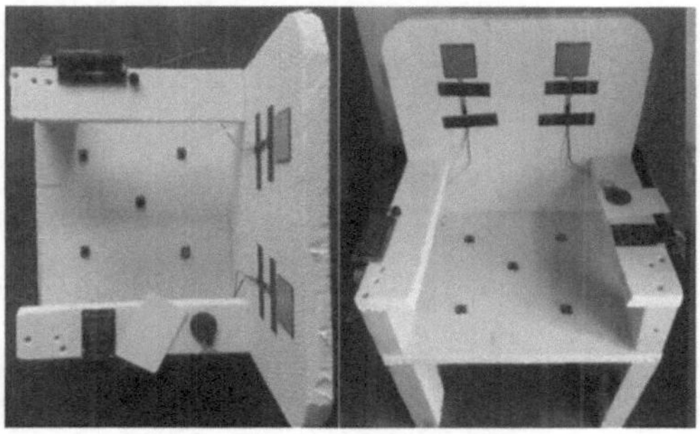

Figure 2. Miniature prototype.

It should be emphasized that the prototype was only built in children-size, and therefore no real test to the chair could be realized, that is noone was able to sit on the prototype and test the response of the chair to different positions. However, the prototype's sensors and code was evaluated.

3 RESULTS

Although the prototype could not be tested, a small experiment was still conducted to see if this prototype would be useful.

Thinking about the combination of the different sensors on the prototype and what positions it would be able to detect with a real-size chair, thirteen positions, one correct and twelve incorrect, were proposed.

As an example, two positions (positions 5 and 9) are visible in Figure 3. Both positions are incorrect; in position 9 we assumed that the angle of the knees is incorrect, and in position 5 the angle is, again, incorrect, as well as the back not being supported by the chair plus the user is not occupying the full length of the sit of the chair.

Besides those thirteen positions, position 14 was also added as a non identified position.

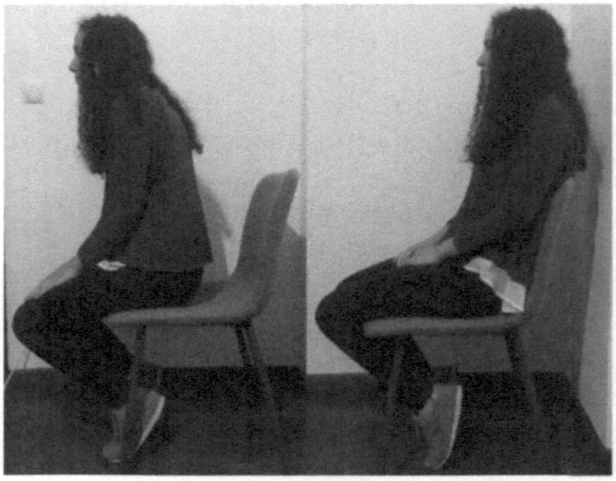

Figure 3. Example of positions. Position 5 (left) and position 9 (right).

Throughout an afternoon, the postural tendencies of a few subjects was analysed and, making use of the positions previously covered, the pie chart below was designed, which regards the positions adopted and the amount of time in each on.

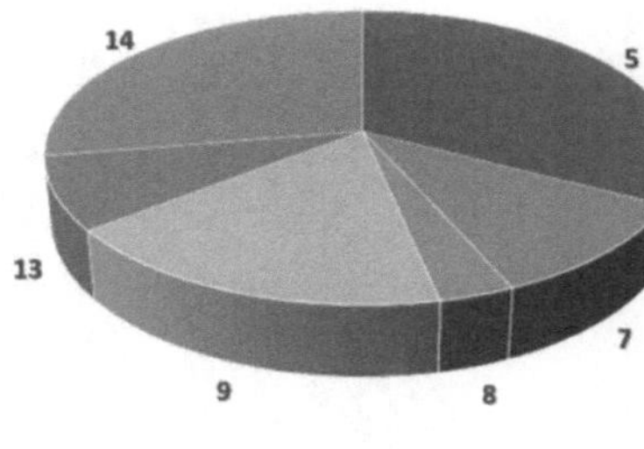

Figure 4. Pie chart for the duration in each position.

At the end of the experiment, the pie chart above was obtained. There it can be observed that the subjects sat in six different positions: position 5, 7, 8, 9, 13 and 14. Five of those positions were considered incorrect in this experiment, and one of those is the unidentified position (position 14). Not once, for all the duration of the experiment, the subjects sat in a correct position.

4 DISCUSSION

Through the results previously showed, it can be concluded that, for a long period of time, the subjects tend to sit in a incorrect position that may have consequences for the subjects' back health in the future. Altought the subjects changed positions over time, they never once sat in a correct position. This shows that these subjects already have an internalized tendency to position in a incorrect way. Thus, measures should be taken into action to change this postural tendency.

In this sense, the prototype described in this article could become one of those measures, by detecting incorrect positions and alerting the users so that they can change their position and establish better siting habits.

5 CONCLUSION

There is an increasing amount of time that the general population spents sitting and, most importantly, sitting in wrong positions, which causes back problems to arise. Such fact was corroborated throughout the experiment, where it was observed that the subjects involved adopted several incorrect positions during the study period. Therefore, the prototype developed in this article can become an asset to combat this issue by detecting and alerting to wrong positions.

However, there is still some work to be done in the future before this prototype can be considered. Foremost, a real-size chair is needed to conduct tests on it so we are able to assess its worth as a mean to improve sitting posture. Moreover, there is a need to evaluate the chosen positions according to biomechanical principles, and evaluate the sensors' placement in the chair to properly detect those positions.

Nevertheless, we believe that the prototype designed is a beginning in creating a helpful tool to combat the growing tendency of bad posture and has attractive future perspectives.

REFERENCES

Fialho, P. B., Souza, A. P. D., Minette, L. J. & Silva, J. D. C. 2007. Avaliação ergonômica de cadeiras residenciais fabricadas no pólo moveleiro de Ubá, MG. *Revista Árvore*, 31, 887–896.

Flutur, G., Movileanu, B., Károly, L., Danci, I., Cosovanu, D. & Stan, O. P. Smart Chair System for Posture Correction. 2019 22nd Euromicro Conference on Digital System Design (DSD), 28-30 Aug. 2019 2019. 436–441.

Marques, N. R., Hallal, C. Z. & Gonçalves, M. 2010. Características biomecânicas, ergonômicas e clínicas da postura sentada: uma revisão. *Fisioterapia e Pesquisa*, 17, 270–276.

Martins, L., Lucena, R., Belo, J., Santos, M., Quaresma, C., Jesus, A. & Vieira, P. 2013. *Intelligent Chair Sensor Classification of Sitting Posture*.

Mengjie, H., Ian, G. & Rui, Y. 2017. Smart Chair for Monitoring of Sitting Behavior. *KnE Engineering*, 2.

Muppavram, S., Patel, N. & Nadeem, M. Posture Alert. 2018 IEEE Region Ten Symposium (Tensymp), 4–6 July 2018 2018. 213–218.

Pulzatto, F., Dias, T. L., Daniele, K. & Silva, F. D. 2017. Análise Ergonômica de Setores Administrativos de Uma Instituição de Ensino Superior.

Shiri, R., Falah-Hassani, K., Heliövaara, M., Solovieva, S., Amiri, S., Lallukka, T., Burdorf, A., Husgafvel-Pursiainen, K. & Viikari-juntura, E. 2019. Risk Factors for Low Back Pain: A Population-Based Longitudinal Study. *Arthritis Care & Research*, 71, 290–299.

Advances and Current Trends in Biomechanics – Belinha et al. (Eds)
© 2022 Copyright the Author(s), ISBN: 978-1-032-10806-3

Monitoring plantar pressure profile using a smart sensor shoe

F. Campos, R. Lopes, A. Flores & F. Direito
INEGI, Porto, Portugal

A. Marques
ISEP-P. Porto, School of Engineering, Polytechnic of Porto, Porto, Portugal

ABSTRACT: This work describes a new and innovative prototype to measure feet plantar pressure so that we can provide means for people that suffer from peripheral neuropathy, like diabetic people, to monitor any sudden increase of pressure and prevent future serious injuries. The presented prototype is wireless, communicating by Bluetooth Classic connection with an Android App, which was also developed specifically to the purpose of displaying all values in an interactive and intuitive way. The sensing mechanism includes seven FSR sensors placed in the most critical areas of pressure increments and a temperature and humidity sensor. All FSR sensors were calibrated and then gait tests were performed. The prototype performance and results were analyzed and validated through the comparison of results obtained with a commercialized product (PEDAR). The smart sensor shoe proved to be capable of measuring and detecting plantar pressure increases.

1 INTRODUCTION

Footwear plays the essential role of being the main interface between human foot and ground during gait. The foot, being the end part of human body's lower limb, supports all the weight and allows movement and locomotion. In that way, poor adjustment between shoe and foot can be responsible for several pathological conditions, such as, deformities in the fingers, metatarsalgia (metatarsal pain), and ulcers due to, peripheral neuropathy (diabetes) (Patil et al., 2009), which are some of the main foot pathologies related to high pressure and overload of our body. In the case of professional/work shoes, as it's mandatory to wear them for many hours a day, the consequences of using inappropriate shoes can be serious in terms of foot health and safety, especially in the long term (Ochsmann et al., 2016). Therefore, it's important to study and develop footwear that can contribute to prevent pathologies and help people according to their needs and footwear use. This work was focused in the development of a solution targeting people with peripheral neuropathies (due to diabetes), and shoes for specific applications, namely for professional use. This way, the main goal was to develop an instrumented shoe that allows plantar pressure and motion monitoring, with an Android app as the user interface, which integrates several features, namely alerting the user whenever a high-pressure event occurs. This solution should contribute to help people to understand how much pressure is exerted by the foot, or a specific part of the foot, alerting for the excessive pressure increase in a certain area, therefore allowing the person to modify his/her foot support in order to prevent injuries. The application works as an auxiliary device in the prevention and alert of bad posture, plantar pressure and load distribution conditions.

DOI: 10.1201/9781003217152-46

2 MATERIALS AND METHOD

In order to carry out the system specifications, questionnaires were delivered to the companies involved in this area, so as to ascertain which characteristics are most appreciated by them and reach a consensus of specifications necessary for the development of smart footwear. At the same time a search was made for possibilities of sensors on the market that could be selected and classification of their characteristics. Then, with the promoter's response already acquired, establish the necessary requirements for the sensor. Also, at this stage, a survey was carried out on what functionalities and features the application should incorporate, with the purpose of define the general layout of the application in the future, considering the needs and specifications defined.

It is described in the requisites of this project the need to specify the transmission, the storage and data processing. However, it was considered that the project should respect some guidelines that should be included into the system, namely:

- The device has to be mobile and weight less than 300 grams and should make use of the smallest number of wires in order to reduce the complexity of the system.
- The chosen sensor is thin and flexible;
- Low electric consumption and lowest possible cost.

Therefore, the hardware selected was considered in terms of controller, pressure sensor, data transmission, temperature and humidity and data storage, which will be described below.

The system was specified to have two distinct parts. One containing the pressure sensors and the humidity, which should have low thickness in order to be inserted between the sole and the shoe insole, with the sensor measurement area being chosen so that seven sensors were possible to mount in the expected highest-pressure zones. The other part corresponds to a housing that will contain the electronic circuitry, having reduced dimensions of interlocking with the aim of allowing its connection to the shoe. Figure 1 represents the system described previously.

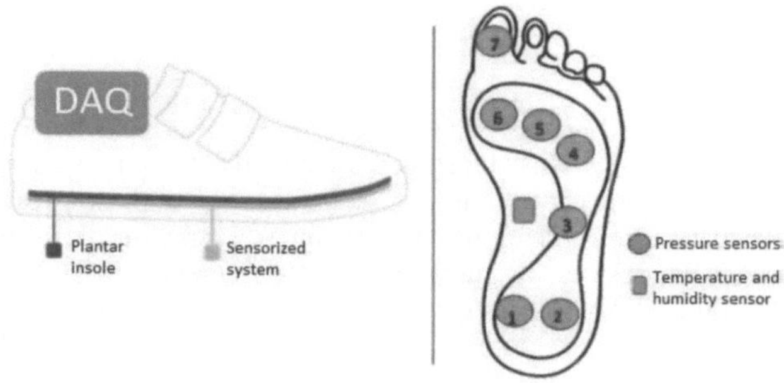

Figure 1. Solution design.

The chosen pressure sensor follows was considered in terms of hysteresis; linearity; temperature sensitivity; sensitive area; load capacity; operating frequency; repeatability; creep. In this way, after studying the different types of existing sensors on the market it was decided to select Flexiforce A301 piezoresistive sensors, placed as shown in Figure 1. Since it was desirable that the developed system should be used in a wide variety of applications, and that the load capacity was not a limiting factor of the system, the use of the A301 sensor that allows a load of up to 445 N, equivalent to 6 MPa in the sensitive area (Abdul Razak et al., 2012).

For the temperature and humidity sensor, the choice fell on AM2320, being a sensor with high performance, fast response and strong anti-interference capability. A flexible PCB was used to build the sensitive part, with Eagle software being used for this purpose, so that it was facilitated the sensors placement in the predetermined positions. This assembly is represented in the Figure 2.

As the power consumption was a critical point of the system, it was decided to use Bluetooth communication protocol. After conducting a market study and considering the technologies and sensors mentioned before, it was decided that the processing should be done through an ESP32-WROOM-32. This MCU has 18 analog inputs, also offering I2C and Bluetooth communication. Later after performing some initial tests, it was noticed this MCU ADC wasn't converting accurately the signals from the pressure sensors. In order to solve this issue, it was added an external ADC in the electronics. The electronic circuit would also include the components for the amplification and processing of the signals and for the power circuit. All these components were soldered in a PCB designed exclusively for this application.

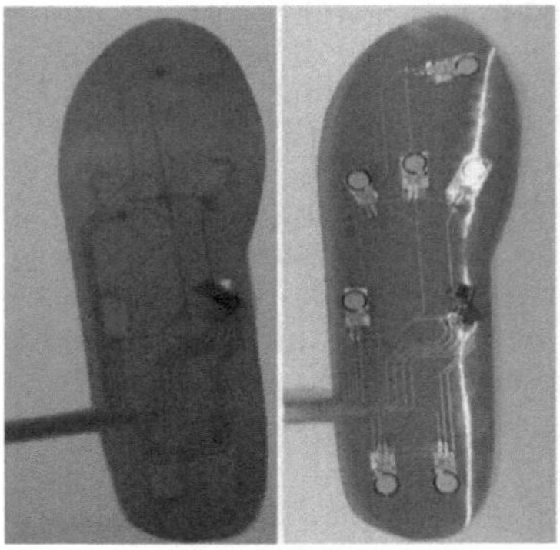

Figure 2. Flexible PCB used to fix sensors.

Regarding the Android app, it was developed specifically for this system and its primary objective is to enable the monitoring and/or visualization of plantar pressures, and it must have the ability to alert the user in different circumstances. Based on the systems that are commercially available and the results of questionnaires and meetings held, the functions and information to be available through the app were determined as follow: Pressure center line; Foot period of contact; Number of steps; Cadence; Weight distribution; Maximum pressure measured by each sensor; Temperature and humidity.

3 TESTS AND RESULTS

After the prototype was conceived, it was necessary to carry out tests and trials, on one hand to calibrate the pressure sensors that were applied to the system, on the other hand to assess whether the overall behavior of the system respected the requirements that had been defined initially. Therefore, the first step was to evaluate the response of the pressure sensors individually, and subsequently, tests were carried out to evaluate the functioning of the system when compared to other reference systems.

The selected pressure sensors, despite having a known behavior, needed to be calibrated according to the supply and acquisition system assigned to it. For this purpose, it was used a test equipment that allows the application of pre-established loads in a dynamic and static way, the EletroPuls E1000, from Instron. Each sensor was tested individually, and the signals were collected through the prototype DAQ. With the results obtained for each sensor, in terms of Voltage vs Force it was possible to draw calibration curves characterizing the behavior of the sensor. In order to obtain a curve with a greater approximation to the real behavior, third degree polynomial trend curves were drawn, as shown in Figure 3. With the definition of the calibration equations, it is possible to process the electrical signals from the sensors and converting them into force values, which in turn, can be converted in pressure values, since the area of the sensors is constant ($A = 71.4\ mm^2$).

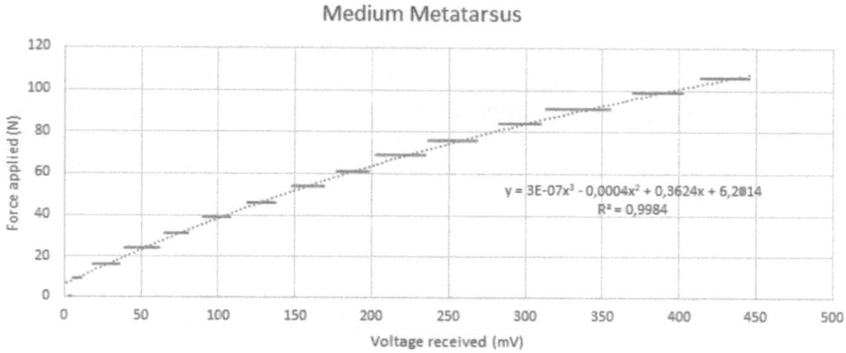

Figure 3. Calibration curve of one of the sensors.

In order to validate the calibration and operation of the plantar pressure measurement prototype, gait analysis tests were carried out at Porto Biomechanics Laboratory (LABIOMEP). These tests had the purpose of comparing the prototype results obtained with the commercial plantar pressure measurement system Pedar-X. This system is made up of sensitive insoles, composed by 99 sensors and is used both in academic and research studies, being considered as the standard of comparison for validating new systems. The test performed consists on simulating a normal gait situation, along about 10 meters with the prototype developed and Pedar-X not simultaneously placed in the shoe. The results for one of the sensors for the prototype and Pedar-X test are shown in Figure 4.

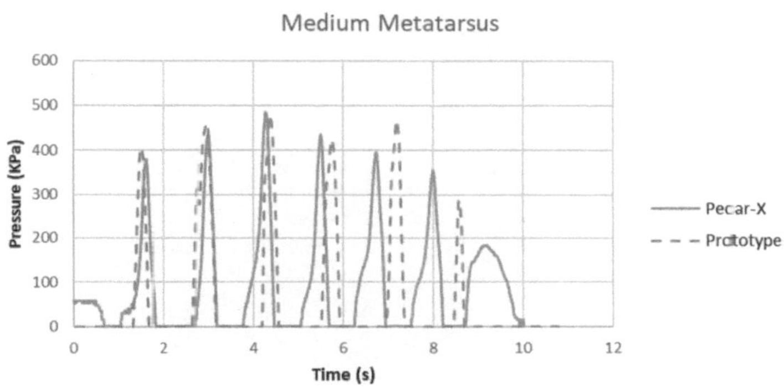

Figure 4. Gait results for prototype and Pedar-X.

4 DISCUSSION

Regarding the calibration test, it can be observed a 3rd degree polynomial trend curve can almost perfectly reply the behavior of the Flexiforce sensor with a good correlation coefficient ($R^2 = 0.9984$). As can be seen, this trend is almost linear for forces higher than 20 N, which will be the range of vertical forces observed in the foot.

Regarding gait tests, as it can be seen, there is a small deviation between the two curves, nevertheless, the behavior of the two systems is very similar, both presenting similar variations (with an error less than 10%) related to each phase of the gait cycle.

5 CONCLUSION

A method to monitor plantar pressures was developed. The results show that the system can be used to measure pressures on the 7 most critical areas considered. Like it was foreseen the system meets all the requirements proposed when the project initiated, being developed a compact DAQ, total absence of electric wires of the prototype, being its energy consumption low (autonomy of about 4 hours in continuous operation) and cost production of about 250€.

ACKNOWLEDGEMENTS

The authors wish to thank Programa COMPETE which supported this work through the funding of PROJECT FAMEST – Project POCI-01-0247-FEDER-024529.

REFERENCES

Abdul razak, A. H., Zayegh, A., Begg, R. K. & Wahab, Y. 2012. Foot plantar pressure measurement system: A review. *Sensors*, 12, 9884–9912.

Ochsmann, E., Noll, U., Ellegast, R., Hermanns, I. & Kraus, T. 2016. Influence of different safety shoes on gait and plantar pressure: a standardized examination of workers in the automotive industry. *Journal of occupational health*, 58, 404–412.

Patil, S., Thatte, M. A. & Chaskar, U. 2009. Development of planter foot pressure distribution system using flexi force sensors. *Sensors & Transducers*, 108, 73.

Advances and Current Trends in Biomechanics – Belinha et al. (Eds)

Mechanical hand remotely controlled by exo-glove

A. Marques, A. Monteiro, P. Gregório & P. Guimarães.
ISEP-P. Porto, School of Engineering, Polytechnic of Porto, Porto, Portugal

ABSTRACT: By observing what happens around us, it is unrealistic to say that technology does not influence our lives. From the smallest memory cards on mobile phones, to robots that operate in factories, it is present and creating new directions. Consequently, the human being remains exposed to several risks that, in a way, can affect their quality of life and even change it drastically. With this approach in mind the present work aims to illustrate the development of a robotic hand to be used in several environments, such as therapeutic, rehabilitation or surgical, focused to help the patient's recovery.

1 INTRODUCTION

The impact of robotics on medicine and other faculties is undeniable today. As our understanding of how to implement robotics services in healthcare matures, a deeper understanding of how to create smarter and more capable surgical robots continues to develop, with the topic of economics also present (Shademan, et al., 2016). The focus of this work was to create a robotic hand capable of reproducing the movements made by a human hand. The project consisted in the design and assembly of a mechanical hand remotely connected to a glove, which is used by the user, being able to reproduce the gestures performed by the glove user.

This idea was enforced by the need to fight the high cost associated with manipulators or robotic components, representing thus an alternative to adopt new physiotherapy strategies, combined with robotic limbs or human parts. This option differs from the usual therapeutic procedure because it allows greater flexibility for the physiotherapist, that is, allowing for higher degree of controlling patient's movements while watching him and adapting different therapeutic designs according to specific patient's needs. The proposed mechanical hand is remotely connected to a glove, which, in turn, acts as an emitter of data that is captured by the mechanical hand, thus representing the receiver of this system. The data sent through this system is obviously visualized by the movement reproduced in the mechanical hand through the information captured and sent by the glove. Considering telemedicine applications, this work can be further improved in a IoT perspective.

2 MATERIALS AND METHODS

This project aimed at the development of a prototype of a sensorized glove capable of controlling a mechanical hand remotely. For this, a glove with resistive sensors was used, capable of interpreting the movements of the fingers of a human hand and converting them into electrical signals that are sent to a central controller, responsible for reproducing the movements in a separate mechanical hand. The signals of the sensed glove are interpreted by an Arduino Uno and by radio frequency the data are transmitted to a second Arduino that controls five servo motors in the mechanical hand, therefore performing the five independent movements of the human fingers.

DOI: 10.1201/9781003217152-47

To use the glove as a control device, a mechanical hand was designed and its parts were manufactured by 3D printing using PLA (lactic polyacid plastic), as well as the glove structure, due to its biodegradability and biocompatibility, as well as the fact that it has a very low toxicity and a high mechanical performance, moldable to needs (Younes & Cohn, 1988). A wooden box was used as a support and fixation structure for the hands and servo motors. In addition, fluor-carbon fishing wires were used as connectors and rubber bands in the joints. The variation of the input signal (coming from the glove) allows the variation of the position of the servo motor, in this way the wires are connected from the motors to the fingertips of the hand and are placed in such a way as to generate a tension according to the angular movement the engine. So, when a tension is generated in the wires, generated by the motors, the fingers flex and the elastics stretch, accumulating potential elastic energy. On the other hand, when there is no tension applied to the wires, the elastic potential energy is converted into kinetic energy allowing the extension of the fingers of the mechanical hand (Carlucci & Santos, 2016).

The communication between the glove and the mechanical hand is carried out remotely by radiofrequency through an RF 433MHz module. The module consists of two units, a transmitter, attached to the glove and a receiver connected to the hand. The following block diagram represents the various functions performed by each component of the circuit, as well as the signal flow.

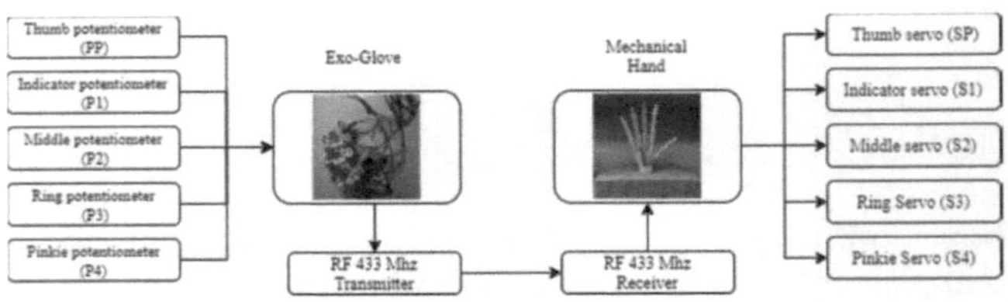

Figure 1. Block diagram representing the signal flow in the described project.

The *exo-glove*, where the various respective potentiometers are allocated to each finger, is the initiator of the process. By measuring the angular variation of the various potentiometers, it detects these changes and communicates with the Arduino by supplying the signal to the transmitting RF. This transmitted signal is captured by the RF receiver, which, through communication with the servomotors of each finger of the mechanical hand, causes their movement in it, depending on the value transmitted to it.

In the development of this project, several electronic equipment was used, as well as a specific printing material that will be addressed below.

- Arduino UNO Boards
  In this project, two Arduino boards were used for the processing and control of all data (Cavalcante, et al., 2011). In the Arduino of the transmitting circuit, the information from the *exo-glove* potentiometers is processed, in which a significant movement of the fingers and filtering is considered so that there is greater accuracy in the data emission and the consequent removal of errors. On the other hand, in the Arduino of the receiving circuit, the connection is monitored via radio frequency, the control of the data transmitted by the previous circuit and its processing for the activation of the servomotors.
- Radio Frequency Module RF 433 MHz
  The application of this module, in the present equipment, has the objective of making the wireless connection between the circuits, to be possible, the remote control of the mechanical hand using the *exo-glove*. These can reach up to 200 meters without obstacles, serving their purpose in possible telemedicine applications.
- Potentiometers
  The *exo-glove* resembles an external skeleton joined by hand, to read its movements. Its structure has a set of joints made up of its respective potentiometer that allows a quick and

accurate reading of the movement of all fingers. In fact, the idea is to rotate the potentiometers, which implies the variation of the electrical resistance (Gussow, 1997) through the flexion and extension of the fingers. In this way, it is possible to transform the mechanical movement into a variation of electrical voltage at the output of the potentiometer, which can be used to control the mechanical hand from a distance.

- Servomotors
  A servomotor is a rotary/linear actuator, which is used to allow precise control of a certain angular/linear position (Silva, 2017). The variation of the input signal, coming from the glove, allows the variation of the position of the servomotor. In this way, the wires connected with the hand are placed to generate a tension according to the angular movement of the motor. Therefore, when tension is generated in the wires by the motors, the fingers flex and the elastics found in the joints of the fingers extends, which generates an elastic tension. On the other hand, through the opposite movement of the motor, tension is no longer applied to the wires and the previously accumulated elastic energy is released, which allows the extension of the fingers of the mechanical hand, returning them to the original position, that is, extended.

3 RESULTS

The figures below show, respectively, how the connection was made between several components of the glove and the two Arduino boards, since one is responsible for programming the transmitter circuit, connected to the *exo-glove*, and the other is responsible by programming the servomotors, which makes the movement of the mechanical hand possible.

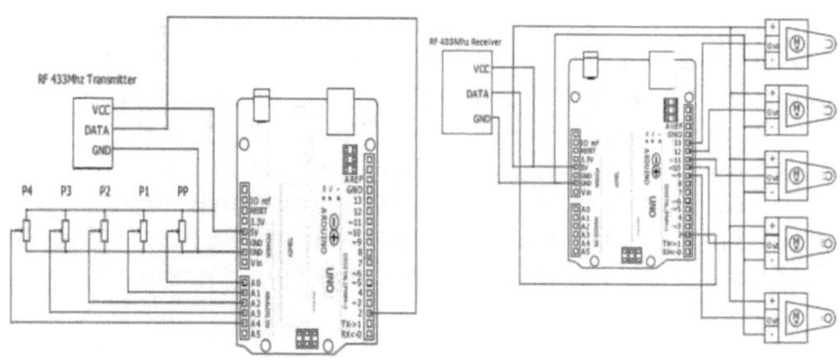

Figure 2. Emitter circuit (on the left) and receiver circuit (on the right).

The thumb (PP), indicator (P1), middle (P2), ring (P3) and pinkie (P4) potentiometers are connected to the analog inputs of the first Arduino A0, A1, A2, A3 and A4, respectively. The data pin of the transmitter RF 433Mhz module is connected to digital pin 2 of the Arduino. The thumb (SP), indicator (S1), middle (S2), ring (S3) and pinkie (S4) servo motors are connected respectively to the digital inputs 9, 10, 11, 12 and 13 of the next Arduino. The data pin of the RF 433Mhz receiver module is connected to digital pin 2 of the Arduino.

After 3D printing the various components and assemble them, the mechanical hand was obtained, composed of 5 fingers like those of human anatomy, allowing a flexion and extension movement. The return to the initial position is guaranteed by elastics, which hold the elastic tension that is generated during the movement. The mechanical hand is connected to the Receiver circuit, while the glove is connected to the Transmitter circuit. The glove is provided with a compartment where the potentiometers are located, and it is intended for it to be used by a subject who performs the movements that will be imitated by the mechanical hand.

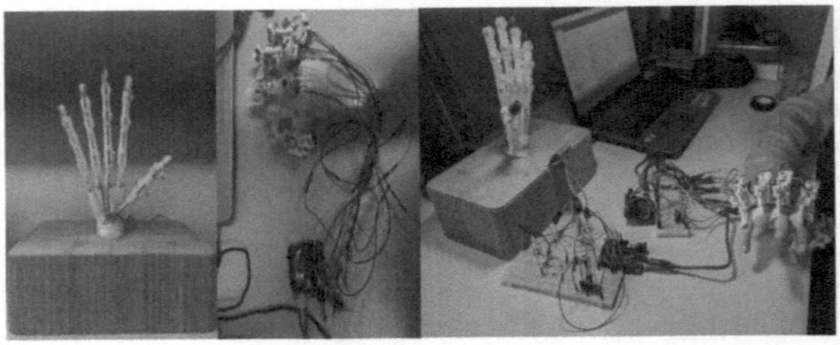

Figure 3. Photographs of the developed prototypes, glove, and robotic hand.

Right after the assembly of both parts, the servomotors were calibrated, defining the initial position of the five fingers, as the alignment of the various segments of the fingers according to an anatomical reference position. This setting allows greater control of articulated movements without disparity from one finger to the other. A set of restrictions were also defined to optimize the movement of the mechanical hand, namely regarding a minimum value of the signal of the potentiometers that determine the movement of the mechanical hand.

4 DISCUSSION

Movement calibration tests were carried out, and some adjustments were found to be necessary. Certain more complex movements, such as flexing the fingertips, are more difficult to achieve with the model of the hand used. This is because a force far greater than that which the servomotors can supply is necessary to contract the whole hand. Another problem found was the bad contact between the several conducting wires, since during the tests, certain movements appeared delayed, uncertain or were non-existent. Thus, after the connections were reworked so that they did not compromise the movements, the hand performed the function correctly. Still associated with the transmission of data between the transmitter and the receiver, a small delay was observed between the transmission of data from the glove and the performance of the movements in the robotic hand, which can be resolved through a device capable of stabilizing or increasing the speed of propagation of the data, thus reducing the observed delay.

5 CONCLUSIONS

Although it is a task with some degree of complexity, it is possible to replicate the movements made by the hand of the human being and, as technology and scientific knowledge evolve, the refinement of movements allows for a closer to real movement. Looking at the current market, the developed prototype has several advantages: it is a low-cost and environmentally friendly prototype since it is printed on PLA (Al-Mulla, et al., 2010), and easy to use for a patient without much preparation. It should also be noted that, in the case of use in a therapeutic environment, its use by the patient, allows the therapist to have greater control of the movements that are performed. However, as this is a first prototype, there is still room for future improvements, namely in terms of the ability to identify movements in space, through the incorporation of inertial sensors.

REFERENCES

Al-Mulla, E., Yunus, W., Ibrahim, N. & Rahman, M., 2010. Journal of Materials Science. *Properties of epoxidized palm oil plasticized polytlactic acid*, 9 january, p. 1942–1946.

Carlucci, F. & Santos, I., 2016. ETIC - Encontro Toledo de Iniciação Científica. *Grandezas Físicas Aplicadas A Uma Catapulta:Energia Potencial Elástica*, september.

Cavalcante, M., Tavolaro, C. & Molisani, E., 2011. Revista Brasileira de Ensino de Física. *Física com Arduino para iniciantes*, december, p. 4503.
Gussow, M., 1997. *Eletricidade Básica*. 2nd ed. São Paulo: Pearson Makron Books.
Shademan, A. et al., 2016. Science translational medicine. *Supervised autonomous robotic soft tissue surgery*, 4 may, p. 337.
Silva, D., 2017. Development of a Servo Motor Optimised for Robotic Applications. *Faculdade de Engenharia da U.P.*, 13 february, p. 68.
Younes, H. & Cohn, D., 1988. Phase separation in poly(ethylene glycol)/poly(lactic acid) blends. *uropean Polymer Journal*, pp. 765–773.

Advances and Current Trends in Biomechanics – Belinha et al. (Eds)
© 2022 Copyright the Author(s), ISBN: 978-1-032-10806-3

Decline angle effects on the lower limb biomechanics of the squat rehabilitation exercise: A systematic review

N. Lomba
Research Centre in Physical Activity, Health and Leisure (CIAFEL), Faculty of Sport and Porto Biomechanics Laboratory (LABIOMEP-UP), University of Porto, Porto, Portugal
Interdisciplinary Centre for Health Sciences (CICS) of ISAVE – Higher Education Institute of Health, Amares, Portugal

J.P.
Centre of Research, Education, Innovation and Intervention in Sport (CIFI2D), Faculty of Sport and Porto Biomechanics Laboratory (LABIOMEP-UP), University of Porto, Porto, Portugal

P. Roriz
University Institute of Maia – ISMAI, Maia, Portugal
Research Center in Sports Sciences, Health Sciences and Human Development (CIDESD), Portugal
Porto Biomechanics Laboratory (LABIOMEP-UP), University of Porto, Porto, Portugal

ABSTRACT: The eccentric squat exercise performed with a 25° decline angle has been suggested as an effective treatment method for the knee joint. However, this clinical strategy lacks a well-founded scientific analysis about its biomechanical effects on the lower limb, and the standardized use of a 25° slant angle. A systematic review was carried out to assess the biomechanical influence on the lower limb (joint moments, muscle activity, load on the patellar tendon and on the femoral patellar joint) from different decline angles during the eccentric squat exercise. Nine studies were identified according to the inclusion criteria. The results suggest that knee extensors muscle activity and knee moment increase as the decline angle also increases. However, the activity of the calf muscles and patellofemoral joint overload increase significantly above 15°. Thus, a 15° angle slant angle seems to provide greater mechanical advantage and safety for knee rehabilitation.

1 INTRODUCTION

Decline squats performed on a 25° slant platform are well accepted as a clinical assessment tool and as an effective rehabilitation exercise (Freke et al., 2020). Originally developed from the traditional eccentric squats performed on flat floor, decline squats are recommended as a more effective strategy to reduce pain and improve function in patients with patellar tendinopathy (Visnes and Bahr, 2007). The principles behind this approach are based on biomechanical changes of the lower limb when forcing the ankle to plantar flexion, where a reduced ankle work and calf muscles activity should be observed (Richards et al., 2008). Consequently, a more selective and higher knee extensors work is expected (Zwerver et al., 2007). Nevertheless, it is important to understand the biomechanics behind this approach, specifically, whether it provides any significant benefit compared with standard eccentric squats performed on flat floor, and if there is an evidence for an optimum angle of decline that could provide a mechanical advantage for the knee joint (Richards et al., 2016). The aim of this study was to identify how

DOI: 10.1201/9781003217152-48

different decline angles in squat exercises change the biomechanical variables of the lower limb during the eccentric phase of the squat and their implications on knee rehabilitation.

2 METHODS

An extensive and comprehensive search was conducted in three major electronic bibliographic databases (*PubMed*, *Scopus* and *Web of Science*) to locate eligible studies. The inclusion criteria were papers written in English or Portuguese, studies conducted in laboratory setting with healthy volunteers comparing the biomechanical variables of the lower extremity (joint moments, muscle activities, load on the patellofemoral joint and on the patellar tendon) during the eccentric squat exercise performed on the horizontal plane and with decline plane, specifying the angle used and squat type (unilateral or bilateral). The search strategy included all articles resulting from the keyword "decline squat" and in addition, the approach "decline squat AND (exercise OR knee OR biomechanics OR rehabilitation)". Articles were screened by title and abstract by both the first and third authors to be assessed for full-text examination. In case of disagreement, a third evaluator (the second author) would decide whether to be included in the selection or not. The same procedure was adopted for the final inclusion. From a total of 619 papers, 301 articles were identified according to the inclusion criteria after removing the duplicates. From these, ten articles were identified for full-text analysis to assess eligibility. One article (Freke et al., 2020) was excluded, as the outcomes observed on the control healthy group were not analyzed in this investigation. Therefore, nine studies were eligible for this systematic review. Quantitative analysis was not possible due to the differences observed between the study protocols.

3 RESULTS

Nine articles compared lower extremity biomechanics between declined with horizontal planes during the descending phase of the squat exercise. Six studies analyzed the biomechanical activity of the lower limb during the traditional eccentric squat exercise performed on the horizontal plane (0°) with the exercise performed with a decline of 25° and 20°, while three studies investigated the influence of different decline angles. Table 1 resumes the studies characteristics.

Kongsgaard et al. (2006) compared unilateral standard squats and unilateral squats on a 25° decline board with the dominant leg. Patellar tendon tension was significantly higher during the squat position on the decline surface compared to the standard surface. Normalized mean electromyography (EMG) amplitudes of the knee extensor muscles were significantly higher during the decline compared to the standard squats, while hamstring and calf muscle mean EMG did not differ. *Tibialis anterior* mean EMG had significantly higher amplitude in standard squat compared to the decline. Frohm et al. (2007) included an assessment of unilateral and double limb squats with a submaximal free weight on a decline board (25°) compared to a horizontal plane. Eccentric work, mean and peak patellar tendon force, and angle at peak force were higher for squats on decline board. Angular excursions at the knee and ankle were larger with the decline board. Mean EMG for *gastrocnemius* and *vastus medialis* was higher on the decline board. Zwerver et al. (2007) evaluated single-leg eccentric squats on a board at different decline angles (0°, 5°, 10°, 15°, 20°, 25° and 30°). An increased knee moment was observed at decline angles of 15° and higher, while ankle and hip moment decreased. Maximum knee and ankle angles increased with steeper decline. Both patellar tendon and patellofemoral forces increased with higher decline angles but excessive patellofemoral load was observed when knee flexion was beyond 60°. Richards et al. (2008) analyzed single-limb squats at four decline angles (0°, 8°, 16° and 24°). Ankle moments decreased as the decline angle increased, and the knee moment increased, particularly at 16°. An increase was seen in the knee moments and integrated EMG in the *rectus femoris* as the decline angle increased, however knee flexion should be limited to 60° to prevent excessive patellofemoral load. A reduction was seen in the ankle moments as the decline angle increased, but EMG activity in the *gastrocnemius* increased between 16° and 24°.

Table 1. Summary of studies elected and included for systematic review.

Year	Authors	Sample Size	Angle comparison	Squat type	Outcomes
2006	Kongsgaard et al.	n=13 (Males: 7)	0° and 25°	Single-limb	*Patellar tendon strain Muscle activity Joint kinematics*
2007	Frohm et al.*	n=14 (Males: 14)	0° and 25°	Unilateral	*Joint kinetics and kinematics* & bilateral *Patellar tendon force Muscle activity*
2007	Zwerver et al.	n=5 (Males: 2)	0°, 5°, 10°, 15°, 20°, 25° and 30°	Single-limb	*Joint kinetics and kinematics Patellar tendon force Patellofemoral contact force*
2008	Richards et al.	n=10 (Unknown)	0°, 8°, 16° and 24°	Single-limb	*Joint kinetics and kinematics Muscle activity*
2009	Alves et al.	n=8 (Males: 3)	0° and 25°	Bilateral	*Muscle activity*
2010	Leporace et al.*	n=8 (Males: 6)	0° and 25°	Unilateral	*Muscle activity*
2015	Lee et al.	n=26 (Unknown)	0° and 25°	Single-limb	*Muscle activity*
2016	Yoo	n=9 (Males: 0)	0° and 20°	Single-limb	*Muscle activity*
2016	Richards et al.	n=18 (Males: 9)	0°, 5°, 10°, 15°, 20° and 25°	Double-limb	*Joint kinetics and kinematics Patellofemoral load Patellar tendon load*

* Include only data comparing decline with horizontal in the free weight tests or with no extra overload

Alves et al. (2009) found no differences in EMG activity (*vastus medialis oblíquus*, *vastus lateralis*, *bíceps femoris*, *soleus* e *tibialis anterior*) and muscle activity patterns were similar during standard and declined squats at 25°. Leporace et al. (2010) reported a higher activation of the quadriceps (*rectus femoris* and *vastus lateralis*) generated in the decline squat at 25° compared to the standard unilateral eccentric squat on a flat surface. Lee et al. (2015) reported a significantly higher muscle activity of the *rectus femoris* and significantly lower muscle activity of the *tibialis anterior* and *gastrocnemius lateralis* during squat exercises on the decline board (25°) compared to flat floor. Yoo (2016) reported that declined wedge squat at 20° was more effective in activating the *rectus femoris* than traditional squat on the horizontal plane. Richards et al. (2016) analyzed the lower extremity biomechanics during double limb squats at six angles of declination (0, 5, 10, 15, 20 and 25 degrees). The joint moments showed an increase at the knee up to 25° and a decrease at the ankle up to 15°. A progressive increase of the load in the patellar tendon and patellofemoral joint was reported, with a corresponding reduction in loading of the ankle. An increase was observed in the *rectus femoris* activity up to 15° declination and a decrease in the *tibialis anterior* activity. As the decline angle increased above 15 degrees the g*astrocnemius* and *biceps femoris* activity also increased.

4 DISCUSSION

Seven studies (Frohm et al., 2007, Kongsgaard et al., 2006, Lee et al., 2015, Leporace et al., 2010, Richards et al., 2016, Richards et al., 2008, Yoo, 2016) reported that muscle activity of the knee extensors (*rectus femoris*, *vastus medialis or vastus lateralis*) was significantly higher during squat exercises on the decline board 24°-25° compared to flat floor. Two studies (Richards et al., 2016, Richards et al., 2008) that evaluated different decline angles also noted a progressive increase of the knee extensors activity with the increase of the decline angle. Only one study (Alves et al., 2009) reported no significant difference in knee extensors EMG activity between declined squats at 25° and standard squats. For calf muscle activity (*gastrocnemius medialis* and/or *soleus muscle*) two studies (Kongsgaard et al., 2006, Leporace et al., 2010) found no significant differences during standard and declined squats at 25°. Lee et al. (2015) observed a lower muscle activity of *gastrocnemius lateralis* during squat exercises on the decline board (25°) compared to flat floor, while two studies (Richards et al., 2016, Richards et al., 2008) reported a decrease on *gastrocnemius* activity until at 15° and 16°, but also a significant increase at 20°, 25°, and 24°. Three studies (Lee et al., 2015, Richards et al., 2016, Richards et al., 2008) reported a decrease on *tibialis anterior* EMG activity during squat exercises on the decline board (25°) compared to flat floor. Richards et al. (2016) pointed out that this is particularly observed up to 15°. Alves et al. (2009) reported no significant difference in *tibialis anterior* activity comparing squats at 0° and 25°. Two studies (Kongsgaard et al., 2006, Leporace et al., 2010) found no differences in hamstring muscles (*biceps femoris, medial hamstring*) EMG activity between standard squats and decline at 25°. However, Richards et al. (2016) reported a decreased on *biceps femoris* until 15° and a significant increase on EMG activity as the decline angle increased above 15°. Four studies (Kongsgaard et al., 2006, Richards et al., 2016, Richards et al., 2008, Zwerver et al., 2007) reported a significantly higher patellar tendon tension and patellofemoral joint load at 24° and 25°. Richards et al. (2016) and Zwerver et al. (2007) suggested that it was particularly observed at higher decline angles, but with possible risks of excessive overloading in the patellofemoral joint. Three studies (Richards et al., 2016, Richards et al., 2008, Zwerver et al., 2007) reported an increased knee moment in decline squats, and two of them (Richards et al., 2008, Zwerver et al., 2007) a significant increased observed at 16° and 15° or higher, by 30% and 40% respectively, while Richards et al. (2016) reported an increase up to 15°-20°. A decreased ankle moment was reported by two studies (Richards et al., 2008, Zwerver et al., 2007) as the decline angle increased up to 25°, while Richards et al. (2016) noted a significant decreased only up to 15°. Zwerver et al. (2007) reported a decrease in hip moment at decline angles of 15° or higher.

No studies evaluated potential differences between double and single limb squat, which may explain some contrasting results. Trunk control during squatting was observed in Frohm et al. (2007) research, and could influence outcomes. Only Zwerver et al. (2007) analyzed the hip biomechanics, although it encompasses a significant role as the proximal joint of the lower limb. Additional studies are needed to determine the efficiency of different decline angles in squat exercises for specific rehabilitation aims.

5 CONCLUSION

Squats exercises performed on a decline board appear to be an effective strategy to selective increase the muscle activity of the knee extensor muscles, as the knee moments increases and ankle moment decreases. Using a decline angle around 15° seems to be the most beneficial and safest strategy to the knee joint, as beyond this range an excessive increase of the load in the patellar tendon and patellofemoral joint may occur along with a high calf muscle activity. Caution is advised when applying these results to specific clinical rehabilitation programs.

REFERENCES

Alves, F. S. M., Oliveira, F. S., Junqueira, C. H. B. F., Azevedo, B. M. S. & Dionísio, V. C. 2009. Analysis of electromyographic patterns during standard and declined squats. *Revista Brasileira de Fisioterapia*, 13, 164–172.

Freke, M. D., Crossley, K., Russell, T., Sims, K. J. & Semciw, A. 2020. Changes in Knee and Trunk Alignment in People With Hip Pain and Healthy Controls When Using a Decline Board During Single-Leg Squat. *J Sport Rehabil*, 29, 886–896.

Frohm, A., Halvorsen, K. & Thorstensson, A. 2007. Patellar tendon load in different types of eccentric squats. *Clin Biomech (Bristol, Avon)*, 22, 704–11.

Kongsgaard, M., Aagaard, P., Roikjaer, S., Olsen, D., Jensen, M., Langberg, H. & Magnusson, S. P. 2006. Decline eccentric squats increases patellar tendon loading compared to standard eccentric squats. *Clin Biomech (Bristol, Avon)*, 21, 748–54.

Lee, D., Lee, S. & Park, J. 2015. Impact of decline-board squat exercises and knee joint angles on the muscle activity of the lower limbs. *J Phys Ther Sci*, 27, 2617–9.

Leporace, G., Pereira, G. R., Carmo, R. C. R., Silva, A. C., Cabral, R. P., Silva Filho, N., Pasqualini, H. E. C. & Batista, L. A. 2010. Specificity of the myoelectrical activity on the eccentric decline squat at 25° and standard squat with different overloads. *Revista Brasileira de Medicina do Esporte*, 16, 205–209.

Richards, J., Selfe, J., Sinclair, J., May, K. & Thomas, G. 2016. The effect of different decline angles on the biomechanics of double limb squats and the implications to clinical and training practice. *J Hum Kinet*, 52, 125–138.

Richards, J., Thewlis, D., Selfe, J., Cunningham, A. & Hayes, C. 2008. A biomechanical investigation of a single-limb squat: implications for lower extremity rehabilitation exercise. *J Athl Train*, 43, 477–82.

Visnes, H. & Bahr, R. 2007. The evolution of eccentric training as treatment for patellar tendinopathy (jumper's knee): a critical review of exercise programmes. *Br J Sports Med*, 41, 217–23.

Yoo, W. G. 2016. Comparison of the hamstring/quadriceps ratio in females during squat exercise using various foot wedges. *J Phys Ther Sci*, 28, 2379–80.

Zwerver, J., Bredeweg, S. W. & Hof, A. L. 2007. Biomechanical analysis of the single-leg decline squat. *Br J Sports Med*, 41, 264–8; discussion 268.

Advances and Current Trends in Biomechanics – Belinha et al. (Eds)
 ISBN: 978-1-032-10806-3

Biomechanical analysis of lumbar spine degeneration

R.M. Fragoso
Escola Superior de Tecnologia de Setúbal, Instituto Politécnico de Setúbal, Portugal

A.P.G. Castro
IDMEC, Instituto Superior Técnico, Universidade de Lisboa, Portugal
Escola Superior de Tecnologia de Setúbal, Instituto Politécnico de Setúbal, Portugal

ABSTRACT: Pathologies of the lumbar region of the spine are increasing in society. As this is a region constantly subjected to loading and of great mobility, the incidence of these problems takes on a great weight in the health systems of the countries. Thus, its study allows the formulation of new methods for diagnosis and treatment. This study aimed at describing the development of a finite element model of the lumbar spine region (L1-L5), as well as the biomechanical response of the spine to successive forces. This was possible with ITK-Snap®, Meshlab®, SolidWorks® and ABAQUS® software packages. The model consists of the vertebrae, intervertebral discs and ligaments that make up this lumbar spine segment. The studies carried out considered physiological efforts of flexion and extension through implementation of 5, 7.5 and 10Nm moments, with no preload, or alternatively with a preload of 50N or 100N. It was shown that moments above 7.5Nm were exaggerated and thus may lead to degeneration of the lumbar spine.

1 INTRODUCTION

The spine is a biomechanically complex anatomical structure, associated with lesions and degenerative pathologies, that result from complex interactions of several factors. Low back pain has become increasingly common in the current population. Annually, about 3.63% of the world population is diagnosed with degenerative disease of the lumbar spine (Ravindra *et al.*, 2018; Real, 2019). The realization of biomechanical tests that involve all the structures associated with the spine allows the development of new mechanisms that provide more effective and specific treatments, as well as the perception of what is at the origin of the different pathologies associated. (Castro *et al.*, 2014). Currently, the literature on spinal problems focuses mainly on the investigation of treatments and solutions, so the main objective of this study is the reconstruction finite element (FE) modeling of a L1-L5 lumbar spine segment and subsequent simulation of the conditions that this structure is subject throughout an individual's life. This study aims to clarify the biomechanical behavior when the healthy conditions of the ligaments are altered, providing the possibility to characterize the changes that happen and determining the influence that they have on the degenerative process of the lumbar spine.

2 METHODOLOGY

In this study, a FE model of the lumbar segment (L1-L5) of the spine was developed, as it is the region that supports the greatest load and rotation movements during daily life, making

DOI: 10.1201/9781003217152-49

problems like fractures or degeneration processes more likely to occur precocious (Carvalho, 2019). Thus, this study uses a 40-year-old female, without any sign of lesion or degeneration. The choice of this patient considered both age and physiological condition, towards obtaining a FE model as realistic as possible.

2.1 *Reconstruction and modeling*

Computational tomography (CT) images of the abovementioned individual were obtained from the xVertSeg database (Korez *et al.*, 2015) and segmented using ITK-Snap program (Yushkevich *et al.*, 2006), which allows the different types of tissue to be separated. Through the thresholding method, the density tissue band is selected, and then the 3D model is generated. In Meshlab® (Cignoni *et al.*, 2008), the number of vertices of the mesh has been reduced (to the order of hundreds) and different filters (such as "Simplification: MC Edge Colapse", "Smoothing Fairing and Deformation: Laplacian Smooth" and "Remeshing, Simplification and Reconstruction: Close Holes") were applied to obtain a smooth and continuous structure.

Segmentation of soft tissue would only be effective with the segmentation of magnetic resonance (MR) images instead of CT, due to the different densities between the bone (vertebrae, VB) and the soft tissue that constitutes the intervertebral discs (IVD). As so, SolidWorks® (Dassault Systèmes SolidWorks Corp., USA) was used for soft tissue inclusion, i.e., the IVD were manually reconstructed by the interpolation of the adjacent VB (L1-L2, L2-L3, L3-L4, L4-L5), in which procedures are combined over three different plans. The first plane corresponds to the upper limit of the disk, followed by an intermediate plane, and finally a plane corresponding to the lower region. The upper and lower face of the IVD should coincide with the limits of the adjacent VB. The reconstruction process of the IVD in SolidWorks® was performed in an iterative way: the configuration of the adjacent VB was not compatible with ABAQUS® due to duplicated surfaces. As so, it was necessary to do inverse extrusions in the upper and lower plane of each VB, with a minimum thickness to ensure the existence of a single face. However, this type of configuration is an approximation since each VB has a characteristic anatomical concavity.

When importing the enhanced 3D model (VB plus IVD) into ABAQUS® (Dassault Systèmes Simulia Corp., USA), a set of material properties was assigned to its different components: i) VB (linear elastic), E=12000MPa, v= 0.3 and ii) IVD (linear elastic), E= 2000MPa, v= 0.4 (Heuer *et al.*, 2007; Carvalho, 2019). It must be highlighted that both VB and IVD were simplified to single layer materials (no internal split), in order to reduce simulation time, hence the further simplification to linear elastic materials.

The mesh assignment to the different types of anatomical parts in the FE model was made with two types of elements: i) the VB and IVD were meshed with linear tetrahedral elements (C3D4), because they are automatically generated on the CT segmentation; ii) truss elements (T3D2) with no response to compression (only tensile strength) were attributed to the ligaments because of their simpler geometry and biomechanical behavior.

The human lumbar spine has seven different types of ligaments, usually denominated by the following acronyms: ALL, PLL, FL, ITL, ISL, SSL and CL (Abreu, 2018; Naserkhaki *et al.*, 2018). The placement in the lumbar spine was made accordingly to the anatomical insertion points. Based on the literature (Carvalho, 2019), each ligament is composed of several bars arranged in parallel with a specific cross-sectional area. In addition, only ALL and PLL ligaments are modeled in series because of their curvilinear shape that assumes the anatomy of the IVD.

2.2 *Simulation*

At this stage of the work, loads, boundary conditions and contacts in the FE model were defined. The loads and boundary conditions intended to simulate natural flexion and extensions movements. Moments were applied to the upper body of the top lumbar VB (L1), through a designated reference point to guide the movement of the whole model. Boundary conditions are applied to the lower region of the bottom VB (L5) to restrict all degrees of freedom. The interactions between the parts of the model (VB and IVD) were defined as rigid links. Contacts were also added in the facets region in a surface to surface interaction, so that there is no penetration between different components of the model, but only a given slippage between facets (which is what happens) (Carvalho, 2019). The range of physiological moments (5Nm, 7.5Nm and 10Nm) were first applied without preload and later combined with a preload of 50N and 100N, added to simulate the compressive forces of structures that aren't included in the FE simulation (e.g., muscles).

Regarding the outputs, vertical stress values (S33) were reported from designated elements on the upper face of L3-L4 and L4-L5, providing an average stress value recorded on the IVD. Additionally, this allows for the comparison with the reference study of Carvalho (2019), in which a L3-L5 FE model was simulated under a moment of 7.5Nm and a preload of 500N. This model was validated against experimental studies. The superior value of the preload is mainly because that model used cortical bone properties in the two VB (i.e., it is more rigid than the present model, in which an average cortical-trabecular stiffness was defined). In this case the preload must be consistent with the material properties used, hence the lower values.

3 RESULTS

Three out of four cases with preload and 10Nm moment were above the limits of the FE model, i.e., the simulation did not converge and so it was not possible to draw any outputs. With 100N preload, three of six simulation cases were also not concluded. All the simulations without preload were concluded. The comparison with the work of Carvalho (2019) is reported in Figure 1.

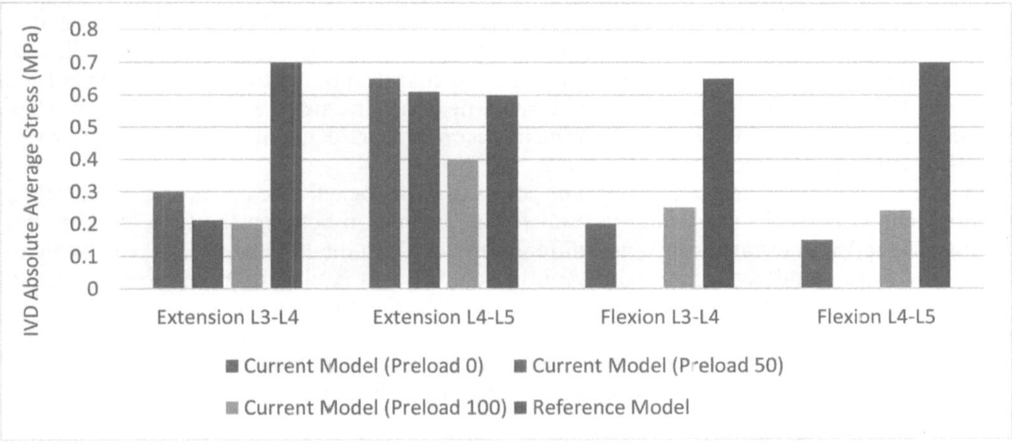

Figure 1. Values of absolute average stress (MPa) from the reference and the current model, with different preload conditions. Based on S33 absolute values extracted from ABAQUS®.

4 DISCUSSION

The simulation outcomes have shown that, for the flexion movement, the behavior is mostly compressive in the IVD, in contrast to the extension movement where there is traction. This is aligned with the natural biomechanics of the spine, and it means that the FE model is working as expected. However, there is a case of compression under an extension movement (moment of 10Nm), meaning that the effort to the IVD is excessive. The differences in stress absolute values seen in Figure 1 can derive from the fact that the load applied in the reference model is distributed only in 3 levels (L3 to L5), in contrast to what happens in the current model (with 5 levels). Nevertheless, the current model is aligned with the reference and therefore with experimental spine studies.

Additionally, when considering this L1 to L5 model, the higher is the applied preload, the less likely was the simulation to be successful. This means that a more complete lumbar spine model is able to respond to movements with a more realistic load distribution over the components.

Regarding the limitations of this work, one can mention the wide anatomical variability from individual to individual (only one set of images was used) as well as the fact that this model does not contain important anatomical structures such as muscles. In addition, the lack of distinction between the types of bone (cortical and trabecular) and IVD components also limits the constitutive modeling of the materials. Finally, when the preload was not considered, the biomechanical behavior of the ligaments may have not been accurately considered as these structures do not resist to compression, meaning that other pre-conditioning efforts could be considered.

5 CONCLUSION

The objectives of this study were the development of a method for reconstruction and modeling of the lumbar spine as well the demonstration and comparison of outcomes obtained through realistic FE simulations. Both were successfully achieved: the 3D model was generated from medical images and modeled according to literature, while the simulation of physiological movements proved to be aligned with computational (and experimental) references. Regarding the results, when increasing the loads and moments, the model's ability to perform the physiological movements of extension and flexion was reduced. Plus, considering the anatomical placement of the ligaments mainly in the posterior region of the model and the lordotic characteristic of the lumbar region, it is expected that during the extension movement the ligaments are less active. On the contrary, when a flexion movement occurs, the ligaments are forced to stretch more so the flexion movement becomes more demanding. Such behavior was recorded with this FE model.

As future work, there are several relevant points that can benefit the deepening of this study and the achievement of more accurate results. The distinction between the type of bone that constitutes the VB and the allocation of different materials in the IVD would make possible to study how these are affected internally. Another change that could bring the results closer to reality would be the insertion of elements that simulate the muscular structures that interact with the spine or the biomechanical response that they represent in the spine. By using MR images, instead of just CT, it would be possible to directly include the soft tissues in the model. Nevertheless, this study produced a L1-L5 model with potential to be used in the study of lumbar spine degeneration and treatment.

REFERENCES

Abreu, J. N. M. (2018) *Efeito da Aplicação do Espaçador Interespinhoso na Biomecânica da Coluna Vertebral*. Faculdade de Engenharia da Universidade do Prto.

Carvalho, V. (2019) *Finite element modelling of innovative interspinous process spacers for the lumbar spine Biomedical Engineering*. Universidade de Lisboa.

Castro, A. P. G. *et al.* (2014) 'Long-Term Creep Behavior of the Intervertebral Disk: Comparison between Bioreactor Data and Numerical Results', *Frontiers in Bioengineering and Biotechnology*, 2, pp. 1–8. doi: 10.3389/fbioe.2014.00056.

Cignoni, P. *et al.* (2008) *MeshLab: An open-source mesh processing tool, 6th Eurographics Italian Chapter Conference 2008 - Proceedings*. doi: 10.2312/LocalChapterEvents/ItalChap/ItalianChapConf2008/129-136.

Heuer, F. *et al.* (2007) 'Stepwise reduction of functional spinal structures increase range of motion and change lordosis angle', *Journal of Biomechanics*, 40(2), pp. 271–280 doi: 10.1016/j.jbiomech.2006.01.007.

Korez, R. *et al.* (2015) 'A Framework for Automated Spine and Vertebrae Interpolation-Based Detection and Model-Based Segmentation', *IEEE Transactions on Medical Imaging*, 34(8), pp. 1649–1662. doi: 10.1109/TMI.2015.2389334.

V. M. Ravindra et al. (2018) 'Degenerative Lumbar Spine Disease: Estimating Global Incidence and Worldwide Volume', *Glob. Spine J.*, 8, pp. 784–794.

Naserkhaki, S. *et al.* (2018) 'Effects of eight different ligament property datasets on biomechanics of a lumbar L4-L5 finite element model', *Journal of Biomechanics*. Elsevier Ltd, 70, pp. 33–42. doi: 10.1016/j.jbiomech.2017.05.003.

Real, J. (2019) *Biomechanical Analysis of Fixation Devices for Lumbar Interbody Fusion*. Instituto Superior Técnico, Universidade de Lisboa.

Yushkevich, P. A. *et al.* (2006) 'User-guided 3D active contour segmentation of anatomical structures: Significantly improved efficiency and reliability', *NeuroImage*, 31(3), pp. 1116–1128. doi: 10.1016/j.neuroimage.2006.01.015.

Advances and Current Trends in Biomechanics – Belinha et al. (Eds)
© 2022 Copyright the Author(s), ISBN: 978-1-032-10806-3

EXOSHOULDER—upper limb exoskeleton for a child with humerus osteomyelitis

Leonor Jud, Joana Grilo, Sérgio Barroso Gonçalves & Miguel Tavares da Silva
IDMEC, Instituto Superior Técnico, Universidade de Lisboa, Lisboa, Portugal

Maria José Costa & Alexandra Castro
Hospital D. Estefânia, Centro Hospitalar Universitário de Lisboa Central, Lisboa, Portugal

ABSTRACT: The EXOSHOULDER project aims the application of a general product development methodology based on advanced manufacturing processes, namely 3D scanning, geometric modelling and additive manufacturing, to the production of an articulated upper limb passive exoskeleton. The validation of the methodology involves its application in the development of a physical prototype for the case study of a child without the humeral diaphysis of the right arm, as a consequence of osteomyelitis. The methodology was applied in the production of a proof of concept for a healthy subject with no history of upper limb disorders. Functional and long-term tests were conducted for assessment of its function and comfort. The obtained results were analyzed and fed into the product development cycle for improvement of the initial concept. Obtained results prove that the implementation of the proposed methodology enables the development of highly customized and functional orthotic solutions swiftly and economically.

1 INTRODUCTION

Advanced production processes, namely 3D scanning, geometric modelling and additive manufacturing, enable the development and production of customized solutions in an efficient way and at a low cost. The application of a methodology based on these processes is useful in the field of the development of assistive devices given that it is not always possible to find general commercial solutions for very specific cases (Baronio et al. 2016). Furthermore, these processes allow for a high customization and parameterization of the end product, resulting in solutions with better function and comfort for the user (Dombroski et al. 2014). The present project aimed to validate the application of a methodology of product development to the manufacturing of an upper limb passive exoskeleton, bespoke for a seven-year-old child who has lost her right humeral diaphysis as a consequence of osteomyelitis (see Figure 1).

Osteomyelitis is an infection caused by bacteria that leads to bone tissue destruction (Lew & Waldvogel 2004). In this particular case, the absence of the humerus diaphysis results in structural and functional changes of the arm, namely a decrease in the articular function of the elbow and shoulder joints and a shortening of the referred limb (see Figure 1).

After a comprehensive literature review and analysis of the existent market solutions, no records of any type of orthotic devices were found to correct or attenuate the referred condition. Thus, the development of a functional and customized solution, adapted to the dimensions and anthropometric characteristics of the patient in study, became necessary.

DOI: 10.1201/9781003217152-50

(a) (b) (c)

Figure 1. Case study: Child with humeral osteomyelitis. a) Posterior view (shortening and lack of structural support of the arm segment). b) Sagittal view (abnormal function of the elbow and shoulder joints with contralateral arm compensatory action). c) X-ray image showing loss of humeral diaphysis.

2 MATERIALS AND METHODS

Considering the goal of developing a functional upper arm orthotic device, i.e., an assistive device that improves the stability of the upper limb and assures the physiological movement of the elbow joint, the product development methodology presented in Figure 2 was applied, based on the approach proposed by Ulrich & Eppinger (2015).

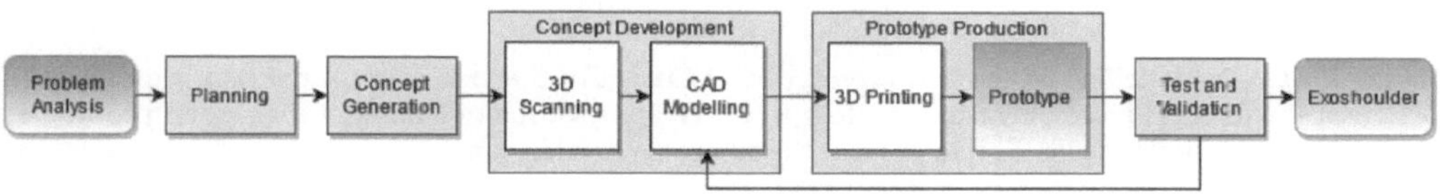

Figure 2. Workflow of the product development methodology. Grey blocks represent the main stages for the design and development of the prototype. Pink blocks represent methodology outcomes.

This methodology can be divided in five different stages: planning, concept generation, concept development, prototype production, and test and validation. The first stage, planning, is where project needs and key features are defined. With this in mind, the analysis of the project needs identified the reestablishment of the upper limb function, the prevention of the arm shortening, and the elbow flexion promotion and hyperextension prevention as main objectives. Moreover, project specifications, such as to have a solution adaptable to growth, comfortable to wear, light and compact (allowing for its use under clothing), and with fast and affordable production, were also defined.

The concept generation stage involves the tasks of analysis of requirements and general concept definition. Regarding this stage, the key points identified for the product development were that the orthosis should have an exterior structure that provides stiffness to the arm segment. This external structure should be composed by two pieces articulated at the elbow, it should have joint stops at the elbow joint to prevent its hyperextension, and use hook-and-loop straps for a comfortable and efficient adjustment of the device to the underlying biological structures. For the purpose of this work it was decided that, for simplicity reasons, the prototype to be produced as proof of concept should present a fixed shoulder joint, as further research on a compact, yet movable solution for this joint is necessary. In what concerns biomechanical and technological requirements, for the former it was defined that the range of motion (ROM) of the elbow joint should vary between 0° to 145° and that the rotation axis of the orthosis should be aligned with the biological axis of the elbow joint, while for the later it was decided that the use of additive manufacturing, through Fused Deposition Modelling (FDM), with low-cost thermoplastics (e.g., PLA, ABS or PETG) should be adopted.

The concept development stage includes the acquisition of the surface shape of the anatomical segments for the customization of the solution to the anthropomorphic features of the user, and the development of the associated 3D geometric models using computer-aided design (CAD) software. The proof of concept was developed for a twenty-year-old volunteer with no history of upper limb disorders. The trunk and right upper limb segments were acquired using the 3D scanner *EinScan Pro* and *EinScan Pro v2.6* software (SHINING3D®, Hangzhou, China), leading to the generation of a point cloud with high dimensional precision. Geometric modelling steps were conducted, using the *MeshLab* (ISTI – CNR, Pisa, Italy) and *NX Siemens* (Siemens PLM Software, Plano, Texas) software packages (Lopes et al. 2017, Ferreira et al. 2019), to adapt the CAD model to a mesh model representative of the body segments (see Figure 3).

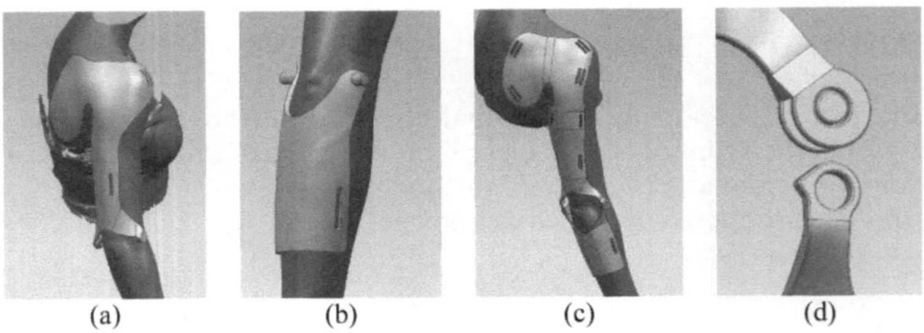

Figure 3. CAD Modelling of the two prototypes. a) Design of the arm piece of the first prototype. b) Design of the forearm piece of the first prototype. c) Assembly of the two parts of the second prototype. d) Close-up of the joint mechanism to prevent the elbow hyperextension.

The prototype production stage consists of using additive manufacturing technologies to obtain a functional prototype. The software *Ultimaker Cura* (Ultimaker, Utrecht, The Netherlands) and the 3D printers *Ultimaker 3* (Ultimaker, Utrecht, The Netherlands), for the forearm piece, and *Sharebot XXL* (Sharebot, Nibionno, Italy), for the arm piece, were used for the additive manufacturing of the prototype, using PLA. The parts were printed using the *Ultimate Cura* fine settings, a gyroid infill structure of 50% and a wall thickness of 1 mm.

The test and experimental validation stage considers the experimental analysis of the developed solution, keeping in mind its possible improvement or validation. After production, the first prototype was tested according to the needs and specifications defined for the project. Several positive aspects were noted, namely, the ergonomics and adaptability of the structure to the upper limb, prototype stiffness and functionality, the alignment of the device joint axis with the biological rotation axis of the elbow joint, and the fixation of the device to the arm and forearm using hook-and-loop straps. However, some aspects were also identified for improvement, namely, the overall weight and volume of the solution that can be further reduced without compromising its function, the volume of the shoulder joint whose bulky design prevents the use of the device under clothing, and, finally, a more efficient fixation of the device to the trunk.

After making the referred changes, the second prototype was printed. The identified improvements were implemented, namely the elbow joint was redesigned in a more compact way, with joint stops included to avoid elbow hyperextension. The device fixation to the trunk was reinforced with an additional strap, as well as the fixation to the lower arm. Overall, the orthotic device became lighter (weighting only 142 grams), more comfortable and more ergonomic.

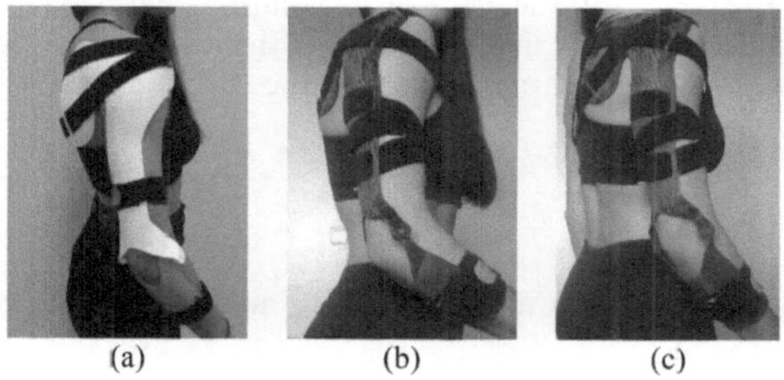

Figure 4. The EXOSHOULDER prototypes: a) First prototype. b) Second prototype – slimmer design, more compact elbow joint, extra strap support on the upper arm and shoulder, overall weight and volume reduction (side view). c) Second prototype (back and side view).

3 RESULTS

From the concept generation phase, a two-part elbow articulated device was envisaged. The upper part was designed to replace the structural function of the humerus, providing stiffness to the arm segment, while the lower part was designed to be attached to the forearm, allowing for the rotation of this segment at the elbow joint level.

After the first testing phase, possible improvements, related with structure ergonomics and elbow function, were identified. Specifically, mechanical joint stops were included in the device joint with the goal of preventing the hyperextension of the biological elbow joint, and new anchor points were added to improve the device fixation to the biological structure, decreasing the pressure points.

4 DISCUSSION

Between the first and second prototypes, the weight was reduced by 30% and there was also a reduction of overall volume. The obtained product allows the normal ROM of the elbow for a healthy subject. Moreover, the changes in the structural parts, namely the redesign of the elbow joint and shoulder interface, resulted in increased comfort during long-term tests.

The designed model is easily scalable, which is a characteristic of particular relevance, since the model can be quickly adapted to the child's growth. Nevertheless, it is important to note that a fully customized solution, based on a 3D scanning of the user, is, in general, the best option, since the solution mimics in more detail his anthropometric characteristics.

The prototype was successfully printed using PLA, a low-cost thermoplastic. The overall printing time (of the two pieces) was close to 40 hours, considering fine printing settings. Due to the low cost of PLA and its large availability, the exoskeleton components can be easily substituted and on a budget, not requiring specific printers or brands.

5 CONCLUSION

The methodology proposed in this work allows for the generation of a fully customized solution, fulfilling the specifications defined in the concept generation phase. The acquisition of the body segments geometry enables the adaptation of the solution to the anthropometric characteristics of the user, guaranteeing a long-term comfort and correct mobility of the elbow joint. Lastly, by using advanced manufacturing processes, such as 3D printing, it is possible to produce customized solutions fast and at a low cost.

The next step of the EXOSHOULDER project is modelling the shoulder joint, without compromising the compactness of the developed solution. Moreover, to more accurately understand the implications of this methodology in the case study, the proof of concept should be validated in the child, guaranteeing that the orthotic device provides the desired support and function.

ACKNOWLEDGEMENTS

This work was supported by FCT, through IDMEC, under LAETA, project UIDB/50022/2020.

REFERENCES

Baronio, G., Harran, S. & Signoroni, A. 2016. A critical analysis of a hand orthosis reverse engineering and 3D printing process. In: *Applied bionics and biomechanics*, 2016(8347478), pp. 1–7.

Dombroski, C.E., Balsdon, M.E.R. & Froats, A. 2014. The use of a low cost 3D scanning and printing tool in the manufacture of custom-made foot orthoses. In: *BMC Research Notes*, 7(1), 443, pp. 1–4.

Ferreira, I., Castro, A., Costa, M.J., Leite, M. & Silva, M.T. 2019. Design and Manufacturing of a Customized Prosthetic Foot for Amputee Children. In: *8º Congresso Nacional de Biomecânica, Covilhã, Portugal, 15–16 February 2019.*

Lew, D.P. & Waldvogel, F.A. 2004. Osteomyelitis. In: *The Lancet*, 364(9431), pp. 369–379.

Lopes, V., Cardoso, R., Leite, M. Silva, M.T., Costa, M.J. & Castro, A. 2017. Concepção e Desenvolvimento de uma Prótese do Membro Superior. In: *7º Congresso Nacional de Biomecânica*, Guimarães, *Portugal, 10–11 February 2017.*

Ulrich, K.T & Eppinger, S.D. 2015. *Product design and development* 6th Ed. New York: McGraw-Hill Education.

Advances and Current Trends in Biomechanics – Belinha et al. (Eds)
© 2022 Copyright the Author(s), ISBN: 978-1-032-10806-3

Development of 3D-printed (FFF) devices to fight COVID-19 transmission

A. Pais, V. Pires, C. Ferreira, J Silva & J.L. Alves
INEGI, Faculty of Engineering, University of Porto, Portugal

J. Bastos & J. Belinha
ISEP, School of Engineering, Polytechnic of Porto, Portugal

ABSTRACT: This work consists on the description of the development process of several devices to prevent COVID-19 spread in common places such as door handlers and sanitizer dispensers. In order to obtain functional prototypes as fast as possible, the Fused Filament Fabrication (FFF) process was used. The chosen material to print the devices was Polylactic acid (PLA), as well as other anti-bacterial polymers. The mechanical functioning of the devices was validated using numerical simulation with the finite element method. The solutions obtained proved to be efficient, functional, and to present a lower cost than already available solutions.

1 INTRODUCTION

All over the world, additive manufacturing (AM) revealed to be a powerful tool in mitigating shortages (Choong *et al.*, 2020; Tino *et al.*, 2020), being face shields for medical professionals one of the most developed designs (Bishop and Leigh, 2020; Celik *et al.*, 2020), as well as 3D-printed masks (Swennen, Pottel and Haers, 2020; Tarfaoui *et al.*, 2020). All these developments were made possible with the use AM, or 3D printing, which bypasses the need for expensive tooling to rapidly develop functional prototypes. Anti-bacterial polymers are also useful for printing these devices. With the growth of AM in biomedical engineering, these materials are gaining more popularity for printing medical devices and implants. Thus, it is shown how these materials present potential for printing clean devices for personal use or to equip public spaces. One of the possible methods to obtain antibacterial polymers is with the addition of metal nanoparticles, such as copper. The antibacterial polymers used in this work were obtained through this method.

The work being proposed is focused on developing devices that do not require medical approval or certification to facilitate daily lives under the circumstances where touching any surface that other people may have touched may be considered hazardous. It has been studied that the novel coronavirus, SARS-CoV-2 can persist in plastic or stainless steel surfaces for up to 72 hours (Van Doremalen *et al.*, 2020). Doorknobs, elevator buttons, handrails are some of the most touched surfaces. If these are adapted to avoid touch, contagion can be reduced, since it is not as common to touch the face with the forearm or elbow. Therefore, some devices are proposed to reduce touching frequent surfaces with one's hand. The developed devices include: a personal hook, a fix door hands-free (forearm) door opener, a hands-free and also electronics-free hand sanitizer dispenser also elbow activated, and also a finger cover for button pressing which is stored in a personal hook.

DOI: 10.1201/9781003217152-51

2 MATERIALS AND METHODS

The additive manufacturing technology chosen to construct the parts was fused filament fabrication (FFF). This process consists of the deposition of material in the form of a filament, layer-by-layer. To make the larger parts, the chosen material was PLA (E = 3500MPa, σ^{max} = 53MPa)(FilamentPM, no date). In addition, antibacterial PLA and TPU were used to print smaller parts which come in direct contact by the user. The parts do not present very intricate detail and thus it is not necessary to have maximum resolution. Thus, it was opted to use standard printing parameters from the slicing software with 0.2mm layer height. The only exception to this layer definition was the clean TPU finger cover which, due to detail, was printed with a resolution of 0.15mm. The remainder printing parameters, namely printing speed and temperatures are dependent on the building material.

To avoid transmission in doorknobs, a device, nicknamed of HANDGENIC was designed to open doors without using one's hand. This device was designed to fit the standard doorknobs found in FEUP which are double curved tubular doorknobs. This device includes a top cover made of anti-bacterial PLA so the area being touched by the user is made of anti-bacterial polymer. TPE grips help adjusting the device to the door handle and a flexible PLA grip improves contact with the door. For the finite element method (FEM) analysis, it was considered that the part would be assembled on a 20mm diameter doorhandle. In addition, two loads were added, to simulate the force applied to the device when opening the door, a vertical load to simulate the activation of the doorhandle and a horizontal load to simulate the pulling of the door. This part is printed with 100% infill since it is subject to demanding and frequent loading.

The elbow-activated sanitizer dispenser was developed to avoid situations where one will touch the dispenser which might be broken or empty and thus is touching a dirty surface. This device is fixed on the wall using adhesives. The FEM analysis considered only the support which holds the sanitizer bottle and so, it was defined as fixed on the wall as well as vertical loading to simulate the weight of the sanitizer bottle.

A portable solution to open doors was also created, the HYHOOK, which includes a thimble-like finger cover made of antibacterial TPU. This finger cover is kept in a casing integrated in the hook so that the "dirty" device does not have to be in contact with the user's belongings. For the FEM analysis of this part, it was considered that the hook was fixed in the extremity which holds the casing and a vertical load directed outwards. Figure 1 shows the three developed devices.

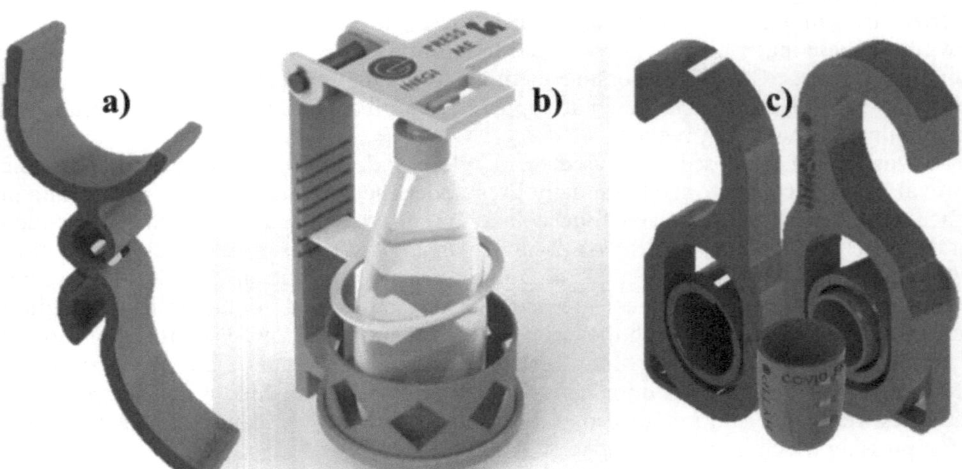

Figure 1. Devices developed a) HANDGENIC (main body), b) elbow activated sanitizer dispenser, c) HYHOOK and finger case.

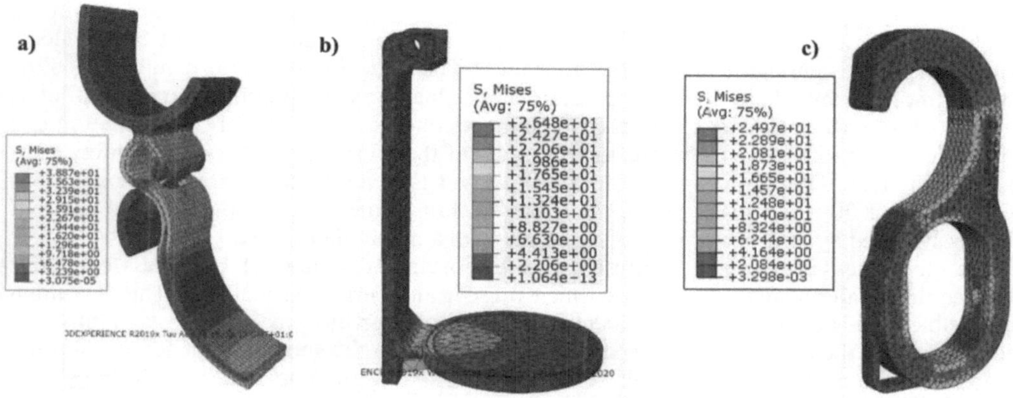

Figure 2. Results of the analysis done in ABAQUS a) handgenic, b) sanitizer dispenser support, c) hook part of the hyhook.

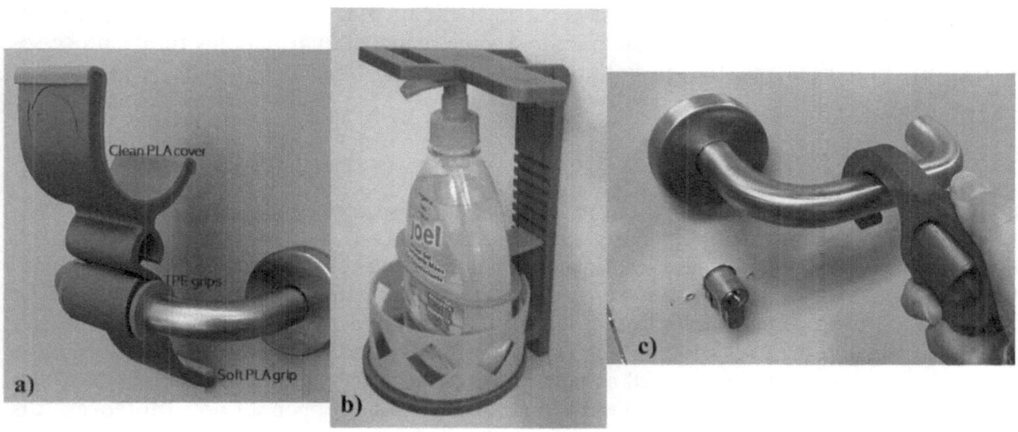

Figure 3. Printed devices in functioning a) handgenic assembled on the door with the accessory parts, b) elbow activated sanitizer dispenser, c) hyhook.

3 RESULTS

The results of the FEM simulation for the three devices subject to more complex loading are shown in Figure 2. All parts were modelled with PLA material. In addition, for safety, it was considered a maximum allowable stress of 40MPa. For the HANDGENIC, a maximum vertical load of 44N and a maximum horizontal load of 22N was obtained. The maximum weight that the sanitizer dispenser can withhold is 284N. The maximum load supported by the HYHOOK can go up to is 320N. Figure 3 shows the three devices assembled and in usage.

4 DISCUSSION

Numerical simulation was conducted to attest the capability of each device as well as defining its maximum admissible load. The parts which are subject to higher loading are made in PLA. The parts were analysed in the elastic domain. For most applications, the parts can withstand the necessary load without deforming. For the case of the HANDGENIC, the maximum stress occurs in the tweezer mechanism because of assembling the component on the door. Moreover, 44N is enough to rotate the doorknob. The critical part of this component is the tweezer mechanism, where there is a constant stress due to assembling in the door and intermittent stresses when the door is opened. The main concerns are, therefore, deformation due to stresses, and fatigue due to the frequent opening of the door. For that reason, a safety coefficient was used. The door handle solution proposed is quite different from other solutions (Chen *et al.*, 2020; François *et al.*, 2020). It can be transformed into an antibacterial device if PLA is switched by anti-bacterial PLA, however with a much higher cost.

The elbow activated dispenser presents the important advantage of being less expensive than commercially available solutions which present an average cost of 20€. The maximum admissible load is enough to hold a 500ml sanitizer container. It is necessary to take into consideration that the adhesive must be strong enough to keep the support attached to the wall. Moreover, it is possible to make this part safer by building the upper piece with antibacterial PLA.

Regarding the HYHOOK, even though it is larger, this solution presents a double purpose, unlike any other solution found in open sources. The FEM analysis showed that the maximum admissible load is more than sufficient to open a door and that the most critical area is at the hook part where the section is minimum, and stress is mostly tensile. Therefore, this part can be printed with some infill to save material, obtain a lighter part and reduce the overall cost. However, if this part is manufactured using conventional manufacturing processes such as injection moulding, the interior infill is not possible.

5 CONCLUSIONS

The presented devices show different approaches to avoid surface contact and present interesting solutions to the university buildings and avoid surface contagion. Additive manufacturing allowed to fully develop all these solutions in a short period. The sanitizer dispenser, besides being elbow activated, presents a lower unit cost than commercially available solutions. The hook solution is a compact and effective solution to distribute among students and university personnel. Finally, the doorknob solution proved effective and ergonomic, and sturdier than some solutions found online, as well as being a single part instead of an assembly. PLA was used as a resistant material to withstand the necessary loads to open a door. Moreover, it is possible to replace some of these components with anti-bacterial polymer however at a higher cost. Finally, some solutions can be adapted to allow large series manufacturing through injection moulding, namely, the HYHOOK, and the individual components of the sanitizer dispenser.

ACKNOWLEDGEMENTS

The authors acknowledge the funding provided by Ministério da Ciência, Tecnologia e Ensino Superior – Fundação para a Ciência e a Tecnologia (Portugal), by project FCT/RESEARCH4-COVID-19/205_596864527 “Assisting the prevention and control of covid-19 with 3D printing solutions” and the funding provided by LAETA with project UIDB/50022/2020.

REFERENCES

Bishop, E. G. and Leigh, S. J. (2020) ‘Using Large-Scale Additive Manufacturing as a Bridge Manufacturing Process in Response to Shortages in Personal Protective Equipment during the COVID-19 Outbreak’, *International Journal of Bioprinting*, 6(4).

Celik, H. K. *et al.* (2020) ‘Design and additive manufacturing of a medical face shield for healthcare workers battling coronavirus (COVID-19)’, *International Journal of Bioprinting*.

Chen, K.-L. *et al.* (2020) 'Novel design for door handle—a potential technology to reduce hand contamination in the COVID-19 pandemic'. Elsevier.

Choong, Y. Y. C. *et al.* (2020) 'The global rise of 3D printing during the COVID-19 pandemic', *Nature Reviews Materials*, pp. 3–5. doi: 10.1038/s41578-020-00234-3.

Van Doremalen, N. *et al.* (2020) 'Aerosol and surface stability of SARS-CoV-2 as compared with SARS-CoV-1', *New England Journal of Medicine*, 382(16), pp. 1564–1567.

FilamentPM (no date) *PLA*. Available at: https://shop.filament-pm.com/pla.

François, P.-M. *et al.* (2020) '3D-printed contact-free devices designed and dispatched against the COVID-19 pandemic: The 3D COVID initiative', *Journal of stomatology, oral and maxillofacial surgery*.

Swennen, G. R. J., Pottel, L. and Haers, P. E. (2020) 'Custom-made 3D-printed face masks in case of pandemic crisis situations with a lack of commercially available FFP2/3 masks', *International Journal of Oral and Maxillofacial Surgery*.

Tarfaoui, M. *et al.* (2020) '3D Printing to Support the Shortage in Personal Protective Equipment Caused by COVID-19 Pandemic', *Materials*, 13(15), p. 3339.

Tino, R. *et al.* (2020) 'COVID-19 and the role of 3D printing in medicine'. Springer.

Advances and Current Trends in Biomechanics – Belinha et al. (Eds)
© 2022 Copyright the Author(s), ISBN: 978-1-032-10806-3

Optimized bone scaffold using a bio-inspired remodelling algorithm (BIRA) and additive manufacturing

A. Pais & J.L. Alves
INEGI, Faculty of Engineering, University of Porto, Portugal

J. Belinha
ISEP, School of Engineering, Polytechnic of Porto, Portugal

ABSTRACT: This work proposes a methodology for the design of prosthetic parts with improved mechanical behaviour, by functional gradients of the gyroid shape with the objective of mimicking bone properties through gyroid scaffold. The gyroid infill was first characterized to obtain a homogenous constitutive model which correlates the elasticity modulus and ultimate stress with the apparent density. The bio-inspired remodelling algorithm (BIRA) was tested with two numerical methods, the FEM and a meshless method. The density distribution obtained proved to be coherent with the density distribution of bone tissue. Then, based on the obtained density distribution, prototypes and experimentally tested. The obtained results can be extrapolated to other materials for inexpensive prototyping.

1 INTRODUCTION

For a long time, functionally graded structures are recognized as an important characteristic in the behaviour of living tissue (Pompe *et al.*, 2003). The development of additive manufacturing facilitated the production of bone tissue scaffolds (Qu, 2020) and the particularities of each different manufacturing process bring different advantages. In the literature, Triply Periodic Minimal Surfaces (TPMS) have often been used as a possibility for bone scaffold (Feng *et al.*, 2018).

The easily controlled density, and the consequent control of elastic modulus, permits to avoid stress shielding, which occurs when there is an abrupt difference in elasticity modulus (Ma et al., 2019). Moreover, gyroid scaffolds are, according to (Melchels et al., 2010; Santos et al., 2020), more easily wetted and intruded, which results in a more homogeneous cell distribution and deeper cell colonization. As pointed out by (Harrysson et al., 2008), optimization algorithms can be used to obtain bone implants with improved properties to avoid stress shielding. Therefore, this work will aim at producing an optimized femur-shaped scaffold using bio-inspired remodelling algorithms that creates an optimized density map. Then, based on that density map, a density gradient with the gyroid infill is created.

DOI: 10.1201/9781003217152-52

2 MATERIALS AND METHODS

Initially, a homogenized gyroid law was developed through tensile and compressive tests on gyroid infill specimens. The gyroid infill was defined from PrusaSlicer v2.2.0. Polylactic acid (PLA) in color "Premium PLA Robotic Grey" (E = 3145 MPa, ρ=1.25g/cm^3) (FormFutura, 2020) was used as the material for construction of the prototypes as well as the infill specimens which were printed using an Ender3 FFF printer. The resultant gyroid material law is:

$$E = -482.65\rho_{app}^3 + 938.34\rho_{app}^2 + 27.693\rho_{app} \quad (1)$$

$$\sigma_c = -18.853\rho_{app}^3 + 43.021\rho_{app}^2 + 12.242\rho_{app} \quad (2)$$

where E is the apparent Young's modulus expressed in MPa, σ_c is the ultimate compressive stress expressed in MPa and ρ_{app} is the average apparent density expressed in g/cm^3.

For the remodelling analyses, a voxelized femur head model with 2.5x2.5x2.5mm^3 voxels was used. The boundary conditions from the human body (Beaupré, Orr and Carter, 1990; Belinha, 2014) were simplified for easier implementation. The model and the boundary conditions are shown in Figure 1a. The model was clamped at the base and two loads were applied, corresponding to the weight and the muscular load.

The remodelling analysis on the gyroid infill model considered a bone density increase ratio of 0.00 and a bone density decrease ratio of 0.075. Moreover, each point was allowed a maximum number of remodels of 2. The maximum density was defined as 1.06 g/cm^3 (which corresponds to an infill density of 85%) and minimum density was defined as 0.125 g/cm^3 (which corresponds to 10% infill density). To compare the resultant density distribution with the density distribution of bone tissue, a similar model was tested with a density increase ratio of 0.00, a density decrease ratio of 0.05 and a maximum number of times that a point can remodel of 2. The maximum density was defined as 2.1 g/cm^3, the minimum density was defined as 0.1 g/cm^3 and the cortical/trabecular transition was defined as 1.3 g/cm^3. Two numerical methods were used for the remodelling analysis, namely, the finite element method (FEM) and a meshless method, the Natural Neighbour Radial Point Interpolation Method (NNRPIM) (Belinha, 2014).

Following the remodelling analysis, the prototype models were defined. The density distribution was replicated by dividing the model into 3 density levels corresponding to a maximum, medium and minimum density. Each of these density levels corresponds to a solid which is then infilled at its average apparent density on the slicing software. The printed prototypes correspond to the remodelling results at 50% and 33% of the initial mass, for remodelling analysis using both methods. A load case similar to the simulated load case was recreated in order to conduct experimental testing. Only the compressive load was applied to the specimen as it is the most prominent because the tensile load is 1/3 of the compressive load. Moreover, in order to keep the prototype in place, the base of the specimen was fixed. The numerical approximation experimental boundary conditions are shown in Figure 1b and the test setting is shown in Figure 1c. To compare the numerical and experimental results, the obtained density distributions were re-tested numerically according to the experimental conditions.

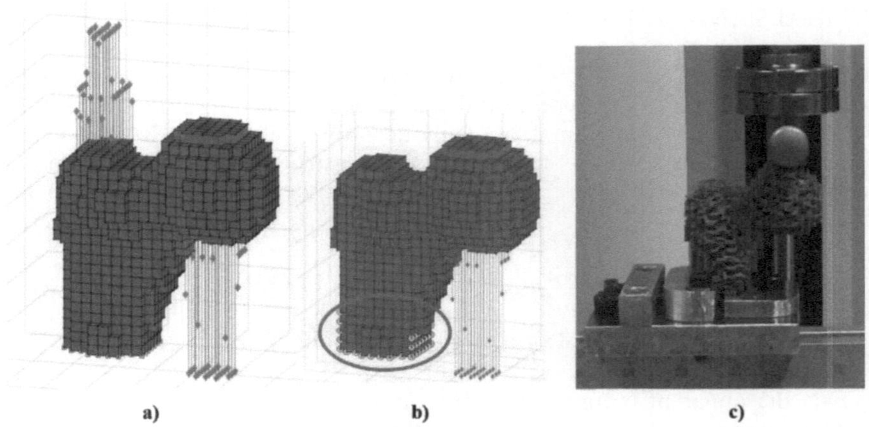

Figure 1. A) Design boundary conditions; b) approximated boundary conditions for experimental testing; c) experimental test conditions.

3 RESULTS

The remodelling analyses allowed to compare the density distributions obtained with the gyroid material law and the bone tissue material law. Figure 2a compares the results obtained for both materials and numerical methods, at 50% of the initial mass, in order to demonstrate how both laws present a similar behaviour. In order to evaluate the mechanical behaviour of the gyroid structure, stiffness K was measured as a function of the apparent density, shown in

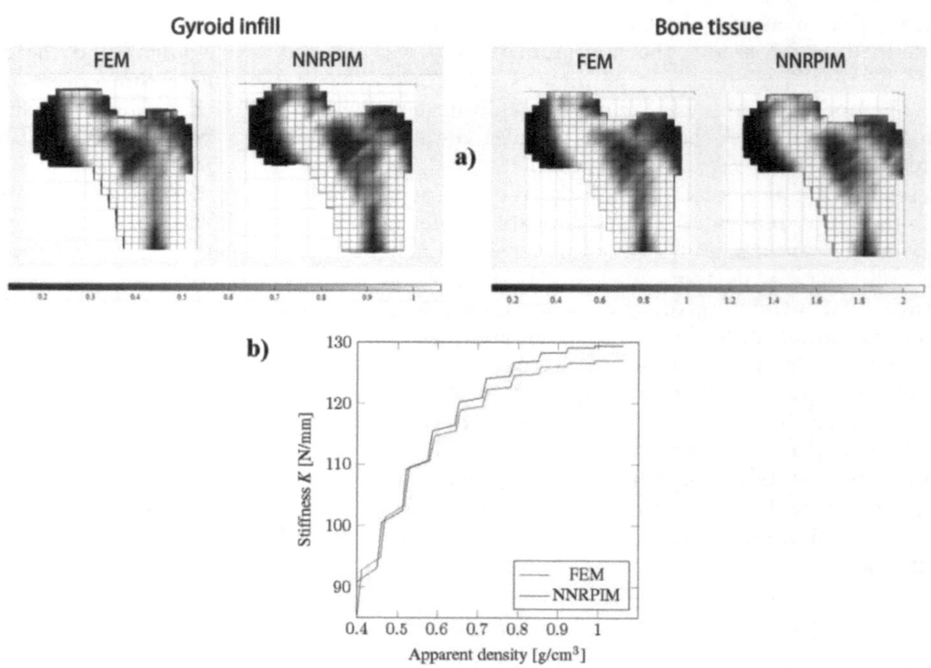

Figure 2. A) Density distribution at the centre cross section for the gyroid model and bone tissue model, at 50% of the initial mass; b) stiffness evolution of the gyroid model for both methods tested.

Table 1. Summary of the experimental tests results and comparison with numerical results for validation of the model.

	vf	m [g]	K [N/mm] Experimental	Numerical Design BC	Numerical Adjusted BC	Error
FEM	33%	13.8	130	85	165	21%
	50%	20.3	285	115	220	30%
NNRPIM	33%	12.5	160	91	162	1%
	50%	18.9	320	116	218	47%

Figure 2b. The results of the experimental tests run on prototypes at 33% and 50 % of the initial mass (vf), and its comparison to numerical results are shown in Table 1.

4 DISCUSSION

With respect to the remodelling analyses, the obtained density distribution for the gyroid infill model obtained a similar density distribution to the model tested with a bone tissue constitutive law. Thus, it can be proved that for equivalent volume fractions, the gyroid infill can mimic the behaviour of cortical or trabecular bone. Moreover, the obtained density distribution is coherent with the literature for analysis ran on realistic femur models, which validates the voxelized model as an approximation. When comparing the differences between the results obtained with the two numerical methods, it is not possible to come to any conclusion, as a very coarse mesh/nodal distribution was used. It would be necessary to use a much finer discretization in order to attest for any significant differences.

Regarding the evolution of stiffness along the remodelling process, there is a slight decrease until 0.8g/cm^3 and a more abrupt decrease until 0.4g/cm^3. Both methods predicted similar displacement values. Once more, the very coarse discretization achieved similar density distributions and thus, similar mechanical behaviour.

The experimental study proved the difficulties when testing the loads in the human body. A similar test setting to (Harrysson *et al.*, 2008) was replicated. The disparity of relative error results was caused by the difficulty in repeating the test conditions exactly for all tests. However, the results can be used as guidance in the iterative design process. In addition, the results on the PLA structures can be extrapolated to other materials to inexpensively develop scaffold prototypes.

5 CONCLUSIONS

The gyroid constitutive model implemented in the remodelling analysis obtained promising results in describing bone distribution for equivalent volume fractions. Both the FEM and meshless method obtained similar and coherent results with what is described in the literature. The experimental results can be extrapolated to other materials suitable for the fabrication of prosthetic parts. Even though the PLA prototype is easy to manufacture through the simplification of the density gradient, experimental testing is difficulted by the complexity of loading conditions in the human body.

ACKNOWLEDGEMENTS

The authors acknowledge the funding provided by Ministério da Ciência, Tecnologia e Ensino Superior – Fundação para a Ciência e a Tecnologia (Portugal), by project FCT/

RESEARCH4COVID-19/205_596864527 "Assisting the prevention and control of covid-19 with 3D printing solutions" and the funding provided by LAETA with project UIDB/50022/2020.

REFERENCES

Beaupré, G. S., Orr, T. E. and Carter, D. R. (1990) 'An approach for time-dependent bone modeling and remodeling—theoretical development', *Journal of Orthopaedic Research*, 8(5), pp. 651–661.

Belinha, J. (2014) *Meshless Methods in Biomechanics - Bone Tissue Remodelling Analysis*. Edited by J. M. R. S. Tavares and R. M. N. Jorge. Porto, Portugal: Springer.

Feng, J. *et al.* (2018) 'Porous scaffold design by solid T-splines and triply periodic minimal surfaces', *Computer Methods in Applied Mechanics and Engineering*, 336, pp. 333–352.

FormFutura (2020) *Premium PLA Robotic Grey*. Available at: https://www.formfutura.com/shop/product/premium-pla-robotic-grey-288?category=175 (Accessed: 28 April 2020).

Harrysson, O. L. A. *et al.* (2008) 'Direct metal fabrication of titanium implants with tailored materials and mechanical properties using electron beam melting technology', *Materials Science and Engineering C*, 28(3), pp. 366–373. doi: 10.1016/j.msec.2007.04.022.

Ma, S. *et al.* (2019) 'Mechanical behaviours and mass transport properties of bone-mimicking scaffolds consisted of gyroid structures manufactured using selective laser melting', *Journal of the Mechanical Behavior of Biomedical Materials*, 93(January), pp. 158–169. doi: 10.1016/j.jmbbm.2019.01.023.

Melchels, F. P. W. *et al.* (2010) 'Mathematically defined tissue engineering scaffold architectures prepared by stereolithography', *Biomaterials*, 31(27), pp. 6909–6916.

Pompe, W. *et al.* (2003) 'Functionally graded materials for biomedical applications', *Materials Science and Engineering A*, 362(1–2), pp. 40–60. doi: 10.1016/S0921-5093(03)00580-X.

Qu, H. (2020) 'Additive manufacturing for bone tissue engineering scaffolds', *Materials Today Communications*, 24(November 2019), p. 101024. doi: 10.1016/j.mtcomm.2020.101024.

Santos, J. *et al.* (2020) 'On the permeability of TPMS scaffolds', *Journal of the Mechanical Behavior of Biomedical Materials*, 110(March), pp. 1–7. doi: 10.1016/j.jmbbm.2020.103932.

Advances and Current Trends in Biomechanics – Belinha et al. (Eds)

Review: Study of erythrocyte cells

N. Gomes & J. Belinha
School of Engineering, Polytechnic of Porto (ISEP), Porto, Portugal

M.I.A. Barbosa
Institute of Science and Innovation in Mechanical and Industrial Engineering (INEGI), Porto, Portugal

ABSTRACT: Erythrocytes are blood cells without a nucleus and with a shape of a biconcave disk. These cells have the main function of transporting oxygen through hemoglobin. Factors such as genetic alterations have consequences in these cells, causing a decrease in red blood cell deformability and permeability, and reducing red blood cell survival. Hereditary disorders of the erythrocyte membrane are caused by mutations in genes that encode various transmembrane or cytoskeletal proteins of red blood cells. These disorders include a set of hemolytic anemias that differ greatly in terms of clinical, morphological, laboratory, and molecular characteristics. This review work presents a summary of the main characteristics of erythrocytes and their pathophysiological changes, in order to allow a better basis for diagnosis and treatment.

1 INTRODUCTION

With the development of nanotechnology, the investigation of the characteristics and mechanical properties of materials thinner than a nanometer has aroused interest in the areas of engineering, biology, and medicine (Suresh, 2006). The erythrocyte, also called red blood cell, has one of the most simple and unique molecular architectures among different types of cells (Peng, Asaro and Zhu, 2010) (Mohandas and Gallagher, 2008). In addition to not having a nucleus, a mature erythrocyte has a cytoplasmic matrix closed by a highly flexible and strong membrane (Peng, Asaro and Zhu, 2010). Moreover, its membrane is the only structural constituent that guarantees its diverse antigenic, transport, and mechanical characteristics (Mohandas and Gallagher, 2008). In 1968, Dutch biologist Jan Swammerdam analyzed and described red blood cells for the first time. However, Antonie van Leeuwenhoek was the first to publish a description of the unique characteristics of red blood cells in the *Philosophical Transactions of the Royal Society* in 1675. Currently, new and unpredictable functions of the red blood cell membrane continue to emerge (Mohandas and Gallagher, 2008). An important criterion for simulating the flow of red blood cells in microcirculation is their mechanostructural characteristics. In the past few decades, many researchers have attempted to describe the micromechanics of red blood cells, producing various mathematical and numerical models (Ju *et al.*, 2013). Throughout this article, the structure of erythrocytes will be addressed, as well as the mechanical properties and pathologies associated with red blood cells.

DOI: 10.1201/9781003217152-53

2 ERYTHROCYTES

Red blood cells are the most important constituents of blood due to their biological functions and their direct effect on hemodynamics. The main function of erythrocytes is to provide oxygen, in a process that occurs mainly in the microvascular network of arterioles, venules, and capillaries (Ju *et al.*, 2013).

The erythrocyte membrane consists of a double layer of lipid fluid (Smith, 1987). This membrane is linked to an intracellular cytoskeleton by protein-protein and lipid-protein interactions that attribute the erythrocyte shape, stability, and deformability. The lipid bilayer acts as a barrier for the retention of cations and anions within the red cells while allowing water molecules to pass freely (Andolfo *et al.*, 2016).

These red blood cells do not have a nucleus and inside them there is a concentrated solution of hemoglobin, which is responsible for the transport of oxygen throughout the body and for the elimination of carbon dioxide (Belinha *et al.*, 2018)(Secomb, 1992).

With the contribution of hemoglobin, carbonic anhydrase, and the exchange of anions of band 3 protein, red blood cells are able to transport the oxygen supply to tissues optimally, depending on their metabolic activity (Hamasaki and Yamamoto, 2000). The average concentration of hemoglobin in the cell is directly related to the cytoplasmic viscosity and affects the deformability of this cell. In a hypertonic medium, these cells lose water, causing an increase in the average hemoglobin concentration and, consequently, a loss of deformability. However, the lack of erythrocyte water contributes to a local increase in the cytoplasmic viscosity of red blood cells (Kim, Lee and Shin, 2015).

The membrane of red blood cells has in its constitution spectrin, actin, 4.1R protein, ankyrin, as well as other proteins related to actin, which bind to the lipid bilayer through band 3, which provide it with excellent biophysical characteristics, such as high elasticity and the ability to quickly respond to shear stresses imposed externally in the circulatory medium (Lux, 2016)(Narla and Mohandas, 2017). Its structural organization facilitates rapid and reversible deformations in high numbers, preserving its mechanical integrity during its useful life (Narla and Mohandas, 2017).

Geometrically, erythrocytes are 6 to 8 μm in diameter in the circular discoid plane and 2 μm in average thickness in the biconcave radial plane. Due to their circular biconcave shape, these cells are considered highly deformable (Ju *et al.*, 2013). This conformation provides a deformable and resilient surface, allowing it to pass through capillaries with half its diameter, which allows it to flow in microvessels during the transport of oxygen and carbon dioxide (Ballas and Krasnow, 1980) (Kim, Lee and Shin, 2015). Several factors cooperate for the deformability of red blood cells, such as the geometric characteristics of the cell, the rheological properties of the intracellular fluid, and the rheological properties of the cell membrane (Chien, 1987).

3 PATHOLOGIES

Genes that encode transmembrane or cytoskeletal proteins cause hereditary disorders of the erythrocyte membrane. As a consequence of these disorders, a reduction in cell deformability and permeability is observed, and a reduction in red blood cell survival (Andolfo *et al.*, 2016) (Iolascon, Andolfo and Russo, 2019). The main pathologies noted in this context include a group of hemolytic anemias caused by altered structural organization of the membrane, such as hereditary spherocytosis, hereditary elliptocytosis and ovalocytosis in Southeast Asia, or by deficient membrane transport function (Iolascon, Andolfo and Russo, 2019).

Thus, anomalies in the shape of red blood cells and in other characteristics of red blood cells can provide significant information in determining a differential diagnosis (Ford, 2013). The changes in the mechanical properties of red blood cells, such as natural shape and characteristics, have serious consequences on their behavior, leading to hereditary defects or disease states (Belinha *et al.*, 2018) (Buys *et al.*, 2013). Therefore, it is considered important to study the mechanical properties of red blood cells.

As disease management is limited by limited public funding due to reduced economic resources, timely diagnosis and treatment of erythrocyte diseases are essential in order to be effective in reducing the clinical, social, and economic burden (Lippi and Mattiuzzi, 2019).

Hereditary spherocytosis is considered a non-immune hereditary hemolytic anemia, with a prevalence of 1:2000 to 1:5000 births (Iolascon, Andolfo and Russo, 2019). This disease affects all ethnicities, however, it is more common in people of northern European descent (1 in 3000) (Mohandas and Gallagher, 2008). In general, the disease is associated with dominant inheritance (75%), although non-dominant inheritance (25%) is also common (Mohandas and Gallagher, 2008). This disease is a disorder of blood cells, more specifically a mutation of ankyrin and/or other membrane protein genes (Suresh, 2006). The red blood cells of a patient with this disorder show a loss of membrane surface and, as a consequence, a change in the shape of the cells from discocytes to stomatocytes and spherocytes, a spherical shape with a diameter smaller than that of healthy red blood cells (Mohandas and Gallagher, 2008) (Suresh, 2006). The shape of red blood cells prevents them from crossing small blood vessels and microcapillaries (Suresh, 2006).

Hereditary elliptocytosis originates from defects in the proteins of the junctional complex, which impair horizontal cytoskeletal interactions, thus decreasing the mechanical stability and deformability of red blood cells (Iolascon, Andolfo and Russo, 2019). Although this disease is found worldwide, it is usually more dominant in malaria-endemic areas in West Africa, with a prevalence of approximately 2% (Mohandas and Gallagher, 2008). This pathology is an autosomal dominant inherited disease, with the majority of patients being asymptomatic. However, roughly 10% of patients show symptoms of moderate to severe anemia, including some cases of fetal hydrops (Mohandas and Gallagher, 2008).

All forms of hereditary elliptocytosis exhibit an unstable membrane resulting from the progressive transformation of the discocyte to elliptocyte form, and in severe cases, there is fragmentation of the membrane and cells with reduced membrane surface area (Mohandas and Gallagher, 2008). The cause of this condition is attributed to mutations in the α-spectrin, β-spectrin, glycophorin C, or protein 4.1 genes, which are more dominant among people of Mediterranean and African descent, and may offer some resistance to malaria (Suresh, 2006).

In Southeast Asian ovalocytosis, as in hereditary elliptocytosis, heterozygous carriers of this disorder have the advantage of survival against all forms of malaria (Iolascon, Andolfo and Russo, 2019). Ovalocytosis is very common in Southeast Asia, with a prevalence between 5% and 25% (Mohandas and Gallagher, 2008). This is determined by the presence of erythrocytes rounded or ovaloid (Tomaiuolo, 2014). Ovalocyte membranes are 4 to 8 times less elastic than normal membranes (Mohandas and Gallagher, 2008).

4 BIOMECHANICAL STUDIES

In recent years, the study of the mechanical properties of red blood cells has attracted the attention of researchers, since many disorders and diseases of hemolytic anemia are highly relevant to the deformation characteristics of these cells. So, it is important to have a perception of the mechanical properties and deformation behaviors of these cells (Wang, Guo and Su, 2014). Peng et al. (Peng, Asaro and Zhu, 2010) applied the finite element method to simulate the membrane of red blood cells as two distinct layers. Thus, using the multiscale modeling technique, they studied the mechanical response induced by mechanical loads, as well as the interaction force between the protein skeleton and the lipid bilayer. With this analysis, it is possible to approximate/simulate the mechanisms of induced membrane morphological changes and to predict the occurrence of normal or defective erythrocyte remodeling. Hansen et al. (Wang, Guo and Su, 2014) determined a continuous model that predicts the mass elastic torsion module and the membrane area expansion module of the red blood cells. Boey et al. (Wang, Guo and Su, 2014) studied the compression module in the plane and the torsion module of the red blood cell spectrin network in the context of a continuous structure. In their research work, three structural models were suggested, the tension-free model, the prestressing model, and the condensed model. Ademiloye et al. (Ademiloye, Zhang and Liew,

2017) studied the effects of *Plasmodium falciparum* infection and maturation on the malformation behavior of red blood cells infected with malaria using a three-dimensional multiscale meshless method.

5 CONCLUSION

Red blood cells are the most important constituents of blood due to their biological functions. These are non-nucleated cells that have the shape of a biconcave disk. Disorders associated with the membrane of red blood cells can be inherited due to mutations in the genes encoding proteins in the membrane or cytoskeleton, as well as for transporters or transmembrane channels. The main consequences of these genetic changes are the decrease in deformability and permeability of red blood cells, and the early removal of erythrocytes from the bloodstream. The diagnosis of these changes is based on the assessment of clinical characteristics, family history, and laboratory data. Thus, it is intended that with the use of new technologies it should be possible to improve the quality and reduce the time of diagnosis of these pathologies.

ACKNOWLEDGMENTS

The authors truly acknowledge the funding provided by Ministério da Ciência, Tecnologia e Ensino Superior – Fundação para a Ciência e a Tecnologia (Portugal), under grant: SFRH/BD/146272/2019 and the funding provided by LAETA, under project UIDB/50022/2020.

REFERENCES

Ademiloye, A. S., Zhang, L. W. and Liew, K. M. (2017) 'Element-Free Multiscale Modeling of Large Deformation Behavior of Red Blood Cell Membrane With Malaria Infection', *bioRxiv*, pp. 316–319.

Andolfo, I. *et al.* (2016) 'New insights on hereditary erythrocyte membrane defects', *Haematologica*, 101 (11), pp. 1284–1294.

Ballas, S. K. and Krasnow, S. H. (1980) 'Structure of erythrocyte membrane and its transport functions', *Annals of Clinical and Laboratory Science*, 10(3), pp. 209–219.

Belinha, J. *et al.* (2018) *Biodental Engineering V: Proceedings of the 5th International Conference on Biodental Engineering*, *Conference on Biodental Engineering*.

Buys, A. V. *et al.* (2013) 'Changes in red blood cell membrane structure in type 2 diabetes: A scanning electron and atomic force microscopy study', *Cardiovascular Diabetology*, 12(1), pp. 1–6.

Chien, S. (1987) 'Red cell deformability and its relevance to blood flow', *Annual Review of Physiology*, 49, pp. 177–192.

Ford, J. (2013) 'Red blood cell morphology', *International Journal of Laboratory Hematology*, 35(3), pp. 351–357.

Hamasaki, N. and Yamamoto, M. (2000) 'Red Blood Cell Function and Blood Storage', *Vox Sanguinis*, 79(4), pp. 191–197.

Iolascon, A., Andolfo, I. and Russo, R. (2019) 'Advances in understanding the pathogenesis of red cell membrane disorders', *British Journal of Haematology*, 187(1), pp. 13–24.

Ju, M. *et al.* (2013) 'A review of numerical methods for red blood cell flow simulation', *Computer Methods in Biomechanics and Biomedical Engineering*, 18(2), pp. 130–140.

Kim, J., Lee, H. and Shin, S. (2015) 'Advances in the measurement of red blood cell deformability: A brief review', *Journal of Cellular Biotechnology*, 1(1), pp. 63–79.

Lippi, G. and Mattiuzzi, C. (2019) 'Updated Worldwide Epidemiology of Inherited Erythrocyte Disorders', *Acta Haematologica*, 143(3), pp. 196–203.

Lux, S. E. (2016) 'Anatomy of the red cell membrane skeleton: Unanswered questions', *Blood*, 127(2), pp. 187–199.

Manciu, S., Matei, E. and Trandafir, B. (2017) 'Hereditary spherocytosis - diagnosis, surgical treatment and outcomes. A literature review', *Chirurgia*, 112(2), pp. 110–116.

Mohandas, N. and Gallagher, P. G. (2008) 'Red cell membrane: past, present, and future', *Blood*, 112 (10), pp. 3939–3948.

Narla, J. and Mohandas, N. (2017) 'Red cell membrane disorders', *International journal of laboratory hematology*, 39(1), pp. 47–52.

Peng, Z., Asaro, R. J. and Zhu, Q. (2010) 'Multiscale simulation of erythrocyte membranes', *Physical Review E - Statistical, Nonlinear, and Soft Matter Physics*, 81(3), pp. 1–11.

Secomb, T. W. (1992) 'Red blood cell mechanics and Capillary Blood Rheology', *Cell Biophysics*, 18(3), pp. 231–251.

Smith, J. E. (1987) 'Erythrocyte Membrane: Structure, Function, and Pathophysiology', *Veterinary Pathology*, 24(6), pp. 471–476.

Suresh, S. (2006) 'Mechanical response of human red blood cells in health and disease: Some structure-property-function relationships', *Journal of Materials Research*, 21(8), pp 1871–1877.

Tomaiuolo, G. (2014) 'Biomechanical properties of red blood cells in health and disease towards microfluidics', *Biomicrofluidics*, 8(5), p. 051501.

Wang, X., Guo, X. and Su, Z. (2014) 'A quasi-continuum model for human erythrocyte membrane based on the higher order cauchy-born rule', *Computer Methods in Applied Mechanics and Engineering*. Elsevier B.V., 268, pp. 284–298.

Advances and Current Trends in Biomechanics – Belinha et al. (Eds)

Study of the structural behavior of the clavicle using meshless methods

N. Gomes & J. Belinha
School of Engineering, Polytechnic of Porto (ISEP), Porto, Portugal

M.I.A. Barbosa
Institute of Science and Innovation in Mechanical and Industrial Engineering (INEGI), Porto, Portugal

ABSTRACT: The bone of the clavicle is long and of double curvature, composed of a single direct connection between the axial and appendicular skeleton. In this study, responses on the clavicle bone were analyzed when different loads were applied using the Finite Element Method (FEM), and the Radial Point Interpolation Method (RPIM). For this, a clavicle bone was segmented, based on an anonymous set of CT images, and then discretized with a 3D tetrahedral element mesh. Several load cases were assumed, and the corresponding variable fields were obtained: displacement, strain and stress fields. According to the results obtained, the highest stress is achieved when the load is applied to the middle zone of the clavicle.

1 INTRODUCTION

The clavicle bone is a double curvature long bone, composed of a single connection between the axial and appendicular skeletons (Burnham, Kim and Kamineni, 2016). Due to this change in conformation, the rate of fractures examined in this area is higher (Calisal and Ugur, 2018). The finite element method is a well-known discretization technique and is a very effective tool for modeling and simulation in the field of biomedical engineering (Belinha, 2016). In recent years, other numerical techniques, such as meshless methods, have shown relevant developments (Wang and Liu, 2002). With this study, it is intended to better comprehend the structural response of the clavicle bone through computer systems. Thus, the objective of this work is to evaluate cases of vertical and horizontal impacts applied to a clavicle using FEM and RPIM. For each load case analyzed, principal stresses and von Mises stress, distributions were analyzed.

2 MATERIALS AND METHODS

First, the clavicle bone was segmented based on a set of anonymous CT images, and, later, it was discretized into a solid mesh with linear tetrahedral elements. The analysis performed with both FEM and RPIM using FEMAS v2020.1 software (cmech.webs.com), an academic freeware. As a simplification, the clavicle model was considered as fully cortical bone. In this study, the material was assumed as elastic, isotropic, and homogeneous, presenting the following mechanical properties: Young's modulus $E = 17000MPa$ and Poisson's coefficient $\upsilon = 0.3$ (Zhang *et al.*, 2020). The natural and essential boundary conditions were imposed in FEMAS following the literature suggestion (Jarvis, 2016). The white dots at the ends of the clavicle in Figure 1 (b) indicate nodes fixed in all degrees of freedom. To study the effects of impact loads on this bone, six load cases with an equal magnitude value ($F = 700\ N$) were simulated, with three of the loads applied along the Oz axis, and the other three applied along the Ox axis, as shown in Figure 1 (a). It is important to note that the loads were all applied independently of each other, and their analysis was carried out separately.

DOI: 10.1201/9781003217152-54

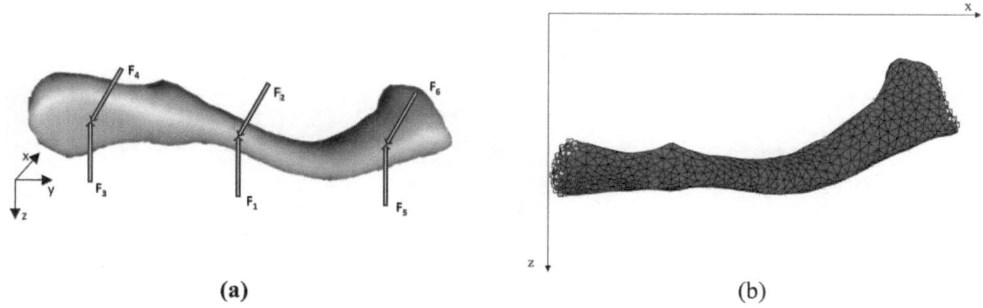

Figure 1. (a) Schematic representation of the load cases; (b) Boundary conditions implemented.

3 RESULTS

Figure 2 illustrates the obtained color maps of the von Mises equivalent stress fields, for each load case. Additionally, Figure 3 and Figure 4 illustrate the color maps of principal stresses σ_1 and σ_3, respectively. Regarding figure interpretation, FEM-F# means that it was analysed load F# using the finite element method and RPIM-F# means that it was analysed load F# using the radial point interpolation method.

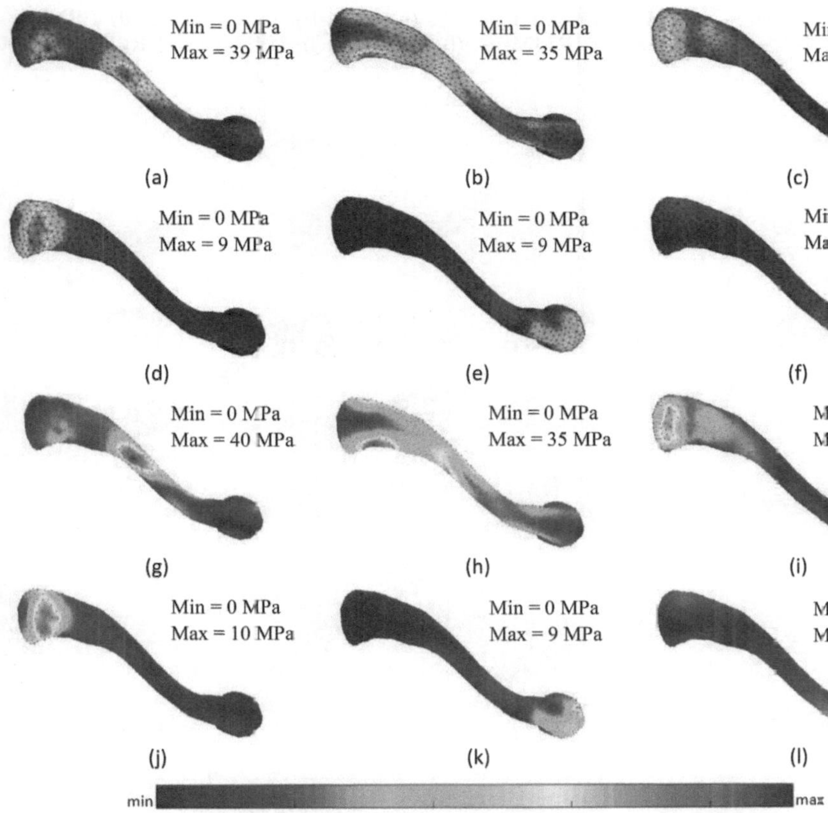

Figure 2. Von Mises stress (MPa) for (a) FEM-F_1, (b) FEM-F_2, (c) FEM-F_3, (d) FEM-F_4, (e) FEM-F_5, (f) FEM-F_6, (g) RPIM-F_1, (h) RPIM-F_2, (i) RPIM-F_3, (j) RPIM-F_4, (k) RPIM-F_5, and (l) RPIM-F_6.

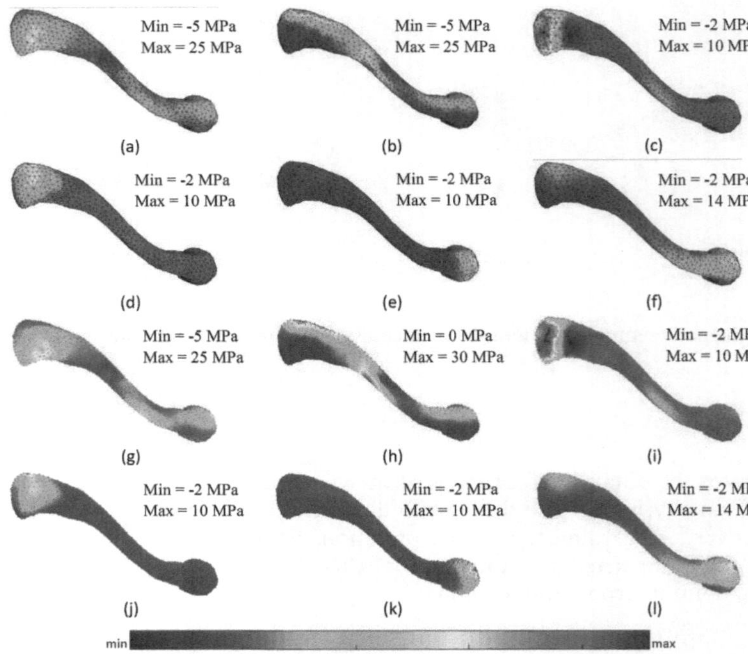

Figure 3. Principal stress σ_1 (MPa) for (a) FEM-F_1, (b) FEM-F_2, (c) FEM-F_3, (d) FEM-F_4, (e) FEM-F_5, (f) FEM-F_6, (g) RPIM-F_1, (h) RPIM-F_2, (i) RPIM-F_3, (j) RPIM-F_4, (k) RPIM-F_5, and (l) RPIM-F_6.

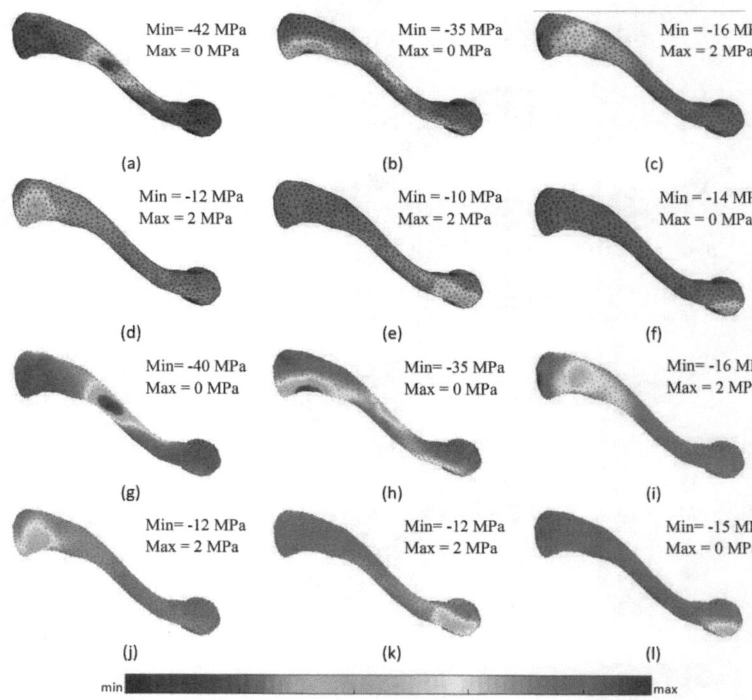

Figure 4. Principal stress σ_3 (MPa) for (a) FEM-F_1, (b) FEM-F_2, (c) FEM-F_3, (d) FEM-F_4, (e) FEM-F_5, (f) FEM-F_6, (g) RPIM-F_1, (h) RPIM-F_2, (i) RPIM-F_3, (j) RPIM-F_4, (k) RPIM-F_5, and (l) RPIM-F_6.

4 DISCUSSION

According to the results obtained, the highest value of von Mises effective stress, in both methods, is reached when the F_1 force is applied, as illustrated in Figure 2. On the other hand, the lower load cases of these stresses are reached when the F_5 force is applied. In the case of the principal stress σ_1, the forces F_1 and F_2, in both methods, induce the highest tensile stress, as shown in Figure 3. However, even for the principal stress σ_1, the lowest values are achieved when the forces F_3, F_4 and F_5 are applied. Also, in both methods, for the principal stress σ_3, the force F_1 gives rise to the highest compressive stress and the force displays the lowest values for this stress, as shown in Figure 4. Thus, according to these results, the F_5 force does not induce high stress levels in the bone, while the F_1 force appears to be the most severe load case. As expected, both F_1 and F_2 forces induce higher stresses in the middle region of the clavicle bone.

5 CONCLUSION

According to the results, regardless the method, the highest von Mises stress is obtained for loads are applied at the middle zone of the clavicle. For both methods, F_1 and F_2 are the forces that induce the highest principal stress σ_1 and F_1 force leads to the highest principal stress σ_3. To conclude, this study allows to better understand the structural response of the clavicle bone to distinct load cases and shown that FEM and RPIM produce similar results.

ACKNOWLEDGMENTS

The authors truly acknowledge the funding provided by Ministério da Ciência, Tecnologia e Ensino Superior – Fundação para a Ciência e a Tecnologia (Portugal), under grant: SFRH/BD/146272/2019 and the funding provided by LAETA, under project UIDB/50022/2020.

REFERENCES

Belinha, J. (2016) 'Meshless Methods: The Future of Computational Biomechanical Simulation', *Journal of Biometrics & Biostatistics*, 07(04).

Burnham, J. M., Kim, D. C. and Kamineni, S. (2016) 'Midshaft Clavicle Fractures A Critical Review', *Orthopedics*, 39(5), pp. e814–e821.

Calisal, E. and Ugur, L. (2018) 'Evaluation of the plate location used in clavicle fractures during shoulder abduction and flexion movements: A finite element analysis', *Acta of Bioengineering and Biomechanics*, 20(4), pp. 41–46.

Jarvis, P. J. (2016) 'Impact Finite Element Analysis of a Detailed Clavicle Implant Model', pp. 1–10.

Wang, J. G. and Liu, G. R. (2002) 'A point interpolation meshless method based on radial basis functions', *International Journal for Numerical Methods in Engineering*, 54(11), pp. 1623–1648.

Zhang, F. *et al.* (2020) 'Finite element analysis of dual small plate fixation and single plate fixation for treatment of midshaft clavicle fractures', *Journal of Orthopaedic Surgery and Research*. Journal of Orthopaedic Surgery and Research, 15(1), pp. 1–7.

Advances and Current Trends in Biomechanics – Belinha et al. (Eds)
© 2022 Copyright the Author(s), ISBN: 978-1-032-10806-3

Bone fracture healing: A review of the remodeling models that simulate bone callus

A. Oliveira & J. Belinha
School of Engineering, Polytechnic of Porto, Porto, Portugal

M. Peyroteo
Institute of Science and Innovation in Mechanical and Industrial Engineering, Porto, Portugal

ABSTRACT: Bone fractures are the most common form of hospitalized trauma. The process of bone healing is complex and involves both biological and mechanical aspects. For this purpose, during the last decade, simulation and modeling of bone healing have been used to understand the details and mechanisms involved and to allow the development of more accurate and rational strategies for the treatment of fractures and to accelerate healing. Many computational models have been established to predict callus formation due to mechanobiological rules. Advances in the mechanobiology of bone enable the understanding of the mechanical response to loading by individual cells, as well as tissue differentiation during events such as fracture healing. The present paper aims to review some of these models of bone regeneration and remodeling.

1 INTRODUCTION

Fractures are one of the most frequent injuries of the musculoskeletal system and the most common form of hospitalized trauma. Unlike other tissues, the "perfect" regeneration capability of the bone is unique, new bone is formed and continuously remodeled until its mechanical properties are restored and the original injured site can hardly be recognized. However, the risk of mal-union, delayed union and non-union remains high, up to 5-10%, contributing to a considerable patient disability, reduced quality of life and significant treatment costs (Ekegren *et al.* 2018). Therefore, a reliable assessment of bone healing is essential for the successful treatment of fractures. Although the bone healing process has been studied experimentally for several decades (McKibbin 1978, Morgan *et al.* 2011, Einhorn & Gerstenfeld, 2015), experimental work involves real scenarios and requires very controlled conditions, state-of-the -art equipment, high precision and cost, and other factors, such as unknown subjects' backgrounds, comorbidities, and genetic variation can influence the outcomes (Moran *et al.* 2016). Additionally, due to the complexity of bone fracture healing, purely experimental approaches are incapable of fully describe the underlying biochemical and biomechanical mechanisms leading to fracture healing (Pivonka & Dunstan 2012). Bone fracture healing involves cell differentiation and tissue remodeling, both of which are influenced by the mechanical environment. Mechanical stimulation can direct the healing process, i.e., tissue differentiation is mechanobiological regulated (Claes & Heigele 1999).

Mechanobiology describes the mechanisms by which biological processes are regulated by signals to cells that are induced by mechanical loads (Van der Meulen & Huiskes 2002). It is possible to observe tissue adaptation in response to changing mechanical stimulation in different tissue types and anatomical locations. Particularly, this is all quite significant in bone,

DOI: 10.1201/9781003217152-55

which is an adaptive material that employs a complex biological system of remodeling in response to mechanical stimulation. Researchers have turned to *in silico* modeling techniques to clarify the stimulation that these cells are subjected to *in vivo* and to determine mechanobiological rules for these adaptive responses. Thus, when the mechanisms of mechanically-regulated tissue formation are well defined, then physiological conditions and pharmacological agents can be patient-specifically developed and applied to prevent non-unions and, furthermore, to promote and help accelerate fracture repair and restore optimal function.

Computational modeling has increasingly grown in bone mechanobiology, stimulating the development of this field, and is recognized as a powerful tool in terms of mechanical interactions at the cellular scale, as well as for the prediction of important resulting tissue-level phenomena, such as cell proliferation and differentiation, tissue growth, adaptation, and maintenance (Van der Meulen & Huiskes 2002). However, some approaches contain some limitations, needing to be refined and validated in order to achieve clinically relevant results.

This review will summarize the work developed over time on mechanobiological regulation and the mechanical factors influencing biological processes and cellular bone healing activities, with the respective models of bone regeneration and remodeling that simulate bone callus.

2 BONE HEALING

Fracture healing is a natural process to reconstruct injured tissue and restore its original function and form. The principles of bone healing are very similar to other processes of bone formation and growth during fetal development. Primary fracture healing (direct healing or intramembranous ossification) involves direct cortical remodeling without any external tissue (callus) formation, and two bone fragments are connected together and healed directly by osteoclasts and osteoblasts activities (Claes, Recknagel & Ignatius 2012).

In contrast, secondary healing (indirect healing or endochondral ossification) is the process by which most fractures heal naturally and occurs in the presence of some interfragmentary movement between the extremities of the fractured bone. It involves a sequential process of tissue differentiation whereby bone fragments are first stabilized by an external callus (Claes, Recknagel & Ignatius 2012). The callus stabilizes the fracture by increasing its stiffness through tissue differentiation and enlarging its cross-sectional area. Ultimately, a bridge is made between the bone fragments, reducing the interfragmentary movement to such a low level that bone formation can occur in the gap. The process of bone healing is divided into three overlapping phases: inflammation, resulting in granulation tissue; repair, a cartilaginous soft callus and then a hard bony callus develops surrounding the fracture site and, finally, remodeling, the ossified callus is restructured until the final bone structure is achieved (Claes, Recknagel & Ignatius 2012).

3 COMPUTATIONAL MECHANOBIOLOGY OF BONE HEALING

More recently mechanobiological models have been used to investigate the role of various mechanical and biological factors in fracture healing (Pivonka & Dunstan 2012, Ghiasi *et al.* 2017). Mechanical stimulation is applied to a model geometry, and the local mechanical environment is calculated, using a numerical method. Mechanical factors such as strain, pressure, stability, and fluid velocity are important parameters that act as stimuli for tissue formation during bone healing. Additionally, biological factors, and chemical stimuli will influence biological processes and cellular activities, such as mesenchymal stem cell (MSC) migration, tissue differentiation, angiogenesis, and growth factor secretion, which in turn influence and regulate the bone healing process (Ghiasi *et al.* 2017). Using models that describe these activities, the processes are simulated adaptively. *In silico* bone mechanobiology has adopted two-pronged investigations: the first related to mechanoregulation algorithms to predict mechanobiological changes in bone tissue, and the second to models investigating cell mechanobiology.

The work done by Pauwels in 1926 was an important milestone for the current knowledge about the regulatory effect of mechanical forces in tissue differentiation (Pauwels 1960). In Pauwels' work, the mechanical environment within a healing fracture callus was analyzed and hypothesized that hydrostatic stress and shear strain are two distinct stimuli that stimulate fibrous tissue and cartilage, respectively, guiding the cells to differentiate into connective tissue. Later, Carter et al. expanded Pauwels' theory by using a finite element (FE) model to explain how mechanical loading guides cell differentiation in a fracture callus (Carter, Blenman & Beaupré 1988). They modelled the tissue in the callus as a single solid (linear elastic) phase, investigating endochondral ossification under low stresses and strains during fracture healing, as well as a developing joint and tissue differentiation around orthopedic implants. The model predicted realistic tissue patterns consistent with biological observations, but without quantification of the magnitude of mechanical stimuli with tissue formation. Also, Carter's studies highlighted the importance of a good blood supply for bone formation, since a poor blood supply promotes cartilaginous tissue formation. On the other hand, Claes and Heigele established certain thresholds for the local stress and strain magnitudes to determine whether endochondral or intramembranous ossification takes place (Claes & Heigele 1999). The developed FE model properly predicted tissue differentiation in the callus at three phases of the healing process, through the validation by histological images from *in vivo* experiments. This algorithm was also combined with other rules of bone healing, applying an iterative FE model controlled by 'fuzzy logic' to investigate trabecular bone fracture healing (Ament & Hofer 2000, Shefelbine *et al.* 2005).

Prendergast et al., presented a poroelastic FE model of a bone implant interface to predict tissue differentiation (Prendergast, Huiskes & Søballe 1997). With this model it was possible to show that the biophysical stimuli experienced by the tissue at the implant interface were generated by both the tissue matrix and the drag force of the interstitial fluid flow. This indicated the need for dynamically loaded, biphasic models, because these effects could not be examined with static or linear elastic representations. Therefore, a biphasic FE model of poroelastic connective tissues was developed composed of two biophysical stimuli: octahedral shear strain in the solid phase and fluid velocity in the interstitial fluid phase as the mechanotransduction variables, where high magnitudes of either, favours fibrous tissue, and only when both stimuli are low enough, can ossification occur. With this study, Isaksson et al. compared the mechanobiological models of Carter, Claes and Heigele, and Prendergast in a fracture healing study, and concluded that the concept based on strain and fluid velocity as stimuli correlated best with experimental results (Isaksson *et al.* 2007). Therefore, over the last two decades, computational models have evolved from simple tools capable of analyzing fracture healing at only one time point to current models that can predict tissue differentiation and remodeling over time. Prendergast model was adapted by Lacroix et al. to describe fracture healing in a time-dependent manner by adding, for the first time, diffusion equations to model the concentrations of progenitor cells in the callus (Lacroix & Prendergast 2002). Although diffusion is not the mechanism of cell migration and proliferation, Lacroix's theory has influenced many researchers to include cell activities in their mechano-regulatory algorithms. Their adaptive poroelastic model was able to simulate direct periosteal bone formation, endochondral ossification in the external callus, stabilisation when bridging of the external callus occurs, and resorption of the external callus. Later, Isaksson et al., were the first to directly link cell phenotype to mechanical stimulation by including four distinct cell types, such as MSCs, fibroblasts, chondrocytes, and osteoblasts in the model (Isaksson *et al.* 2008). These cells were able to migrate, proliferate, differentiate, or die, based on both mechanical stimulation and the behavior of adjacent cells, allowing spatial and temporal predictions of bone fracture healing phases. However, the modeling of cell activities was carried out on an element basis, and anisotropy in cell movement was not considered. Consequently, Perez and Prendergast developed a 2D FE lattice model representing both cells and extracellular matrix in the callus (Pérez & Prendergast 2007). A 'random walk' model was included to represent cell migration both with and without a preferred direction, which implies anisotropic proliferation and migration of cells. Byrne adapted this model into 3D and later implemented it experimentally in fracture healing of a human tibia under muscle loading, predicting healing beyond the repair phase (Byrne, Lacroix & Prendergast 2011). Additionally, it was presented by Checa and Prendergast another lattice approach, incorporating angiogenesis in the modulation of cell phenotype (Checa & Prendergast 2009). Simulations of the early stages of bone healing were not frequently modelled and may be partly due to mechanical factors playing a minor role in this phase, while biological and chemical factors are much more significant (Ribeiro *et al.* 2015). Nevertheless, a more recent study by Ghiasi et al., developed a FE-based model to simulate different levels of diffusion coefficient for MSC migration, Young's modulus of granulation tissue, callus thickness and interfragmentary gap size to understand the modulatory effects of these initial phase parameters on bone healing (Ghiasi *et al.* 2019). However, further experimental investigations on the biological and mechanical factors in early stage of healing are required to develop more robust and predictive models to simulate healing from the beginning to the end.

4 CONCLUSION

In silico modeling has increasingly contributed to the field of bone mechanobiology, helping to understand how mechanical factors interact with biological processes, and considers both of these aspects simultaneously, aiding in the interpretation of experimental findings and identifying new avenues of investigation. However, there are still parameters that need to be incorporated into modeling efforts, such as growth factors, the mechanism of cellular activities and angiogenesis in bone healing. In the future, the evolution in both techniques and computational power will move the field towards patient-specific models in the clinic.

ACKNOWLEDGMENTS

The authors acknowledge the funding from Ministério da Ciência, Tecnologia e Ensino Superior – Fundação para a Ciência e a Tecnologia, under grant SFRH/BD/133105/2017 and LAETA project UIDB/50022/2020.

REFERENCES

Ament, C. & Hofer, E. P. 2000. A fuzzy logic model of fracture healing, *Journal of Biomechanics* 33(8): 961–968.

Byrne, D. P., Lacroix, D. & Prendergast, P. J. 2011. Simulation of fracture healing in the tibia: Mechanoregulation of cell activity using a lattice modeling approach, *Journal of Orthopaedic Research* 29(10): 1496–1503.

Carter, D. R., Blenman, P. R. & Beaupré, G. S. 1988. Correlations between mechanical stress history and tissue differentiation in initial fracture healing, *Journal of Orthopaedic Research* 6(5): 736–748.

Checa, S. & Prendergast, P. J. 2009. A mechanobiological model for tissue differentiation that includes angiogenesis: A lattice-based modeling approach, *Annals of Biomedical Engineering* 37(1): 129–145.

Claes, L. E. & Heigele, C. A. 1999. Magnitudes of local stress and strain along bony surfaces predict the course and type of fracture healing, *Journal of Biomechanics* 32(3): 255–266.

Claes, L., Recknagel, S. & Ignatius, A. 2012. Fracture healing under healthy and inflammatory conditions, *Nature Reviews Rheumatology* 8(3): 133–143.

Einhorn, T. A. & Gerstenfeld, L. C. 2015. Fracture healing: Mechanisms and interventions, *Nature Reviews Rheumatology* 11(1): 45–54.

Ekegren, C. L. *et al.* 2018. Incidence, costs and predictors of non-union, delayed union and mal-union following long bone fracture, *International Journal of Environmental Research and Public Health* 15 (12).

Ghiasi, M. S. *et al.* 2019. Computational modeling of human bone fracture healing affected by different conditions of initial healing stage, *BMC Musculoskeletal Disorders* 20(1): 1–14.

Isaksson, H. *et al.* 2007. Bone regeneration during distraction osteogenesis: Mechano-regulation by shear strain and fluid velocity, *Journal of Biomechanics* 40(9): 2002–2011.

Isaksson, H. *et al.* 2008. A mechano-regulatory bone-healing model incorporating cell-phenotype specific activity, *Journal of Theoretical Biology* 252(2): 30–246.
Lacroix, D. & Prendergast, P. J. 2002. A mechano-regulation model for tissue differentiation during fracture healing: Analysis of gap size and loading, *Journal of Biomechanics* 35(9): 1163–1171.
McKibbin, B. 1978. The biology of fracture healing in long bones, *Journal of Bone and Joint Surgery* 60 (2): 150–62.
Van der Meulen, M. C. H. & Huiskes, R. 2002. Why mechanobiology? A survey article, *Journal of Biomechanics* 35(4): 01–414.
Moran, C. J. *et al.* 2016. The benefits and limitations of animal models for translational research in cartilage repair, *Journal of Experimental Orthopaedics* 3(1): 1–12.
Morgan, E. F. *et al.* 2011. Experimental Model of Skeletal Healing 43(12): 2418–2424.
Pauwels, F. 1960. A new theory on the influence of mechanical stimuli on the differentiation of supporting tissue. The tenth contribution to the functional anatomy and causal morphology of the supporting structure, *Zeitschrift fur Anatomie und Entwicklungsgeschichte* 121: 478–515.
Pérez, M. A. & Prendergast, P. J. 2007. Random-walk models of cell dispersal included in mechanobiological simulations of tissue differentiation, *Journal of Biomechanics* 40(10): 2244–2253.
Pivonka, P. & Dunstan, C. R. 2012. Role of mathematical modeling in bone fracture healing, *BoneKEy Reports* 1(August): 1–10.
Prendergast, P. J., Huiskes, R. & Søballe, K. 1997. Biophysical stimuli on cells during tissue differentiation at implant interfaces, *Journal of Biomechanics* 30(6): 539–548.
Ribeiro, F. O. *et al.* 2015. In silico mechano-chemical model of bone healing for the regeneration of critical defects: The effect of BMP-2, *PLoS ONE* 10(6): 1–25.
Shefelbine, S. J. *et al.* 2005. Trabecular bone fracture healing simulation with finite element analysis and fuzzy logic, *Journal of Biomechanics* 38(12): 2440–2450.

Advances and Current Trends in Biomechanics – Belinha et al. (Eds)

Stress analysis in tibias with bone callus: A numerical study

A. Oliveira & J. Belinha
School of Engineering, Polytechnic of Porto, Porto, Portugal

M. Peyroteo
Institute of Science and Innovation in Mechanical and Industrial Engineering, Porto, Portugal

ABSTRACT: Human tibial fractures are complex injuries that usually result in long periods of hospitalization and rehabilitation. During the slow healing process, the bone callus gradually in-creases its mechanical properties. This work aims to evaluate the structural influence of the mechanical properties of a bone callus due to a transverse tibial fracture. Therefore, a 2D tibia model was constructed, considering distinct geometries and load cases simulating a normal alignment and malalignment conditions (valgus and a varus knee). Each model was analyzed with the finite element method (FEM) and the meshless method, particularly the Radial Point Interpolation Method (RPIM). The results show that both malalignment conditions induced the highest stresses in the model compared to the aligned condition. The maximum stresses were observed in the tibia model with the E=250 MPa callus and gradually decreased with the increase of the mechanical properties of the callus.

1 INTRODUCTION

The prediction of fracture healing remains one of the major challenges (Sridevi et al. 2015). The conventional method of assessing fracture healing in long bones is by a combination of passive manipulation of the injured limb, to access the strength of the fracture, along with medical imaging to determine the volume and state of fracture-binding tissues, referred to as callus; however, this assessment method is subjective and unanimity lacks among clinicians (McClelland et al. 2007). Thus, in research community, there has been a need for objective, quantitative and simulation methods that allow monitoring the consolidation of the fracture, assisting the diagnosis and treatment of delayed unions and non-unions, as well as helping to identify the endpoint of healing, avoiding prolonged treatments. In addition, computer modeling is having a profound effect on scientific research as many biological processes, including bone healing, are so complex that physical experimentation is either too expensive or too time consuming and requires very controlled conditions (Moran et al. 2016). The healing process strongly depends on the mechanical actions on the callus, which determine the relative movement of the bone fragments. It is a complex process involving cellular differentiation, strongly stimulated by mechanical loading (Flores et al. 2017). In this process, the bone goes through three main phases: inflammation, soft and hard callus formation, and bone remodeling. Mechanical stresses or strains in fracture sites are typically estimated using computational models. This work aims to study how the variation of the Young's modulus of the bone callus due to a transverse tibial fracture influences the stress field in the healing fracture vicinity. Thus, the mechanical response of the tibia model, with a callus possessing distinct mechanical properties, is analyzed using the finite element method (FEM) and Radial Point Interpolation Method (RPIM). Furthermore, it is investigated how the geometry of the tibia, combined with the presence of a bone callus, influences the stress magnitude and

DOI: 10.1201/9781003217152-56

distribution. For this purpose, three models of the tibia were constructed, each one possessing its own load case: a normal tibia showing a normal alignment and two malalignment conditions, such as a valgus and a varus knee.

2 MATERIALS AND METHODS

Three 2D models of the tibia bone were constructed in FEMAP© software (student version). The 2D models were obtained from anonymized radiographs of an adult male healthy individual, an adult male individual possessing the valgus malalignment condition and an adult male individual possessing the varus malalignment condition. The three 2D models possess the same number of nodes and elements: 2461 nodes and 4574 triangular elements (constant strain elements). After the construction of the models, the corresponding node/element mesh was imported to the "Finite Element and Meshless Methods Analysis Software", FEMAS©, an academic freeware that runs on MATLAB© (cmech.webs.com). The material properties and the essential and natural boundary conditions are declared in FEMAS. The mechanical properties of the materials, Young's modulus (E) and Poisson's coefficient (ν) were obtained from the literature (Lacroix & Prendergast 2002). For the homogenized bone tissue (cortical bone): E=16.7GPa and ν=0.3. For the bone callus, four tissue types were considered: E1=0.25GPa (cartilage), E2=1.5GPa (intermediate bone), E3=4GPa (mature bone), E4=6GPa (mature bone), and a Poisson's coefficient ν=0.4999 was assumed in all of them. The intention is to have the representative behavior of each healing stage, in a range of elasticity values that vary between the soft callus, until a phase that represents the hard callus stage. All materials were assumed to be homogeneous, isotropic, and linear elastic. Regarding the essential boundary conditions (displacements constrains), the model was fixed in the distal extremity, with no displacement allowed in any of the directions of the two degrees-of-freedom. Concerning the natural boundary conditions (external forces), this work intends to analyze a peak load during walking for an approximately 80kg human being. Therefore, assuming that the individual is standing in one leg, all the load is applied to the proximal tibia distributed into a 10x10 mm^2 surface. For the 2D analysis, it was considered the plane strain deformation theory. Thus, only a 2D slice with 1mm thickness was analyzed. The load on that 1mm thickness was distributed at the proximal tibia (only 1/10 of the total load). In this work, three 2D models were constructed, each one representing a health condition: normal alignment (case 1), 10° valgus malalignment (case 2) and 10° varus malalignment (case 3). Each one of these models will possess its own load. For the normal alignment (case 1) it was considered: $\vec{F}_1 = \{0, 80\}^T N$, for 10° valgus malalignment (case 2) it was assumed: $\vec{F}_2 = \{13.89, -78.78\}^T N$ and for 10° varus malalignment (case 3) it was considered: $\vec{F}_3 = \{-13.89, -78.78\}^T N$.

3 NUMERICAL RESULTS

The models described previously were analyzed with FEM and RPIM, and the stress fields of the von Mises effective stress (σ_{ef}) and the principal stresses (σ_1 and σ_2) were calculated for each bone callus and geometric model as shown in Figures 1 and 2.

4 DISCUSSION

In this work, the main goal was to understand the mechanical behavior of the tibia with a callus of different mechanical properties that simulate the healing process. Furthermore, the influence of the geometry was also investigated, by analyzing models presenting normal alignment and malalignment conditions, such as a valgus and a varus knee. Twelve models were analyzed with FEM and RPIM (three distinct geometries in which the Young's modulus of the bone callus was varied four times between 250 MPa and 6000 MPa). The analysis allowed to obtain the stress fields of each model, corresponding to the von Mises effective stress, σ_{ef}, and the principal stresses, σ_1, and σ_2.

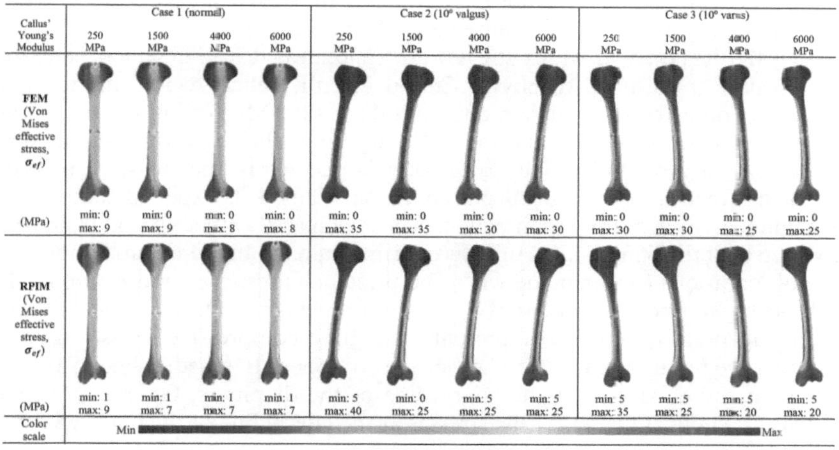

Figure 1. Von Mises stress maps obtained for each model and callus case with both FEM and RPIM.

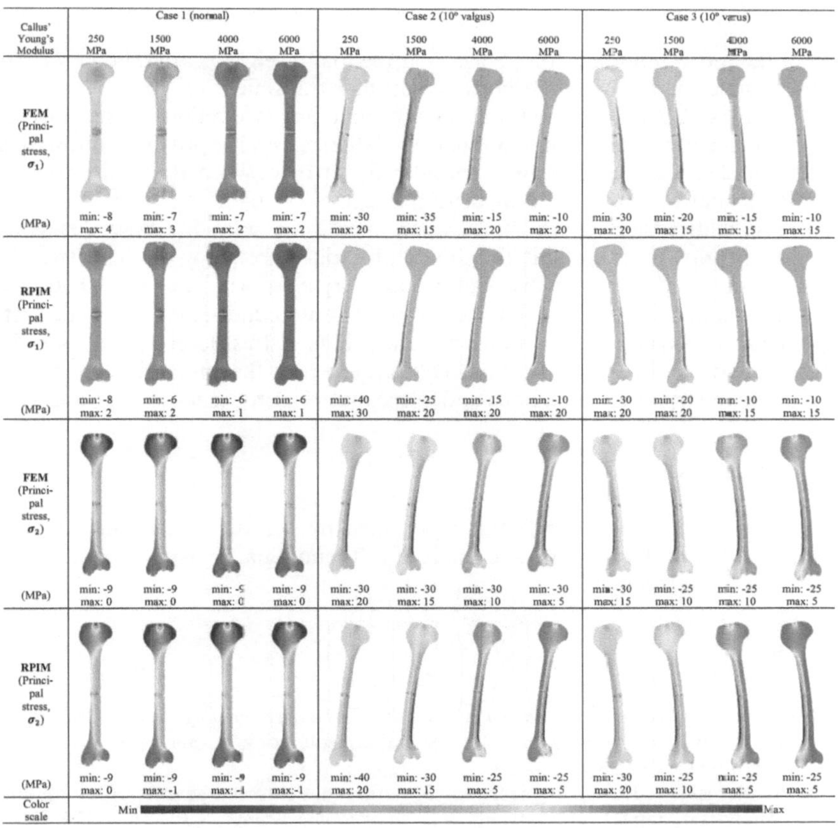

Figure 2. Principal stress (σ_1 and σ_2) maps obtained for each model and callus case with both FEM and RPIM.

The color maps in Figures 1 and 2 show that the results for both methods were similar. However, it was possible to analyze the stresses in the callus' region in more detail with

RPIM. In the meshless methods, the nodes discretizing the problem domain can be arbitrarily distributed without any connectivity requirement (Belinha 2014).

Additionally, the highest von Mises stresses are concentrated in different regions, being the higher values found at the tibial diaphysis. Considering the callus' region, the maximum von Mises stresses were observed in the tibia model with the E=250 MPa callus and then, it gradually decreased with the increase of the bone callus Young's modulus. This observation is valid for three 2D models analyzed. Both valgus knee (case 2) and varus knee (case 3) induce stresses 350%-400% higher than a normal alignment condition. Thus, it is expected that malalignment conditions require larger immobilization periods when compared with a normal alignment condition (higher stresses at the bone callus can induce further damage and re-inflammation).

The same observations can be made when the principal stresses σ_1 and σ_2 are analyzed. In a 2D model, the highest positive values of σ_1 represent the highest tensile stress and the highest negative value (absolute value) of σ_2 represent the highest compressive stress. Figure 2 shows that the valgus knee (case 2) induces higher tensile stresses at the medial line of the diaphysis and higher compressive stresses at the lateral line of the diaphysis. On the other hand, the varus knee (case 3) induces higher tensile stresses at the lateral line of the diaphysis and higher compressive stresses at the medial line of the diaphysis.

5 CONCLUSION

Tibia is one of the most important long bones and the main weight-bearing bones of the human leg, being responsible for receiving the body weight and transmit it to the foot when a human being stands or walks. A 2D simulation was performed aiming to obtain the stress distribution in the tibia. In order to perform the computational simulation, several approximations were assumed to simplify the model, e.g., the bone was considered isotropic, when it is anisotropic, and the FEM and RPIM analyses were performed assuming the 2D plane strain deformation theory (PSDT). These simplifications only allow to approximate the solution. Nevertheless, since the PSDT comes as a simplification of the full 3D deformation theory for the observed plane, it is expected that the obtained 2D stress distributions are very close with the real 3D stress distributions in the observed plane. Thus, the obtained results allow to understand the stress distribution in the tibia bone and predict which locations are under higher stress levels. This work also shows how a material discontinuity, such as a bone callus, influences the stress distribution, as well as variations in the geometry of the tibia bone and in the material properties of the callus bone.

ACKNOWLEDGMENTS

The authors truly acknowledge the funding provided by Ministério da Ciência, Tecnologia e Ensino Superior - Fundação para a Ciência e a Tecnologia (Portugal), and by LAETA, under project UIDB/50022/2020.

REFERENCES

Belinha, J. 2014. *Meshless Methods in Biomechanics - Bone Tissue Remodelling Analysis*. Springer.

Flores, A. *et al.* 2017. Bone Immobilization devices and consolidation mechanisms: Impact on healing time, *Procedia Structural Integrity* 5: 34–39.

Lacroix, D. & Prendergast, P. J. 2002. A mechano-regulation model for tissue differentiation during fracture healing: Analysis of gap size and loading, *Journal of Biomechanics* 35(9): 1163–1171.

McClelland, D. *et al.* 2007. Fracture healing assessment comparing stiffness measurements using radiographs, *Clinical Orthopaedics and Related Research* 457: 214–219.

Moran, C. J. *et al.* 2016. The benefits and limitations of animal models for translational research in cartilage repair, *Journal of Experimental Orthopaedics* 3(1): 1–12.

Sridevi, M. *et al.* 2015. Tibia Fracture Healing Prediction Using First-Order Mathematical Model, *Computational and Mathematical Methods in Medicine* 2015.

 ISBN: 978-1-032-10806-3

An elastostatic structural analysis of a proximal tibia using a 3D FEM and meshless RPIM

T. Sousa
School of Engineering, Polytechnic of Porto, Portugal

A. Guerra
INEGI, Faculty of Engineering, University of Porto, Portugal

J. Belinha
Department of Mechanical Engineering, School of Engineering, Polytechnic of Porto, Portugal

ABSTRACT: Obesity is related to several clinical conditions and diseases including bone related diseases. Bones support the body's soft tissues and can handle intense loads, yet they can damage or fracture under high loads. Computational mechanical behavior analysis of bones is crucial considering that it allows to understand the structural characteristics of the tissues and materials forming bone tissue and predict its mechanical limits. Finite element and meshless methods using radial point interpolators are well-known discretization techniques that can be used to analyze the mechanical response of structures and materials. This work, using the FEM and a meshless method, the proximal tibia is analyzed under two distinct load cases, corresponding to a healthy and an obese individual. An elastostatic analysis was performed, and stress and strain fields were obtained. The results show that higher strain and stress occur near the tibial tuberosity and that increasing the load magnitude increases the highest strain and stress values.

1 INTRODUCTION

Obesity is related to several diseases such as diabetes, hypertension, coronary heart disease, and several kinds of cancer (Aronne, 2002). Bone issues related to obesity, such as osteoarthritis, are also very common since more weight means more pressure to the bone. The study of bone structures behavior when subjected to different loads, caused by obesity, is of extreme relevance and importance. Finite element analysis (FEA) is a widely adopted method to investigate mechanical bone behavior (Ruffoni and Van Lenthe, 2017). The Radial Point Interpolation Method (RPIM), although not as popular as FEA for mechanical bone behavior analysis, recently, in a mechanical bone remodeling study, demonstrated smoother and more accurate results when compared to the Finite Element Method (FEM), while also being more organic (Peyroteo *et al.*, 2018). Using computed tomography (CT) data is possible to generate models of the bone that provide accurate information regarding the bone size and geometry which, along with the tissue's properties, influence its behavior. This work aims to develop a functional 3D model of the tibia via segmentation of CT images, as well as the study of the mechanical behavior of this bone when forces of different intensities (simulating a weight increase) are applied to the tibial condyles. This work also compares the results obtained by the FEM and the RPIM.

DOI: 10.1201/9781003217152-57

2 METHODS

This work evaluates the stress and strain in the human tibia during the midstance phase of the gait cycle where only one leg is supporting the weight of the individual. Forces of different magnitudes that represent the weight of the individual were applied in the tibial condyles. Two force cases were studied. The force magnitude for load case 1 (LC1) was of 689.62 N, corresponding to 75Kg and for load case 2 (LC2) was 919.40 N (100Kg). In both load cases the force was divided and distributed in the surfaces of both tibial condyles. The computational model developed consists only of the proximal tibial bone. The reconstruction of the tibial structure was made using an anonymized CT image of a knee joint obtained from "The biomedical 3D printing community". The spongy bone was considered homogeneous, isotropic and linear elastic and it was the only material in the model. The values for the Young's modulus (E) and Poisson's ratio (υ) were obtain from the literature (Vaverka, 1997; Kubíček and Florian, 2009). Thus, in this work, the following values were considered: E = 500MPa and υ = 0.3. The final model consisted of 2909 nodes and 12998 elements (3D constant strain tetrahedral elements of 4 nodes).

3 RESULTS

Regarding the principal strains and stresses, in a 3D model, the highest positive values of ε_1 and σ_1 represent the highest tensile strain and the highest tensile stress, respectively. In contrast, the highest negative value (absolute value) of $\varepsilon_3$and σ_3 represent the highest compressive strain and the highest compressive stress, respectively. Furthermore, $\bar{\varepsilon}$ and $\bar{\sigma}$, represent equivalent effective strain and equivalent von Mises stress, respectively. The model was fixed (all degrees of freedom were constrained) at the plane *Oxy* ($z = 0$). Thus, strain and stress maps obtained at the plane *Oxy* ($z = 0$) for both load cases and both methods are represented in Figure 1. The colored bar indicates the maximum and minimum value of each distribution map. Figure 1. allows to visualize the locations, in the base plane, possessing the highest

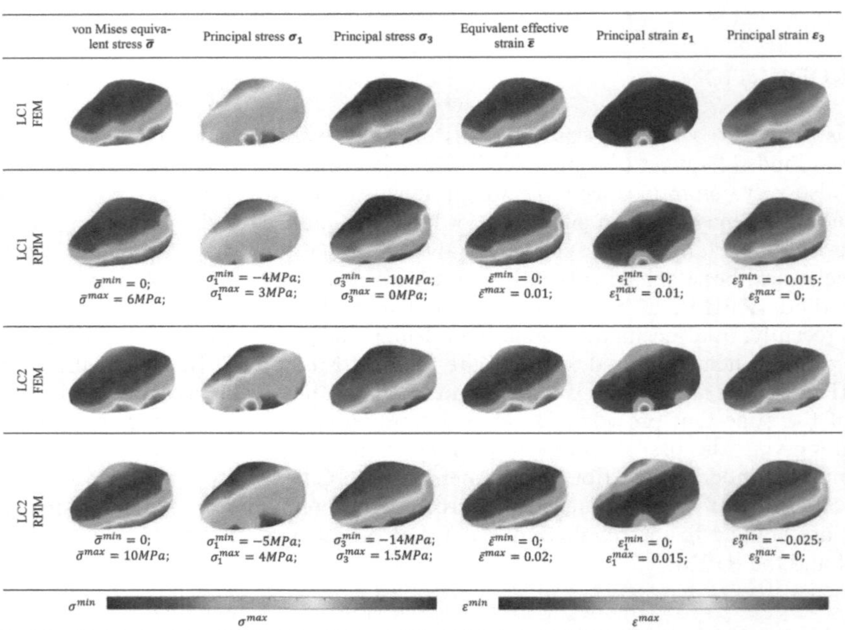

Figure 1. Stress and strain distributions maps for LC1 and LC2. Base plane. FEM and RPIM.

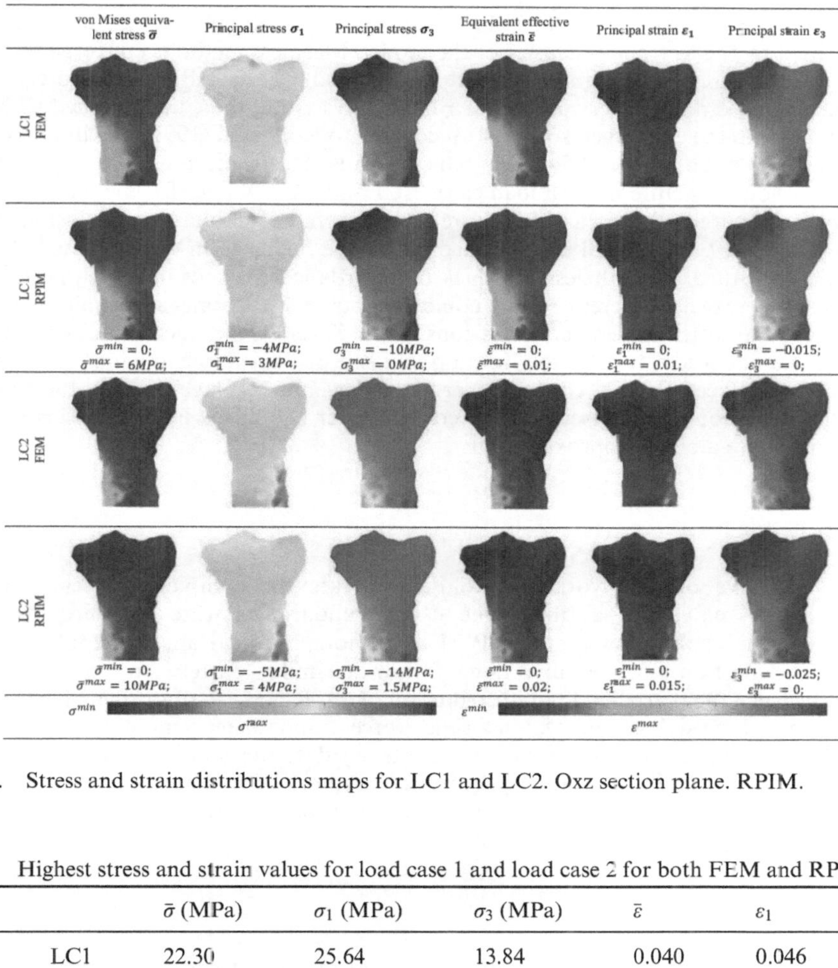

Figure 2. Stress and strain distributions maps for LC1 and LC2. Oxz section plane. RPIM.

Table 1. Highest stress and strain values for load case 1 and load case 2 for both FEM and RPIM.

		$\bar{\sigma}$ (MPa)	σ_1 (MPa)	σ_3 (MPa)	$\bar{\varepsilon}$	ε_1	ε_3
FEM	LC1	22.30	25.64	13.84	0.040	0.046	0.021
	LC2	32.88	38.30	19.44	0.060	0.069	0.029
RPIM	LC1	8.461	4.707	13.47	0.017	0.005	0.021
	LC2	12,73	7.112	20.29	0.026	0.011	0.031

values of strains and stresses. To understand how the stress is distributed within the model, the 3D model was sectioned by a *Oxz* plane and the corresponding stress and strain fields were captured. The results for both load cases are shown in Figure 2. for the FEM and the RPIM. Table 1 contains the highest values registered for every evaluated stress and strain for each method and load case.

4 DISCUSSION

The force cases had the expected impact on the stress/strain variable field since an increase in the force's intensity resulted in more deformed condyles as well an increase in the maximum strain and stress. Both von Mises stress and effective strain distribution maps allow to clearly identify the regions with higher valued stress/strain, which is the clamped boundary. Due to the location of the application of the external forces, this observation was expected. The values for the von Mises stress, equivalent effective stress, tensile strain and stress obtained with the RPIM, were lower when compared to the FEM. On the other hand, values for compressive strain and stress were similar in both methods. The difference in the obtained values can be justified by the low nodal density of the model. Meshless methods converge faster than mesh methods. Therefore, possibly, the results obtained by the RPIM are more accurate, because the analyzed mesh is very coarse for FEM analysis and, thus, the obtained FEM solution is still far from the converged solution. According to Røhl et al (1991), the ultimate tensile stress of the spongy bone is of 2.54 MPa, which means that in the present analysis suggests that the bone would fracture in both load cases LC1 and LC2. Values for the ultimate tensile/compressive strain are of 0.0155 and 0.0111 respectively and of ultimate compression stress of 2.22 MPa (Røhl *et al.*, 1991). All the obtained values are higher than the ultimate values presented in the literature. Nevertheless, this lack of accordance does not indicate that the developed model is incorrect. Different results could be obtained if distinct mechanical properties were assumed, and different geometry was considered. Several other factors influence the outcome in the present study, such as the location and size of the surface in which external force is applied, the absence of a thin cortical layer, condyles, and cartilage. In fact, the trabecular bone in the tibia should be denser and the cortical layer should be included. This would be sufficient to achieve a better approximation.

5 CONCLUSION

The main objective of this work was to demonstrate the complete process from CT-images to stress analysis of a tibia bone using standard freeware software, while also demonstrating the capabilities of the RPIM as a biomechanical analysis tool. The stress and strain distribution on the tibia bone follow a similar pattern for both FEM and RPIM. However, the variable field distribution with RPIM is smoother. Furthermore, differences in the model geometry, materials, forces, and displacement constrains could lead to distinct results. So, the obtained results need to be studied and interpreted to understand their reliability. Although distinct authors obtained distinct values, for a similar analysis, it is not possible to indicate which one is the most accurate one. The final validation requires experimental tests.

ACKNOWLEDGMENT

The authors truly acknowledge the funding provided by Ministério da Ciência, Tecnologia e Ensino Superior - Fundação para a Ciência e a Tecnologia (Portugal), under grant SFRH/BD/133894/2017. Additionally, the authors acknowledge the funding provided by LAETA, under project UIDB/50022/2020.

REFERENCES

Aronne, L. J. (2002) 'Classification of obesity and assessment of obesity-related health risks', *Obesity Research*, 10(SUPPL.2), pp. 105–115. doi: 10.1038/oby.2002.203.

Kubíček, M. and Florian, Z. (2009) 'Stress Strain Analysis of Knee Joint', *Engineering Mechanics*, 16(5), pp. 315–322.

Peyroteo, M. M. A. *et al.* (2018) 'Mechanical bone remodelling : Comparative study of distinct numerical approaches', *Engineering Analysis with Boundary Elements*. Elsevier Ltd, 000(July 2017), pp. 1–15. doi: 10.1016/j.enganabound.2018.01.011.

Røhl, L. *et al.* (1991) 'Tensile and compressive properties of cancellous bone', *Journal of Biomechanics*, 24(12), pp. 1143–1149. doi: 10.1016/0021-9290(91)90006-9.

Ruffoni, D. and Van Lenthe, G. H. (2017) '3.10 Finite element analysis in bone research: A computational method relating structure to mechanical function', in *Comprehensive Biomaterials II*. Elsevier, pp. 169–196. doi: 10.1016/B978-0-12-803581-8.09798-8.

Vaverka, F.1938- (1997) *Fyzikální základ biomechaniky/*. 3. vyd. Vydavatelství Univerzity Palackého,.

Advances and Current Trends in Biomechanics – Belinha et al. (Eds)

Angiogenic biomaterials

T. Sousa
School of Engineering, Polytechnic of Porto, Portugal

A. Guerra
INEGI, Faculty of Engineering, University of Porto, Portugal

J. Belinha
Department of Mechanical Engineering, School of Engineering, Polytechnic of Porto, Portugal

ABSTRACT: Angiogenesis is the formation of new blood vessels from pre-existing ones, and it is the process by which new capillaries are formed. It is a central mechanism to healthy tissue maintenance but also a fundamental process in tumor growth. Tissue engineering aims for the replacement of damaged tissues such as bone and skin. The functional vascularization of the produced tissue is essential in its integration and function. Thus, a safe and controlled induction of angiogenesis is required. Cancer therapies, on the other hand, require localized angiogenic inhibition. Effectively control the angiogenic process is fundamental for the development of safe inductive or inhibitory therapies. Biomaterials and bioactive materials can be used to achieve such control. A conjugation of adequate cell types, biomarkers, scaffolds, and growth factors delivery/inhibition are the key to controlled angiogenic therapies.

1 INTRODUCTION

Tissue replacement with engineered constructs to combat medical disorders is the basis of tissue engineering (TE). A bottleneck in TE is recreating functional vascularization to ensure appropriate integration and function of substitute bioengineered structures. The clinical applicability of TE is limited as a result of large volumes of implanted tissue being unable to stimulate the formation of blood vessels required for their survival. A vasculature is a critical element of most multicellular organisms to nourish organs with vital nutrients and oxygen, as well as to excrete waste metabolic products (Liu *et al.*, 2020). Angiogenesis, the growth of new mature blood vessels from pre-existing vessels, is a crucial event in various biological processes including wound healing and organ development (Risau, 1997). It is a central mechanism to healthy tissue maintenance albeit also being a fundamental process in tumor growth (Mastrullo *et al.*, 2020). Understanding the mechanisms that regulate angiogenesis, physiological and pathological, is fundamental to recreate functional vascularization and integration of tissue engineered constructs. Similarly to natural reparative angiogenesis, biomaterial-driven therapeutic angiogenesis aims at integrating a network of functional vessels to engineered tissue replacements. Traditionally, biomimetic angiogenesis is stimulated by growth factors or proteins that generally lead to immature disorganized blood vessels, reminiscent of uncontrolled tumoral angiogenesis (Liu *et al.*, 2020; Mastrullo *et al.*, 2020). Biomaterials can aid in effectively achieving control over angiogenesis.

DOI: 10.1201/9781003217152-58

2 BIOMATERIALS FOR ANGIOGENIC INDUCTION

Promoting angiogenesis is an extremely beneficial therapy in diseases such as ischemia, which compromises the blood supply to tissues and organs (Jalilian, Elkin and Shin, 2020). Different types of cells can be used to promote angiogenesis. Vascular endothelial cells are often used in therapies since they can organize themselves into a functional capillary network and are of easy integration in tissues, however, there are some limitations to their usage. There are majorly three types of cells that are accepted for angiogenic purposes: somatic, stem and induced pluripotent stem cells (Saberianpour *et al.*, 2018). Stem cells show high vascularization potential since they can differentiate into endothelial cells after exposure to angiogenic growth factors. Nevertheless, there are still concerns regarding their heterogenic nature and high individual variability (Jalilian, Elkin and Shin, 2020). The growth and differentiation of cells is triggered by several specific conditions, which can be activated by scaffolds. Scaffolds are temporary three-dimensional structures that simulate a biological environment to deliver, align and maintain cell connection that favors angiogenesis (Saberianpour *et al.*, 2018). Scaffolds are required to possess several properties, namely: biocompatibility (for cell adherence and avoid rejection by the body), biodegradability (for the eventual replacement of the scaffold by the body's own cells) and mechanical properties that allow for tissue's support and surgical handling (O'Brien, 2011). These structures are fabricated of natural materials such as collagen, fibrin and alginate, or synthetic materials like carbon fiber, bioactive glasses, and Teflon (Nikolova and Chavali, 2019). The choice of which material to use is made based on the tissue's properties where the scaffold is to be integrated. For instance, the use of bioactive glasses is extremely common in TE, especially in hard tissues (bones) due to their mechanical resistance, angiogenic and osteogenic capabilities (Kargozar *et al.*, 2016). Furthermore, they can be embedded with metallic ions, like Cu2+ which upregulates vascular endothelial growth factor (VEGF) expression and promotes neovascularization while also having antibacterial capabilities (JACOBS *et al.*, 2020). Delivery of angiogenic molecules through scaffolds is a key strategy to stimulate angiogenesis. Other than copper, angiogenic ions such as cobalt, silicate, strontium and magnesium are especially useful in bone regenerative scaffolds (Lee *et al.*, 2020). Angiogenic proteins that are delivered include VEGF, platelet-derived growth factor, fibroblast growth factor and transforming growth factor β1. Normally, for vascularization purposes, scaffolds need to be porous and mimic the ECM, which is essential when delivering angiogenic growth factors, since they all naturally bind to it (Martino and Hubbell, 2010). The use of scaffolds, natural or synthetic, allows for spatial and temporal control of angiogenesis, controlled release of growth factors and bioactive molecules and cell phenotype definition. Regardless, they are constructs of extreme complexity that need to be carefully engineered. From their surface topography and stiffness to pore structure, every detail must be meticulously tailored, otherwise their effectiveness is compromised.

3 BIOMATERIALS FOR ANGIOGENIC INHIBITION

Cancer therapies require angiogenesis inhibition rather than its induction. Kim *et al* (1993) firstly demonstrated in vivo that VEGF inhibition successfully suppressed tumor growth, proving the therapeutical effects of VEGF inhibition. As of January 2021, the U.S Food and Drug Administration (FDA) has approved 14 different drugs that predominantly target VEGF or its receptors. Inhibiting VEGF will prevent new tumorous vessel formation which compromises the nutrients and oxygen delivery, therefore "starving" the tumor. Although promising, the current drugs still present problems such as inefficacy and serious side effects (Ribatti, 2017). Reports of antiangiogenic drugs that targeted the VEGF pathway promoting metastasis and invasion were made in 2009 (Ebos *et al.*, 2009; Pàez-Ribes *et al.*, 2009). VEGF inhibitors efficiency is compromised by genetic mutations, vascular changes, up-regulation of other angiogenic promoters and metastasis (Ribatti, 2010b). VEGF biomarkers can help identifying subsets of patients who probably will respond better to a certain antiangiogenic therapy, identifying early clinical benefit and/or treatment resistances, ultimately providing a personalized cancer treatment to each individual (Ribatti, 2010a). In the last few year there has been progress in identifying and validating biomarkers that predict overall survival and treatment effectiveness in cancer patients (Hein *et al.*, 2015; Balikova *et al.*, 2019), however, there is still a long way to go for personalized cancer therapies (Liang *et al.*, 2019). Moreover, the growing number of antiangiogenic agents that are being approved, intensifies the need for biomarkers that assess safety and efficacy (Mukherjee, 2018). As an alternative to conventional cancer therapies, nanomedicine and nanoparticles arise as potential candidates. Nanoparticles have been widely investigated in the biomedical field due to their many applications from targeted drug delivery and optical bioimaging to biosensors and immunoassays (Abdalla *et al.*, 2018). The unique properties of nanomaterials can be used not only to diagnose cancer but also to develop new therapeutic approaches for its treatment. When compared to current cancer diagnostic tools, several nanoparticles-based assays displayed a far superior selectivity and sensitivity (Zhang *et al.*, 2019). Due to their large surface area to volume ratio, nanoparticles can more efficiently capture cancer biomarkers (Ferrari, 2005). Regarding antiangiogenic cancer therapy, nanoparticles such as gold (Vines *et al.*, 2019; Siddique and Chow, 2020), silica (Setyawati and Leong, 2017), copper (Benguigui *et al.*, 2019) and silver (Vlăsceanu *et al.*, 2016; Huy *et al.*, 2019), have demonstrated antiangiogenic capabilities and promising results for new combinatory therapies. The future of nanomedicine is promising and offers several new opportunities for the creation of therapies to fight and diagnose cancer. Nevertheless, additional knowledge of material properties, cancerous cells metabolic properties and physiological barriers is required to improve antiangiogenic nanoparticle therapies (Abdalla *et al.*, 2018). As the demand and need for knowledge increases, the recent advances in customizable biomaterials pose an interesting solution to create physiologically-relevant tumor models that will aid in developing relevant studies and discovery of new anti-angiogenic drugs and therapies (Nguyen and Murphy, 2018)

4 CONCLUSION

Angiogenesis is an essential process that allows for the body's tissue vascularization and oxygenation. Angiogenic induction is critical in tissue engineering since it allows for the vascularization of the new tissues and their integration in the body. Biomaterials are versatile materials that can be integrated onto different angiogenic therapies. The conjugation of the adequate cell type, scaffold, delivery method and active biomolecule, accelerate the vascularization rate and can promote controlled angiogenesis. Anti-angiogenic therapies using biomarkers and nanoparticles show high potential, however, the challenge of solely targeting cancer cells without side effects remains.

ACKNOWLEDGMENT

The authors truly acknowledge the funding provided by Ministério da Ciência, Tecnologia e Ensino Superior - Fundação para a Ciência e a Tecnologia (Portugal), under grant SFRH/BD/133894/2017. Additionally, the authors acknowledge the funding provided by LAETA, under project UIDB/50022/2020.

REFERENCES

Abdalla, A. M. E. *et al.* (2018) 'Current challenges of cancer anti-angiogenic therapy and the promise of nanotherapeutics', *Theranostics*. Ivyspring International Publisher, pp. 533–549. doi: 10.7150/thno.21674.

Balikova, I. *et al.* (2019) 'Genetic biomarkers in the VEGF pathway predicting response to anti-VEGF therapy in age-related macular degeneration', *BMJ Open Ophthalmology*. BMJ Publishing Group, 4 (1). doi: 10.1136/bmjophth-2019-000273.

Benguigui, M. *et al.* (2019) 'Copper oxide nanoparticles inhibit pancreatic tumor growth primarily by targeting tumor initiating cells', *Scientific Reports*. Nature Publishing Group, 9(1), pp. 1–10. doi: 10.1038/s41598-019-48959-8.

Ebos, J. M. L. *et al.* (2009) 'Accelerated Metastasis after Short-Term Treatment with a Potent Inhibitor of Tumor Angiogenesis', *Cancer Cell*. Cancer Cell, 15(3), pp. 232–239. doi: 10.1016/j.ccr.2009.01.021.

Ferrari, M. (2005) 'Cancer nanotechnology: Opportunities and challenges', *Nature Reviews Cancer*. Nature Publishing Group, pp. 161–171. doi: 10.1038/nrc1566.

Hein, A. *et al.* (2015) 'Genetic variants in VEGF pathway genes in neoadjuvant breast cancer patients receiving bevacizumab: Results from the randomized phase III GeparQuinto study', *International Journal of Cancer*. Wiley-Liss Inc., 137(12), pp. 2981–2988. doi: 10.1002/ijc.29656.

Huy, T. Q. *et al.* (2019) 'Recent Advances of Silver Nanoparticles in Cancer Diagnosis and Treatment', *Anti-Cancer Agents in Medicinal Chemistry*. Bentham Science Publishers Ltd., 20(11), pp. 1276–1287. doi: 10.2174/1871520619666190710121727.

JACOBS, A. *et al.* (2020) 'Biological properties of copper-doped biomaterials for orthopedic applications: a review of antibacterial, angiogenic and osteogenic aspects', *Acta Biomaterialia*. Elsevier BV. doi: 10.1016/j.actbio.2020.09.044.

Jalilian, E., Elkin, K. and Shin, S. R. (2020) 'Novel cell-based and tissue engineering approaches for induction of angiogenesis as an alternative therapy for diabetic retinopathy', *International Journal of Molecular Sciences*, 21(10), pp. 1–20. doi: 10.3390/ijms21103496.

Kargozar, S. *et al.* (2016) 'Synthesis, physico-chemical and biological characterization of strontium and cobalt substituted bioactive glasses for bone tissue engineering', *Journal of Non-Crystalline Solids*. Elsevier B.V., 449, pp. 133–140. doi: 10.1016/j.jnoncrysol.2016.07.025.

Kim, K. J. *et al.* (1993) 'Inhibition of vascular endothelial growth factor-induced angiogenesis suppresses tumour growth in vivo', *Nature*, 362(6423), pp. 841–844. doi: 10.1038/362841a0.

Lee, J. H. *et al.* (2020) 'Materials roles for promoting angiogenesis in tissue regeneration', *Progress in Materials Science*. Elsevier Ltd, p. 100732. doi: 10.1016/j.pmatsci.2020.100732.

Liang, X. *et al.* (2019) 'An update on biomarkers of potential benefit with bevacizumab for breast cancer treatment: Do we make progress?', *Chinese Journal of Cancer Research*. Chinese Journal of Cancer Research, 31(4), pp. 586–600. doi: 10.21147/j.issn.1000-9604.2019.04.03.

Liu, W. *et al.* (2020) 'Insights into the angiogenic effects of nanomaterials: Mechanisms involved and potential applications', *Journal of Nanobiotechnology*. BioMed Central, 18(1), pp. 1–22. doi: 10.1186/s12951-019-0570-3.

Martino, M. M. and Hubbell, J. A. (2010) 'The 12th–14th type III repeats of fibronectin function as a highly promiscuous growth factor-binding domain', *The FASEB Journal*. Wiley, 24(12), pp. 4711–4721. doi: 10.1096/fj.09-151282.

Mastrullo, V. *et al.* (2020) 'Angiogenesis in Tissue Engineering: As Nature Intended?', *Frontiers in Bioengineering and Biotechnology*, 8(March), pp. 1–13. doi: 10.3389/fbioe.2020.00188.

Mukherjee, S. (2018) 'Recent progress toward antiangiogenesis application of nanomedicine in cancer therapy', *Future Science OA*. Future Medicine Ltd. doi: 10.4155/fsoa-2018-0051.

Nguyen, E. H. and Murphy, W. L. (2018) 'Customizable biomaterials as tools for advanced anti-angiogenic drug discovery', *Biomaterials*. Elsevier Ltd, pp. 53–66. doi: 10.1016/j.biomaterials.2018.07.041.

Nikolova, M. P. and Chavali, M. S. (2019) 'Recent advances in biomaterials for 3D scaffolds: A review', *Bioactive Materials*. KeAi Communications Co., 4, pp. 271–292. doi: 10.1016/j.bioactmat.2019.10.005.

O'Brien, F. J. (2011) 'Biomaterials & scaffolds for tissue engineering', *Materials Today*. Elsevier Ltd, 14 (3), pp. 88–95. doi: 10.1016/S1369-7021(11)70058-X.

Pàez-Ribes, M. *et al.* (2009) 'Antiangiogenic Therapy Elicits Malignant Progression of Tumors to Increased Local Invasion and Distant Metastasis', *Cancer Cell*. Cancer Cell, 15(3), pp. 220–231. doi: 10.1016/j.ccr.2009.01.027.

Ribatti, D. (2010a) 'Biomarkers of response to angiogenesis inhibitors: An open and unsolved question', *European Journal of Cancer*. Elsevier Ltd, pp. 6–8. doi: 10.1016/j.ejca.2009.10.022.

Ribatti, D. (2010b) 'The inefficacy of antiangiogenic therapies', *Journal of Angiogenesis Research*, p. 27. doi: 10.1186/2040-2384-2-27.

Ribatti, D. (2017) 'Sales for anti-angiogenic drugs', *Oncotarget*. Impact Journals LLC, pp. 38080–38081. doi: 10.18632/oncotarget.17373.

Risau, W. (1997) 'Mechanisms of angiogenesis', *Nature*, pp. 671–674. doi: 10.1038/386671a0.

Saberianpour, S. *et al.* (2018) 'Tissue engineering strategies for the induction of angiogenesis using biomaterials', *Journal of Biological Engineering*. BioMed Central Ltd., p. 36. doi: 10.1186/s13036-018-0133-4.

Setyawati, M. I. and Leong, D. T. (2017) 'Mesoporous Silica Nanoparticles as an Antitumoral-Angiogenesis Strategy', *ACS Applied Materials and Interfaces*. American Chemical Society, 9(8), pp. 6690–6703. doi: 10.1021/acsami.6b12524.

Siddique, S. and Chow, J. C. L. (2020) 'Gold Nanoparticles for Drug Delivery and Cancer Therapy', *Applied Sciences*. MDPI AG, 10(11), p. 3824. doi: 10.3390/app10113824.

Vines, J. B. *et al.* (2019) 'Gold nanoparticles for photothermal cancer therapy', *Frontiers in Chemistry*. Frontiers Media S.A., p. 167. doi: 10.3389/fchem.2019.00167.

Vlăsceanu, G. M. *et al.* (2016) 'Silver nanoparticles in cancer therapy', in *Nanobiomaterials in Cancer Therapy: Applications of Nanobiomaterials*. Elsevier Inc., pp. 29–56. doi: 10.1016/B978-0-323-42863-7.00002-5.

Zhang, Y. *et al.* (2019) 'Nanotechnology in cancer diagnosis: Progress, challenges and opportunities', *Journal of Hematology and Oncology*. BioMed Central Ltd., p. 137. doi: 10.1186/s13045-019-0833-3.

Advances and Current Trends in Biomechanics – Belinha et al. (Eds)
© 2022 Copyright the Author(s), ISBN: 978-1-032-10806-3

Acid effect on PMMA microhardness and surface roughness in fixed prosthodontics

T. Soares & E.R. Carneiro
Faculty of Dental Medicine, University of Porto, Portugal

J. Lino Alves
LAETA, INEGI, Faculty of Engineering, University of Porto, Portugal

P.J. Almeida, S.J. Oliveira & M.H. Figueiral
INEGI, Faculty of Dental Medicine, University of Porto, Portugal

ABSTRACT: The use of temporary restorations is an important step of the prosthodontic treatment of the patient. There is a great diversity of materials, including acrylic and bis-acrylic resins. As for the confection technique, restorations can be performed directly on the prepared teeth or, indirectly, through CAD/CAM technology. The oral environment causes degradation and aging of dental restorations, due to constant contact with saliva, food components and drinks. The aim of this work was to evaluate the microhardness and surface roughness of different materials used in temporary restorations, when in contact with acidic agents found in the diet. In the experimental conditions tested, the only erosive agent affecting the surface roughness of polymethyl-methacrylate (PMMA) was Red Bull®, with Manual PMMA samples showing increased surface roughness when i.n contact with this agent. With regard to microhardness, no statistically significant differences were found.

1 INTRODUCTION

Temporary restorations on fixed prosthodontics are an intermediate stage of the treatment, between tooth preparation and placement of the final indirect restoration (Abdullah et al. 2016). An adequate temporary restoration must fulfil biological, mechanical and aesthetic principles, respecting the following requirements: pulp protection, dimensional stability, favoring oral hygiene and shaping of the gingival margins (Strassler 2009).

There is a great diversity of provisional materials, such as acrylic resins and bis-acrylic composite resins (Perry & Magnuson 2012). Temporary restorations can be performed directly on prepared teeth or, indirectly, using CAD/CAM methodologies involving digital impression and 3D printing (Regish et al. 2011).

With the widespread consumption of acidic foods and drinks, exposure of tooth structures and restorative materials to acidic environments is increasingly common, making acid erosion in the oral cavity a growing reality (de Carvalho Sales-Peres et al. 2007).

This work aimed to evaluate the microhardness and surface roughness of different materials of temporary restorations used in fixed prosthodontics, when in contact with acidic agents found in the diet.

DOI: 10.1201/9781003217152-59

2 MATERIALS AND METHODS

50 specimens of PMMA acrylic resin (TAB 2000®, Kerr, Italy) were produced by direct mixing and manipulation (Manual PMMA) and 50 specimens of PMMA (Temp Basic – Zirkonzahn SRL®, Italy) were produced by laboratory CAD/CAM. All specimens were subjected to the same polishing conditions, with drills and polishing discs, in order to simulate the in-office polishing protocol. The specimens were divided into subgroups as described in Table 1.

Table 1. Distribution of the specimens according to the contaminating solution.

Resin of manual PMMA (TAB2000®)	Control group – Distilled water (n= 10) Soft drink – Coca-Cola® (n= 10) Red wine (n= 10) Natural orange juice (n= 10) Energy drink – Red Bull® (n= 10)
Resin of PMMA for CAD/CAM (Temp Basic - Zirkonzahn SLR® A2-B2 95H16)	Control group – Distilled water (n= 10) Soft drink – Coca-Cola® (n= 10) Red wine (n= 10) Natural orange juice (n=10) Energy drink – Red Bull® (n= 10)

The specimens were stored in an oven at 37°C for 7 days, after being randomly distributed by the different acid solutions and distilled water (control group). The pH of the storing solutions was measured at the beginning and at the end of the experiment using pH test strips (Figure 1).

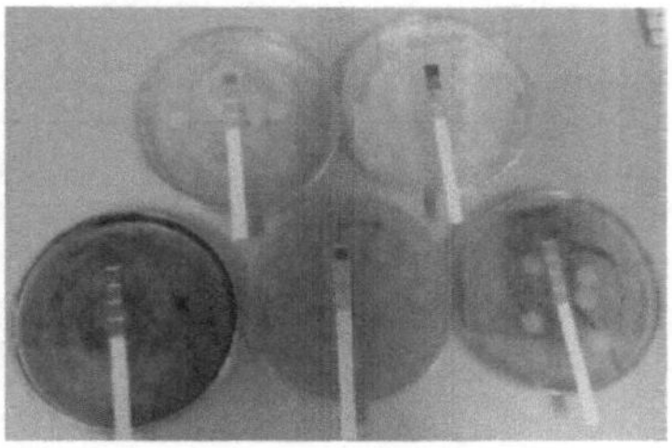

Figure 1. pH measurement after storage of the specimens for 7 days in an oven at 37°C.

To assess the surface roughness, Optical Interferometry was used [determination of arithmetic mean roughness parameters (Ra) and maximum heights of the roughness profile (Rz)]; and to assess the surface microhardness, the Vickers hardness test was performed. The data were collected and stored in a database created using the Excel® software and the statistical analysis was performed using SPSS®, considering a significance level of 0.05 ($\alpha = 0.05$).

3 RESULTS

3.1 *Surface roughness analysis*

A total of 100 specimens was subjected to surface roughness testing (Table 2).

Table 2. Descriptive statistics of Ra and Rz for each group.

Material	Solution	Mean Ra	Std deviation Ra	Median Ra	Mean Rz	Std deviation Rz	Median Rz
Manual PMMA	Distilled H_2O	0.46	0.32	0.33	7.06	6.53	3.19
	Red wine	0.45	0.22	0.37	7.78	5.92	6.72
	Orange juice	0.38	0.13	0.40	2.99	1.17	2.79
	Red Bull®	1.03	0.86	0.84	18.89	14.26	17.58
	Coca-Cola®	0.36	0.14	0.35	2.78	0.97	2.94
CAD/CAM PMMA	Distilled H_2O	1.15	0.87	1.02	17.25	15.68	15.17
	Orange juice	0.67	0.22	0.61	9.83	4.39	8.83
	Red Bull®	0.55	0.52	0.39	5.62	8.93	2.71
	Coca-Cola®	0.54	0.15	0.51	5.11	2.67	3.95
	Red wine	0.54	0.45	0.39	5.41	8.78	2.72

With regard to surface roughness, the Kruskal-Wallis test revealed statistically significant differences when all groups were compared. After performing the Mann-Whitney post hoc test, only the specimens of Manual PMMA stored in Red Bull® solution showed statistically significant differences ($p=0.034$ for Ra and Rz), exhibiting increased surface roughness in comparison with the control samples. In CAD/CAM PMMA specimens, there were no statistically significant differences between the groups.

3.2 *Surface microhardness analysis*

The microhardness test was carried out in 30 samples (3 specimens per group; Table 3).

Table 3. Descriptive statistics of microhardness for each group.

Material	Solution	Mean	Std deviation	Median
Manual-PMMA	Distilled H_2O	10.88	0.50	10.75
	Coca-Cola®	11.32	0.48	11.3
	Orange juice	11.53	0.52	11.75
	Red Bull®	10.27	0.02	10.25
	Red wine	10.70	0.13	10.8
CAD/CAM PMMA	Distilled H_2O	10.27	0.03	10.3
	Coca-Cola®	10.35	0.03	10.35
	Orange juice	10.25	0.10	10.3
	Red Bull®	10.32	0.02	10.3
	Red wine	9.97	0.14	10.05

After performing the Kruskal-Wallis test, no statistically significant differences were found concerning the microhardness (Manual PMMA: $p=0.177$ and CAD/CAM PMMA: $p=0.096$), nor in terms of manufacturing protocol neither storage conditions.

4 DISCUSSION

The current work aimed at assessing the possible changes in surface roughness and microhardness of PMMA specimens, when in contact with acidic agents found in the diet.

During the experimental stages, the pH values of the storing solutions did not change. However, a limitation of this study relates with the absence of all the factors present in the oral cavity, namely the buffer capacity of saliva, which gradually increases the pH values, and the fermentation of carbohydrates, which decreases the pH values (Wongkhantee et al. 2006, Alencar et al., 2016).

According to the literature, temporary crowns produced with CAD/CAM technologies have less roughness and porosity because they are manufactured under controlled conditions, in contrast to temporary crowns obtained by the direct technique (Rosentritt et al. 2017, Rayyan et al. 2015). In this work, there were statistically significant differences in the roughness of the specimens, and those manufactured by the direct technique showed higher roughness values. Moreover, significantly increased surface roughness was only observed for Manual PMMA specimens stored in Red Bull® solution, when compared with control conditions. These findings may arise from the in-office polishing performed in all specimens. Microscopically, however, the manual polishing of the specimens proved to be insufficient, producing a very irregular and unclear surface that made difficult the surface roughness and microhardness measurements.

Nicodemo et al. compared the changes in the microhardness of PMMA used in provisional crowns, manufactured by different techniques, after pH variations. The authors concluded that, although there was a decrease in the hardness of the material, no statistically significant differences between the different types of processing were found (Nicodemo et al. 2013). In our experimental conditions, with regard to microhardness, when comparing groups of the same material or the same acidic agent, there were no statistically significant differences. Moreover, despite the erosive potential of the contaminants, the hardness of the materials tested did not changed significantly.

5 CONCLUSIONS

Compared to the provisional restorations of Manual PMMA, the specimens produced by CAD/CAM appear to have better performance when subjected to dietary acids, as their surface roughness and microhardness were not significantly affected by those agents.

REFERENCES

Abdullah, A. O., Tsitrou, E. A. & Pollington, S. 2016. Comparative in vitro evaluation of CAD/CAM vs conventional provisional crowns. *Journal of Applied Oral Science*, 24(3): 258–63.

Alencar, C. R., Mendonça, F. L., Guerrini, L. B., Jordão, M. C., Oliveira, G. C., Honório, H. M., Magalhães, A. C. & Rios, D. 2016. Effect of different salivary exposure times on the rehardening of acid-softened enamel. *Brazilian Oral Research*, 30(1): e104.

de Carvalho Sales-Peres, S. H., Magalhães, A. C., de Andrade Moreira Machado, M. A. & Buzalaf, M. A. 2007. Evaluation of the erosive potential of soft drinks. *European Journal of Dentistry*, 1(1): 10–3.

Nicodemo, C. A. Z., Rezende, C. E. E., Moretti-Neto, R. T. & Rubo, J. H. 2013. Micro-hardness of Acrylic Resin utilized for Provisional Crowns: Effect of different polymerization techniques and pH-Cycling. *Brazilian dental science*, 16(2): 44–50.

Perry, R. D. & Magnuson, B. 2012. Provisional materials: key components of interim fixed restorations. *Compendium of continuing education in dentistry*, 33(1): 59–60, 62.

Rayyan, M. M., Aboushelib, M., Sayed, N. M., Ibrahim, A. & Jimbo, R. 2015. Comparison of interim restorations fabricated by CAD/CAM with those fabricated manually. *Journal of Prosthetic Dentistry*, 114(4): 414–9.

Regish, K. M., Sharma, D. & Prithviraj, D. R. 2011. Techniques of fabrication of provisional restoration: an overview. *International Journal of Dentistry*, 2011: 134659.

Rosentritt, M., Raab, P., Hahnel, S., Stöckle, M. & Preis, V. 2017. In-vitro performance of CAD/CAM-fabricated implant-supported temporary crowns. *Clinical Oral Investigations*, 21(8): 2581–2587.

Strassler, H. E. 2009. In-office Provisional Restorative Materials for Fixed Prosthodontics. *Inside Dentistry*, 5(4).

Wongkhantee, S., Patanapiradej, V., Maneenut, C. & Tantbirojn, D. 2006. Effect of acidic food and drinks on surface hardness of enamel, dentine, and tooth-coloured filling materials. *Journal of Dentistry*, 34(3): 214–20.

Advances and Current Trends in Biomechanics – Belinha et al. (Eds)

Force-velocity relationship in overarm throwing—an exploratory study

H. Lobo & L. Estriga
Faculdade de Desporto, Universidade do Porto, Porto, Portugal

ABSTRACT: Research about the force-velocity relationship in multi-joint movements is growing. In this research, we developed a specific methodology aiming to assess the force-velocity relationships in handball. Nineteen handball players of the Portuguese first division were organized in sets of four players to perform the throws randomly with four modified handballs (size 3 with different weights). The participants were asked to replicate a 7-meter throw with their maximal shooting velocity. The developed method appears to be relatively simple to use and ecologically valid to assess force-velocity and power-velocity capabilities in handball throwing, but still needs to be improved and tested with a larger sample and a wider range of balls' weight.

1 INTRODUCTION

Throwing velocity is a factor of major importance in high-level handball. Furthermore, velocity in handball throwing is influenced by power and strength (Ortega-Becerra et al., 2018) and, thus, the study of the power-force-velocity relationship seems fundamental.

The force-velocity relationship has been deeply studied both in its physiological basis and, more recently, in its applications to specific sports. These applied studies showed that not only it is possible to obtain an individual profile through the evaluation of performance in different loading conditions (Jaric, 2015, Samozino et al., 2012) but also that training according to each athlete's profile is more effective to improve performance in ballistic movements (Jimenez-Reyes et al., 2017).

The results and importance of force-velocity testing and profiling seem promising, but its results are specific to each movement. When it comes to the upper limbs, the literature is scarcer, but some studies have researched this relationship in both bench press throws (Sreckovic et al., 2015) and with medicine balls (Sreckovic et al., 2015). Nevertheless, if our goal is to study the force-velocity capabilities in such a complex movement as the handball throw, a new method to do this assessment with more specific procedures is needed. Regarding this approach, to our knowledge, there are almost no studies with such methodology. (Tillaar and Ettema, 2004) studied overarm throwing with different ball weights (0.2 to 0.8 kg) and found a linear relationship between force and velocity and no change in the global coordination patterns of throwing across the entire range of balls. However, even though they tested handball players, it is important to note that smaller balls were used (circumference 0.3m). In our exploratory study, we used only weighted official size #3 handballs to evaluate the force-velocity relationship with minimal disturbance of the grip and throw patterns.

DOI: 10.1201/9781003217152-60

2 METHODOLOGY

Nineteen adult male handball players (age: 21.74±2.96 years, body mass= 92.09 ± 12.10 kg, stature= 1.86 ± 0.07 m) were tested. All of them were playing in the Portuguese first division. All recruited subjects had no shoulder pain at the time of the testing or any previous shoulder injury or pathology.

For the data collection, four official handballs (size 3) were used. One was unmodified (433g) while the other three were disassembled, weighted (546 g, 607 g and 747 g) and reassembled by a professional ball manufacturer, without any changes to the exterior shape or surface. The added weights were solidly fixed and uniformly distributed inside the balls.

Due to logistical reasons, the tests were conducted in two different sessions and locations. In one session, fifteen players from the same team performed the tests during a training session, after 20 minutes of a standard warming up, with throwing tasks included. Because of time constraints, each subject did only two shooting attempts with the same ball, consisting of a total of 8 throws per each. In another session, four participants agreed to perform the testing in the Sport Faculty facilities, where there was no time limit for the testing. In this case, three shooting attempts were done. These subjects also performed a similar warming-up protocol. In all situations, a set of four players were organized to perform the throws randomly with each of the four different balls, with one minute and thirty seconds of rest between attempts. The participants were asked to replicate a 7- meter throw (and thus, without lifting one of their feet) with their maximal shooting velocity.

Maximal velocity was measured with a radar gun (Stalker Radar) with an accuracy of ± 0.028 m/s. All throws were within the central 10° arc of the radar gun that was placed behind the goal.

From the energy/work equivalence, one obtains the average force exerted by the player on the ball (F_{ave}):

$$\frac{1}{2} \cdot m \cdot v_{throw}^2 - \frac{1}{2} \cdot m \cdot v_{initial}^2 = F_{ave} \cdot s \tag{1}$$

The initial velocity of the ball is null and so the average throwing force is just the ball's final kinetic energy over the acceleration length (s):

$$F_{ave} = \frac{m\ v_{throw}^2}{2\ s} \tag{2}$$

The acceleration length precise determination requires the usage of a 3-dimensional video system or similar technologies. We assumed that it will change little from each player's throw to throw and so we used a fixed (average value) of 0.5m. As long as the intra-subject constancy of s is true, even if its precise value is not accurate, the determination of the optimal weight and power can still be done exactly the same.

Finally, the average (spatial) throwing power is computed according to:

$$P_{ave} \propto F_{ave} v_{throw} \tag{3}$$

The full details of this elementary derivation are:

$$P_{ave} = (F \cdot v)_{ave} = \frac{1}{s}\int_0^s m\frac{dv}{dt} v dx = \frac{m}{s}\int_0^T \frac{dv}{dt} v^2 dt = \frac{m}{s}\int_0^{v_{throw}} v^2 dv =$$

$$= \frac{m}{s}\left[\frac{v^3}{3}\right]_0^{v_{throw}} = \frac{2m\ v_{throw}^2}{3\ \ 2}\frac{v_{throw}}{s} = \frac{2}{3}(F_{ave}s)\frac{v_{throw}}{s} = \frac{2}{3}F_{ave}v_{throw} \tag{4}$$

3 RESULTS

As the aim of this study was to develop a new methodology for the individual evaluation of force-power-velocity capabilities in handball, individual analysis was carried for each athlete. During this analysis, we realized that the data of some of the subjects was statistically weak (low coefficient of determination, R^2). For that reason, we focused our analysis on a more restrict group of seven subjects with higher R^2 that are presented in Table 1, together with their throwing velocity, force and power with the official ball, in comparison with the same measures with the theoretical optimal ball weight (for power maximization).

In the Figure 1, we can observe two different power-velocity profiles for handball throwing. Additionally, we computed a theoretical optimal ball weight that would maximize power for that athlete (apex of the power-velocity curve).

Table 1. Throwing velocity, force and power with both the official ball and the theoretical optimal ball weight.

	Official ball				Optimal ball weight			
Subject (ID)	Throw velocity (m/s)	$Force_{mean}$ (N)	P_{max} (W)	R^2, N	m_{opt_ball} (kg)	Throw velocity (m/s)	Force (N)	P_{max} (W)
2	22.18	221.46	4912.87	0.45, 8	0.968	17.59	299.69	5272.09
7	27.42	338.33	9276.97	0.45, 8	0.825	22.73	426.37	9690.10
10	25.80	299.58	7729.80	0.61, 8	1.052	20.02	421.39	8434.57
16	23.75	253.80	6027.45	0.56, 12	0.694	18.91	341.20	6450.69
17	25.14	280.47	7050.12	0.50, 12	0.444	25.02	281.75	7049.98
18	26.64	319.2	8506.19	0.65, 12	0.605	24.22	354.78	8592.01
19	21.46	207.23	4447.11	0.42,12	0.755	18.25	251.50	4589.27

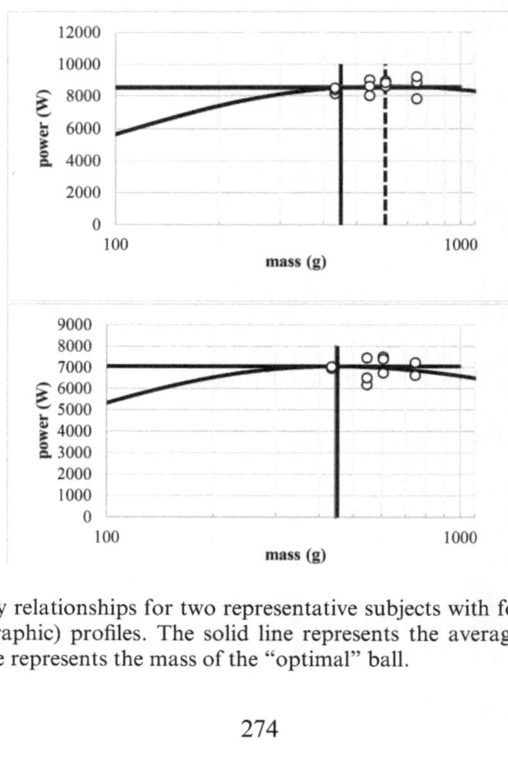

Figure 1. Power-velocity relationships for two representative subjects with force-oriented (top graphic) and balanced (bottom graphic) profiles. The solid line represents the average mass of the official ball (450g) and the dashed line represents the mass of the "optimal" ball.

4 DISCUSSION

In the top graphic of Figure 1 we can observe that, for a specific subject, the theoretical most suited ball (ball weight with which the athlete would produce more power) is a lot heavier (605g) than the official ball (425-475g). These results indicate that this athlete has a more force-oriented profile and might have a speed deficit. These observations are in line with the ideas of (Morin and Samozino, 2015) that consider that an evaluation of individual force-velocity (in our study, the consequent power-velocity) relationships can help identifying imbalances and optimize strength training, according to individual necessities. Furthermore, to strengthen this perspective, Jimenez-Reyes et al. (2017) showed that training addressing own imbalances and deficits (studied with force-velocity assessments in squat jump) is more effective than training following "one-size-fits-all" programs. If the subject on the top graphic of Figure 1 wants to improve is throwing velocity, then the focus should be on minimizing this speed deficit. According to the same study, the subject on the bottom graphic presents a balanced power-velocity profile and should search all-round improvements in the force-velocity curve.

One should be aware that, even though these individual assessments appear to offer some valuable insights about the athlete's needs, the methodology used in this study still has several limitations. One aspect that should be assessed in order to have more control over the variables of the throw is the throwing angle. As we know, handball throwing technique is not the same between athletes and even between throws. An approach similar to the one used by Tillaar and Ettema (2004), that used high velocity cameras and reflective markers to record the movement could be one way to reduce the effect of the variation in throwing technique and to know the ball's propulsion distance (until the point of ball release). The direct measure of acceleration would also allow a more accurate determinarion of the force-velocity and power-velocity relationships. Additionally, increasing the range of ball weights or increasing the number of throws could result in a higher R^{2}.

5 CONCLUSIONS

Our findings revealed that the majority of the evaluated players would produce higher standing-throw power if the ball was heavier than the official size 3, meaning that they have a speed deficit and a more force-oriented profile. The methodology can still be improved with the assessment of acceleration and joint angles.

REFERENCES

Jaric, S. 2015. Force-velocity Relationship of Muscles Performing Multi-joint Maximum Performance Tasks. *International journal of sports medicine*, 36.

Jimenez-reyes, P., Samozino, P., Brughelli, M. & Morin, J.-B. 2017. *Effectiveness of an Individualized Training Based on Force-Velocity Profiling during Jumping*, Frontiers in Physiology.

Morin, J.-B. & Samozino, P. 2015. Interpreting Power-Force-Velocity Profiles for Individualized and Specific Training. *International journal of sports physiology and performance*, 11.

Ortega-becerra, M., Pareja-blanco, F., JimÉnez-Reyes, P., Cuadrado-peÑafiel, V. & GonzÁlez-badillo, J. J. 2018. Determinant Factors of Physical Performance and Specific Throwing in Handball Players of Different Ages. *Journal of strength and conditioning research*, 32, 1778–1786.

Samozino, P., Rejc, E., Enrico di prampero, P., Belli, A. & Morin, J.-B. 2012. Optimal Force–Velocity Profile in Ballistic Movements—Altius. *Medicine and science in sports and exercise*, 44, 313–22.

Sreckovic, S., Cuk, I., Djuric, S., Nedeljkovic, A., Mirkov, D. & Jaric, S. 2015. *Evaluation of force–velocity and power–velocity relationship of arm muscles.*

Tillaar, R. & Ettema, G. 2004. *A Force-Velocity Relationship and Coordination Patterns in Overarm Throwing.*

Advances and Current Trends in Biomechanics – Belinha et al. (Eds)
© 2022 Copyright the Author(s), ISBN: 978-1-032-10806-3

Thermographic study of cranio-cervico-mandibular complex after oral rehabilitation

A. Moreira & R. Batista
Faculdade de Medicina Dentária, Universidade do Porto, Portugal

J. Mendes
INEGI, Faculdade de Engenharia, Universidade do Porto, Portugal

S.J. Oliveira, M. Sampaio-Fernandes & M.H. Figueiral
INEGI, Faculdade de Medicina Dentária, Universidade do Porto, Portugal

ABSTRACT: An unstable occlusion may be responsible for an unbalanced masticatory muscular activity bringing consequences in terms of vascularization and local temperature. Infrared thermography (IRT) allows to quantify and map different temperatures at the skin's surface corresponding to the under anatomical structures. The main objective of this investigation was to evaluate the influence of a prosthodontic treatment on the cranio-cervico-mandibular complex using IRT. The sample was divided into three groups: control, initial treatment group and the treatment group after oral rehabilitation. The thermographic recording was performed under controlled and protocoled conditions. Our results suggest that masseter and *orbicularis oris* muscles may show thermal changes after oral rehabilitation treatments in complete and partial edentulous patients.

1 INTRODUCTION

The cranio-cervico-mandibular complex (CCMC) is composed of a large number of anatomical structures. The muscles involved in the CCMC can be categorized as: a) muscles responsible for the mandibular movement; b) tongue muscles; c) facial muscles and d) neck muscles (Lucas 2010). The temporomandibular joint (TMJ) is the anatomical region where the mandible articulates with the cranium (Standring 2016). Other component of the CCMC are the teeth found in the alveolar processes of the maxilla and mandible. When the elevator muscles of the mandible contract, together with both TMJ (left and right), mandibular teeth contact the maxillary ones, a phenomenon known as intercuspation. An unstable occlusion may be responsible for an unbalanced masticatory muscles activity, being one of multiple factors for developing temporomandibular disorders (TMD) (Ribeiro et al. 2014). TMD are classified as a subgroup of musculoskeletal disorders that affect the TMJ, the masticatory muscles and all the surrounding structures. These conditions are characterized by an altered mandibular cinematic, myalgia, arthralgia and/or articular sounds. However, the masticatory dysfunction is independent of signs and symptoms of TMD since the human body has a great adaptive capacity. Mammals are dependent of a correct chewing process and bolus formation since it reduces the size of food and increases the area of enzymatic contact in the gastrointestinal tract (Bortoluzzi et al. 2012). The human body is provided with a complex mechanism of temperature maintenance. The sympathetic and parasympathetic nervous system regulate the body temperature controlling the transference of heat with radiation, conduction, convection

DOI: 10.1201/9781003217152-61

and evaporation. Infrared thermography (IRT) allows to quantify and map different temperatures at the skin's surface corresponding to the anatomical structures underneath. IRT has being described as a valid diagnostic tool for TMD and facial pain (Dibai Filho et al. 2013, Pogrel et al. 1996). The present study intends to evaluate the influence of prosthodontic treatments on the cranio-cervico-mandibular complex (temporal, masseter and *orbicularis oris* muscles and TMJ) using IRT. Null hypotheses were as follows: H0a – The ΔT has no differences when comparing the sample groups and H0b – The state (healthy or pathological) has no differences when comparing the sample groups.

2 MATERIALS AND METHODS

The treatment group was composed of adults of both sexes, that underwent a prosthodontic treatment: removable or implant-/tooth-supported fixed dentures, that added at least two posterior occlusal contacts and/or reestablished the vertical dimension of occlusion (complete edentulous or Kennedy class I/II). The control group was composed of adults of both sexes that had never received any prosthodontic treatment, with no more than a single missing posterior tooth, with the exception of third molars. The presence of TMD in both groups was excluded upon the 3Q/RDC questionnaire together with clinical examination. The thermographic analysis to the CCMC was performed with the FLIR i7 infrared camera from Flir® Systems Inc, Oregon, U.S.A., under controlled and protocoled conditions.

The thermal images of the treatment group were obtained before the oral rehabilitation (BOR; n=33) and two months after concluding the oral rehabilitation treatment (AOR; n=19). For the control group (CG; n=33) there was only one moment of thermographic recording.

For each participant and IRT assessment a total of six images, two images per view (right, left and frontal view), was collected (Figure 1). Thus, a total of 510 thermograms was obtained and analyzed with the software Flir Tools®.

For each view and region of interest (ROI) an average temperature value was recorded and organized in a data base in the SPSS v24 software (IBM®, Armonk, NY, USA). In this data base the thermal difference (ΔT) (asymmetry value) and the state of the ROI (healthy/pathological) were also recorded. Considering previous studies, the cut-off point to describe a ROI as pathological corresponds to ΔT equal or superior to 0.4°C. After testing the variables' homogeneity it was concluded that data did not follow a normal distribution. Therefore, only non-parametric statistical tests were performed with a 95% confidence interval to compare and test the null hypotheses.

To compare the ΔT of the treatment group before and after oral rehabilitation (AOR vs BOR) a Wilcoxon test was run. To compare the ΔT between the treatment groups and the control group (AOR vs CG and BOR vs CG) the U-Mann Whitney test was used. To compare the health status of the ROI between the different sample groups (BOR, AOR and CG) the chi-square test was run.

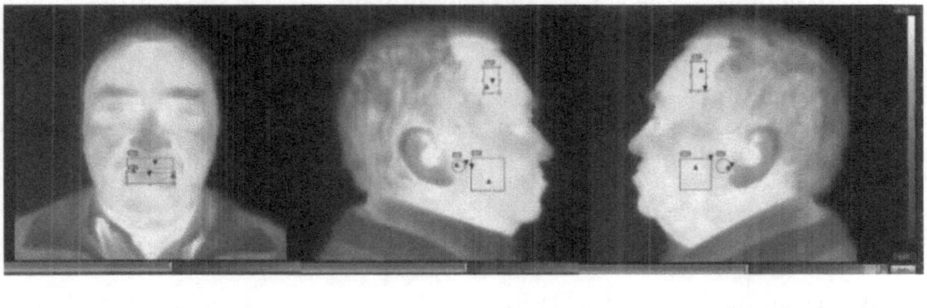

Figure 1. Example of the frontal, lateral right and left thermograms obtained and analyzed with the software Flir Tools ®.

3 RESULTS

Our results showed significant statistical differences when the ΔT and the health status of the *orbicularis oris* muscle were compared between the BOR and the AOR groups (p=0.020 and p=0.003, respectively). Overall, after the oral rehabilitation, patients showed lower ΔT values and pathological ROI comparing to the initial condition (i.e. before oral rehabilitation). When the health status of the ROI was compared between the CG and the BOR groups, only the masseter muscle showed significant statistical differences (p=0.030), with the frequency of pathological ROI being greater in the BOR group. No statistical differences were found for the ΔT and the health status of the ROI for the remaining structures.

4 DISCUSSION

IRT is considered an efficient method to assess the surface temperature of the skin, being used in diverse medical situations (Pogrel et al. 1996). The role of IRT as a diagnostic tool for DTM has been described elsewhere (Pogrel et al. 1996). Nevertheless, very few studies similar to the present one are available in the literature, addressing the potential of IRT in the context of oral rehabilitations. One way to assess the health status of a ROI is comparing the temperature values of ipsilateral structures (ΔT= |left ROI – right ROI|).

Taking into consideration the results of the study, it is only possible to accept partially the null hypotheses, once statistically significant differences were found for the *orbicularis oris* and masseter muscle. In contrast, no statistically significant differences were found for the temporal muscle and temporomandibular joint. In the particular case of TMJ, these results suggest that when participants without TMD are enrolled, the TMJ temperature is not influenced by the presence or absence of teeth and/or an oral rehabilitation. Such behavior should not be expected in patients with intracapsular pathologies, lupus, rheumatoid arthritis, in which the temperature is influenced by the condition and its severity (Gratt et al. 1994, Lasanen et al. 2015). Similarly, the temporal muscle showed no significant statistical differences independently of the groups analyzed.

The electromyographic behavior of masticatory muscles has been studied during the prosthodontic treatment process. Usually, when the retention, stability and occlusion are improved, oral rehabilitated patients show an increased muscular activity (Fueki et al. 2007). It is important to highlight that in the present study all the thermograms were collected with the participants in a rest position and without any chewing exercise prior to the examination. Moreover, it should be emphasized that even though minimum, electromyography activity remains positive with the masticatory muscles in a rest position, maintaining the basal muscle tone (Oncins et al. 2014).

5 CONCLUSIONS

A prosthodontic treatment appears to have minimum or null effect on the ΔT or health status of some CCMC structures, namely TMJ and temporal muscle. On the contrary, the masseter and *orbicularis oris* muscles showed significant thermal changes when participants underwent oral rehabilitation including two or more posterior contacts and/or the restoration of the vertical dimension of occlusion. This pattern should be validated by further studies.

Our preliminary data suggest that IRT may be an important tool to evaluate the quality of prosthodontic treatments while assessing thermal parameters of the orofacial structures.

REFERENCES

Bortoluzzi, M. C., Traebert, J., Lasta, R., Da Rosa, T. N., Capella, D. L. & Presta, A. A. 2012. Tooth loss, chewing ability and quality of life. *Contemporary clinical dentistry*, 3(4), 393–397.

Dibai filho, A. V., Packer, A. C., Costa, A. C. & Rodrigues-bigaton, D. 2013. Accuracy of infrared thermography of the masticatory muscles for the diagnosis of myogenous temporomandibular disorder. *Journal of Manipulative and Physiological Therapeutics*, 36(4), 245–52.

Fueki, K., Kimoto, K., Ogawa, T. & Garrett, N. R. 2007. Effect of implant-supported or retained dentures on masticatory performance: a systematic review. *Journal of Prosthetic Dentistry*, 98(6), 470–7.

Gratt, B. M., Sickles, E. A., Wexler, C. E. & Ross, J. B. 1994. Thermographic characterization of internal derangement of the temporomandibular joint. *Journal of Orofacial Pain*, 8(2), 197–206.

Lasanen, R., Piippo-savolainen, E., Remes-pakarinen, T., Kröger, L., Heikkilä, A., Julkunen, P., Karhu, J. & Töyräs, J. 2015. Thermal imaging in screening of joint inflammation and rheumatoid arthritis in children. *Physiological Measurement*. 36 (2), 273–82.

Lucas, P. W. 2010. *Dental Functional Morphology - How teeth work*, Cambridge University Press.

Oncins, M. C., Vieira, M. M. & Bommarito, S. 2014. Eletromiografia dos músculos mastigatórios: análise em valor original e RMS. *Revista CEFAC*, 16(4), 1215–1221.

Pogrel, M. A., McNeill, C. & Kim, J. M. 1996. The assessment of trapezius muscle symptoms of patients with temporomandibular disorders by the use of liquid crystal thermography. *Oral Surgery, Oral Medicine, Oral Pathology, Oral Radiology, and Endododontology*, 82(2), 145–51.

Ribeiro, J. A., de Resende, C. M., Lopes, A. L., Farias-Neto, A. & Carreiro, A.D. F. 2014. Association between prosthetic factors and temporomandibular disorders in complete denture wearers. *Gerodontology*, 31(4), 308–13.

Standring, S. 2016. *Gray's anatomy: the anatomical basis of clinical practice*. Elsevier Limited. ISBN: 978-0-7020-5230-9.

Advances and Current Trends in Biomechanics – Belinha et al. (Eds)

Implementation and assessment of an autonomous control for human supervision replacement in a robotic walker toward an adaptive shared control

A. Pereira, J. Lopes & J. Afonso
Center for Microelectromechanical Systems, University of Minho, Portugal

L. Costa
ALGORITMI, University of Minho, Portugal

J. Figueiredo & C.P. Santos
Center for Microelectromechanical Systems, University of Minho, Portugal

ABSTRACT: Robotic-walker-aided rehabilitation of patients with gait disorders has shown improvements over classical methods. However, it requires constant supervision from a medical expert so that collisions are avoided. In this paper, an autonomous controller capable of guiding the patient to a destination while avoiding collisions is proposed as a replacement for human supervisors. The association of a LiDAR sensor and Google's Cartographer algorithm creates the map, registering obstacles. The global planner A* plans a path to meet the user's selected destination. The local planner Dynamic Window Approach was also implemented, which considers the dynamic constraints. The autonomous control was validated with 3 healthy volunteers using a walker under 5 validation scenarios, assessing its suitability of use as a supervisor in a shared control. Results show a mapping and localization accuracy of 0.38 m. The walker never registered any collision and always reached the destination, highlighting its applicability as human supervisor replacement.

1 INTRODUCTION

Gait disorders usually lead to a poor quality of life due to a decrease in physical, psychological, and social health indexes (Mahlknecht et al. 2013). Robotic-walker-aided rehabilitation has proven to be capable of offering increased physical support for the patient over classical methods, while removing the need for the medical expert to physically carry the patient (Sierra. et al. 2019). Moreover, these actively drivable walkers allow the patient to independently determine the locomotion direction, simulating more accurately daily life activities, possibly improving the rehabilitation's efficiency (Sierra, et al. 2019).

Despite the increased autonomy given by the walker when compared to classic methods of rehabilitation, the constant supervision by a medical expert is still mandatory, so that the patient's safety and rehabilitation are ensured. Patient's safety is compromised when unintended falls occur due to three main factors: i) bad quality of gait pattern, ii) walker-patient interface compliance, and iii) rehabilitation scenario complexity (presence of obstacles) (Sierra et al. 2019). While monitoring the risk of fall, the human supervisor does not only need to plan how to decrease this risk, but also when it is beneficial to interfere.

The full dependency of a medical expert turns the whole rehabilitation procedure inefficient from a medical institution's staff allocation point-of-view. State-of-the-art approaches deploy

DOI: 10.1201/9781003217152-62

shared controllers which can be divided into 3 sections: i) a patient intention detector, ii) an autonomous supervisor, and iii) a prioritizer (Sierra et al. 2019). The patient intention detector is responsible for converting the user's intention into information understandable by the walker, i.e., velocity commands. The autonomous supervisor is always aware of the user's risks of collision with either permanent or dynamic obstacles, and it is responsible for creating a safe trajectory that avoids obstacles and reaches a certain goal. The prioritizer is the entity responsible for contextualizing the commands suggested by both patient intention detector and autonomous supervisor, choosing which one is the most suitable from a safety point-of-view.

The deployment of autonomous driving systems is common among the automotive and generic robotics community; however, literature show a scarce amount of information when implemented in biomedical devices. Before any type of driving is achieved, the robot needs to read, to perceive, and to interpret its surroundings, so that it can locate itself, assess collision risk, and determine the best route to meet the destination. To acquire environment data, LiDAR, SoNAR, RaDAR, GPS, video cameras, and inertial motion sensors are usually used (Woo et al. 2019). However, the LiDAR sensor is the one that presents best range and resolution. To locate the robot, Global Positioning Systems and Simultaneous Localization and Mapping (SLAM) algorithms are commonly used, being SLAM more beneficial in indoor settings considering its high availability and no need for external setup. Nevertheless, it first needs to fully scan the environment before any type of use.

After the autonomous controller is fully aware of its surroundings, global planners are used to determine which path can reach the destination more efficiently. Dijkstra's algorithm and A*, usually applied to shortest path solving of graphs, determines a set of sequential positions capable of meeting the destination (Yurtsever et al. 2020). These global planners only consider spatial constraints, e.g., the presence of obstacles (either static or dynamic). Local planners, on the other hand, use kinodynamic constraints to plan paths, translating the spatial trajectory into velocity commands which can be read by robot controllers. The Dynamic Window Approach (DWA) is an efficient but simple local planner which predicts multiple possible trajectories based on a "dynamic window" which is regulated based on the robot's properties such as maximum accelerations and velocities. The best trajectory is chosen based on the quality parameterization of the users, involving variables such as: proximity to obstacles, destination, and initially planned path from the global planner.

In this paper, we implemented an identic framework as an autonomous driving system for a rehabilitation robotic walker. It uses a LiDAR sensor and Google's Cartographer SLAM algorithm as a localization component, and the A* and DWA as global and local planners, respectively. Its accuracy to reach the destination and the ability of avoiding obstacles was assessed so that it can be used as the supervisor of a shared control.

2 MATERIALS AND METHODS

In this work, the WALKit (previously name ASBGo*) walker is used as the rehabilitation tool. It was specially designed to attain the necessities and difficulties of cerebellar ataxia patients, that present a strongly affected gait, poor motor coordination and postural control. Its architecture is based on, a powerful software for robots' control.

To implement the autonomous control and acquire environment data, a LiDAR URG-04LX-UG01 (HOKUYO, Japan) sensor was used. This sensor was chosen since it provides 5 m of range, 0.325° of angular resolution, it requires low power, and it is of small size. All these characteristics are ideal for its integration into the robotic walker due to not requiring any significant structural and functional changes.

To map the environment and localize the walker, Google's Cartographer SLAM algorithm was used since it can map small environments accurately and also correlates multiple small maps, turning them coherent between each other. Moreover, the Cartographer does not need any odometry algorithm to bootstrap the SLAM process, which is an advantage when compared to others described in literature (Hess *et al.* 2016).

Figure 1. The position of the LiDAR sensor in the walker (a), and the ArUco marker detection for validation (b).

To validate the LiDAR-SLAM pair, it is essential to measure the accuracy of localization. To objectively measure the map quality and trajectory accuracy, the proposed map must be compared to ground-truth maps and literature-validated positioning systems. However, these are difficult to obtain, and so an alternative approach was devised. The SLAM system is recursive, i.e., when mapping, the algorithm needs the walker's pose, and to determine the walker's pose, it needs a definition of the environment; thus, if the trajectory's error is low, the mapping must be accurate enough to allow for good pose inferences. Therefore, measuring the trajectory error and maps quality are one and the same. Ten ArUco markers were position in the environment, as illustrated in Figure 1, along with an Orbbec Camera (Orbbec, USA) installed in the robotic walker. The ArUco marker detection package allowed the walker to determine its relative position to the markers, creating a sequence of relative positions through time and throughout the environment. This trajectory created by the marker-camera association can used to determine the relative position of the markers to the LiDAR-SLAM world frame, and the distance between markers (in the LiDAR-SLAM frame) measured. These measurements were compared to the ones measured manually with a ruler of precision 0.05 m, using the root mean squared as a validation metric.

Regarding the walker's control, the global planner A* was chosen due to its high presence in literature, simplicity, and efficiency. After the global plan is calculated, the local planner DWA converts the first path into velocity commands. These commands are vector comprising two scalars: i) the linear velocity: velocity aligned with the heading of the walker, and ii) the angular velocity.

To validate the global and local planners, 5 validation scenarios were performed involving 3 healthy volunteers. These validation scenarios comprise a set of environment states that reflect the possible configurations met in gait rehabilitation, namely: i) destination chosen straight ahead with no obstacles present, ii) destination straight backward with no obstacles, iii) contouring a 90° corner to the right and left (static obstacle present), and iv) driving straight forward with a dynamic obstacle temporarily crossing in front of the initially planned path. The number of collisions and destinations reached were used as validation metrics, in a way of measuring how suitable the controller is in keeping the user safe.

3 RESULTS

In terms of mapping and localization, the proposed LiDAR-SLAM structure created maps with all 100th percentiles around 0.21 m and outliers of 0.38 m of root mean squared error. In terms of global and local plan, no collision was observed with any static or dynamic obstacles and the walker was always able to reach the set destination. Figure 2 illustrates the velocity commands and distance to the closest obstacle.

4 DISCUSSION

To assess the use of this autonomous control as the supervisor of a shared control, the map and localization accuracy was measured. Moreover, the ability of the walker to avoid

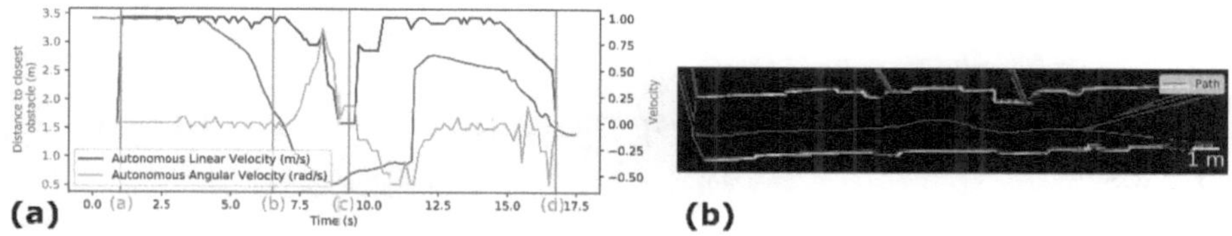

Figure 2. The plot of the velocity commands and distance to the closest obstacle in the validation trial involving a dynamic obstacle (a) and a map obtained (b).

obstacles was verified to determine if its use is suitable so that the user's safety is prioritized. Regarding the localization errors, we verified a low RMSE, with and average of 0.21 m and maximum outliers of 0.38 m. Therefore, the LiDAR-SLAM system proved to be able to produce quality maps as evidenced in Figure 2 (b).

From Figure 2 (a), we can notice the walker's behavior changes as the distance whenever an obstacle is detected, mainly when the distance to the closest obstacle is not lower than 1.5 m, suggesting that the walker was able to detect obstacles and plan a trajectory accordingly to avoid them. Moreover, the walker always reached its destination, guarantying the task's execution, and no collisions were observed.

5 CONCLUSIONS

In this paper, an autonomous control was implemented and assess to be used as a supervisor in a shared control of a robotic walker. The controller was able to adequately guide the user to the destination, possibly serving as a valid replacement for the human supervisor. Results showed that the autonomous controller can be used as the supervisor in the shared control of the robotic walker since the walker always reached its destination without any collision with both permanent and dynamic obstacles. Future work addresses the extensive validation of the autonomous controller along with the shared control and the study regarding the biomechanical impact along with patients in a way of assessing the direct benefits to gait performance.

ACKNOWLEDGEMENTS

This work has been supported by the FCT - Fundação para a Ciência e Tecnologia - with the reference scholarship 2020.05708.BD and under the national support to R&D units grant, through the reference project UIDB/04436/2020 and UIDP/04436/2020.

REFERENCES

Hess, W. et al. (2016) Real-time loop closure in 2D LIDAR SLAM, Proceedings - IEEE International Conference on Robotics and Automation. doi: 10.1109/ICRA.2016.7487258.

Mahlknecht, P. et al. (2013) 'Prevalence and Burden of Gait Disorders in Elderly Men and Women Aged 60-97 Years: A Population-Based Study', PLoS ONE, 8(7). doi: 10.1371/journal.pone.0069627.

Sierra, S. D. et al. (2019) 'Remote-Operated Multimodal Interface for Therapists During Walker-Assisted Gait Rehabilitation: A Preliminary Assessment', in 2019 14th ACM/IEEE International Confer-ence on Human-Robot Interaction (HRI). IEEE, pp. 528–529. doi: 10.1109/HRI.2019.8673099.

Woo, A., Fidan, B. and Melek, W. W. (2019) 'Localization for Autonomous Driving', in *Handbook of Position Location*. Wiley, pp. 1051–1087. doi: 10.1002/9781119434610.ch29.

Yurtsever, E. et al. (2020) 'A Survey of Autonomous Driving: Common Practices and Emerging Technologies', IEEE Access, 8, pp. 58443–58469. doi: 10.1109/ACCESS.2020.2983149

Advances and Current Trends in Biomechanics – Belinha et al. (Eds)

Health 4.0 technologies to develop customized external breast prostheses

F.C.N. Paula, M.E. Kunkel, T.S. Bina & T.V. Ribeiro
Institute of Science and Technology, Federal University of Sao Paulo, Sao José dos Campos, Brazil

R.T. Stramandinoli-Zanicotti
Rehabilitation Hospital Center of Parana, Curitiba , Brazil

J.C. Leme
Sport Science Department, State University of Londrina, Londrina, Brazil

ABSTRACT: The quality of life of mastectomized women has been extensively studied. Health 4.0 technologies, such as digitalization, 3D modeling, and additive manufacturing, have been explored in the virtual planning of prostheses production. Fusion and deposition modeling (FDM) is the most accessible process. This study presents a prototype-making method for the manufacturing of customized silicone external breast prostheses. The breast of a mannequin was scanned, and the design of the breast mold of two prostheses was created with the Meshmixer and Tinkercad software. A virtual mastectomy was performed to create the prosthetic counter mold to achieve a flush fit and to facilitate adherence. The prostheses molds were manufactured through the FDM process using low-cost 3D printers. The evaluated technologies of industry 4.0 have made it possible to produce customized external breast prostheses of good quality, reducing the number of steps of the conventional method.

1 INTRODUCTION

Globally, the most incidental cancer in women is breast cancer, representing 24.2% of the cases in 2018, with approximately 2.1 million new cases (WHO, 2020). In Brazil, breast cancer is the second most incidental in women. In patients under 40, there are ten deaths per 100,000 cases and from 60 years of age on the risk is ten times greater (INCA, 2019). From the biopsychosocial point of view, the diagnosis of breast cancer has a negative impact on the woman's life during the therapeutic and survival phase (Ciesla & Polom, 2010). The main option for treating breast cancer is surgical with or without radiotherapy and drugs to prevent cancer cells from spreading to other tissues. The quality of life of women surviving breast cancer has been extensively studied (Miyashita et al, 2015; Chen et al, 2016). Breast loss has great significance for women, for this reason, breast reconstruction is part of the recovery of mastectomized women. An external breast prosthesis is an option that does not involve a surgical procedure, and it can be used at all times. The market offers prostheses of affordable value; however, the models are generic with a design that does not provide the breast's natural tactile sensation (Gallagher et al, 2010).

Industry 4.0 technologies, such as 3D scanning, computer simulation, and additive manufacturing, have been explored in the virtual planning of prostheses' production, optimizing their production (Artioli et al, 2019). Fused deposition and modeling (FDM) is the most affordable process due to the low-cost of 3D printers and materials. This article presents

DOI: 10.1201/9781003217152-63

a method to produce personalized external breast prostheses using industry 4.0 tools such as the FDM process of additive manufacturing. Some studies have studied the feasibility of producing external silicone breast prostheses from molds made with additive manufacturing (Eggbeer & Evans, 2011; Shin et al, 2019). However, due to the variety of processes and materials, many experiments need to be carried out to find a more suitable method. This study presents a prototype-making method for the manufacturing of customized silicone external breast prostheses using the FDM process.

2 MATERIAL AND METHODS

A life-size plastic mannequin was used as a standard female breast model to produce two customized external breast prostheses prototypes. The 3D scanning of the breast region was performed with the iSense Ultra HD 3D Scanner (3D Systems Inc, EUA) developed for use on iPad (Apple Inc., Cupertino, California, EUA) (Figure 1a). The 3D scanning of the trunk of the mannequin was performed in a place with enough space to make the complete circumference around the mannequin with an approximate distance of 1 meter with natural lighting (Figure 1b). First, the scanner recognized the mannequin, and then the image capture is done in a slow circular motion.

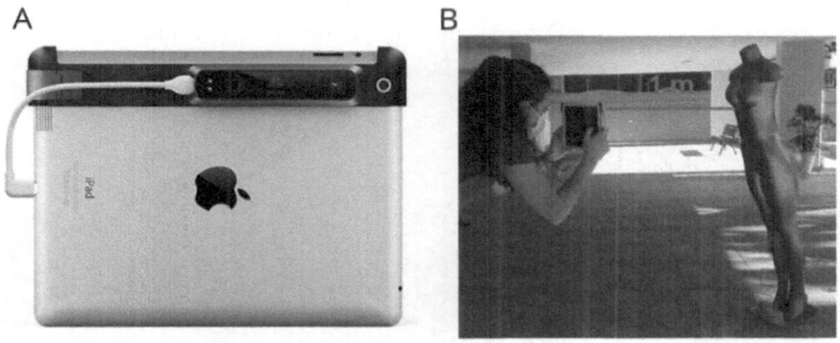

Figure 1. 3D Scanning of the trunk of the mannequin. (A) iSense Ultra HD 3D scanner and (B) Operator positioning during 3D scanning of the trunk.

The 3D model of the breast captured in the 3D scanner was used to make a mold of the breast prostheses that was manufactured by 3D printing. Subsequently, the 3D printed mold receives the medical grade silicone to create the customized prosthesis. The 3D mold of the breast and the breast prostheses were created in the computer-aided design (CAD) software Meshmixer and Tinkercad. The 3D models were saved with the *.stl* extension and were sliced virtually with the free software Ultimaker Cura 4.7 (Ultimaker USA Inc., Utrecht, the Netherlands) to create a Gcode file, standard code used in 3D printers. A virtual mastectomy simulation was performed to create the prosthesis1 counter mold model to promote better fit and adhesion. A virtual cut was made in the breast region of the reconstructed mannequin, and the irregular surface was used to define the negative mold model's shape.

The manufacture of two types of molds was done using the FDM process on two types of 3D printers and material: The 3D GTMax3D CoreA3 printer (GTMax, Brazil) with acrylonitrile butadiene styrene (ABS) filament and on the 3D printer Stella 2 (Boa impressão 3D, Brazil) with PLA (Lactic polyacid) filaments. The 3D printing configuration used were 10% density, 100 μm resolution, 0.1 mm layer thickness, and 60 mm/s print speed. The prostheses were made of medical-grade Silpuran 2420 silicone (Polisil, Brazil).

3 RESULTS

The 3D scanning of the mannequin's trunk took 5 min. Figure 2 shows the production process of custom breast prosthesis mold. The virtual mastectomy simulation was carried out to create the prosthetic counter mold. The counter mold in the posterior part of the second prosthesis had the exact shape of the mastectomized region (Figure 3). In the 3D printing of the ABS prosthesis mold, 52 g of the filament was used, and in the PLA mold and counter mold, 59 g and 56 g, respectively. Each mold was scaled to 50% of the breast size and manufactured in approximately 5 hours. After the silicone curing, the two prostheses showed fine lines resulting from the printed layers of the mold and counter mold (Figure 4).

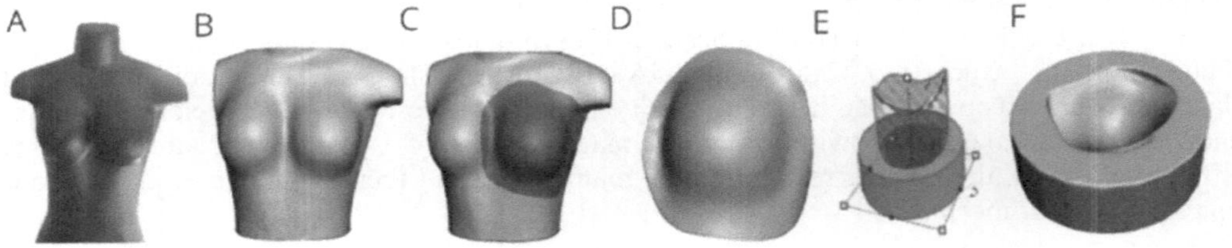

Figure 2. Production process of custom a breast prosthesis mold. (A) Mannequin; (B) 3D scanning of the trunk; (C) Breast selection; (D) CAD breast model; (E) 3D mold creation and (F) 3D model of the breast prosthesis mold.

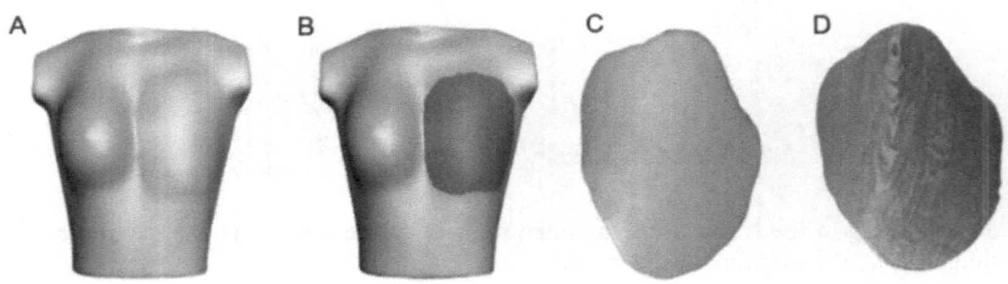

Figure 3. Simulation of mastectomy on the mannequin's digital image for the production of the counter mold (negative part). (A) 3D model of the mannequin with breast extraction; (B) Selection of the supposed mastectomized area; (C) 3D model of the mastectomized area (D) 3D printed mastectomized area on PLA.

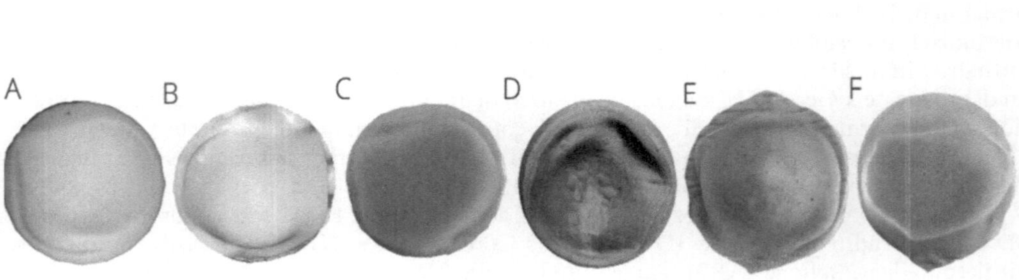

Figure 4. Personalized silicone breast prostheses. A) ABS mold; B) and C) anterior and posterior views of the prosthesis produced in the ABS mold without counter mold; D) PLA mold; E) and F) anterior and posterior views of the prosthesis produced in the PLA mold with counter mold.

4 DISCUSSION

The material cost of the silicone prostheses made from molds manufactured in material ABS and PLA scaled to 50% of the breast size was the US $ 5.83 and US $ 7.81, respectively. Little esthetic difference was noticed in the use of ABS and PLA filament for mold's manufacture of molds. However, the prosthesis prototypes made with PLA mold showed better results with the counter mold, which delimits the prosthesis fitting region of the simulated mastectomy area.

The mold printed in ABS showed a more uniform appearance of the layers compared to the mold printed in PLA. Therefore, at first, it was suspected that the silicone prosthesis made in the ABS mold would have an aesthetic appearance closer to the breast's natural image. However, after the production of both prostheses, no differences in aspect were observed, except for the posterior portion of the second prosthesis, which presents the prosthesis fitting area in the supposed mastectomized area. Compared to the conventional method to produce breast prostheses, the digital method has many advantages, as it does not require the presence of the user at all stages of the process and does not require the professional proficiency for the prosthesis′ manual impression.

The methodology developed in the current research can also be used to create biomodels for surgical planning. For example, in cases of breast reconstruction surgeries, it is possible to scan the natural breast before its surgical removal, to then be able to fabricate a prosthesis exactly the original size of the patient's breast. 3D printing technology has become a trending issue in phantom constructions for medical applications, and other areas as maxillofacial prosthesis, with its advantages of being customizable and cost-efficient (He et al, 2019).

The results of our research indicate that it is possible to manufacture a good quality breast prosthesis mold using the FDM additive manufacturing process that is more accessible. In the next phase of the research, the best prosthetic attachment to the mastectomized region will be investigated. Technological innovations in the area of prosthesis development can bring many benefits to the prosthesis user, helping in a general way in the rehabilitation process. One of the advantages of 3D modeling is the reduction of direct contact with the patient during the production process, as the anatomical structure can be acquired from medical images or 3D scanning. Another advantage of using 3D printing is that from a single 3D digital model, it is possible to produce as many prostheses as needed.

5 CONCLUSION

The 3D Scanning and FDM process can be used to produce personalized external breast prostheses with good quality using fewer steps than in the conventional process. These factors can cooperate to reduce the final overall cost of external breast prostheses. The study raised several questions that may be considered in prosthesis esthetics improvement and clinical trials.

ACKNOWLEDGMENT

The authors thank National Council for Scientific and Technological Development, the International Development Research Centre (IDRC) and the GDS Program.

REFERENCES

WHO. World Health Organization. International Agency for Research on Cancer. 2020. Available in: <gco.iarc.fr/today/>. Accessed on July 23, 2020.

INCA. Instituto Nacional de Câncer. E-book: Estimativa 2020, Incidência de Câncer no Brasil. RJ: INCA, 2019. Available in: <www.inca.gov.br/sites/ufu.sti.inca.local/files/media/document/estimativa-2020-incidencia-de-cancer-no-brasil.pdf>. Accessed on September 23, 2020.

Ciesla, S. & Polom, K. 2010. The effect of immediate breast reconstruction with Becker-25 prosthesis on the preservation of proper body posture in patients after mastectomy. *European Journal of Surgical Oncology* (EJSO), 36(7), 625–631.

Miyashita, M., Ohno, S., Kataoka, A., Tokunaga, E., Masuda, N., Shien, T. ... & Takahashi, M. 2015. Unmet information needs and quality of life in young breast cancer survivors in Japan. *Cancer nursing*, 38(6), E1.

Cheng, K.K.F., Wong, W.H. & Koh, C. 2016. Unmet needs mediate the relationship between symptoms and quality of life in breast cancer survivors. Supportive care in cancer, 24(5), 2025–2033.

Gallagher, P., Buckmaster, A., O'carroll, S., Kiernan, G. & Geraghty, J. 2010. External breast prostheses in post-mastectomy care: women's qualitative accounts. *European journal of cancer care*, 19(1), 61–71.

Artioli B.O., Kunkel M.E., Mestanza S.N. 2019. Feasibility study of a methodology using additive manufacture to produce silicone ear prostheses. In World Congress on Medical Physics and Biomedical Engineering 2018 (211–215*). Springer*, Singapore.

Eggbeer, D. & Evans, P. 2011. Computer-aided methods in bespoke breast prosthesis design and fabrication. Proceedings of the Institution of Mechanical Engineers, Part H: *Journal of Engineering in Medicin*e, 225(1), 94–99.

Shin, K. 2019. Thermal and moisture comfort mastectomy bra and prosthesis. *Textile Research Journal.*

He, Y., Liu, Y., Dyer, B.A., Boone, J.M., Liu, S., Chen, T. ... & Qiu, J. 2019. 3D-printed breast phantom for multi-purpose and multi-modality imaging. *Quantitative imaging in medicine and surgery*, 9(1), 63.

Advances and Current Trends in Biomechanics – Belinha et al. (Eds)
© 2022 Copyright the Author(s), ISBN: 978-1-032-10806-3

Challenges in finite element modelling of bioabsorbable stents for the treatment of Aortic Coarctation

F.J. Santos, B.A. Hernandez & M.P.D. Souza
UNESP, São Paulo State University, Campus of Bauru, Brazil

A.J.P. Andrade
Adib Jatene Foundation, Dante Pazzanese Institute of Cardiology, São Paulo, Brazil

E.A.C. Sousa
UNESP, São Paulo State University, Campus of Bauru, Brazil

ABSTRACT: Metallic stents are used for the treatment of Aortic Coarctation (CoA). However, their application in new-borns is limited, mostly due to the patient's continuous growth, demanding new-debilitating stent expansion surgeries. An alternative could be the use of bioabsorbable stents (BS). The Finite Element Method (FEM) has been widely used in the development of new medical equipment, mostly due to its capacity for design exploration. Nonetheless, the modelling of stents is not trivial, as many and different variables will affect the model's performance. The aim of this study was to evaluate the challenges in finite element simulation of bioabsorbable stents. A new BS model was proposed, modelled, and analysed using the FEM. The analysis showed that, among other parameters, the use of cylindrical-coordinates instead of Cartesian, and radial-displacement instead of pressure as loading input parameters presented the best results in the modelling of BS, which could be used as default parameters for future modelling.

1 INTRODUCTION

Aortic Coarctation (CoA) is a Congenital Heart Disease (CHD) in which the aortic artery is constricted, decreasing in the available cross-sectional area (lumen) for the blood flow (Doshi, 2012). According to Brown et al., (2013), it occurs more frequently in men, and, according to Nguyen et al., (2015), it is the seventh most common congenital heart disease (8% of the total CHDs). It usually develops close to the aortic arch, between the left subclavian artery and the arterial ligament (Doshi et al., 2012). Approximately 51,000 new-borns per year worldwide have some degree of coarctation of the aorta (Brown et al., 2013). This is the equivalent of one in every 2,500 births, according to some studies (Van der Linde et al., 2011).

The treatment for CoA is usually through surgery and using metallic stents, which are deformable thin circular structures that use their radial deformation and geometry to expand and re-open the coarctation (Torki et al., 2020). However, metallic stents are not a viable solution in the treatment of new-borns, as new surgeries for stent dilation would be necessary due to the patient's constant growth. An alternative would be the use of bioabsorbable stents, as these devices remain temporarily in the patient's body. According to Qiu et al., (2018), this kind of stent is generally made of magnesium or polymer alloys, with PLLA (Poly-L-Lactide Acid) and PLA (PolyLactic Acid) being the most used polymers.

DOI: 10.1201/9781003217152-64

Numerical methods, especially the Finite Element Method (FEM), have been widely used in the analysis of new cardiovascular equipment and in the development and optimisation of new stents (Torki et al., 2020). A main advantage of FEM is that several designs can be explored at the same time and it is considerably less expensive than experimental procedures.

Despite the advantages of FEM, the modelling of bioabsorbable stents is not a trivial activity, as it involves many variables, such as element type and size (for both the stent and the artery), large displacements, plasticity, distortion of the elements, contact problems, lower material strength, etc.

Therefore, this study aimed to evaluate which parameters will have an influence in the modelling of bioabsorbable stents and which is the best strategy to achieve convergence.

2 MATERIAL AND METHODS

A new stent geometry was proposed by the Adib Jatene Foundation (Dante Pazzanese Institute of Cardiology, São Paulo, Brazil). It has an external diameter of 6.25mm and a wall thickness of 0.25mm. Two types of artery geometries were developed by our research group, the Centre for Simulation in Bioengineering, Biomechanics and Biomaterials (CS3B), at UNESP, campus of Bauru, Brazil. The first geometry is a healthy artery with an external diameter ranging from 10 to 14mm. The wall thickness was set as 10% of the outside diameter. The second geometry was an artery with coarctation, with an external diameter of 13mm in the healthy region and an internal diameter of 6.8mm in the unhealthy region, Figure 1. Both the stent and artery were modelled in SpaceClaim software (ANSYS, Inc., Canonsburg, PA USA).

The material used for the stent was PLLA (Poly-L-Lactide Acid), with Young's modulus of 2,800 MPa, Poisson's ratio of 0.3, and with a stress-strain curve given by Qiu et al., (2018). For the artery, a hyperelastic behaviour was assumed according to a second-order Mooney-Rivlin material model, with $C_{10} = 0.174$ MPa, $C_{20} = 1.881$ MPa, and d = 1.149 MPa (Simsek et al., 2015). The steel material available in the Ansys software library was used in the initial simulations.

Quadratic solid element type (Solid 186) was chosen for the stent. For the artery, shell elements (Shell 181) were initially picked but changed later to quadratic solid elements (Solid 186).

The first modelling challenge was to achieve the model's convergence. Simulations were initially carried out with the stent set as made of steel, without establishing contact with the artery. After this, a healthy artery was included and contact between stent and artery was calibrated. Finally, an artery with coarctation was introduced. The steps taken to achieve model convergence are depicted in Table 1.

3 RESULTS

An optimum 0.1mm element size was achieved for the stent model, while it was 0.4mm for the artery. A mesh convergence study for the stent is presented in Table 2.

Figure 1. An artery with coarctation.

Table 1. Steps to achieve model convergence.

Simulation/ Stage number	Stent Material	Artery
1	Steel	No
2	PLLA	No
3	Steel	Yes
4	PLLA	Yes
5	PLLA	Yes and with coarctation

Table 2. Element convergence study.

Simulation/Stage number	Element Quantity	Size (mm)	Principal Stress (MPa)	Difference (%)
S1	19051	0.16	109.57	-
S2	29418	0.14	104.66	4.48
S3	43378	0.12	102.88	1.70
S4	70393	0.1	101.66	1.19

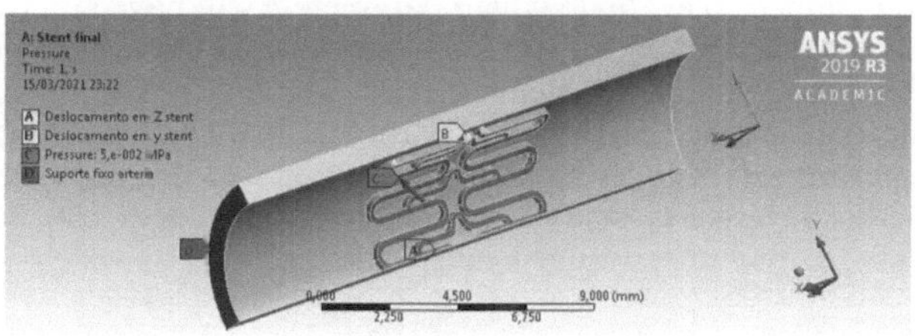

Figure 2. Initial boundary conditions.

Stage 4 (Table 1) presented a computational model with the highest degree of nonlinearity, due to the stent complex geometry, artery material property, and contact proprieties. Consequently, it was the stage with the highest computational efforts. Initially, boundary conditions were set as shown in Figure 2: the artery was set with shell elements and its extremities were fully restrained. Pressure loading was applied in the internal side of the stent, and the stent was restrained in y and z directions, with a symmetry plane on the x-axis. In order to improve convergence, an artery with an inner diameter close to the stent's external diameter was introduced, causing immediate contact with the stent. The results with these boundary conditions were not encouraging, as the stent maximum expansion was only 0.25mm.

Finally, after several interactions, a change of coordinates system from Cartesian to cylindrical, the use of solid elements for the artery, and the application of radial displacement instead of pressure loading on the internal surface of the stent presented the best results, Figure 3.

4 DISCUSSION

Metallic stents have been successfully used for the treatment of aortic coarctation in adult patients. However, their application in new-borns and young children is limited, mostly due to

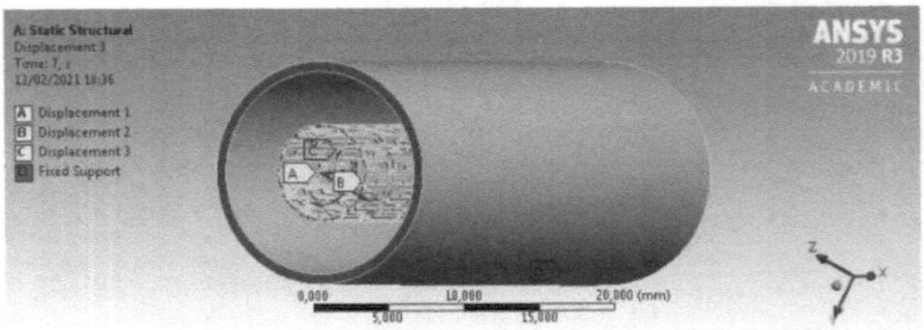

Figure 3. Final boundary conditions.

the patient's continuous growth, which demands new and debilitating stent expansion surgeries. An alternative to the metallic type is the use of bioabsorbable stents, as these devices, usually made of organic polymer materials such as PLLA or PLA, remain only temporarily in the patient's body. Still, some challenges have to be overcome to make this type of stent a reality, such as improving its low structural strength.

The Finite Element Method (FEM) has been extensively used to develop and analyse new stents, cardiac devices, and medical equipment (Torki et al., 2020). The main advantage of FEM is that several and different designs can be explored before clinical trials, which would save resources and time as an optimum design could be achieved at the beginning of the development process. Nonetheless, the finite element modelling of stents is not a trivial activity as it involves many variables, such as type and size of the element (for both the stent and the artery), large displacements, plasticity, element distortion, contact problems, and, especially in bioabsorbable stents, the low material strength.

Therefore, this study aimed to evaluate which parameters will have an influence in the modelling of bioabsorbable stents, and which is the best strategy to achieve convergence.

The first parameter assessed was the element type. The use of shell elements in the artery greatly increased the convergence difficulty due to the challenge in identifying the correct stent-artery contact parameters. Changing it to solid eliminated this problem.

Sequentially, a mesh convergence study was conducted and, as it can be seen in Table 2, a mesh size of 0.1mm for the stent resulted in an error of 1.19%, which was chosen for the modelling.

The initial boundary conditions, the use of internal pressure loading and Cartesian coordinate system, resulted in element distortion and poor stent performance, i.e., small stent opening. An alternative was a change of coordinates system, from Cartesian to cylindrical, and in the load type, from the pressure to radial displacement on the internal surface of the stent. These two modifications lead to an improvement in stent performance.

Another change was in the number of the load steps and time substeps. In the first case, the load steps had to be increased for a slow load (or displacement) application, with 10 loadsteps being the optimum value. Likewise, the number of substeps had raised to improve contact between the stent and artery. The optimum value was 200.

Finally, the inclusion of an artery with a coarctation, Figure 1, occurred without major problems.

5 CONCLUSION

This study aimed to analyse the challenges in finite element modelling of bioabsorbable stents. The finite element method proved to be an important tool in the development of new cardiac devices. A new bioabsorbable stent geometry was proposed and its performance in the treatment of an artery with coarctation was evaluated.

The use of a cylindrical coordinate system and the application of displacement directly on the internal surface of the stent are the main findings of this study. They improved convergence and the model's performance. The next steps of this study will be to conduct a full evaluation of stent performance and to validate the model.

REFERENCES

Brown, M.L., Burkhart, H.M., Connolly H.M., Dearani, J.A., Cetta, F., Li, Z., Oliver, W.C., Warnes, C.A. and Schaff, H.V. 2013. Coarctation of the aorta: Lifelong surveillance is mandatory following surgical repair. *Journal of the American College of Cardiology* 62(11): 1020–1025.

Doshi, A.R. 2012. Coarctation of Aorta-Management Options and Decision Making. *Pediatrics & Therapeutics* 01(S5): 1–10.

Nance, J.W., Ringel, R.E. and Fishman, E.K. 2016. Coarctation of the aorta in adolescents and adults: A review of clinical features and CT imaging. *Journal of Cardiovascular Computed Tomography* 10 (1): 1–12.

Nguyen L. and Cook, S.C. 2015. Coarctation of the Aorta. Strategies for Improving Outcomes. *Cardiology Clinics* 33(4): 521–530.

Qiu, T.Y., Song, M. and Zhao, L.G. 2018. A computational study of crimping and expansion of bioresorbable polymeric stents. *Mechanics of Time-Dependent Materials* 22(2): 273–290.

Simsek, F.G., Kwon, Y.W. 2015. Investigation of material modelling in fluid–structure interaction analysis of an idealized three-layered abdominal aorta: aneurysm initiation and fully developed aneurysms. *Journal of Biological Physics* 41(2): 173–201.

Torki, M.M., Hassanajili, S. and Jalisi, M.M. 2020. Design optimizations of PLA stent structure by FEM and investigating its function in a simulated plaque artery. *Mathematics and Computers in Simulation* 169: 103–116.

Van der Linde, D., Konings, E.E., Slager, M.A., Witsenburg, M., Helbing, W.A., Takkenberg, J. J. and Roos-Hesselink, J. W. 2011. Birth prevalence of congenital heart disease worldwide: A systematic review and meta-analysis. *Journal of the American College of Cardiology*, 58(21): 2241–2247.

Advances and Current Trends in Biomechanics – Belinha et al. (Eds)
© 2022 Copyright the Author(s), ISBN: 978-1-032-10806-3

Development of a 3D-printed body-powered prosthesis with flexible materials

A.M. Oliveira
Physics Department, NOVA School of Science and Technology, NOVA University of Lisbon, Portugal

C. Quaresma
Instrumentation, Biomedical Engineering and Radiation Physics Laboratory (LIBPhys-UNL), Physics Department, NOVA School of Science and Technology, NOVA University of Lisbon, Caparica, Portugal

B.A.R. Soares
UNIDEMI, Department of Mechanical and Industrial Engineering, NOVA School of Science and Technology, NOVA University of Lisbon, Caparica, Portugal

ABSTRACT: The research on 3D printed prostheses has been increasing, as they appear to be a good solution to solve some of the issues of common prostheses. However, despite of their growth, these prostheses have a high rejection rate, especially in children, due to its artificial appearance. The main goal of this paper is to present the development of an improved 3D printed body-powered prosthesis, both in aesthetical and functional terms. Body casting and 3D scanning were the methods used to capture the anatomical features of the child's healthy hand. The whole prosthesis was designed using the Fusion 360 CAD software and printed using PLA and Filaflex filaments. Different assembly approaches are now being tested to achieve the best functional performance possible. The prosthesis possesses a high anthropomorphic level, with the fingers possessing good bending characteristics. However, tests of the full prosthesis are not finalized due to the world situation.

1 INTRODUCTION

Hands are one of the most important anatomical structures in the human body, as they are used in most day-to-day tasks. The absence of this anatomical structure, either partial or complete, compromises the quality of life of upper limb amputees, decreasing their autonomy level and limiting the ability to perform a wide range of activities (Cordella *et al.*, 2016). However, prosthetic rehabilitation can help re-establishs function and consequently improve the quality of the patients' lives (Raichle *et al.*, 2008).

Congenital factors are the main cause of upper extremity absence in children, however, for many children with congenital pathologies it is easier to perform their activities of daily living without a prosthesis (James *et al.*, 2006). To get the children used to prostheses, The Association of Children's Prosthetic and Orthotic Clinics recommends prosthetic fitting with a passive device at the age of six months, when the child learns to sit (Nelson *et al.*, 2006). The introduction of a prosthesis at early ages contributes to psycho-motor development, promotes the use of both limbs in prehension activities and crawling, ensures body symmetry during the child's growth, helps in the self-body perception. In addition, it also increases the chance of accepting more sophisticated prosthetic devices in the future, such as body-powered and

DOI: 10.1201/9781003217152-65

electronic prostheses (Afonso, Coelho and Cadete, 2009; Egermann, Kasten and Thomsen, 2009; Mano, Fujiwara and Haga, 2018).

. Prosthetic devices have a huge lack of functional benefits which contributes for their rejection. Besides the cosmetic appearance, comfort and functionality, children's prostheses need to have a higher resistance, and be easily repairable or replaced. Lower costs are also desired as children's fast growth requires a new prosthesis in a short time span. Adjustments are usually done every three to six months and the replacement of the prosthesis is typically done every one two years. (Nelson *et al.*, 2006; Burn, Ta and Gogola, 2016).

Three-dimensional(3D)-printed prostheses may be the solution to these problems as they are functional, low-cost, light, and easy-to-fix (Burn, Ta and Gogola, 2016). However, despite the huge growth of 3D-printed prostheses, through e-NABLE community efforts, these devices present a high rejection rate in children. Although the similarity between these prostheses and hero hands was expected to improve their acceptance rate for children, 3D-printed prostheses are rejected after a few hours of play, as many children consider them toys (Zuniga *et al.*, 2015).

This paper presents the development of a highly customised prothesis by leveraging 3D printing capabilities, which ameliorates the problems of existing prostheses. Functional and aesthetical improvements were the focus of this study.

2 MATERIALS AND METHODS

This study was based on a single clinical case: a four-year-old child with a transverse metacarpal total deficiency. A highly customised prothesis, both in terms of size and appearance was developed. The methodology was divided in two stages: the assessment of the anatomic features of the child and the prosthesis' development.

The acquire the shape of the limbs' extremities of the child body casting with alginate and plaster was used, followed by 3D scanning using an EinScan-SE 3D scanner. This allowed the virtualization of the both the healthy limb and the stump. Furthermore, some additional anatomical measures of the child's limbs were taken. These assessments were then used to develop the prothesis.

The prosthesis was designed using © Autodesk's Fusion 360 CAD software. The designed models were printed in a Prusa i3 MK3S, using ©PrusaSlicer as slicing software. The final model was printed with a combination of rigid and flexible filaments: PLA and ©Recreus' Filaflex 82A, respectively.

The design of the prosthetic device was made considering the Design for Additive Manufacturing principles, to achieve the best results when printing the different components that compose the prosthesis. The entire design was made considering the extrusion width and height, so the design walls could be a multiple of these values. The printing position was also a considered factor as well as the existence of support structures, that were avoided whenever possible. The design of the idealized prosthesis was divided in three main sections: fingers, metacarpal region and wrist. Furthermore, all parts were scaled to 110% during their design to prevent the child's growth.

The fingers were the first part to be addressed and consisted of an adaptation of the child's healthy fingers. The final design of the index was used as a reference for the remaining fingers design, except for the thumb, due to the anatomical similarity. All the designed fingers have a natural curvature. In addition, air chambers were added in the interphalangeal joints' zone so the finger could bend properly. In addition to these features, inner cylinders were added to simulate the existence of phalanges. During the fingers' design, several printing tests were made to find the ideal printing quality. In parallel to these tests, pull tests were performed to select the best printing settings and finger design that allowed a reasonable printing quality and small flexion forces.

The metacarpal region was constructed using the Lofts tool from the Fusion 360 software. This tool connects different sketches located on different planes. This connection was oriented by guidelines to give the metacarpal region the desired shape. In similarity to the fingers, the metacarpal region also contains an air chamber located in the metacarpophalangeal joints' zone, with the aim of allowing the hand to bend. The prosthesis-stump interface was customized according to the child's stump form and has a ventilation system to let the child's skin breathe.

Regarding the wrist mechanism, it consists of an adaptation of a prosthesis available on the e-NABLE online platform. The chosen model was the Phoenix Hand v2 (Phoenix Hand – Enabling The Future, no date). This part, as well as the thread's ring, were printed with PLA.

The fingers and the metacarpal region were printing using Filaflex filament. The metacarpal region needed several supports during its printing process. Different infill percentages were used to simulate the soft tissues and rigid parts such as the phalanges. In the metacarpal region, different infill percentages were also used, with the aim of simulating carpometacarpal joint flexion when flexing the thumb.

As any other body-powered prosthesis, the design device works through a tensioning cables system. The fingertips have additional chambers where rigid rings, tied to the tendon cables, rest. The threads are inserted in the fingers through a hole in the fingertips and go through the fingers and continues through the metacarpal region. The cables are tensioned by a whippletree mechanism, which gives the prosthesis the ability to adapt to the object during the grasping performance.

The fingers fit into the metacarpal region through pressure. After docked, the fingers contain a mechanism to prevent the fingers from breaking apart from the metacarpal region body.

3 RESULTS

The aim of study was to develop a highly customised 3D-printed body-powered prosthesis and improve the functional and cosmetic features of the existing prosthetic devices. The resulting device is displayed in Figure 1a.

The air chambers that simulate the interphalangeal joints are a conic void structure with a 60° opening and variable depth correspondent to 90% of the transversal section length (L) where the chambers vertexes were inserted. These values are a result of the pull tests made, in which index needed approximately 12 N to flex entirely. The PLA rings were inserted in the fingertips chambers through elliptic holes located in anterior region of the fingertips. These chambers are connected to a 2.1 mm diameter hole that goes lengthwise through the whole finger. The hole's centre is located at approximately 0.2 L from the anterior region of the finger and follows the finger's curvature. Figure 1b presents the design of the index finger.

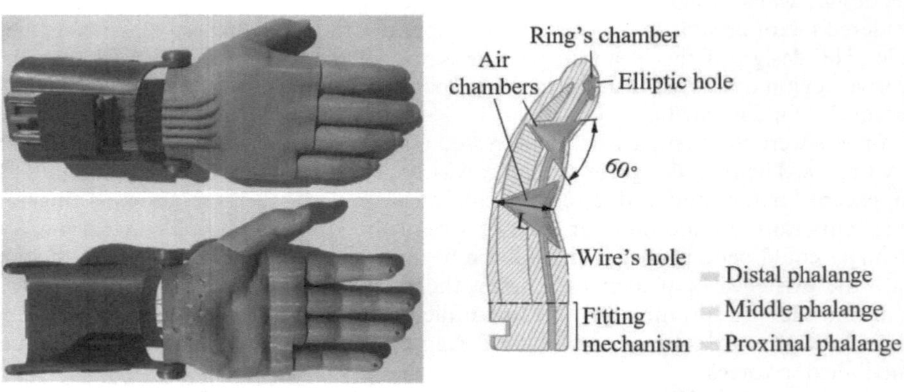

Figure 1. a) Flexible prosthesis, b) prosthesis finger design.

The resulting design of the metacarpal region has an anthropomorphic shape. The air chamber that simulated the metacarpophalangeal joints has a 30° opening and is centred 20 mm above the top of metacarpal region component. The wires holes start in anterior region of the metacarpal and end at the posterior region of the wrist, forming a smooth curved path for the tendon cables go through. The ventilation mechanism is composed by diamond shaped holes.

According to the measurements of the child's limb, the Phoenix Hand's components that compose the present prosthesis such as the whippletree mechanism, the gauntlet and the respective pins didn't need to be scaled. However, the gauntlet and its pins had to be reformulated to meet the design of the developed prosthesis in this study.

4 DISCUSSION

In this study, a 3D-printed prosthesis was developed by using a combination of rigid and flexible filaments. Body casting and 3D scanning allowed to obtain a high customization level. Unlike other 3D-printed body-powered prostheses, the developed device matched the sound hand of the child in terms of size and anatomical features, eliminating the childish appearance that leads to the rejection of these type of prostheses.

During the design of the fingers, several printing tests were made to obtain the ideal printing quality. Nevertheless, it was difficult to achieve a good printing quality that led to small flexion forces. A load of 12N was required to bend each finger, and when bending the prosthesis, it teared in the connection between the metacarpal region and the gauntlet.

Therefore, new assembly approaches are now being tested to solve these problems. In the future, after the prosthesis is completely functional, the whole prosthesis will be evaluated in terms of mechanical and functional performance. The results of these tests must be the starting point for further improvements so this methodology must be applied to other subjects.

Despite the functionality issues, this study is a good starting point to improve the functionality and aesthetics of 3D-printed prostheses. By enhancing these features, guaranteeing all the other user's needs and customising these devices, 3D-printed prostheses may become the key to decrease the prostheses' rejection rate.

5 CONCLUSION

3D-printed hand prostheses have shown a lot of benefits when compared with other prostheses. However, their childish appearance and stiffness are decreasing their overall acceptance rate. This paper presents an improved 3D-printed body-powered prosthesis, both in aesthetical and functional terms, aiming to increase the adhesion to these types of prostheses. Different assembly approaches are now being made as the resulting device revealed some functionality issues. However, the device shown good results, especially regarding cosmetic appearance. After solving these complications, the whole prosthetic device will be evaluated, and the testing results should be the starting point for further improvements.

ACKNOWLEDGEMENTS

The authors would like to thank the support of the FCT, through UNIDEMI, project UID/EMS/00667/2020.

REFERENCES

Afonso, C., Coelho, I. J. P. and Cadete, I. A. 2009. Amputações e Malformações Congénitas do Membro Superior na População Pediátrica - Revisão de 27 Anos. Revista da Sociedade Portuguesa de Medicina Física e de Reabilitação 17(1): 1–4.

Burn, M. B., Ta, A. and Gogola, G. R. 2016. Three-Dimensional Printing of Prosthetic Hands for Children. The Journal of Hand Surgery 41(5): e103–e109.
Cordella, F. et al. 2016. Literature Review on Needs of Upper Limb Prosthesis Users. Frontiers in Neuroscience 10(209).
Egermann, M., Kasten, P. and Thomsen, M. 2009. Myoelectric hand prostheses in very young children. International Orthopaedics, 33(4): 1101–1105.
James, M. A. et al. 2006. Impact of Prostheses on Function and Quality of Life for Children with Unilateral Congenital Below-the-Elbow Deficiency. The Journal of Bone & Joint Surgery, 88(11): 2356–2365.
Mano, H., Fujiwara, S. and Haga, N. 2018. Adaptive behaviour and motor skills in children with upper limb deficiency. Prosthetics and Orthotics International 42(2): 236–240.
Nelson, V. S. et al. 2006. Limb Deficiency and Prosthetic Management. 1. Decision Making in Prosthetic Prescription and Management. Archives of Physical Medicine and Rehabilitation 87(3): 3–9.
Enabling The Future. no date. Phoenix Hand. Viewed 16 March 2021. <http://enablingthefuture.org/phoenix-hand/>
Raichle, K. A. et al. 2008. Prosthesis use in persons with lower- and upper-limb amputation. Journal of rehabilitation research and development 45(7): 961–972.
Zuniga, J. et al. 2015. Cyborg beast: a low-cost 3d-printed prosthetic hand for children with upper-limb differences. BMC Research Notes 8(1).

Advances and Current Trends in Biomechanics – Belinha et al. (Eds)
© 2022 Copyright the Author(s), ISBN: 978-1-032-10806-3

Is flexibility associated with body composition? Bioimpedance analysis in children and adolescents

N. Azevedo
CICS, ISAVE, Faculdade de Desporto, Universidade do Porto, Portugal

J.C. Ribeiro
CIAFEL, Faculdade de Desporto, Universidade do Porto, Portugal

L. Machado
CIFI2D, LABIOMEP, Faculdade de Desporto, Universidade do Porto, Portugal

ABSTRACT: Flexibility is an important parameter of musculoskeletal assessment. It is related to sports performance, but also to health conditions. Flexibility is part of the characterization of the global physical fitness of children and adolescents, as is the assessment of body composition. There is a scarce literature reporting the association between the levels of flexibility and the variables related to body composition.

To understand the existence of an association between body composition and flexibility, a cross-sectional study in children and adolescents was carried out using valid instruments to assess flexibility and body composition. The results suggest an association between bioimpedance values and flexibility, while observing differences between genders and ages.

Keywords: Flexibility, children, adolescents, physical fitness, body composition

1 INTRODUCTION

Physical fitness analysis in children and adolescents is extremely important because it may serve as a predictor of future health in children, both in terms of cardiovascular diseases as well as musculoskeletal diseases, especially diseases of the spine (Ruiz *et al.*, 2009). In Portugal, the benchmark for assessing physical fitness in the education system is through the FITescola® program. This program is divided into 3 specific areas, namely aerobic fitness, body composition and neuromuscular fitness. Flexibility is assessed in the neuromuscular fitness area. Flexibility is an important parameter for sports evaluation (Arnason *et al.*, 2004), but also for musculoskeletal health (Martim and Simas, 2012). The benefits of a good physical fitness level are not only limited to the physical component but also to the quality of life in children (Garber *et al.*, 2011). There are several factors that contribute to the different levels of flexibility, including the parameters of body composition (Eler, 2018). Studies have found differences between genders and levels of flexibility (Haugen, Høigaard and Seiler, 2014; Eler, 2018), but there is little evidence of the association that body composition parameters have with flexibility in children and adolescents (Ganley *et al.*, 2011).

DOI: 10.1201/9781003217152-66

2 MATERIALS AND METHODS

A cross-sectional study was carried out with children from some schools in the north of Portugal, in Braga district, between October and December 2019. As instruments, a sample characterization questionnaire, a body composition analyser (InBody 230) and the FITescola® test battery was used. The test battery is divided into three areas, aerobic condition, neuromuscular fitness, and body composition. Flexibility was included in the neuromuscular fitness and was assessed by the seat and reach test (assessing the flexibility of the lower limbs). Three categories were created according to the performance of each student (low flexibility, healthy/normal flexibility, and athletic/high flexibility). The categorization was done by age and according to the FITescola® guidelines. The application of all instruments took place at the same moment of evaluation, during a physical education class. Of a total of 1907 children whose caretakers freely signed the informed consent, 1491 (78.2%) agreed to participate in the study, that included boys and girls aged 9 to 19 years.

3 STATISTICAL ANALYSIS

Descriptive statistics was used to characterize the study sample. Normality of the data was tested, and Mann–Whitney U-test was used to estimate differences in the studied variables between the two groups (gender), and multinomial logistic regression was used to calculate the odds ratio for body composition and flexibility. The statistical significance was set as $\alpha=0.05$. IBM SPSS Statistic version 26 was used.

4 RESULTS AND DISCUSSION

In the analysis, significant differences were observed regarding all variables between genders, except for age (p=0.261). Results shown in Table 1.

Table 1. Sample characterization by gender.

	Female			Male			
	N	Mean	SD	N	Mean	SD	p.value
Age	729	13.97	2.432	762	13.84	2.391	0.261
Flexibility	711	26.22	9.358	750	21.68	8.513	0.000
Body Fat Mass	725	16.88	8.362	755	10.76	6.952	0.000
Fat Free Mass	725	38.02	6.785	755	46.23	12.43	0.000
Skeletal Muscle Mass	725	20.50	4.066	755	25.57	7.545	0.000
Body Mass Index	725	21.89	4.217	755	20.91	3.645	0.000
Percent Body Fat	725	29.31	8.061	755	18.46	8.807	0.000
Basal Metabolic Rate	725	1191	146.5	755	1368	268.3	0.000
Waist-Hip Ratio	725	0.87	0.064	755	0.81	0.059	0.000

Mann–Whitney U-test: (level of significance 95%)

To calculate the Odds Ratio (OR), a multinomial logistic regression was performed considering flexibility as the dependent variable (3 categories: low flexibility profile - LFP, normal flexibility profile – NFP, and athletic flexibility profile - AFP) and the remaining variables as independent variables. The reference category is the normal flexibility profile.

Through the observation of the sample characterization data, we can see that there are significant differences between genders in the studied variables, but this difference does not exist in the age variable.

Table 2. Distribution of the subjects by group for the studied variables.

	Female			Male		
	Low	Normal	High	Low	Normal	High
Flexibility	54.4%	26.9%	18.7%	45.9%	42.3%	11.9%
Body Fat Mass	11.3%	51.2%	37.5%	32.5%	40.9%	26.6%
Fat Free Mass	14.8%	65.9%	19.3%	16.7%	74.0%	9.3%
Skeletal Muscle Mass	34.8%	58.6%	6.6%	34.0%	58.0%	7.9%
Body Mass Index	9.4%	62.2%	28.4%	16.7%	66.0%	17.4%
Percent Body Fat	4.4%	35.2%	60.4%	15.6%	48.9%	35.5%
Basal Metabolic Rate	46.6%	51.6%	1.8%	24.9%	71.8%	3.3%
Waist-Hip Ratio	0.3%	38.3%	61.4%	48.1%	43.2%	8.7%

When comparing the two genders in relation to the percentage in the variables in the study (except age), dividing them into three groups according to the reference values obtained by the bioimpedance system (low profile, normal profile and high profile), we can see that there are also differences between genders. (Table 2).

These gender differences are evident, for example, in individuals with normal flexibility, where the male gender has a higher percentage of individuals with normal flexibility (42.3%) compared to the female gender (26.9%). The male gender also has a higher percentage (32.5% - about three times more) than the female gender (11.3%) in the low-profile group of body fat mass.

At Percent Body Fat this difference is more pronounced in the high-profile group of the female gender (60.4%) compared to the male gender (35.5%). This relationship is also evident in the waist-hip ratio, in which the female gender has a higher percentage (61.4%) compared to the male gender (8.7%) in the high-profile group.

Table 3. Logistic regression with flexibility and body composition. Considering normal flexibility as reference.

	Low flexibility profile				Athletic flexibility profile			
		95% CI				95% CI		
	Odds Ratio	Lower Bound	Upper Bound	Sig.	Odds Ratio	Lower Bound	Upper Bound	Sig.
Intercept				**0.000**				**0.000**
Body Fat Mass	1.028	0.729	1.449	0.876	1.054	0.657	1.692	0.827
Fat Free Mass	0.898	0.643	1.254	0.527	1.534	0.975	2.414	0.064
Skel. Musc. Mass	**0.610**	0.449	0.828	**0.002**	0.848	0.564	1.277	0.431
Body Mass Index	0.718	0.496	1.039	0.079	0.935	0.557	1.569	0.799
Percent Body Fat	1.033	0.749	1.426	0.842	1.034	0.667	1.603	0.882
Basal Met. Rate	**0.665**	0.445	0.995	**0.047**	1.416	0.826	2.427	0.205
Waist-Hip Ratio	**1.440**	1.176	1.764	**0.000**	**1.475**	1.111	1.960	**0.007**
Age	**0.819**	0.774	0.867	**0.000**	**1.127**	1.038	1.223	**0.004**

Note: R^2 = .08 (McFadden), .14 (Cox & Snell), .17 (Nagelkerke). Model $\chi^2(1)$ = 226,945, p < .00. * p < .01.

By analysing the data, in Table 3, we can observe that, an increase in skeletal muscle mass (SMM) reduced the probability by 39% (OR: 0.610; CI: 0.449-0.828) of belonging to the LFP, comparatively to be in the NFP.

The same is observed with the increase of basal metabolic rate (BMR), where we can notice that an increase in this variable reduces the probability to be in the LFP, comparatively to be in the reference category.

The waist-hip ratio (WHR) showed a positive relationship in both groups, where increases in WHR increase the probability of belonging to the LFP and AFP when compared to the NFP.

The probability of belonging to the LFP is lower with increasing age (18%; OR: 0.819; CI: 0.774-1.867), however, at older ages the probability of belonging to the AFP increases by about 13% when compared to the reference category (OR: 1.127; CI: 1,038-1.223).

After analysing the variables together, we carry out the evaluation separately by gender.

Table 4. Logistic regression with flexibility and body composition. Data separated between genders: Female.

	Low flexibility profile				Athletic flexibility profile			
		95% CI				95% CI		
	Odds Ratio	Lower Bound	Upper Bound	Sig.	Odds Ratio	Lower Bound	Upper Bound	Sig.
Intercept				**0.004**				**0.000**
Body Fat Mass	**1.883**	1.055	3.361	**0.032**	1.399	0.660	2.967	0.381
Fat Free Mass	0.674	0.428	1.062	0.089	1.010	0.568	1.797	0.973
Skel. Musc. Mass	0.777	0.504	1.198	0.254	1.010	0.587	1.738	0.971
Body Mass Index	**0.467**	0.256	0.853	**0.013**	0.752	0.344	1.644	0.475
Percent Body Fat	0.924	0.561	1.521	0.755	1.136	0.614	2.103	0.685
Basal Met. Rate	0.721	0.386	1.348	0.306	1.689	0.755	3.776	0.202
Waist-Hip Ratio	1.119	0.704	1.778	0.633	0.931	0.516	1.680	0.813
Age	0.947	0.871	1.030	0.206	**1.269**	1.134	1.420	**0.000**

Note: R2 = .07 (McFadden), .13 (Cox & Snell), .15 (Nagelkerke). Model $\chi2(1)$ = 99,611, p < .00. * p < .01.

In the female gender, the increase in body fat mass (BFM) is associated with an increased probability of belonging to the LFP group (88%; OR: 1.883; CI: 1.055-3,361). On the other hand, the higher levels of the body mass index (BMI) increase the probability of belonging to the LFP group by 53% (OR: 0.457; CI: 0.256-0.853).

In females, age increases the probability of belonging to the AFP group by 27% (OR: 1.269; CI: 1.134-1.420).

Table 5. Logistic regression with flexibility and body composition. Data separated between genders: Male.

	Low flexibility profile				Athletic flexibility profile			
		95% CI				95% CI		
	Odds Ratio	Lower Bound	Upper Bound	Sig.	Odds Ratio	Lower Bound	Upper Bound	Sig.
Intercept				**0.000**				**0.003**
Body Fat Mass	0.698	0.438	1.112	0.131	1.201	0.592	2.435	0.612
Fat Free Mass	1.123	0.665	1.896	0.665	**2.701**	1.190	6.128	**0.017**
Skel. Musc. Mass	**0.533**	0.335	0.848	**0.008**	0.788	0.392	1.584	0.504
Body Mass Index	1.044	0.626	1.741	0.868	1.290	0.604	2.756	0.511
Percent Body Fat	1.071	0.685	1.675	0.764	0.762	0.378	1.536	0.447
Basal Met. Rate	0.758	0.425	1.351	0.347	0.995	0.441	2.242	0.989
Waist-Hip Ratio	1.260	0.879	1.805	0.208	0.743	0.441	1.251	0.264
Age	**0.709**	0.646	0.778	**0.000**	1.105	0.961	1.270	0.161

Note: R2 = .12 (McFadden), .20 (Cox & Snell), .24 (Nagelkerke). Model $\chi2(1)$ = 169,10, p < .00. * p < .01.

In males, with the increase in SMM and age variables, the probability of belonging to the LFP group reduced by 47% and 29% respectively (OR: 0.533; CI: 0.335-0.848 and OR: 0.709; CI: 0.646-0.778). when we analyse the probability of belonging to the group of greater flexibility in males, an increase in fat free mass (FFM) increases the probability of belonging to the AFP group when compared to the NFP (OR: 2,701; CI: 1.190-6.128).

5 CONCLUSION

These data indicate that there are differences in the genres regarding flexibility, and that this is mediated by several factors, including parameters obtained through bio-impedance. In the specific work of flexibility, both in school and in sports in children and adolescents, the use of bio-impedance seems to be an important assessment tool.

REFERENCES

Arnason, A. *et al.* (2004) 'Physical Fitness, Injuries, and Team Performance in Soccer', *Med. Sci. Sports Exerc*, 36(2), pp. 278–285. doi: 10.1249/01.MSS.0000113478.92945.CA.

Eler, N. (2018) 'The Relationship between Body Composition and Physical Fitness Parameters in Children', *International Education Studies*. Canadian Center of Science and Education, 11(9), p. 71. doi: 10.5539/ies.v11n9p71.

Ganley, K. J. *et al.* (2011) 'Health-Related Fitness in Children and Adolescents', *Pediatric Physical Therapy*, 23(3), pp. 208–220. doi: 10.1097/PEP.0b013e318227b3fc.

Garber, C. E. *et al.* (2011) 'Quantity and quality of exercise for developing and maintaining cardiorespiratory, musculoskeletal, and neuromotor fitness in apparently healthy adults: Guidance for prescribing exercise', *Medicine and Science in Sports and Exercise*, 43(7), pp. 1334–1359. doi: 10.1249/MSS.0b013e318213fefb.

Haugen, T., Høigaard, R. and Seiler, S. (2014) 'Normative data of BMI and physical fitness in a Norwegian sample of early adolescents', *Scandinavian Journal of Public Health*, 42(1), pp. 67–73. doi: 10.1177/1403494813504502.

Martim, J. and Simas, M. (2012) 'Flexibilidade e Lesões no Voleibol'. *1. 48-53.*

Ruiz, J. R. *et al.* (2009) 'Predictive validity of health-related fitness in youth: A systematic review', *British Journal of Sports Medicine*. British Association of Sport and Excercise Medicine, pp. 909–923. doi: 10.1136/bjsm.2008.056499.

Advances and Current Trends in Biomechanics – Belinha et al. (Eds)

Deformation of retaining screws on implants: Preliminary study using microCT

R. Batista, A. Moreira & P. Mesquita
Faculdade de Medicina Dentária, Universidade do Porto, Portugal

R.B. Ruben
CDRsp-ESTG, Politécnico de Leiria, Portugal

J.C. Sampaio-Fernandes & M.H. Figueiral
INEGI, Faculdade de Medicina Dentária, Universidade do Porto, Portugal

ABSTRACT: The screws that attach the column, abutment or implant suprastructure to the implant through a threaded joint are called retaining screws. The stability of the implant-pillar connection is an important factor for the maintenance of the bone crest, and the retaining screw is a fundamental element of this connection. Although the design of the head and its section have evolved from a grooved shape to hexagonal, square and star shapes, there are no studies evaluating the wear and deformation of the retaining screw head section after torque application. In this sense, 4 titanium alloy screws and 4 gold-plated titanium alloy screws with different tightening protocols were evaluated and analyzed by microcomputed tomography (microCT). The results suggest that high torque values and successive grips deform the screw head regardless of the alloy.

1 INTRODUCTION

Fixed prostheses, either screwed or cemented, remain attached to the implant by retaining screws that hold the abutment to implant body (Ferro 2005). After osteointegration, the untightening of the prosthetic screw is the most common problem in oral rehabilitation with implants. These screws are made up of head and body. Both the shape of the head and the body have been the subject of various studies in relation to loosening; deformation of the screw head, however, is poorly addressed in the literature. The torque applied to the retaining screw inside the threaded joint develops a force called preload (McGlumphy et al. 1998). Additionally, tightening the screw causes it stretching due to the tension exerted on the threaded joint, followed by an elastic recovery. The combination of these forces at the joint level results in a tightening force that holds the abutment-implant interface together (Haack et al. 1995). Although the design of the head and its section have evolved from a grooved shape to hexagonal, square and star shapes, there are no studies evaluating the wear and deformation of the retaining screw head section after torque application. The present work aims at quantifying the deformation of the prosthetic retention screw head on hexagonal section implant when subjected to successive tightening, depending on the material, number of grips and magnitude of the torque.

DOI: 10.1201/9781003217152-67

2 MATERIALS AND METHODS

Computerized microtomography (microCT) with the GE Phoenix V Take X M device was used to evaluate the screw head deformation. From a previous study with 34 titanium alloy screws (Ti; n=17) and gold-plated titanium alloy zimmer biomet^{TM} (Gold; n=17), 5 screws of each type were randomly sellected and assigned to the tightening protocols as follows: new screw/control [Ti0 (Ti-new-P1 & Ti-new-P2) and Gold0 (gold-new-P1 & gold-new-P2)], screw tightened 10 times to 20N.m (Ti1 and Gold1), tightened 20 times to 20N.m (Ti2 and Gold2) and tightened 10 times to 30N.m (Ti3 and Gold3). Each screw was subjected to microCT scanning (Figure 1). To determine the internal area of the hexagon of the screws, and thus assess the deformation (corresponding to an increase of the area), a Python program was used, based on the Otsu segmentation method, which, for all scans, used an accuracy of 19.9 μm between pixels and between slices (Figure 2) (Castro et al. 2019). From slice 1 till slice 25 it is possible to observe the head of the screw and the from 25 till 45 it is possible to observe the neck and the threads of the screw.

Figure 1. MicroCT image of retention screw.

Figure 2. Internal area of screw head (in black).

3 RESULTS

The main results of our investigation are summarized in Figures 3 and 4. Regarding the titanium screws, in the region of the screw head between section 1 and section 25, i) lower deformation of Ti2 and Ti1 was observed in comparison to Ti0; ii) a significant difference in screw deformation was found for Ti3 (Figure 3).

Concerning the gold-plated titanium alloy zimmer biomet^{TM} screws, it was observed that in the same sections mentioned above, the screws of groups Gold1 and Gold3 have a significant deformation in relation to the new screws, with Gold2 being the specimen with the highest deformation level (Figure 4).

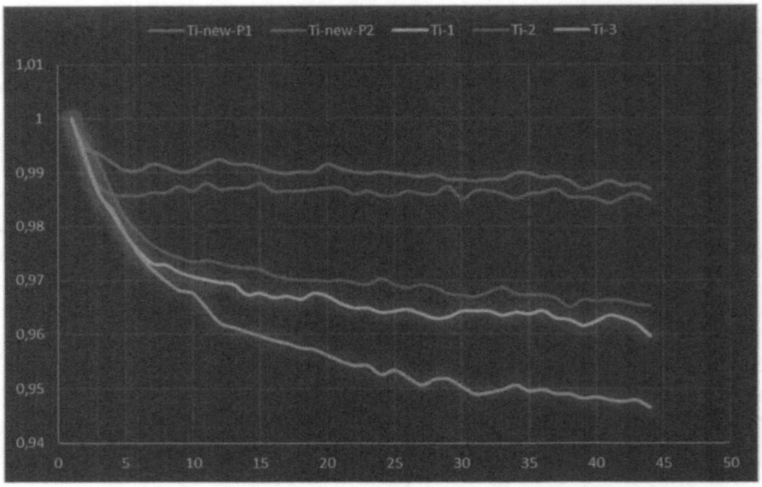

Figure 3. Graph illustrating results of titanium screws. Xx axis with the slices of screws and yy axis with deformation of used screws comparing with new ones (area measured in pixels). Ti new P1 and P2 being Ti0 screws.

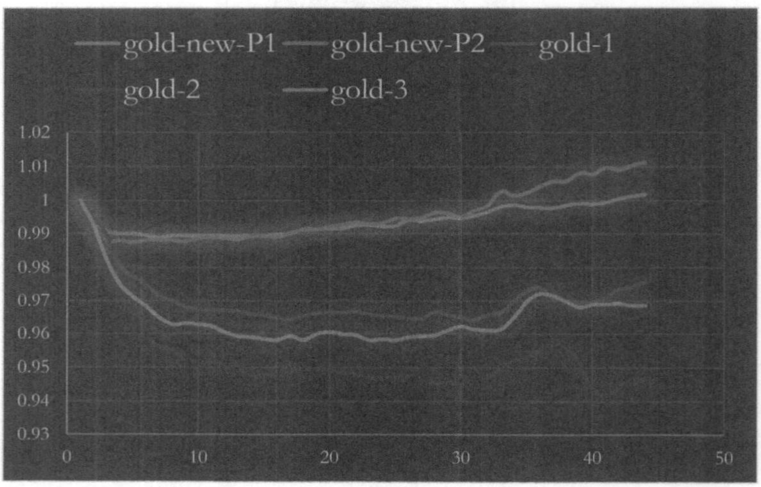

Figure 4. Graph illustrating results of gold-plated screws. Xx axis with the slices of screws and yy axis with deformation of used screws comparing with new ones (area measured in pixels). The gold new P1 and P2 being G0 screws.

4 DISCUSSION

From the analysis of all screws, it is possible to verify that in the first sections corresponding to the area of the screw head, there is a great deformation of the tighten screws, in relation to the new ones. Regarding the titanium alloy, the screw with the largest area was the Ti3, exhibiting a clearly distinct behavior from Ti2 and Ti1. One can therefore assume that concerning the titanium screws the torque applied is the most decisive factor causing deformation. With regards to gold-plated titanium alloy screws, the one with the largest area was the Gold2 followed by the Gold3. Based on this observation, we could assume that the number of successive grips would be the preponderant factor for deformation. However, since only one screw was analyzed per condition, inter-sample variability may have contributed for these results and should not be ruled out, rather demanding additional studies.

Regarding the terminal sections, corresponding to the threads of the screws, our data show that there is also deformation, which can be explained by the increase of the screws' roughness during the tightening procedure (Figure 4).

5 CONCLUSIONS

Based on the microCT analysis we conclude that the screws already tightened have a significantly larger area when compared to the new ones. Regarding the titanium alloy screws, the tightening protocol of 10 grips to 30N.m of torque is the one that induces the largest area and therefore the greatest deformation. In relation to gold-plated titanium alloy, the screw subjected to 20 grips to 20N.m of torque displays the major deformation.

In our experimental conditions i) gold-plated titanium alloy screws had the best behavior; and ii) titanium alloy screws tightened 10 times with 30N.m of torque had the poorest performance.

Overall, our findings suggest that high torque magnitude and successive tightening cause deformation of the screw head regardless of alloy. Although experimentally assessed in this study, the level of deformation cannot be directly extrapolated to the clinical practice. It will be important to invest efforts to confirm whether the deformations observed by microCT may compromise the success of implant-bolted rehabilitations when the retaining screws are subjected to successive tightening and high torque conditions.

REFERENCES

Castro, A. P. G., Ruben, R. B., Gonçalves, S. B., Pinheiro, J., Guedes, J. M., Fernandes, P. M. 2019. Numerical and experimental evaluation of TPMS Gyroid scaffolds for bone tissue engineering. *Computer Methods in Biomechanics and Biomedical Engineering*, 22(6):567–573. doi:10.1080/10255842.2019.1569638. Epub 2019 Feb 18. PMID: 30773050.

Ferro J. K. 2005. The glossary of prosthodontic terms. *Journal of Prosthetic Dentistry*, 94 (1): 10–92. doi: 10.1016/j.prosdent.2005.03.013

Haack, J. E., Sakaguchi, R. L., Sun, T., Coffey, J. P. 1995. Elongation and preload stress in dental implantabutment screws. *The International Journal of Oral & Maxillofacial Implants*,10(5):529–36. PMID:7590997

McGlumphy, E. A., Mendel, D. A., Holloway, J.A. (1998). Implant screw mechanics. *Dental Clinics of North America;*42(1):71–89. PMID: 9421671.

Advances and Current Trends in Biomechanics – Belinha et al. (Eds)
© 2022 Copyright the Author(s), ISBN: 978-1-032-10806-3

Structural characterization of a chlorhexidine delivery system based on reline resins

J. Costa, J. Portugal & C.B. Neves
UICOB, Faculty of Dental Medicine, Universidade de Lisboa, Lisboa, Portugal

M. Franco & N. Alves
CDRSP, Instituto Politécnico de Leiria, Marinha Grande, Portugal

C. Santos
ESTSetúbal, CDP2T, Instituto Politécnico de Setúbal, Setúbal, Portugal
CQE, Instituto Superior Técnico, Universidade de Lisboa, Lisboa, Portugal

A. Bettencourt
iMED.ULisboa, Faculty of Pharmacy, Universidade de Lisboa, Lisboa, Portugal

ABSTRACT: The study evaluates the effect of loading chlorhexidine on the porosity and chemical composition of reline resins. All materials included a control group (0% chlorhexidine) and experimental groups loaded with 2.5% chlorhexidine (w/w) in Kooliner and 5% in Ufi Gel Hard and Probase Cold. The open porosity at surface of the cylindric specimens was evaluated by water immersion method. Micro-CT technology was used for total porosity analysis and FTIR-ATR for chemical composition evaluation. Quantitative data were submitted to the Mann-Whitney tests (α=0.05). Loading with chlorhexidine did not significantly (p>0.05) change the surface porosity of Ufi Gel Hard and Probase Cold, but resulted in higher surface porosity (p=0.008) of Kooliner. Micro-CT images of experimental specimens presented a lower level of total porosity than the controls. FTIR-ATR spectra of all chlorhexidine-loaded specimens showed absorption bands of the drug. In all resins, loading of chlorhexidine decreased total porosity, not changing their chemical composition.

1 INTRODUCTION

The rehabilitation with removable dentures is one of the most common treatment options to restore the masticatory, aesthetic, and phonetic functions in edentulous patients (Alves et al., 2018). Dentures are susceptible to microbial colonization in the oral environment and the most common chronic mucosal lesion associated is denture stomatitis, characterized by erythema and edema of the underlying mucosal areas (Jainkittivong et al., 2010). Despite its multifactorial etiology, the infection by *Candida* species is considered the main etiological factor (Bueno et al., 2015). Conventional therapy with topical and/or systemic antifungals seems to be unsatisfying. Low patient compliance and incomplete elimination of the *Candida* species are the reasons for high relapse rates. A sustained-release drug delivery system (DDS) using chlorhexidine (CHX) incorporated in reline acrylic resins has been suggested for the treatment of denture stomatitis (Rijo et al., 2018). The relining procedure fills the gap between the original denture and the altered tissue with hard or soft lining material,

DOI: 10.1201/9781003217152-68

thus replacing the denture base (Felton et al., 2011, Kreve and Dos Reis, 2019). The use of DDS in the oral cavity inhibits microbial adhesion and growth and are based on a continued, localized controlled release of a drug over a long period, promoting high therapeutic efficiency. It does not require patient compliance and presents minimal adverse risks (Salim et al., 2012). However, DDS can affect the properties of biomaterials when the drug is incorporated, compromising its performance. A previous study concluded that a concentration of 2.5% of CHX in Kooliner and 5% in Ufi Gel Hard and Probase Cold seems to be enough to prevent fungus development and did not negatively affect their reline resins' mechanical properties (Costa et al., 2017). However, the effect of this CHX delivery system on their structural properties has not been investigated. The main purpose of this work is to evaluate the effect of loading three reline acrylic resins (Kooliner, Ufi Gel Hard and Probase Cold) with a previously established concentration of CHX on their porosity and chemical composition.

2 MATERIALS AND METHODS

Three autopolymerizing hard reline acrylic resins, in the powder-liquid form, with different chemical compositions and physical structures were selected. Two were direct reline resins: Kooliner (K) (GC America Inc., Alsip, IL., USA) and Ufi Gel Hard (UGH) (Voco GMbH., Cuxhaven, Germany), and the third was an indirect reline resin: Probase Cold (PC) (Ivoclar Vivadent AG., Liechtenstein). For each material, two groups were settled: a control group with no added CHX; and an experimental group, where a 2.5% (w/w) concentration of CHX was incorporated in the Kooliner and a 5% (w/w) concentration of CHX in both the Ufi Gel Hard and the Probase Cold. The powders of the reline resin and the CHX diacetate monohydrate (Panreac Applichem, Darmstadt, Germany) were each weighed using a precision scale (A&D Company, Limited, Tokyo, Japan). Each specimen was prepared by incorporating CHX and mixing it with the corresponding reline acrylic resins' powder, using a mortar and a pestle for homogenization. Then, the powder was mixed with the liquid monomer, measured using a graduated pipette, and the final mixture was placed into a cylinder-shaped stainless-steel mold and left there during the time recommended by the manufacturer's instructions. A total of 96 specimens (12x6 mm) were prepared. The open porosity at the surface of the specimens (n=5) was measured using the Archimedes' principle. Porosity (P) was calculated according to the following Equation 1:

$$P(\%) = \frac{W_{sat} - W_{dry}}{W_{sat} - W_{sus}} \times 100 \tag{1}$$

where W_{sat} = weight of a specimen saturated with water; W_{dry} = dry weight of a specimen; and W_{sus} = weight of a specimen suspended in water.

The 3D microarchitecture morphology was evaluated by morphometric analysis and 3D model images in a micro-CT equipment (SkyScan 1174; Bruker, Kontich, Belgium). The specimen was mounted on a stub and scanned with the following parameters: 30.1 μm image pixel size; 50 kV; 800 mA; 5500 ms exposure time; 0.9° rotation step; and no aluminum filter. Data were reconstruction with NRecon software (Bruker, microCT, Kontich, Belgium). Fourier Transform Infrared with Attenuated Total Reflectance (FTIR-ATR) spectroscopy was used for the specimens' chemical analysis, using an Alpha-P spectrometer (Bruker, Brussel, Belgium). The tests were performed at room temperature in a spectral range of 400-4000 cm^{-1} at a resolution of 2.00 cm^{-1}. The obtained spectra were averaged over 64 scans. After scanning, characteristic infrared absorption bands of organic functional groups were analyzed using the Opus software 6,5 version. Since normality and homogeneity of variance were not verified in scale variables (Shapiro-Wilk and Levene tests, $p<0.05$), data were submitted to Mann-Whitney non-parametric tests (α=0.05).

3 RESULTS

The CHX incorporation did not significantly change the porosity of Ufi Gel Hard (p=0.690) or Probase Cold (p=0.310). However, loading Kooliner with CHX (0.66%) resulted in significantly (p=0.008) higher porosity values compared to the control group (0.043%) (Table 1).

Selected 3D reconstruction models by micro-CT allowed a qualitative evaluation of total porosity (open and closed) of the specimens, represented by the green color. Micro-CT images of CHX loaded specimens of all materials presented a lower level of porosity when compared to the micro-CT images of the control group (Figure 1). To analyze the chemical composition, FTIR-ATR spectra of the specimens and pure CHX were obtained. The pure CHX spectrum is characterized by the presence of three bands at 1164, 1480 and 1724 cm^{-1}, corresponding to the C-O, C=N and C=O bonds, respectively (Figure 2a). The FTIR-ATR spectrum of CHX-loaded Kooliner specimen shows the characteristic vibration bands of the control specimen. In addition, it is also possible to find two characteristic CHX vibration bands which can be assigned to C-O (at 1150 cm^{-1}), and C=O (at 1733 cm^{-1}) (Figure 2b). The spectrum of Ufi Gel Hard and CHX are almost identical with bands in the same range of wavenumber (Figure 2c). CHX-loaded Ufi Gel Hard spectrum showed no new bands, remaining with the sum of CHX and Ufi Gel Hard bands. In the CHX-loaded Probase Cold spectrum, the three characteristic CHX vibration bands appeared (Figure 2d).

Table 1. Calculated open porosity at the surface of the specimen (%) according to Equation 1.

Experimental Groups	POROSITY (%) M ± SD	MED ± IQR	p value
0% CHX K	$4.3 \times 10^{-2} \pm 2.77 \times 10^{-2}$	$6.2 \times 10^{-2} \pm 4.70 \times 10^{-2}$	p=0.008
2.5% CHX K	$6.6 \times 10^{-1} \pm 9.92 \times 10^{-1}$	$2.5 \times 10^{-1} \pm 1.25 \times 10^{0}$	
0% CHX UGH	$3.7 \times 10^{-2} \pm 3.31 \times 10^{-2}$	$3.1 \times 10^{-2} \pm 4.60 \times 10^{-2}$	p=0.690
5% CHX UGH	$1.8 \times 10^{-2} \pm 6.53 \times 10^{-2}$	$3.0 \times 10^{-2} \pm 9.00 \times 10^{-2}$	
0% CHX PC	$1.2 \times 10^{-1} \pm 8.87 \times 10^{-2}$	$1.5 \times 10^{-1} \pm 1.63 \times 10^{-1}$	p=0.310
5% CHX PC	$1.8 \times 10^{-2} \pm 3.22 \times 10^{-2}$	$3.0 \times 10^{-2} \pm 4.22 \times 10^{-2}$	

M = Mean, SD = Standard deviation, MED = Median, IQR = Interquartile range.

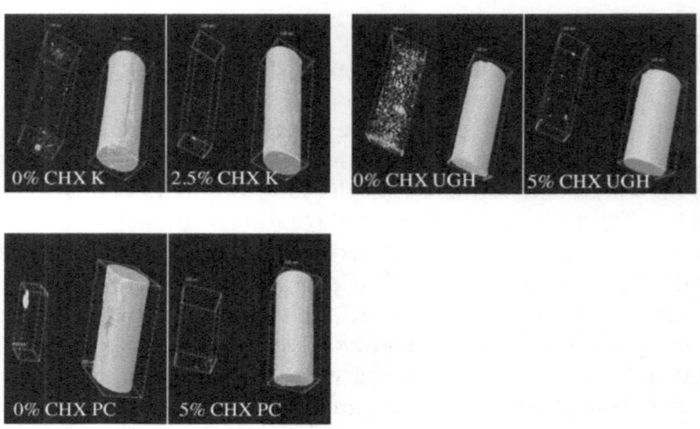

Figure 1. 3D Reconstruction models of the Kooliner (K), Ufi Gel Hard (UGH), and Probase Cold (PC) control and experimental specimens by micro-CT (CTAn and CTVol, Bruker softwares) [CHX – Chlorhexidine].

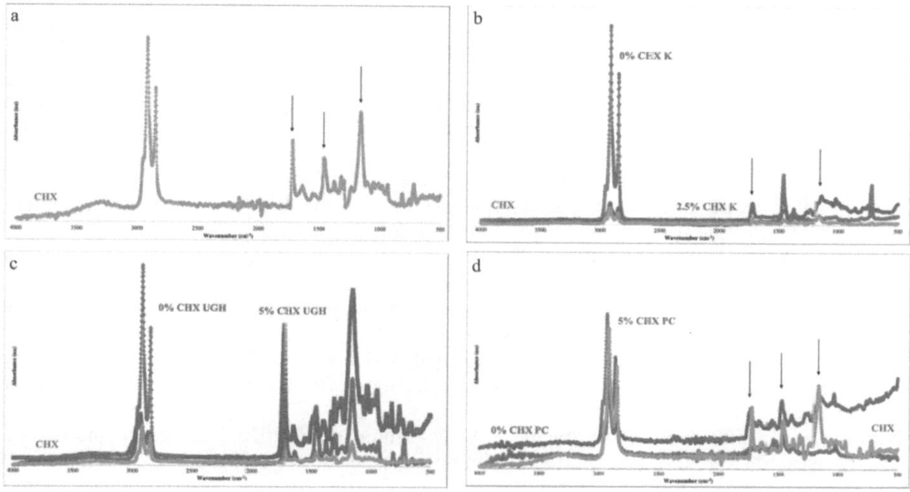

Figure 2. FTIR-ATR spectra of chlorhexidine (a), and Kooliner K (b), Ufi Gel Hard UGH (c), and Probase Cold PC (d) control and experimental specimens [CHX – Chlorhexidine].

4 DISCUSSION

The surface porosity of relining resins was firstly investigated by a classic method that employs the volume of water absorbed by the biomaterial. This method is limited to the evaluation of the open pores at the surface of the biomaterial (Chiang et al., 2010). For the obtained results the incorporation of CHX did not influence the surface porosity of the resins, except for Kooliner. It is well known that Kooliner has a simple polymer structure, less complex than the crosslinking net formed in Ufi Gel Hard, and Probase Cold, in turn, an indirect reline resin with a higher level of monomer-polymer conversion and less distance between particles (Arima et al., 1996). This structural organization of this resin can be responsible for the difference observed in surface porosity. Micro-CT images allowed to verify that CHX loading decreased the total porosity of all the resins. The obtained results suggest that the presence of CHX particles inside resin can fill the pores and reduce the global porosity. Additionally, from FTIR-ATR results, it was observed that adding CHX particles to the resin do not change the main absorption bands of the resin neither CHX, suggesting that CHX does not react chemically with the resins.

5 CONCLUSIONS

In all studied resins, loading of CHX decreased the total porosity without changing their chemical composition. Considering the results, CHX delivery systems based on reline acrylic resins of removable dentures appears to be a potential approach in the prevention or treatment of denture stomatitis. However, further studies are needed to assess the potential of their application so as not to affect the behavior of the biomaterials.

REFERENCES

Alves, A. C., Cavalcanti, R. V., Calderon, P. S., Pernambuco, L. & Alchieri, J. C. 2018. Quality of life related to complete denture. *Acta Odontol Latinoam* 31: 91–96.

Arima, T., Murata, H. & Hamada, T. 1996. Analysis of composition and structure of hard autopolymerizing reline resins. *J Oral Rehabil* 23: 346–52.

Bueno, M. G., Urban, V. M., Barberio, G. S., Da Silva, W. J., Porto, V. C., Pinto, L. & Neppelenbroek, K. H. 2015. Effect of antimicrobial agents incorporated into resilient denture relines on the Candida albicans biofilm. *Oral Dis* 21: 57–65.

Chiang, Y. C., Rosch, P., Dabanoglu, A., Lin, C. P., Hickel, R. & Kunzelmann, K. H. 2010. Polymerization composite shrinkage evaluation with 3D deformation analysis from microCT images. *Dent Mater* 26: 223–31.

Costa, J., Alexandre, F., Bettencourt, A., Ribeiro, I., Portugal, J. & Neves, C. 2017. Incorporação de Clorexidina em Resinas de Rebasamento – Estudos Microbiológicos. *Rev Port Estomat Med Dent Cir Maxilofac* 58: e51.

Felton, D., Cooper, L., Duqum, I., Minsley, G., Guckes, A., Haug, S., Meredith, P., Solie, C., Avery, D. & Deal Chandler, N. 2011. Evidence-based guidelines for the care and maintenance of complete dentures: a publication of the American College of Prosthodontists. *J Prosthodont* 20: S1–12.

Jainkittivong, A., Aneksuk, V. & Langlais, R. P. 2010. Oral mucosal lesions in denture wearers. *Gerodontology* 27: 26–32.

Kreve, S. & Dos Reis, A. C. 2019. Denture Liners: A Systematic Review Relative to Adhesion and Mechanical Properties. *ScientificWorldJournal 2019*: 6913080.

Rijo, I., Pedro, D., Costa, J., Bettencourt, A. F., Portugal, J. & Neves, C. B. 2018. Chlorhexidine loading of acrylic reline resins – Microhardness and flexural strength after thermal aging. *Rev Port Estomatol Cir Maxilofac* 59: 154–161.

Salim, N., Moore, C., Silikas, N., Satterthwaite, J. D. & Rautemaa, R. 2012. Fungicidal amounts of antifungals are released from impregnated denture lining material for up to 28 days. *J Dent* 40: 506–12.

Advances and Current Trends in Biomechanics – Belinha et al. (Eds)

Interpretation of the acceleration obtained by an Inertial Measurement Unit (IMU) in countermovement jump, in Athletics' elite athletes

P. Miranda-Oliveira
Federação Portuguesa de Atletismo, Lisboa, Portugal
ESTG, Politécnico de Leiria, Portugal

P. Serra & M. Branco
Escola Superior de Desporto de Rio Maior, Instituto Politécnico de Santarém, Portugal
Centro Interdisciplinar para o Estudo da Performance Humana (CIPER), Faculdade de Motricidade Humana, Universidade de Lisboa, Cruz Quebrada, Dafundo, Portugal

O. Fernandes
Departamento de Desporto e Saúde, Escola de Saúde e Desenvolvimento Humano, Universidade de Évora, Portugal
Comprehensive Health Research Center, Universidade de Évora, Portugal

ABSTRACT: This study aimed to correlate the vertical acceleration obtained by an Inertial Measurement Unit (IMU) and the biomechanical parameters obtained by a force plate (FP). Eight athletes (six males and two females; Aged 26 ± 3.61; body mass: 73.19 ± 8.96Kg; and height: 1.77 ± 0.08m) with international representation by their countries were included in the study. An IMU developed collecting at 275Hz and a Bertec FP collecting at 1000Hz were used to obtain the selected biomechanical parameters. Statistical Pearson and Spearman correlation tests were applied. Several biomechanical parameters obtained a higher correlation with the vertical acceleration obtained by IMU. Through the linear regression models, the vertical acceleration root mean square (RMS) obtained a higher explanation with a $R^2 = 59.2\%$. The proposed model showed that higher vertical acceleration RMS is associated with high force values and high-rate force development, and small temporal values in negative and positive impulse and rate force development.

1 INTRODUCTION

The motor skills of vertical jump have been studied over the years (González-Badillo & Marques 2010). Several studies showed that higher jump height in Countermovement jumps (CMJ) and Squat jumps (SJ) was associated with high power levels (Dowling & Vamos 1993; González-Badillo & Marques 2010; Mandic et al. 2015), high force levels (Dowling & Vamos 1993; Mandic et al. 2015), high vertical velocity during positive impulse (Dowling & Vamos 1993; Ferragut et al. 2003; González-Badillo & Marques 2010), high capacity to accelerate and high vertical impulses (Ferragut et al. 2003). The correlation between the kinetic and kinematic parameters and the jump height allows understanding the VJ execution (Dowling & Vamos 1993; Ferragut et al. 2003;

DOI: 10.1201/9781003217152-69

González-Badillo et al. 2010; Mandic et al. 2015). Inertial Measurement Units (IMU) have emerged because of their sizes and showed a higher validity in CMJ (Nielsen et al. 2019). However, it did not verify studies that associated the IMU's vertical acceleration with the kinematic and kinetic parameters by the force plate (FP) in CMJ. This study aimed to correlate the vertical acceleration obtained by an IMU and the biomechanical parameters obtained by a FP, in an Athletics' elite athletes' group.

2 MATERIAL AND METHODS

Eight athletes, six males (pole vaulters: 5.53 ± 0.13m; decathletes: 7280 ± 183points; aged: 24.33 ± 2.71; body mass (BM): 77.90 ± 8.69Kg; and height (h): 1.80 ± 0.07m) and two females (pole vaulters: 4.47 ± 0.04m; aged: 30.00 ± 2.00; BM: 63.23 ± 1.87Kg; and h: 1.70 ± 0.01m) with international representation by their countries were included in the study. All participants were informed about the study objectives and signed the informed consent approved by the Ethical Council of the *Universidade de Évora*. An instrument developed by the investigation group, encompassing a tridimensional (3D) IMU (size: 42 x 32 x 17mm; weight: 22g) was used to obtain the vertical acceleration (the maximum, minimum, difference between maximum and minimum and RMS). The acceleration data was corrected through the curve slope obtained by the linear regression. The IMU was fixed with adhesive tape on the surface in the 5th lumbar vertebral apophysis (L5), as a reference for the Centre of Gravity (CG) (Nielsen et al. 2019; Camomilla et al. 2018) and in the right calcaneus area (IMU GRF) (Camomilla et al. 2018) as reference for the ground reaction force (GRF). The PF (Bertec, Columbus, USA; com 1.2 x 0.6m) was used to determine the kinetic parameters (the relative, maximum, and minimum force, the force of the negative and active impulse phases and rate force development; the negative and active impulse; and the relative and maximum power), the kinematic parameters (maximum vertical displacement, velocity, and acceleration) and temporal parameters in the negative and active impulse and the rate force development. Each athlete performed three to five CMJ, with a minute of rest between jumps. The Jamovi® was used to perform the statistical analysis. The normality was verified by the Shapiro-Wilk test ($p > 0.05$). Pearson and Spearman correlation between the IMU CG and the FP, and the IMU GRF and the FP was determined. The correlation values (r) suggested by Hopkins et al. (2009) were considered, $r \leq 0.3$, small; r between 0.3 and 0.5, moderated; and $r > 0.5$, high. In predictive models' development, linear regression with enter method was used. The presented model considered the correlations between the dependent variable and the independent variables whose significance degree was $\rho \leq 0{,}05$.

3 RESULTS

The athletes performed a total of 30 CMJ. The mean and standard deviation, the normality analysis for all analyzed parameters were shown in Table 1.

The correlation tests between the vertical acceleration obtained through the IMU in the CG and GRF position and the kinematic and kinetic parameters obtained through the PF were represented in Table 2.

Through the correlation results, were proposed several models, and the vertical acceleration RMS obtained through IMU in CG position had a higher explanation with a $R^2 = 59.2\%$. The negative impulse phase was a significant predictor ($p = 0.022$) (see Table 3).

Table 1. Mean and standard deviation of the vertical acceleration from IMU and kinematics and kinetics parameters from FP.

	Mean	±SD	
IMU			
Acc_y_max_CG	12.9	±3.61	
Acc_y_max_GRF	41.2	±7.85	
Acc_y_min_CG	–26.3	±7.85	
Acc_y_min_GRF	–5.81	±3.52	¥
Acc_y_diff_CG	39.1	±7.93	
Acc_y_diff_GRF	47.0	±11.9	
Acc_y_RMS_CG	7.86	±1.16	
Acc_y_RMS_GRF	8.11	±2.39	
Force plate			
Fy_max_rel	27.7	±2.28	
RFDy_max	13906	±3457	
RFDy_rel	192	±46.5	¥
Acc_y_max	17.9	±2.28	
Fy_Ineg	107	±121	¥
Ineg	–120	±23.2	
t_Ineg	0.325	±0.06	¥
t_Iact	0.423	±0.04	
t_RFDy	0.154	±0.05	¥

¥ non-normal - $p \le 0.05$; Acc_y – vertical acceleration; max – maximum; min – minimum; diff – difference between the maximum and minimum accelerationvalue; RMS – root mean square; CG – centre of gravity; GRF – ground reaction force; rel – relative; Fy – vertical force; RFDy – vertical rate force development; Ineg – negative impulse; Iact – active impulse t – time.

Table 2. Correlation between the vertical acceleration obtained through the IMU in the CG and GRF position and the biomechanical parameters obtained through the PF.

		Vertical acceleration							
		Maximum		Minimum		Difference		RMS	
		CG	GRF	CG	GRF	CG	GRF	CG	GRF
Fy_max_rel_	r	-.130	-.318	-.106	-.260	.046	-.225	.479 ψ	-.277
	p	.494	.087	.578	.165	.811	.232	.007	.138
RFDy_max	r	-.330	-.314	-.208	-.174	.055	-.281	.431	-.266
	p	.075	.091	.270	.357	.771	.133	.017 ψ	.155
RFDy_rel	r	-.088	-.317	-.254	.002	.169	-.317	.602	-.213
	p	.643	.088	.175	.994	.370	.089	< .001 ¥	.256
Acc_y_max	r	-.130	-.318	-.106	-.260	.046	-.225	.479	-.277
	p	.494	.087	.578	.165	.811	.232	.007 ψ	.138
Fy_Ineg	r	.035	.026	.099	-.075	-.084	.080	-.588	-.077
	p	.855	.892	.600	.695	.658	.672	< .001 ¥	.684
Ineg	r	.403	.312	.331	.139	-.144	.291	-.279	.362
	p	.027 ψ	.093	.074	.462	.447	.119	.136	.049 ψ
t_Ineg	r	-.482	.129	.202	-.426	-.401	.277	-.650	.013
	p	.007 ψ	.496	.285	.019 ψ	.028 ψ	.138	< .001 ¥	.944
i_Iact	r	-.432	.092	-.120	-.303	-.078	.133	-.458	-.000
	p	.017 ψ	.628	.527	.103	.683	.485	.011 ψ	.999
t_RFDy	r	.136	.290	.318	-.111	-.209	.337	-.570	.192
	p	.475	.120	.087	.558	.267	.069	.001 ψ	.310

Significance: ψ $p<0.05$; ¥ $p<0.001$; Acc_y – vertical acceleration; max – maximum; RMS – root mean square; CG – centre of gravity; GRF – ground reaction force; rel – relative; Fy – vertical force; RFDy, vertical rate force development; Ineg – negative impulse; Iact – active impulse; t – time.

Table 3. Linear regression model for vertical acceleration RMS in CG position.

Predictor variable	Vertical acceleration RMS in CG position	
	Estimate	p
Intercept	9.0145	0.083
Fy_max_rel_	0.1819	0.381
RFDy_max	0.00002	0.187
RFDy_rel	-0.0171	0.412
Fy_Ineg	-0.000003	0.992
t_Ineg	-15.9886	0.022 ψ
i_Iact	-0.8655	0.898
t_RFDy	-0.0826	0.992

Significance: ψ $p<0.05$; RMS – root mean square; CG – centre of gravity; max – maximum; rel – relative; Fy – vertical force; RFDy - vertical rate force development; Ineg – negative impulse; Iact – active impulse; t – time.

4 DISCUSSION

This study aimed to correlate the vertical acceleration obtained by an Inertial Measurement Unit (IMU) and the biomechanical parameters obtained by a force plate (FP), in an Athletics' elite athletes' group.

When analyzed the acceleration values from the IMU positions, the GRF position obtained higher values than the CG position (see Table 1), it was expected due to the application of the same force to different masses according to the Newtonian definitions of Force.

The significant correlation between the vertical acceleration with the negative impulse phase reinforced the importance of the eccentric and concentric phases during the CMJ execution (Ferragut et al. 2003; González-Badillo & Marques 2010).

The linear regression model suggested higher vertical acceleration, we had higher vertical forces. Dowling & Vamos (1993) and Mandic et al. (2015) obtained a significant association between higher jump height and higher vertical force. During the negative and active phase, several temporal and kinetic parameters were inserted in the model reinforced the importance between the eccentric and concentric phases during the CMJ execution (González-Badillo & Marques 2010). Ferragut et al. (2003) also suggested a positive mechanical impulse as an accuracy predictive variable of the jump height during the CMJ.

5 CONCLUSION

This study agrees with the basic notions of variation of the magnitude of the accelerations as it has been demonstrated with higher values in GRF position than CG position. The significant correlation between vertical acceleration and the negative impulse phase showed the importance of good performance during the eccentric and concentric phases.

These models were important to show that higher vertical acceleration RMS during CMJ execution suggested that the athlete carried out higher vertical forces and rate force development, and with smaller time in negative and active impulse phases and time to achieve the maximum force.

ACKNOWLEDGMENT

To athletes by your professionalism showed during the study and the coaches by to accept conjugate the data collection during your training plan.

To Federação Portuguesa de Atletismo by the given support, in material, space, and time availability to perform the data collection.

To Instituto Português do Desporto e Juventude to allow the use of the requested space.

REFERENCES

González-Badillo, J. J. and Marques, M. C. (2010). Relationship Between Kinematic Factors and Counter-movement Jump Height in Trained Track and Field Athletes. The Journal of Strength & Conditioning Research, 24(12):3443–3447.

Dowling, J. J. and Vamos, L. (1993). Identification of Kinetic and Temporal Factors Related to Vertical jump Performance. Journal of Applied Biomechanics, 9:95–110.

Mandic, R., Jakovljevic, S., and Jaric, S. (2015). Effects of countermovement depth on kinematic and kinetic patterns of maximum vertical jumps. Journal of Electromyography and Kinesiology, 25 (2):265–272.

Ferragut, C., Arteaga-ortiz, R., and Calbet, J. A. (2003). Predicción de la altura de salto vertical: importancia del impulso mecánico y de la masa muscular de las extremidades inferiores. Revista Motricidad, 10:7–22.

Hopkins, W. G., Marshall, S. W., Batterham, A. M., & Hanin, J. (2009). Progressive statistics for studies in sports medicine and exercise science. Medicine and Science in Sports and Exercise, 41(1),3–12.

Nielsen, E. T., Jørgensen, P. B., Mechlenburg, I., and Sørensen, H. (2019). Validation of an inertial measurement unit to determine countermovement jump height. Asia-Pacific Journal of Sports Medicine, Arthroscopy, Rehabilitation and Technology, 16:8–13.

Camomilla, V., Bergamini, E., Fantozzi, S., & Vannozzi, G. (2018). Trends supporting the in-field use of wearable inertial sensors for sport performance evaluation: A systematic review. Sensors (Switzerland), 18(3). https://doi.org/10.3390/s18030873

Advances and Current Trends in Biomechanics – Belinha et al. (Eds)
© 2022 Copyright the Author(s), ISBN: 978-1-032-10806-3

Characterization of walking patterns using a smartphone in patients with peripheral arterial disease

Vítor Filipe
Universidade de Trás-os-Montes e Alto Douro, Portugal

Miguel Correia
Faculdade de Engenharia da Universidade do Porto, Portugal

Hugo Paredes & Bruno Pinto
Instituto de Engenharia de Sistemas e Computadores, Tecnologia e Ciência - INESC TEC, Portugal

Ivone Silva
Instituto de Ciências Biomédicas Abel Salazar da Universidade do Porto - ICBAS, Portugal
Centro Hospitalar Universitário do Porto, Hospital de Santo António, Porto, Portugal

Catarina Abrantes
Centro de Investigação em Desporto, Saúde e Desenvolvimento Humano - CIDESD, Portugal

ABSTRACT: Cardiovascular diseases are the leading cause of global death and have a high impact on the economics and the quality of life of the populations, requiring effective preventive measures. The WalkingPAD project objective is to develop a prevention and treatment program based on supervised walking for patients with intermittent claudication supported by technological self-care tools, as a key to changing long-term behaviors. The identification of intermittent claudication by automatic mechanisms, using the walking patterns identified in the signal collected by smartphone sensors is explored in this study, to overcome the bias associated with patients' verbal reporting. The preliminary results obtained give good indicators about the possibility of using this mechanism to automatically identify the onset of pain in patients with peripheral arterial disease and intermittent claudication.

1 INTRODUCTION

Patients with peripheral arterial disease (PAD) have a severe limitation of their functional capacity concerning the distance walked, driven by the pain resulting from muscle ischemia and defined as intermittent claudication (IC). This limitation promotes a sedentary lifestyle, decreasing the level of physical fitness, enhancing cardiovascular risk factors, and consequent disease progression. Physical exercise programs and increased physical activity are effective options with scientific evidence, of low cost and risk, compared to more invasive therapies. These remotely structured and supervised programs that fit within the residence area are much more motivating and supportive than hospital programs.

The WalkingPAD project aims to develop an intervention strategy centered on the patient and supported by information and communication technologies, aiming to demonstrate the technical and economic viability of an integrated technological solution for personalized medicine framed in the patient residence area and supported in their daily routines. Intending to

DOI: 10.1201/9781003217152-70

increase adherence to walking and the distance covered, the project aims to create a program of supervised outpatient therapeutic physical exercise, for education and co-responsibility of the patients in the management of their disease. One of the technological limitations to the development of an effective solution (Paredes et al. 2021) lies in the identification of the onset of pain moment, which in most situations, introduces a bias in the supervision of exercises.

In this article, an exploratory study is presented to develop automatic mechanisms for identifying the onset of pain based on the analysis of the smartphone sensor signal. The results obtained reveal a potential solution for efficient identification of IC.

2 METHODS

The Walking PAD project aims to collect data from 120 patients with peripheral arterial disease that are being monitored clinically at the Centro Hospitalar e Universitário do Porto, Hospital de Santo António, Portugal. For this exploratory study, data was collected from a walking session on a treadmill of two patients with PAD, both with IC and participants in a supervised physical exercise program for PAD (Renner et al. 2020).

Data was collected during the patients' walk on a treadmill, for 40 minutes. A smartphone, Samsung Galaxy S7, which has a 6-axis IMU with a 3D digital accelerometer with a full-scale acceleration range of ±2/±4/±8/±16g (STMicroelectronics LSM6DS3), was placed on the patients' right iliac crest to record the acceleration values in the three planes of motion: medio-lateral (X-axis), vertical (Y-axis) and anterior-posterior (Z-axis), during the walking exercise. Data collection was performed using the Physics Toolbox Accelerometer (Google Play Store) application, for the Android system, with a sampling rate of 50Hz. The data acquired by the application was saved, in a CSV file, for later offline processing with the acceleration values organized in columns for each axis.

During the exercise, at a 5-minute interval, the patient was asked to indicate whether he felt pain and, if so, in which limb. Also, the minute when the patient felt pain was registered so that it is possible to compare the patterns/variables of walking before and after the sensation of pain.

The acceleration data was filtered with a 4th order Butterworth low-pass filter with a cut-off frequency of 5Hz since in this study the parameters are extracted through the detection of cycles and the upper limit of the normal human step frequency is 4Hz, the relevant data is found below the cutoff point (Fridolfsson et al. 2020). This operation removes the noise that typically affects the accelerometry data while preserving the characteristics necessary to calculate the walking parameters.

In this study the gait patterns of patients with PAD, are described by the following parameters: cadence, step time, stride regularity, step regularity, and step symmetry.

In the acceleration signal, the time between two consecutive steps constitutes a cycle. The detection of local maxima or minima in the acceleration signals can be used to segment it and count the number of steps. A peak detector algorithm, is applied to the vertical acceleration signal to split it into segments from which the temporal parameters of the cadence and time of the step were calculated (Mantyjarvi et al. 2020). An example of peak detection can be found in Figure 1.

The remaining parameters, step and stride regularity, and step symmetry were determined by calculating the unbiased autocorrelation sequence, *A(m)*, with *N* samples, obtained by summing the products of the vertical acceleration signal *z(i)* multiplied by the signal with a delay *m*, Equation 1 (Moe-Nilssen et al. 2020):

$$x' = \frac{1}{N-|m|}\sum_{i=1}^{N-|m|}(i).z(i+m) \quad (1)$$

The first peak of *A(m)*, *Ad1*, expresses the acceleration regularity between two consecutive steps. The second peak, *Ad2*, represents the regularity of two strides. The step symmetry is calculated by the quotient between the step regularity and the stride regularity, *Ad1/Ad2*.

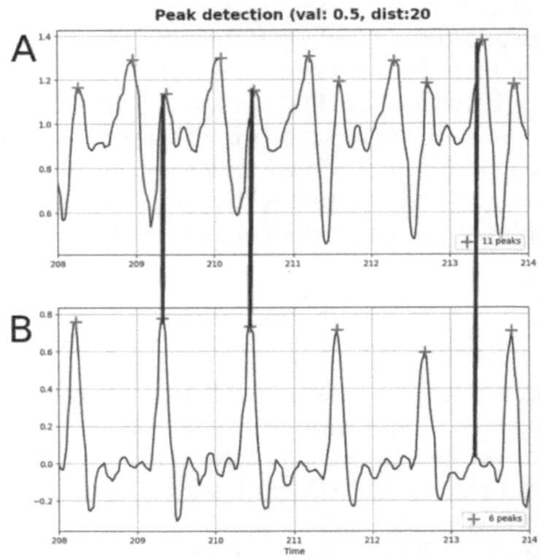

Figure 1. 1A: peak detection of vertical acceleration (every peak represents a step), 1B: peak detection of anterior-posterior acceleration (every peak represents the heel-strike of one leg, separating the data into cycles); x-axis = time [s], y-axis = acceleration [g].

3 RESULTS

As an example of the results obtained, Table 1 shows the walking parameters measured in patient 1, with pain appearing at 5:36 minutes, and Table 2 the parameters of patient 2, where pain appeared at minute 5 for the right leg and between 10 and 15 minutes for the left leg. The measurements are divided into four periods, before the onset of pain and three periods of 10 minutes (10 min, 20 min, and 30 min) after the pain appearance. The major variations in the parameters are observed in the transition of period no pain (Before) to pain (After(10)), especially in cadence and in step time.

Table 1. Mean and standard deviation of walking parameters measured in a patient 1 trial.

	Before	After(10)	After(20)	After(30)
Cadence steps/min	108.50 ± 0.03	102.20 ± 0.04	102.10 ± 0.09	102.20 ± 0.02
Mean Step Time/s	0.56 ± 0.10	0.58 ± 0.08	0.60 ± 0.16	0.59 ± 0.09
Step Regularity	0.94 ± 0.02	0.93 ± 0.01	0.93 ± 0.01	0.92 ± 0.01
Stride Regularity	0.99 ± 0.00	0.98 ± 0.00	0.98 ± 0.01	0.98 ± 0.01
Step Symmetry	0.98 ± 0.01	0.94 ± 0.00	0.95 ± 0.01	0.94 ± 0.01

Table 2. Mean and standard deviation of walking parameters measured in a patient 2 trial.

	Before	After(10)	After(20)	After(30)
Cadence steps/min	108.29 ± 0.04	99.52 ± 2.86	96.25 ± 3.11	96.23 ± 0.01
Mean Step Time/s	0.56 ± 0.03	0.60 ± 0.02	0.62 ± 0.01	0.61 ± 0.03
Step Regularity	0.97 ± 0.02	0.96 ± 0.02	0.99 ± 0.02	0.97 ± 0.01
Stride Regularity	0.99 ± 0.00	0.99 ± 0.00	0.99 ± 0.02	0.99 ± 0.00
Step Symmetry	0.98 ± 0.01	0.97 ± 0.00	0.98 ± 0.01	0.98 ± 0.01

Patient 1 experienced pain 5 minutes after the onset of pain, marked with a cadence decreased of 5.8% compared to the period before the pain. As expected, a decrease in cadence is associated with an increase of about 3.6% of the average step time, after the onset of pain. Once the level of pain becomes stable, the walking parameters also remain stable. This shows that even after 40 minutes of walking, tiredness does not significantly affect gait parameters, unlike the appearance of pain.

Patient 2 felt pain in the right leg first at minute 5 that resulted in decreased of 8.3% in the cadence and an increase of 7.1% in mean step time, compared to the period before the pain. The pain in the left leg appeared between minute 10 and 15 that resulted in a 3% decrease in cadence and an increase of 3.3% in the mean step time, showing that the parameters are affected not only with the onset of pain but also with its intensification. Once again after the stabilization of the pain, the parameters remained stable.

Since the parameters *step regularity* and *stride regularity*, and *step symmetry* do not change much during the walking exercise they became irrelevant for this specific analysis. The fact that the intensity in this exercise is low may explain the small variations observed before and after the occurrence of pain in these parameters. The reference values for step and stride regularity, on a similar age group, show slightly lower values, around 0.8. The high values can also be explained by the fact that the patients have a very symmetrical and regular gait, which seems plausible for this small sample of patients, or the parameters of regularity and symmetry are poorly represented in this experiment. One reason for the high values in the symmetry and regularity parameters may be the experimental configuration of the treadmill. When walking on a treadmill, there are not many changes in the way the steps are taken, facilitating stability.

4 CONCLUSIONS

This exploratory study revealed that the manifestation of IC in PAD patients leads to the alteration of some walking parameters. The results indicate that PAD patients decrease the cadence to ensure continuity of walking while experiencing claudicating pain, which is in line with other studies (Gardner et al. 2020). Additionally, viable walking parameters can be extracted from an accelerometer embedded in a smartphone which provides the opportunity to establish a larger database that can be analyzed to deepen the knowledge of PAD and thus contribute to the introduction of innovative treatments of the disease that modify the disease progression.

For future work, the gait cycle stance and swing phases should be analyzed to determine if a pathological walking pattern arises when claudicating pain manifests. Another interesting aspect would be to observe whether a larger change in the walking parameters occurs with an increase in exercise intensity. Also, comparing the treadmill exercise regime to a home exercise regime can generate interesting insights to develop future exercise programs accordingly.

ACKNOWLEDGMENT

This work was funded by the Foundation for Science and Technology, I.P., within the scope of the WalkingPAD project (PTDC/MEC-VAS/31161/2017). The authors would like to thank Erasmus student Kassandra Renner for her collaboration in the experimental data collection.

REFERENCES

Fridolfsson, J., Börjesson, M., Buck, C., Buck, C., Ekblom, Ö., Ekblom-Bak, E., Hunsberger, M., Lissner, L., Lissner, L., Lissner, L., Arvidsson, D., 2019. Effects of Frequency Filtering on Intensity and Noise in Accelerometer-Based Physical Activity Measurements. Sensors.

Gardner, A., Ritti-Dias, R.M., Stoner, J.A., Montgomery, P.S., Scott, K.J., Blevins, S.M , 2010. Walking economy before and after the onset of claudication pain in patients with peripheral arterial disease. Journal of Vascular Surgery.

Mantyjarvi, J., Lindholm, M., Vildjiounaite, E., Makela, S.-, Ailisto, H.A., 2005. Identifying users of portable devices from gait pattern with accelerometers, in: Proceedings. (ICASSP '05). IEEE International Conference on Acoustics, Speech, and Signal Processing, 2005. Presented at the Proceedings. (ICASSP '05). IEEE International Conference on Acoustics, Speech, and Signal Processing, 2005., p. ii/973-ii/976 Vol. 2.

Moe-Nilssen, R., Helbostad, J.L., 2004. Estimation of gait cycle characteristics by trunk accelerometry. Journal of Biomechanics 37, 121–126.

Paredes, H., Paulino, D., Barroso, J., Abrantes, C., Machado, I., & Silva, I. (2021). Supervised physical exercise therapy of peripheral artery disease patients: M-health challenges and opportunities. In Proceedings of the 54th Hawaii International Conference on System Sciences (p. 3881).

Renner, K., Filipe, V., Torres-Pereira, L., Silva, I., & Paredes, H. (2020). Gait pattern analysis with accelerometer data from a smartphone in PAD patients. In Proceedings of the 8th IEEE International Conference on E-Health and Bioengineering - EHB 2020.

Advances and Current Trends in Biomechanics – Belinha et al. (Eds)

Fit accuracy of RPD framework produced by CAD-CAM—pilot clinical trial

P.R. Conceição, A. Peleija, J.P. Martins, J. Portugal & C.B. Neves
UICOB, Faculty of Dental Medicine, Universidade de Lisboa, Lisboa, Portugal

M. Franco & N. Alves
CDRSP, Instituto Politécnico de Leiria, Marinha Grande, Portugal

ABSTRACT: Aim: To assess the effect of CAD-CAM fabrication on the clinical fit accuracy of removable partial denture (RPD)'s metal frameworks. Methods: Eight patients were selected to participate in this study and twelve RPD were planned. For each RPD ($n=12$) two Cobalt-Chromium frameworks were produced with two alternative protocols: CAD-CAM protocol (experimental group); and conventional lost-wax casting technique (control group). Clinical fit accuracy was assessed by direct evaluation of the gap between the framework clasp assemblies and their dental supports using an endodontic plugger; and by indirect evaluation of the gap between the oclusal rests and their dental seats through a silicone mold that was further digitized and analyzed by micro-CT. Data were analyzed with Wilcoxon and paired t-tests ($\alpha = 0.05$). Results: No statistically significant ($p>0.05$) differences were found between groups, either with direct or indirect evaluation. Conclusions: CAD-CAM technology is a viable method for fabrication of RPD's metal frameworks.

1 INTRODUCTION

Removable partial dentures (RPD) are considered important alternatives to fixed dentures and the correct fit to the oral support structures is fundamental for their success (Nadin et al., 2020). For decades, the RPD metal framework fabrication has been carried out using the lost wax casting technique. This laboratory procedure is susceptible of technical errors and material distortions, which compromise the fit accuracy of the framework's complex structure (Bajunaid et al., 2019).

Computer-aided design and computer aided-manufacturing (CAD-CAM) technology emerged to overcome those disadvantages (Ahmed et al., 2016). This technology was first described for the production of a RPD metal framework in 2004 by Williams and colleagues. Since then, the improvement of additive techniques and the development of framework design softwares innovated the fabrication of RPD metal framework. However, there are still few clinical studies that evaluate the impact of digital technology on the RPD fit accuracy (Ye et al., 2017).

The objective of this pilot clinical trial was to assess the effect of CAD-CAM framework fabrication on the clinical fit accuracy of RPD to the tooth support structures. The null hypothesis tested was that the fabrication protocol does not influence the clinical fit of RPD ($p>0.05$).

2 MATERIALS AND METHODS

After approval by the Institutional Ethics Committee for Health, eight participants were selected from patients who needed RPD rehabilitation according to the following inclusion

DOI: 10.1201/9781003217152-71

criteria: age over 18 years; at least one premolar or molar as a prosthetic abutment; absence of metal allergy; absence of active dental caries or grade II or III tooth mobility in the prosthetic abutments; absence of other oral lesions. All patients have voluntarily signed the written informed consent agreement and a total of twelve RPD clinical cases were planned.

After treatment planning and tooth preparation, an impression was made to each of the twelve clinical case using alginate impression material (Hydrogum 5, Zhermack GmbH, Deutschland) and individual trays. Impressions were subsequently poured with type IV plaster (Elite Rock, Zhermack GmbH, Deutschland) obtaining the master stone casts. From each master cast, two Cobalt-Chromium RPD frameworks were produced with alternative techniques: digital CAD-CAM protocol (experimental group; n=12) and conventional lost wax casting technique (control group; n=12).

In the experimental CAD-CAM group, the master stone cast was previously digitized in a laboratory scanner (S600 Arti, Zirkonzahn GmbH, Italy). The analysis of the insertion axis and the consequent digital design of the metal framework were performed with the Partial Planner software (Zirkonzahn GmbH, Italy) and the final file was sent for fabrication in a commercial laboratory production center (Sineldent, Spain). The digital built up of the metal framework was carried out in Cobalt-Chromium SP2 alloy (EOS GmbH, Germany) with an EOSINT M270 equipment, using the Direct Metal Laser Sintering (DMLS) technique, and submitted to heat treatment for 45 minutes.

In the control group of lost wax technique the master stone cast was duplicated to a refractory cast and wax patterns were used to reproduce the framework design. Patterns were eliminated in an oven (Infinity L30, Jelrus, USA) and then the Cobalt-Chromium alloy (Remanium G 380+, Dentaurum GmbH & Co. KG, Germany) was injected with an induction casting equipment (Ducatron Quattro, Ugin Dentaire, France).

Both metal frameworks were finished with hand drills and rubbers, followed by immersion in an electrolytic bath (Polytherm compact, Dentaurum GmbH & Co. KG, Germany) for 3 minutes and polished with brushes and paste. Finishing and polishing procedures of the two frameworks for each case were carried out by the same technician, using the same master for proofing.

Fit accuracy of the RPD's metal frameworks was assessed by a trained and previously calibrated investigator, using two methods for measuring thickness of the gap between framework structures and their underlying tooth preparations: a direct and an indirect method.

The direct evaluation was carried out, after the placement of the framework in the oral cavity, based in the assessment of the thickness of the gap between the framework assemblies (occlusal or cingular rests and reciprocal arms) and their tooth support structures using an endodontic plugger 35 (with a 350 micrometers tip). Data was recorded using the number of components, per framework, in which a gap higher than 350μm was detected (Figure 1).

The indirect evaluation was performed with a silicone mold (V-Posil Light Fast set, Voco GmbH, Germany) of the gap between the framework occlusal rests and their respective dental seats. After removal of exceeding material with a scalpel blade, the collected specimens were individually digitized by micro computed tomography (micro-CT) equipment (Skyscan 1174, Bruker, Belgium). The projection images reconstructed with the NRecon software (Bruker, Belgium) were uploaded in the DataViewer software (Bruker, Belgium) and the datasets of the silicone molds were reoriented and 3D registered using their counterpart as reference; the same VOI (Volume of Interest) was defined for reference and target datasets. The average thickness of the selected portion of each specimen was calculated in micrometers, after segmentation, using the CT-Analyzer software (Bruker, Belgium). Final thickness data were obtain by the mean of thickness values of all specimens assessed per framework (Figure 2).

The collected data was statistically analyzed (IBM SPSS, v.25) with Wilcoxon tests for the results obtained by the direct method and with paired t-Student test for the data achieved by the indirect evaluation ($\alpha = 0.05$).

3 RESULTS

The results obtained by direct evaluation are shown in Table 1. No statistical significant differences were found between protocols in occlusal rests (p=0.234), cingular rests (p=0.564) or reciprocal arms (p=0.157) (Table 1). Thickness mean values of the silicone molds were 293.4± 84.12μm for conventional protocol and 325.3±112.14μm in CAD-CAM group. No statistical significantly (p=0.412) differences were found between the two fabrication protocols.

Table 1. Mean (standard deviation) number of gaps found by the direct evaluation of the fit accuracy.

Production method	Oclusal resttts	Cingular rest	Reciprocal arms
Conventional protocol	0.67 (±0.65)	0.5 (±0.54)	0.42 (±0.67)
CAD-CAM	1 (±0.85)	0.38 (±0.52)	0.08 (±0.29)
p value	p=0.234	p=0.564	p=0.157

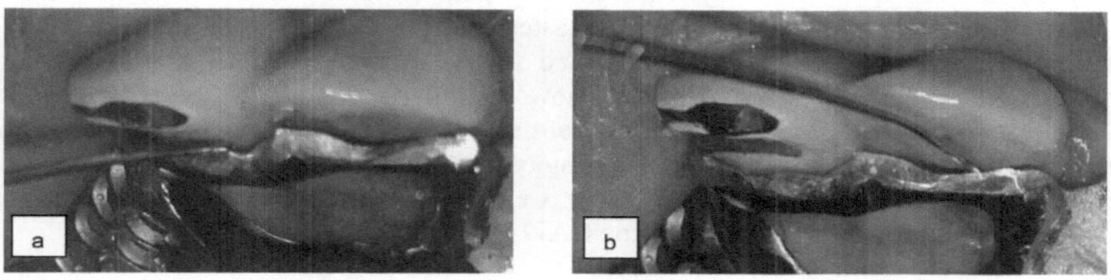

Figure 1. Evaluation of fit accuracy by direct clinical inspection of the gap between surfaces, using an endodontic plugger 35. a- Absence of a gap of 350μm or more in thickness. b- Presence of a gap of 350 μm or more in thickness.

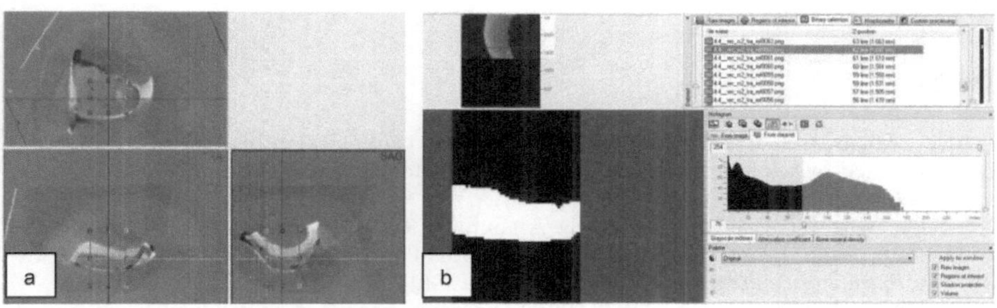

Figure 2. Images of the silicone molds obtained by Micro-CT. a- Overlapping of the datasets, of the two metal frameworks occlusal rests of a dental seat, using the same VOI in the DataViewer. b- Calculation of the average thickness of one silicone mold, after segmentation, in the CT-Analyzer.

4 DISCUSSION

This pilot clinical trial suggests that the CAD-CAM technology used in RPD metal framework production may be a valid alternative to the conventional manufacturing process, since null hypothesis could not be rejected.

One of the most described direct methods to RPD fit accuracy assessment is the visual inspection (Ye et al., 2017). However, due to the complexity of the prosthetic design and to the subjectivity of the human eye evaluation, this method is subject to errors (Ye et al., 2017). For this reason, in our study we used a calibrated instrument to perform a more reproducible identification of gaps, as Keltjens and colleagues (1997) proposed. An endodontic plugger number 35 (with a diameter of 350μm on the tip) allowed to identify gaps close to 311μm, a value that was considered as an acceptable discrepancy due to technical errors (Eggbeer et al., 2005).

Considering the indirect methods of quantitative evaluation of the gap between occlusal rest and the respective dental seat the usage of a silicone mold was considered the most reliable (Ye et al., 2017). In the present study micro-CT was used to quantify the silicone molds thickness. The advantage of this technique over optical microscopy (Ye et al., 2017) and digital microscope measurement (Bajunaid et al., 2019) are sample preservation, higher resolution power and ability to measure the total thickness of the sample (Han et al., 2016).

Thickness mean values of the gaps evaluated with the indirect method was similar to the observed in a previous laboratory study (Bajunaid et al., 2019), where there were also no differences in the fit accuracy of RPD metal frameworks produced with CAD-CAM or conventional technique. However, some differences were found, in a clinical study (Ye et al., 2017), between RPD metal framework produced by CAD-CAM and those produced by conventional protocol. These results can be explained not only by the use of a different quantitative assessment method, as already exposed, but also by the differences in the production protocol. In the present study, to avoid different distortions we used the same master cast for the two production methods, in opposition of the usage of two master casts by Ye and colleagues (2017). Both studies used the same production equipment (EOSINT M270), however different Cobalt-Chromium alloys were used. Instead of Wirebond C+ chosen by the other authors, our study used the EOS SP2, known to create structures of high density, strength and biocompatibility (Ahmed et al., 2016).

Although clinical investigations on this topic are complex, further studies with larger samples are needed to confirm the viability of the CAD-CAM method.

5 CONCLUSION

Considering the obtained results, and other advantages of the digital fabrication, this study suggests that the CAD-CAM protocol is a viable method for RPD's metal framework production.

REFERENCES

Ahmed, M.F., Abbas, F.S. & Omar, S.S. 2016. Clinical and cytological evaluation of removable partial denture fabricated by selective laser sintering additice prototyping technique. *Alexandria Dental Journal* 41: 42–49.

Bajunaid, S.O., Altwaim, B., Alhassan, M. & Alammari, R. 2019. The Fit Accuracy of Removable Partial Denture Metal Frameworks Using Conventional and 3D Printed Techniques: An In Vitro Study. *The Journal of Contemporary Dental Practice* 20(4): 476–481.

Eggbeer, D., Bibb, R.J. & Williams, R.J. 2005. The computer-aided design and rapid prototyping fabrication of removable partial denture frameworks. *Professional Engineering Publishing* 219(3): 195–202.

Han, S.-H., Sadr, A., Tagami, J. & Park, S.-H. 2016. Non-destructive evaluation of an internal adaptation of resin composite restoration with swept-source optical coherence tomography and micro-CT. *Dental Materials* 32(1): e1–7.

Keltjens, H.M., Mulder, J., Käyser, A.F. & Creugers, N.H. 1997. Fit of direct retainers in removable partial dentures after 8. *Journal of Oral Rehabilitation* 24: 138–142.

Nadin, A.H., Mutlu, Ö., Martin, S. & Samir, A. 2020. A digital cast-free clinical workflow for oral rehabilitation with removable partial dentures: A dental technique. *Journal of Prosthethic Dentistry* (5): 680–685.

Williams, R.J., Bibb, R. & Rafik, T. 2004. A technique for fabricating patterns for removable partial denture frameworks using digitized casts and electronic surveying. *J Prosthet Dent* 91: 85–8.

Ye, H. et al. 2017. Preliminary clinical application of removable partial denture frameworks fabricated using computeraided design and rapid prototyping techniques. *The International Journal Prosthodontics* 30: 348–353.

Advances and Current Trends in Biomechanics – Belinha et al. (Eds)

Creation of a 3D printing protocol for the unlimbited arm prosthesis

T.S. Bina, M.E. Kunkel, E.G.R. dos Anjos, R.C. Ribeiro, T.V. Ribeiro & H.D. Silveira
Mao3D, Institute of Science and Technology, Federal University of Sao Paulo, São José dos Campos, Brazil

ABSTRACT: The amputation of a limb leads to difficulties and limitations in daily activities. Mao3D is a social project in São Paulo, Brazil, that produces and donates 3D-printed mechanical prostheses to promote more autonomy to people with upper limb differences. The production of 3D prostheses in Mao3D is made from open source models through fused and deposition modeling (FDM) process of additive manufacturing. Many parameters of this process may generate printing errors and low quality of the manufactured pieces. In this research, the manufacturing of the Unlimbited Arm v2.1 prosthesis using the FDM process and polylactic acid (PLA) filament was carried out. This defined the most appropriate 3D printing configurations and assisted in the creation of a 3D printing quality control protocol. The protocol is available on the website www.mao3d.com.br to assist in producing this 3D prosthesis model of good printing quality.

1 INTRODUCTION

Amputation is the total or partial removal of a body part due to treatment of diseases and accidents or due to genetic factors or even problems during embryonic formation. Prostheses are devices prescribed to replace lost limbs. However, the high costs and the adaptation difficulty limit the improvement in the life quality of low-income people with disabilities (Tanaka & Lightdale-Miric, 2016). Additive manufacturing technology has been used to produce accessible, lightweight, and low-cost prostheses with new designs. This technology allows the production of 3D objects through the material layers deposition, making it possible to create customized low-cost prostheses with adaptable parts of different sizes (Zuninga, et al, 2015).

The manufacturing process of a 3D prosthesis is initiated with the 3D design created in computer-aided design (CAD) software. This design is transformed into a Standard Tessellation Language (STL) file, which contains the data that describes the layout of the 3D object. In order for the STL file to be ready for the 3D printer, it is necessary to send it to a slicing software where the manufacturing parameters used are defined. Examples of these parameters are the position of the part on the print table, scale determination, planning of the contour and filling trajectory and printing time. These configurations may vary according to the part of the prosthesis to be produced. With the settings defined, a software generates a G-code file, the programming language of 3D printers that contains all the instructions necessary for the printer to manufacture the part (Mwema & Akinlabi, 2020). The slicing software's main printing parameters are layer height or thickness, build orientation, raster angle or raster orientation, infill pattern, infill density and infill percentage, perimeters shells, top solid layers, and bottom solid layers.

e-Nable is a global on-line community that provides open-source files for additive manufacturing of upper limb prostheses for free through fused and deposition modeling (FDM)

DOI: 10.1201/9781003217152-72

process (Schull, 2015). Mao3D is a Brazilian social impact initiative that prints 3D prostheses, free of charge for those who need them, using open-source models from e-Nable (Kunkel et al, 2018; Bina et al, 2020). The FDM process using polylactic acid (PLA) filament in low-cost 3D printers requires knowledge of specific parameters to avoid the generation of printing errors and low-quality parts. This research was carried out to analyze the manufacture of prosthetic parts of the Unlimbited Arm.v2.1 model to define the most appropriate printing configurations and creation of a 3D printing quality protocol.

2 MATERIAL AND METHODS

The Unlimbited Arms v2.1 prosthesis, the model most used in the Mao3D, is a mechanical prosthesis actioned by the elbow movement, indicated for cases of amputation or malformation of the transradial region (MakerBot Thingiverse, 2020). This model consists of 32 pieces grouped in six sets for the 3D printing manufacturing process (Figure 1).

For the quality analysis of prothesis' manufactured parts, each part of the Unlimbited Arm model was printed eight times using the FDM process on the Stella 2 3D printer (Boa Impressão 3D, Brazil) and PLA filaments. The common errors that occur during the manufacture of the parts and affect the prosthesis quality and structure were analyzed. The most common errors are (A) Not sticking to the bed, (B) Warping, (C) Under-extrusion, (D) Layer separation and splitting and (E) Poor bridging. Some of these problems were solved with manual actions directly on the 3D printer during the production. Some manual solutions were tested to prevent these errors from occurring. Other problems are solved when the printing parameters are defined in slicing software. The parameters used in the analysis are related to the support configuration, internal fill, infill percentage, layer configurations (bottom, top and perimeter), and print positions of the pieces considering the X, Y, and Z axes (Figure 2). Each manufactured part's print quality was analyzed, a quality standard was defined, and a detailed print quality protocol for each part of the Unlimbited Arm v2.1 prosthesis was created.

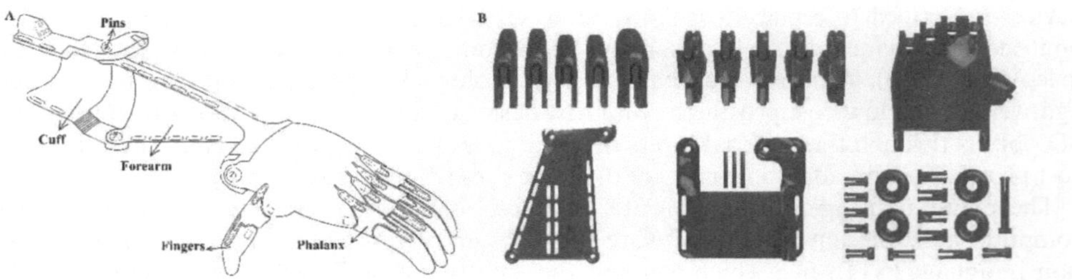

Figure 1. The upper limb mechanical prosthesis Unlimbited Arm v2.1 (A) assembled prosthesis; and (B) sets of pieces grouped for the 3D printing manufacturing process.

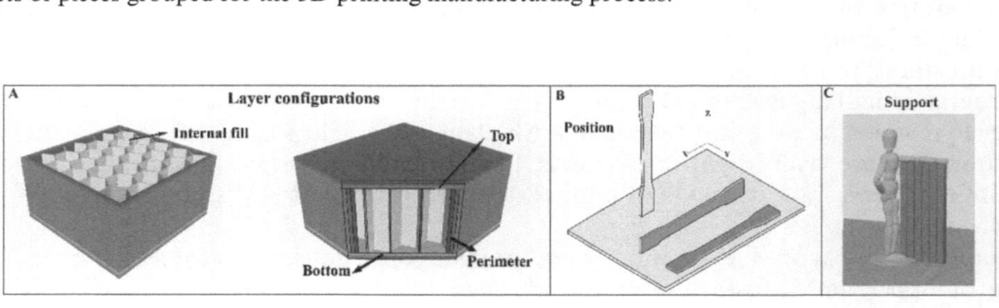

Figure 2. Printing parameters defined in slicing software. (A) Layer configuration; (B) Position; and (C) Support.

3 RESULTS

In the analysis, the printed parts of the Unlimbited Arm v2.1 prosthesis with the best quality control were selected and the parameters used in printing them were added to the protocol. (Figure 3). The print quality protocol is available in Portuguese and English language on the Mao3D website (Mao3D, 2020) (Figure 4).

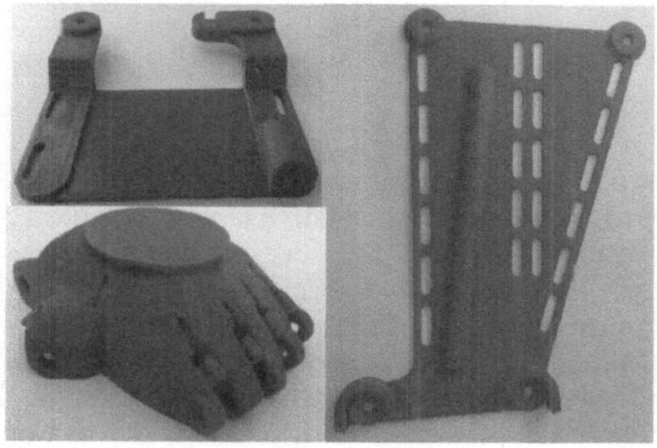

Figure 3. Main parts of the Unlimbited prosthesis model printed using the FDM process that achieved good quality.

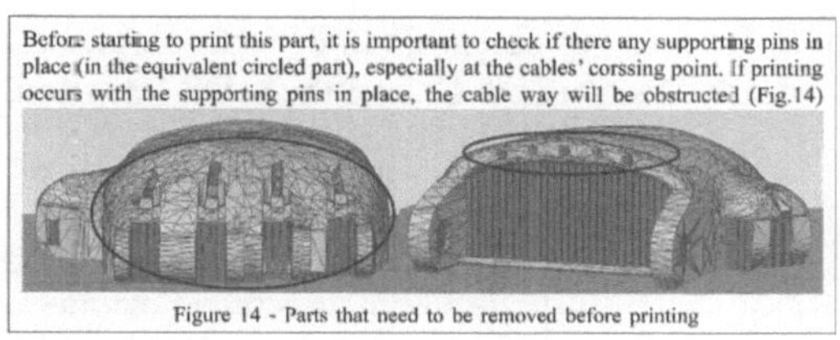

Before starting to print this part, it is important to check if there any supporting pins in place (in the equivalent circled part), especially at the cables' corssing point. If printing occurs with the supporting pins in place, the cable way will be obstructed (Fig.14)

Figure 14 - Parts that need to be removed before printing

Figure 4. Image of the created 3D printing quality protocol.

4 DISCUSSION

During the FDM process, the PLA material is deposited on the printing bed in layers. The way these layers are deposited is defined as external or internal fill, one of the 3D printing parameters defined in the slicing software. In the external fill, the variation occurs mainly for the design of the part externally, the most used design is rectilinear. At the internal fill, it is possible to choose how the material will be deposited. The most well-known forms are honeycomb, rectilinear, triangle, wiggle, and grid. These fill forms can influence the part's resistance and can vary from 0 to 100% (Dey & Yodo, 2019). The quality of a 3D printed piece depends on the maintenance of the equipment, quality of the material and STL file, environmental factors, and the 3D printer parameters (Mohamed et al, 2015). Our research showed that the choice of parameters is extremely important in 3D printing, as any parameter can significantly change the quality and properties mechanics of the produced part. Sheoran & Kumar (2020) have explored and utilized various experimental or statistical Design of Experiment (DOE) techniques for optimizing the FDM process parameters to improve the mechanical properties or part quality or both. They concluded that the parameters used influence efficiency, properties, and quality of the printed product. In the literature, some process parameters such as layer thickness, raster angle and direction construction are much more explored than others, such as fill density and number of perimeters. In addition to the printing parameters, other factors influence the FDM process, such as 3D printer maintenance, filament quality, 3D model quality, and environmental factors. The protocol developed in this research presenting the steps to printing each piece of the Unlimbited prosthesis with the illustration of errors and printed piece with good quality. The protocol aims assist in the producing this prosthesis model, resulting in a good print quality, good mechanical resistance, and functionality to assists in the user's rehabilitation process. The printing quality also results in the prosthesis' esthetics, which influences the user's acceptance process. The quality protocol fits into an essential control item for a production system.

5 CONCLUSION

Nowadays, it is common to find people who have low-cost 3D printers, some of them work in projects to produce 3D prostheses or other assistive technology devices to assist people with disabilities. The analysis of the manufacturing process of the prosthesis Unlimbited Arm.v2.1 allowed the definition of the most appropriate printing configurations. The 3D printing quality protocol produced from this analysis will be useful to assist those who wish to help social projects as e-Nable and Mao3D providing a better esthetic standard of 3D upper limb prostheses. In addition to allowing anyone who owns a 3D printer, including in hospitals or rehabilitation centers, to print 3D prostheses with good quality. It may allow the production of a more functional and resistant prosthesis for the user, improving its rehabilitation and acceptance process of the prosthesis.

ACKNOWLEDGMENT

The authors thank National Council for Scientific and Technological Development, the International Development Research Centre (IDRC) and the GDS Program.

REFERENCES

Bina, T.S., Kunkel, M. E., Ribeiro, et al. 2020. Produção de próteses mecânicas 3D de membro superior para um caso de amputação bilateral infantil: desafios da reabilitação. In: FCCRC Ferrari. (eds), Processos de Intervenção em Fisioterapia e Terapia Ocupacional 2. 1ed. Ponta Grossa: *Atena*, 1, 185–98.

Dey, A. & Yodo, N. 2019. A systematic survey of FDM process parameter optimization and influence on part characteristics. *Journal Manufacturing and Materials Processing*, 3(3), 64.

Kunkel, M.E. Gonçalves, I.T., de Pinheiro, L.M, et al. 2019. Mao3D-Protetização e reabilitação de membro superior adulto com a tecnologia de impressão 3D. In: A.C. Oliveira (ed), A Produção do Conhecimento na Engenharia Biomédica. 1ed. Ponta Grossa: *Atena*, 1, 14–29.

MakerBot Thingiverse. The Unlimbited Arm v2.1 – Alfie Edition by Team Unlimbited. Available in: https://www.thingiverse.com/thing:1672381. Accessed on March 19, 2021.

Mao3D. 2020. 3D printing protocol of upper limb protheses. Available in: https://www.mao3d.com.br/. Accessed on March 19, 2021.

Mohamed, O. A., Masood, S. H. & Bhowmik, J. L. 2015. Optimization of fused deposition modeling process parameters: a review of research and future prospects. *Advances in Manufacturing*, 3(1), 42–53.

Mwema, F.M. & Akinlabi, E.T. 2020. Basics of Fused Deposition Modelling (FDM). In Fused Deposition Modeling (1–15). *Springer*, Cham.

Schull, J. (2015, October). Enabling the future: Crowdsourced 3D-printed prosthetics as a model for open source assistive Technology Innovation and mutual aid. In *Proceedings of the 17th International ACM SIGACCESS conference on computers & accessibility* (pp. 1–1).

Tanaka, K.S. & Lightdale-Miric, N. 2016. Advances in 3D-printed pediatric prostheses for upper extremity differences. *The Journal of Bone and Joint Surgery*, 98(15),1320–1326.

Zuniga, J., Katsavelis, D., Peck, J., Stollberg, J., Petrykowski, M., Carson, A. & Fernandez, C. 2015. Cyborg beast: a low-cost 3d-printed prosthetic hand for children with upper-limb differences. *BMC research notes*, 8(1),1–9.

Advances and Current Trends in Biomechanics – Belinha et al. (Eds)

Fit accuracy assessment of RPD metal framework by digital superimposition

P.R. Conceição, R. Pinto, D. Marques, J. Portugal & C.B. Neves
UICOB, Faculty of Dental Medicine, Universidade de Lisboa, Lisboa, Portugal

M. Franco & N. Alves
CDRSP, Instituto Politécnico de Leiria, Marinha Grande, Portugal

ABSTRACT: Aim: To evaluate the validity of an innovative digital method to assess the fit accuracy of removable partial denture frameworks. Methods: Fit accuracy of seven removable partial denture's metal frameworks was assessed through the evaluation of the gap between the framework occlusal rests and the respective master cast. Digitization of the framework and the respective master cast was performed followed by digital superimposition and direct measurement with Geomagic ControlX (experimental group; n=7). Also, an indirect evaluation of the thickness of the gap was performed through a silicone mold analyzed by micro-CT (control group; n=7). Wilcoxon test and the Spearman correlation test were performed (α=0,05). Results: No statistically significant (p=0.866) differences were observed between the two methods and the correlation found was moderately positive (ρ=0.607). Conclusions: The digital superimposition emerges as a valid direct method of fit accuracy assessment.

1 INTRODUCTION

Removable partial dentures (RPD) allow replacement of teeth loss with an excellent cost-benefit ratio and improve the quality of life of patients by restoring masticatory, phonetic and aesthetic functions (Bajunaid et al., 2019). However, a correct fit to the oral support structures is essential for success, preventing not only denture deformation and fracture but also oral lesions like ulcerations, teeth wear and periodontal disease (Schweiger et al., 2020).

Over the last decades, several methods for assessing the fit accuracy have been described, with emphasis on those that focus on occlusal rests, due to their importance in the stability and function of prostheses (Tasaka et al., 2020). Recently, in terms of biomedical measurement, micro computed tomography (micro-CT) has replaced some analog techniques for its simplicity and preservation of the samples (Ram et al., 2019). In turn, the development of digital technologies, namely those of computer-aided engineering (CAE) has also introduced new methods for the quality control of RPD production, making it possible to identify clinical imperceptible problems and to overcome limitations of analog methods (Soltanzadeh et al., 2019). However, due to the complexity of the prosthetic design, studies that quantitatively assess the fit accuracy of metal framework to oral support structures are still scarce (Arnold et al., 2018).

DOI: 10.1201/9781003217152-73

This laboratorial pilot study aims to evaluate the validity of an innovative digital method to assess the fit accuracy of RPD framework according to the following null hypothesis: the method of evaluation does not influence the fit accuracy assessment.

2 MATERIALS AND METHODS

Seven Cobalt-Chromium RPD frameworks and respective master stones casts were selected according to the following inclusion criteria: frameworks extended through the two quadrants of the dental arch and have at least one occlusal rest in each quadrant. The fit accuracy of the framework occlusal rests was evaluated by measuring total thickness of the the gap between the occlusal rest and the respective master cast seats using two methods by one researcher, previously submitted to calibration.

The direct method of fit accuracy assessment (experimental group, n=7) consisted in the individual digitization of the metal framework followed by the individual digitization of the corresponding master cast using a laboratory scanner (Arti S600, Zirkonzahn GmbH, Italy). The framework was placed in the master cast and digitization of this set was also performed. The digital superimposition was performed with Geomagic ControlX software (3D System GmbH, EUA) using as reference the image of the digitization of the set. The transform alignment and best fit tools were used to correctly positioned each individual image to ascertain the virtual gap between the framework oclusal rests and the cast seats.

With the "plane" tool of software, and considering the shape of each occlusal rest, a grid of ten equidistant virtual sagittal planes was created. The intersection of these planes resulted in nine interior lines that in turn intersected the surface of the occlusal rest at nine points and the surface of the dental seat at another nine points. The mean length of the nine segments obtained was defined as the result for each occlusal rest and the mean of the oclusal rests defined the value for each framework (Figure 1). Results were depicted as mean ± standard deviation.

In the indirect method of fit accuracy assessment (control group, n=7) a silicone mold (V-Posil Light, Voco GmbH, Germany) was created in the gap between the framework occlusal rest and cast seat. Each silicone specimen was then digitized by micro-CT (Skyscan 1174, Bruker, Belgium). After the reconstruction of the projection images with the NRecon software (Bruker, Belgium) and the spatial orientation made with DataViewer software (Bruker, Belgium), a VOI (Volume of Interest) was defined in each specimen according to the area of the points defined in the ControlX (Figure 2). The mean thickness of each silicone molds was calculated in micrometers, after segmentation, using the CT-Analyzer software (Bruker, Belgium). The final data per framework was based in the mean of thickness values of all silicone molds of each framework.

The collected data was analyzed (IBM SPSS, v.25) using the Wilcoxon test ($\alpha = 0.05$) and the Spearman correlation test with analysis of the scatter plot and the box plot.

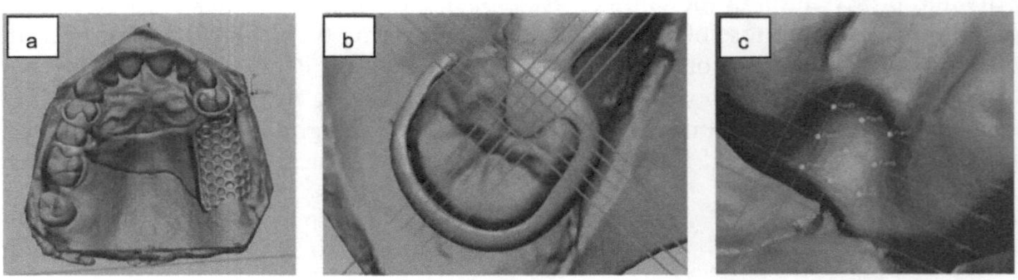

Figure 1. Direct evaluation method by digital superimposition using Geomagic ControlX. a- Final superimposition between the individual images of the master cast and the metal framework. b- Grid resulting from the intersection of the built planes. c- Points projected on the surface of the dental seat by the intersecting lines of the created planes.

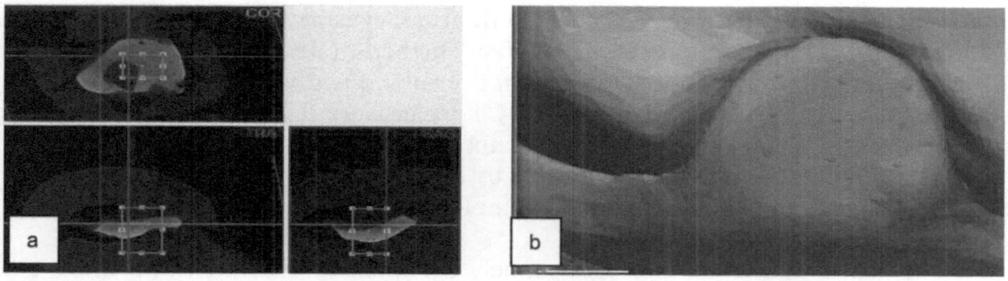

Figure 2. Selection of a Volume of interest in a virtual image of a silicone molds digitized by micro-CT (a) according to the area of the nine points defined in Geomagic ControlX (b).

3 RESULTS

The average thickness between the occlusal rests and the cast seats was 238.7±138.09μm by the direct method of digital superimposition and 253.9±102.57μm by the indirect method of micro-CT (Figure 3a). No statistically significant (p=0.866) differences were found between methods and a moderate positive correlation was verified (ρ = 0.607) (Figure 3b).

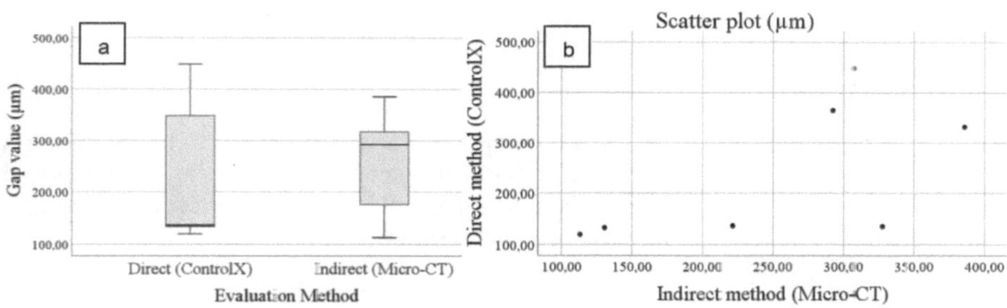

Figure 3. a- Box plots of the gap (in micrometers) by the two evaluation methods under study, with the mean and the interquartile range of data. b- Scatter plot of the mean distances values (in micrometers) of all oclusal rests in each of the seven metal frameworks, to the corresponding dental seat.

4 DISCUSSION

This laboratorial pilot study suggests that the proposed direct digital assessment of the gap between occlusal rests and cast seats is a valid alternative method to assess the fit accuracy of RPD metal framework, since the null hypothesis could not be rejected.

The evaluation of the RPD framework fit accuracy to the oral structures is a fundamental step to ensure prosthetic function and patient's satisfaction. Although there are several laboratorial methods described to assess the RPD fit accuracy, no method is considered as gold-standard (Ye et al., 2017) because the application of each one has advantages and disadvantages which must be contemplated according to each situation. Quantitative assessment of fit accuracy allows capturing much more detailed information than qualitative dichotomy measures, although its application is more complex (Ye et al., 2017).

Methods for quantitative assessment of RPD fit accuracy can be classified according to the information being collected in direct or indirect methods (Ye et al., 2017). The indirect method uses negative substrates of the gap that are subsequently measured (Bajunaid et al., 2019). In the present study, by applying micro-CT to measure the silicone molds, it was possible to overcome the inconveniences of other microscopy techniques. Micro-CT techniques allow measuring the entire specimen contrary to the method applied by Bajunaid and colleagues (2019) as well as preserve the silicone mold integrity as it does not require cuts (Ye et al., 2017).

The development of digital technologies, namely CAE, has revolutionized the direct quantitative assessment of gaps by preserving the evaluated structures and avoiding the cuts implied in the analog techniques (Arnold et al., 2018). The Geomagic Control is a certified software that allows digital superimposition and measurement, being already used in other areas of dentistry (Marques et al., 2019). The superimposition technique exposed in our study is similar to that described in the study by Soltanzadeh and colleagues (2019) as they also used a version of the software to measure gaps after digital superimposition. However, our method differs in the measurement mode, since instead of being applied to the entire structure, is applied to occlusal rests. This possibility, focusing on a specific prosthetic component, is a novelty.

Our study showed that there were no statistically significant differences between the two methods of assessment. Despite higher resolution power (Ram et al. 2019), Micro CT method requires high financial investments that are not compatible with everyday framework quality assessment. The method of superimposition and digital measurement is more accessible to laboratories and many already have similar technology used in our study.

Further studies with bigger sample size are needed to prove the validity of the proposed innovative direct quantitative method for assessing the metal frameworks fit accuracy.

5 CONCLUSION

Considering the obtained results, this study suggests that the innovative digital superimposition and measurement, is a valid method of RPD fit accuracy assessment. This method can be a considerable tool in the production control of the RPD metal frameworks, available to dental laboratories, in order to ensure the prosthetic function.

REFERENCES

Arnold, C., Hey, J., Schweyen, R. & Setz, J.M. 2018. Accuracy of CAD-CAM-fabricated removable partial dentures. *The Journal of Prosthetic Dentistry* 119(4): 586–592.

Bajunaid, S.O., Altwaim, B., Alhassan, M. & Alammari, R. 2019. The Fit Accuracy of Removable Partial Denture Metal Frameworks Using Conventional and 3D Printed Techniques: An In Vitro Study. *The Journal of Contemporary Dental Practice* 20(4): 476–481.

Marques, N. et al. 2019. Soft tissue replication in single unit implant impressions—A three dimensional clinical study. *J Esthet Restor Dent* 31(4): 359–368.

Ram, S.M., Ranadive, N.M. & Nadgere, J.B. 2019. Microcomputed tomography a noninvasive method to evaluate the fit of a restoration as compared to conventional replica technique. *The Journal of Indian Prosthodontic Society* 19: 233–239.

Schweiger, J. et al. 2020. Internal porosities, retentive force, and survival of cobalt–chromium alloy clasps fabricated by selective laser-sintering. *Journal of Prosthodontic Research* 64(2):210–216.

Soltanzadeh, P. et al. 2019. An in vitro investigation of accuracy and fit of conventional and CAD/CAM removable partial denture frameworks. *Journal of Prosthodontics* 28: 547–555.

Tasaka, A. et al. 2020. Accuracy of removable partial denture framework fabricated by casting with a 3D printed pattern and selective laser sintering. *Journal of Prosthodontic Research*, 64(2): 224–230.

Ye, H. et al. 2017. Preliminary clinical application of removable partial denture frameworks fabricated using computeraided design and rapid prototyping techniques. *The International Journal Prosthodontics* 30: 348–353.

Advances and Current Trends in Biomechanics – Belinha et al. (Eds)

Analysing the structural performance of dental restorations using meshless methods

G.A.R. Caldas
Institute of Science and Innovation in Mechanical and Industrial Engineering (INEGI), Portugal

Jorge Belinha
Mechanical Engineering Department, School of Engineering, Polytechnic of Porto (ISEP), Portugal

J.G. Mendes & Renato Natal Jorge
Mechanical Engineering Department, Faculty of Engineering of the University of Porto (FEUP), Portugal

ABSTRACT: In this work adhesive dental bridges are analysed, with the main objective of study the influence of the bridge design, adhesive material, and adhesive thickness on the mechanical resistance of the bridge. Elasto-static numerical analyses were performed and for the structural analysis, three discrete numerical techniques were used: Finite Element Method (FEM), Radial Point Interpolation Method (RPIM) and Natural Neighbour Radial Point Interpolation Method (NNRPIM). With the data obtained was concluded that the use of two retainer wings, each one with 0.1 mm of thickness, appears to be the most favourable combination of parameters. However, the effect of the resin-cement was unclear. In addition, the results obtained using meshless methods are in agreement with the FEM.

1 INTRODUCTION

Adhesive bridges are a type of fixed prosthesis in which the pontic tooth (replaced tooth) is attached to a structure (retainer wing) that is cemented on the abutment teeth (patient's natural teeth) (Walmsley et al., 2007). With this type of restoration, the replaced teeth are sealed primarily to enamel, and therefore the fixed dental prosthesis is supported by natural teeth. This type of restoration distinguishes from the others because it is a conservative method and it is reversible, once it requires minimal tooth preparation and does not compromise the abutment tooth (Walmsley et al., 2007) (Durey, Nixon, Robinson, & Chan, 2011). However, there are still some concerns in using these bridges, which require a more complex geometry design and have a low retention rate. This low retention rate between the retainer wing and the abutment tooth leads to debonding, which is the main cause of failure of this type of dental restoration (Śmielak, Świniarski, Wołowiec-Korecka, & Klimek, 2016) (Vallittu & Sevelius, 2000).

2 MATERIALS AND METHODS

This work proposes to study the biomechanical structural response of an adhesive dental bridge using advanced discretization computational techniques, such as meshless methods, which will allow to show that meshless methods are a solid alternative numerical technique. Also, it was studied the effect of three different factors (adhesive's thickness, bridge design and adhesive material) on the mechanical resistance of a resin-bonded dental bridge.

DOI: 10.1201/9781003217152-74

Table 1. Mechanical properties (Cornacchia, Las Casas, Cimini, & Peixoto, 2010) (Moreira, Belinha, Dinis, & Jorge, 2014) (Della Bona, Donassollo, Demarco, Barrett, & Mecholsky, 2007) (Manicone, Rossi Iommetti, & Raffaelli, 2007).

	Zirconia	Admira (A)	Brilliant Coltene (B)	NC Coltène (NC)
Young's Modulus (E) [GPa]	245	3.6	2.3	2.3
Poisson's Ratio (υ)	0.26	0.3	0.3	0.3

Three different adhesive's materials were tested: Admira Fusion VOCO, Brilliant COLTÈNE and NC COLTÈNE. For the pontic tooth it was assigned the material Zirconia. The materials considered as elastic, homogeneous and isotropic linear behaviour. The mechanical properties are given in Table 1.

Three numerical methods were used: FEM, RPIM and NNRPIM. Thus, the results of all methods were compared, and their performance was evaluated. The 2D model, constructed based on an orthopantomography of an unknown patient, consisted of a mandibular central incisor and canine as abutment teeth, and lateral incisor as pontic tooth. Wings were built on each side of the pontic tooth, thus simulating the presence of the adhesive. In this model, each wing was divided into four parts, each one with 0.1 mm. Thus, the adhesive's thickness was varied from 0.1 mm to 0.4 mm. To simulate the presence of only one wing, to the right-side wing was assigned the following properties: E=0.0841 MPa and υ=0.3.

Therefore, with this generic model and these variations, it was possible to simulate eight different models. As a natural boundary condition, it was assumed a 1N load, with 90° of inclination with the longitudinal axis of the tooth. The bottom nodes of the central incisor and the canine were constrained in both Ox and Oy directions.

Based on the results obtained on the 2D study and due to computational limitations, a 3D model of the restored tooth and the adhesive bridge was built in a CAD software, considering a constant and equal to 0.4 mm adhesive's thickness and the two-retainer design. Therefore, the focus of this 3D study was the resin-cement.

The numerical model was constructed with FEMAP and analysed with FEMAS - Finite Element and Meshless Analysis Software, which is an academic software capable to perform several kinds of computational mechanics analysis, using both the FEM and meshless methods (more details in cmech.webs.com). The final model is represented in Figure 1, shown from a top view. As essential boundary conditions, the nodes marked in red are constrained in Ox, the nodes marked in yellow are constrained in Oy and Ox and the nodes marked in green are constrained in Oy. It was assumed that there was a perfect mechanical and chemical bond between the adhesive and the abutment, hence there are also nodes marked in white, which are constrained in all directions. As a natural boundary condition, it was considered a 1 N load, applied in the z direction.

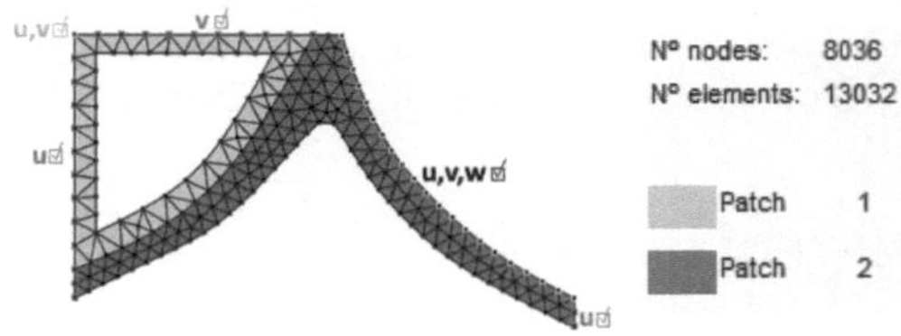

Figure 1. Representation of the global 3D generic model from top view.

3 RESULTS AND DISCUSSION

3.1 *2D study of structural response of an adhesive dental bridge*

In Figure 2 are represented the colour distribution maps of principal stress σ1 for 0.4 mm of adhesive's thickness, the resin-cement Admira and numerical method FEM. As it can be verified, when using the single-retainer design, the maximum principal stress obtained is ten times higher than the one obtained when using the two-retainer design. Moreover, for both designs the higher stresses were found in the interface between the adhesive and the pontic.

It was evaluated the total displacement of a point located on the lower centre part of the pontic tooth. These data are presented in Figure 3 and it can be verified that the total displacement is higher for the single-retainer design, for the 0.4 mm thickness and for the adhesive Brilliant and NC Coltene. The results for the three methods are very similar, being FEM and NNRPIM the closest.

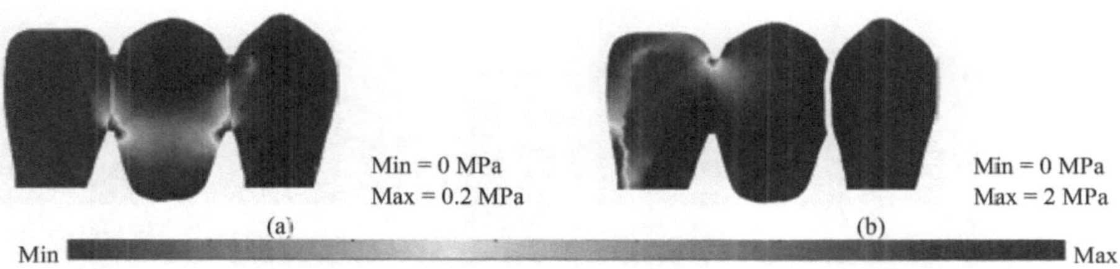

Figure 2. Principal stress σ1 map for Admira, FEM and: (a) two-retainer and (b) single-retainer design.

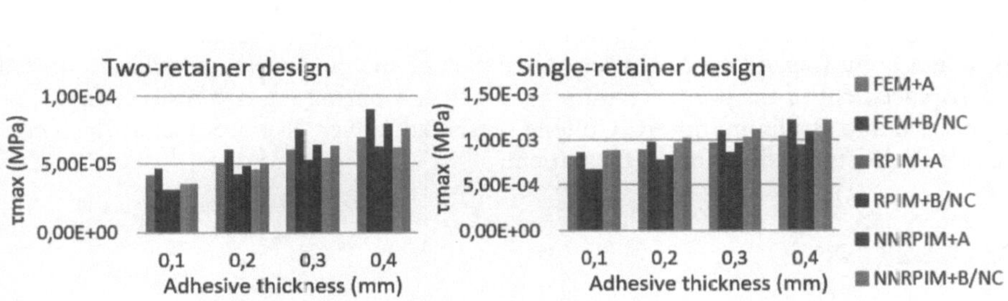

Figure 3. Histograms of total displacement for a node in the inferior portion of the pontic.

3.2 *3D study of the structural response of an adhesive dental bridge*

In Figure 4 are represented the colour distribution maps of principal stress σ1 and shear stress τ_{zx} for resin-cement Admira and FEM. Is possible to observe that the area of the adhesive that is at risk is the one between the abutment and the pontic, and the area of the adhesive in contact with the pontic. The adhesive in contact with the abutment does not exhibit high stresses due to the assumptions made that the adhesive is perfectly bonded to the abutment tooth. Also in Figure 4 are represented the principal stress σ1 for all resin-cements and numerical

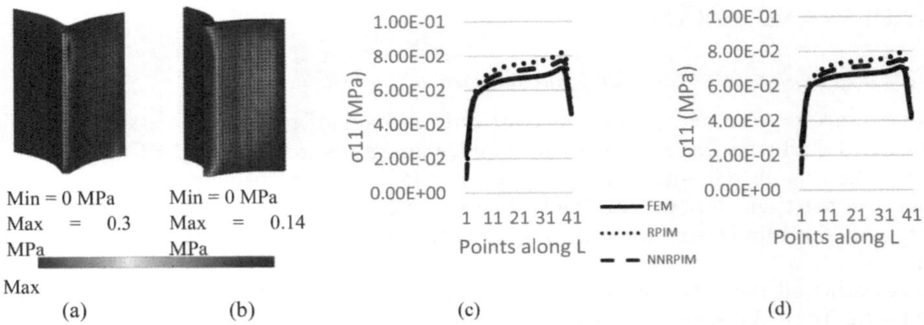

Figure 4. Colour distribution maps of principal stress σ1 (a) and shear stress τ_{zx} (b) for FEM and resin cement Admira. Graphic representation of principal stress σ1 for Admira (c) and Brilliant/NC Coltène (d).

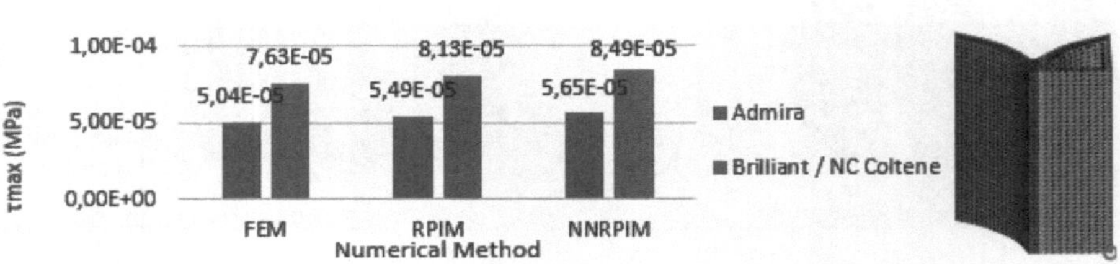

Figure 5. Histogram representation of total displacement.

methods, exanimated in a line at the interior of the adhesive. Is possible to verify that with adhesive Brilliant/NC Coltène are obtained slightly lower tensile stresses.

Analysing a point located on the lower centre part of the pontic was possible to investigate the total displacement of the pontic (Figure 5). When comparing the two resin-cements, is possible to observe that Brilliant and NC Coltène presented higher displacements, which can represent a less comfortable situation to the patient.

4 CONCLUSIONS

It was possible to observe that the area of the adhesive that is at risk is the one between the abutment and the pontic, which can indicate the initial point of crack propagation, leading to breakage of the adhesive and consequently the entire dental prosthesis. The performed elastostatic analysis shown that the single-retainer design increases the risk of debonding, and the adhesive's thickness of 0.1 mm appears to be the most favourable. Regarding the resin-cement materials tested, despite Brilliant and NC Coltène presented lower tensile stresses, which could indicate lower probability to occur debonding, they also presented higher displacements of the pontic, which can represent a less comfortable situation to the patient. Therefore, at this point all resin-cement present as valid solutions. In addition, the results obtained using meshless methods agree with FEM.

ACKNOWLEDGEMENTS

The authors truly acknowledge the funding provided by Ministério da Ciência, Tecnologia e Ensino Superior –Fundação para a Ciência e a Tecnologia(Portugal), under the funding of LAETA-project UIDB/50022/2020,and project "SIMPLIFIED: Easy Tooth Abutment" – POCI-01-0247-FEDER-017982, co-funded by FEDER through COMPETE20220.

REFERENCES

Cornacchia, T. P. M., Las Casas, E. B., Cimini, C. A., & Peixoto, R. G. (2010). 3D finite element analysis on esthetic indirect dental restorations under thermal and mechanical loading. Medical & Biological Engineering & Computing, 48(11), 1107–1113. http://dx.doi.org/10.1007/s11517-010-0661-7

Della Bona, A., Donassollo, T. A., Demarco, F. F., Barrett, A. A., & Mecholsky, J. J. (2007). Characterization and surface treatment effects on topography of a glass-infiltrated alumina/zirconia-reinforced ceramic. Dental Materials, 23(6), 769–775. http://dx.doi.org/10.1016/j.dental.2006 06.043

Durey, K. A., Nixon, P. J., Robinson, S., & Chan, M. F. W. Y. (2011). Resin bonded bridges: Techniques for success. British Dental Journal, 211(3), 113–118. http://dx.doi.org/10.1038/sj.bdj.2011.619

Manicone, P. F., Rossi Iommetti, P., & Raffaelli, L. (2007). An overview of zirconia ceramics: Basic properties and clinical applications. Journal of Dentistry, 35(11), 819–826. http://dx.doi.org/10.1016/j.jdent.2007.07.008

Moreira, S. F., Belinha, J., Dinis, L. M. J. S., & Jorge, R. M. N. (2014). A global numerical analysis of the"central incisor/local maxillary bone" system using a meshless method. Molecular & Cellular Biomechanics : MCB, 11(3), 151–184. http://dx.doi.org/10.3970/mcb.2014.011.151

Śmielak, B., Świniarski, J., Wołowiec-Korecka, E., & Klimek, L. (2016). 2D-finite element analysis of inlay-, onlay bridges with using various materials. International Scientific Journal, 79(2), 71–78.

Vallittu, P. K., & Sevelius, C. (2000). Resin-bonded, glass fiber-reinforced composite fixed partial dentures: a clinical study. The Journal of Prosthetic Dentistry, 84(4), 413–418. http://dx.doi.org/10.1067/mpr.2000.109782

Walmsley, A. D., Walsh, T. F., Lumley, P., Burke, F. J. T., Shortall, A. C., Hayes-Hall, R., & Pretty, I. (2007). Restorative Dentistry. (Elsevier, Ed.) (Second) Edi. Churchill Livingstone. Retrieved from https://www.sciencedirect.com/science/book/9780443102462

Advances and Current Trends in Biomechanics – Belinha et al. (Eds)
© 2022 Copyright the Author(s), ISBN: 978-1-032-10806-3

Elasto-plastic analysis of dental restorations using meshless methods

G.A.R. Caldas
Institute of Science and Innovation in Mechanical and Industrial Engineering (INEGI), Portugal

Jorge Belinha
Mechanical Engineering Department, School of Engineering, Polytechnic of Porto (ISEP), Portugal

J.G. Mendes & Renato Natal Jorge
Mechanical Engineering Department, Faculty of Engineering of the University of Porto (FEUP), Portugal

ABSTRACT: Using advanced meshless discretization computational techniques, in this work it is studied the structural response of adhesive dental bridges, analysing the effect of different resin-cements and the bridge design. For comparison purposes, three discretization techniques were used, the Finite Element Method (FEM) and two meshless methods, the Radial Point Interpolation Method (RPIM) and the Natural Neighbour Radial Point Interpolation Method (NNRPIM). 2D and 3D simulations of adhesive/tooth bio-systems were performed, assuming the adhesive with a nonlinear elasto-plastic behaviour. It was possible to conclude that the resin-cement has a clear influence over the stress concentration along the bridge and the single-retainer design increases the risk of debonding. Furthermore, the obtained results show that meshless methods are capable to obtain solutions very similar with FEM.

1 INTRODUCTION

Lost or extracted teeth compromise functional or aesthetic roles such as chewing and smile, and can cause infections, so it is very important to replace them. Adhesive bridges are a type of fixed prothesis in which the pontic tooth (replaced tooth) is attached to a structure (adhesive) that is cemented on the abutment teeth (adjacent teeth to the space of the missing tooth and that will serve as a support) (Walmsley et al., 2007). The fixed dental prosthesis is supported by patient's natural teeth. This type of prosthesis is distinguished from others due to the conservation of dental structures and their reversibility. However, adhesive dental bridges are associated with failure due to debonding between the retainer wing and the cement or between the cement and the dental surface, to the fracture of the abutment tooth and delamination of the veneering material of the framework (P. K. Vallittu et all., 2000) (L. Shi et all., 2009).

2 MATERIAL AND METHODS

Preliminary elasto-static numerical analyses were performed to understand the influence of different parameters on the mechanical resistance of the dental bridge. However, the effect of the resin-cement was unclear because these analyses do not account for material non-linearity.

DOI: 10.1201/9781003217152-75

Table 1. Mechanical properties (Cornacchia, Las Casas, Cimini, & Peixoto, 2010) (Moreira, Belinha, Dinis, & Jorge, 2014) (Della Bona, Donassollo, Demarco, Barrett, & Mecholsky, 2007) (Manicone, Rossi Iommetti, & Raffaelli, 2007).

	Zirconia	Admira (A)	Brilliant Coltene (B)	NC Coltène (NC)
Young's Modulus (E) [GPa]	245	3.6	2.3	2.3
Poisson's Ratio (υ)	0.26	0.3	0.3	0.3
Yield Stress (MPa)	$12 \cdot 10^{14}$	290	480	260
Plastic Modulus (GPa)	24.5	1.739	2.532	1.111

Therefore, to accurately analyse the resin-cements behaviour, 2D and 3D analysis models were developed, and elasto-plastic numerical analysis were performed using three numerical methods (FEM, RPIM and NNRPIM).

Different adhesive's materials were considered: Admira Fusion VOCO, Brilliant COLTÈNE and NC COLTÈNE, which have a bi-linear elasto-plastic, homogeneous and isotropic behaviour. The mechanical properties are presented in Table 1. However, the resin-cement Brilliant was not analysed because its Plastic modulus is higher than the Young's modulus, which is not supposed to happen. This inconsistency may be due to errors in experimental tests.

2.1 *2D Adhesive dental bridge*

The 2D model was based on preliminary works previously performed. Due to computational limitations, only half model was analysed, being the new model composed by the left adhesive wing and half of the mandibular lateral incisor, as presented in Figure 1. The adhesive's thickness was constant and equal to 0.4 mm. To simulate the presence of one or two retainer wings, different essential boundary conditions were applied, as demonstrated in Figure 1(a) and Figure 1(b), respectively. As a natural boundary condition was considered a 1N load with 90° of inclination with the longitudinal axis of the tooth, applied on the top of the pontic.

2.2 *3D Adhesive dental bridge*

Due to the same limitations previously mentioned, a 3D model of the restored tooth and the adhesive bridge was built considering a constant and equal to 0.4 mm adhesive's thickness and the two-retainer design. The numerical model was constructed with FEMAP and analysed with FEMAS– Finite Element and Meshless Analysis Software, which is an academic software (more details in cmech.webs.com). The final model is represented in Figure 2 from a top view and is possible to observe the essential boundary conditions on the left. It was assumed that there was a perfect mechanical and chemical bond between the adhesive and the abutment,

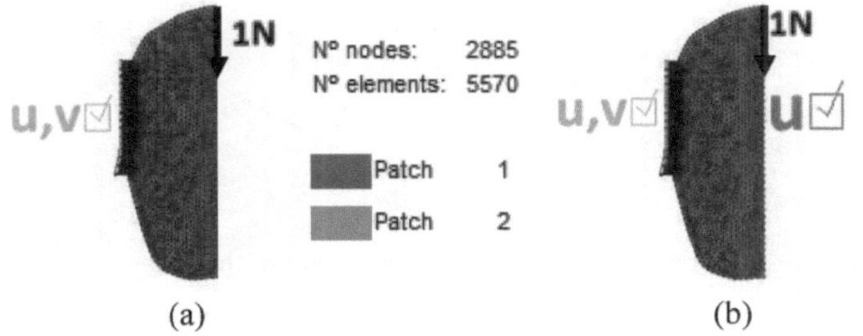

Figure 1. Global 2D geometric model for the (a) single-retainer and (b) two-retainer design.

Figure 2. Representation of the global 3D generic model.

hence there are nodes marked in white, which are constrained in all directions. As a natural boundary condition, it was considered a 1 N load, applied in the z direction.

3 RESULTS AND DISCUSSION

3.1 *2D adhesive dental bridge*

The elasto-plastic behaviour of the adhesive was registered through the graphic representation of stress/strain curve in Figure 3, obtained for resin-cement Admira and all numerical methods, butboth bridge designs. Although for the single retainer design the plastic behaviour was less noticeable, is possible to verify that this design reaches the plastic limit for lower stresses and higher deformations. Figure 3 shows that the three numerical methods overlap almost perfectly.

To evaluate the difference between the two resin-cements, in the graphs depicted in Figure 4 is possible to observe the total displacement of the pontic obtained with the force applied for the two-retainer design, all numerical methods and all resin-cements. NC Coltène reaches the plastic limit for lower forces and higher displacements.

3.2 *3D adhesive dental bridge*

The colour distribution maps of Figure 5 were obtained for the last increment of the non-linear-elasto-plastic analysis, i=10, numerical method FEM and resin-cement Admira. Is possible to observe that the area of the adhesive at risk is the one between the abutment and the pontic and the area in contact with the pontic. Therefore, these could be the initial areas of crack propagation. The adhesive in contact with the abutment does not exhibit stresses due to

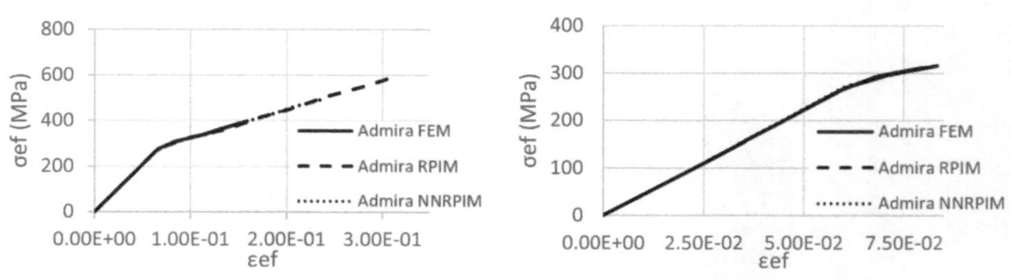

Figure 3. Graphic representation of stress/strain curve in a point of the adhesive for two-retainer design on the left and single-retainer design on the right.

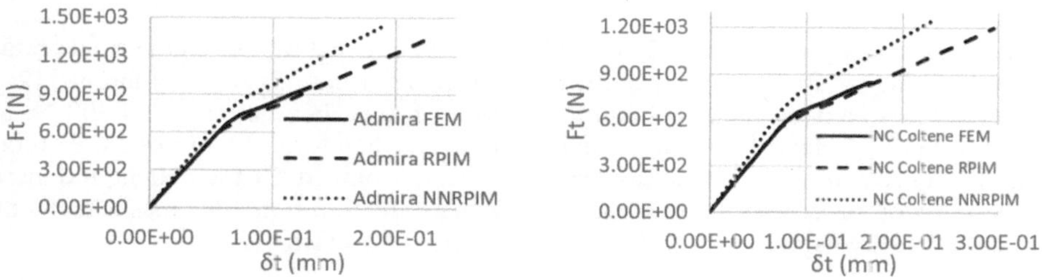

Figure 4. Graphic representation of force/displacement curve in a point of the pontic for two-retainer design.

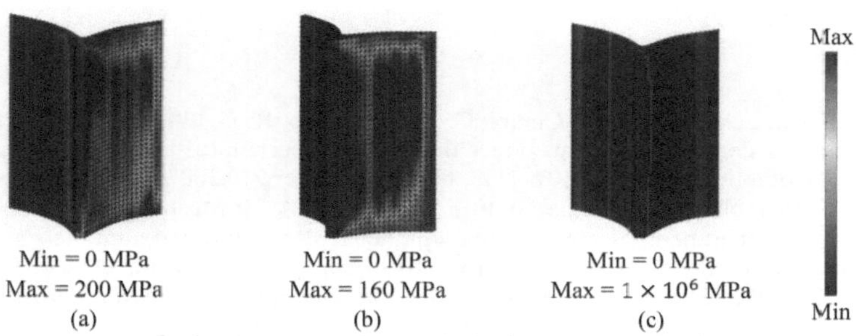

Figure 5. Colour distribution maps of: (a) principal stress σ11, (b) shear stress τ_{zx} and (c) plastic deformation εf_{plast}, for numerical method FEM and resin cement Admira.

the assumptions made that the adhesive is perfectly bonded to the abutment tooth. In Figure 6 is possible to observe that NC Coltène reaches the plastic limit for lower forces and higher displacement of the pontic, which leads to a sensation of discomfort for the patient. Permanent deformations are reached for lower force levels, increasing the probability of prosthesis failure and replacement.

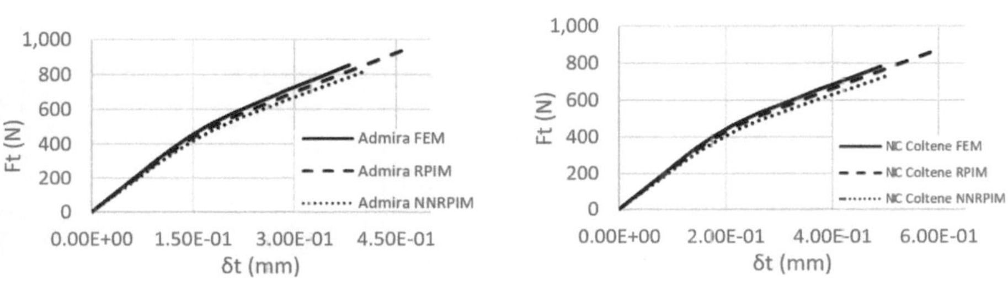

Figure 6. Graphic representation of force/displacement curve in a point of the pontic.

4 CONCLUSIONS

The elasto-plastic analyses performed allowed two relevant conclusions: (1) the single-retainer design leads to higher stress concentration and higher deformations of the adhesive; (2) NC Coltène undergoes irreversible deformations for lower forces, between 200 and 300 N, force levels that are easily reached during the daily activities. Therefore, the use of single-retainer design and resin cement NC Coltène can lead to higher discomfort for the patient and increase the probability of prosthesis failure and replacement. In addition, the meshless methods approximate the FEM solution in such way that the curves overlap.

ACKNOWLEDGEMENTS

The authors truly acknowledge the funding provided by Ministério da Ciência, Tecnologia e Ensino Superior –Fundação para a Ciência e a Tecnologia(Portugal), under the funding of LAETA-project UIDB/50022/2020,and project "SIMPLIFIED: Easy Tooth Abutment" – POCI-01-0247-FEDER-017982, co-funded by FEDER through COMPETE20220.

REFERENCES

Cornacchia, T. P. M., Las Casas, E. B., Cimini, C. A., & Peixoto, R. G. (2010). 3D finite element analysis on esthetic indirect dental restorations under thermal and mechanical loading. Medical & Biological Engineering & Computing, 48(11), 1107–1113. https://dx.doi.org/10.1007/s11517-010-0661-7

Della Bona, A., Donassollo, T. A., Demarco, F. F., Barrett, A. A., & Mecholsky, J. J. (2007). Characterization and surface treatment effects on topography of a glass-infiltrated alumina/zirconia-reinforced ceramic. Dental Materials, 23(6), 769–775. https://dx.doi.org/10.1016/j.dental.2006.06.043

L. Shi and A. S. L. Fok, *"Structural optimization of the fibre-reinforced composite substructure in a three-unit dental bridge,"* Dent. Mater., vol. 25, no. 6, pp. 791–801, 2009

Manicone, P. F., Rossi Iommetti, P., & Raffaelli, L. (2007). *An overview of zirconia ceramics: Basic properties and clinical applications*. Journal of Dentistry, 35(11), 819–826. https://dx.doi.org/10.1016/j.jdent.2007.07.008

Moreira, S. F., Belinha, J., Dinis, L. M. J. S., & Jorge, R. M. N. (2014). A global numerical analysis of the "central incisor/local maxillary bone" system using a meshless method. *Molecular & Cellular Biomechanics : MCB*, *11*(3), 151–184. https://dx.doi.org/10.3970/mcb.2014.011.151

P. K. Vallittu and C. Sevelius, "Resin-bonded, glass fiber-reinforced composite fixed partial dentures: a clinical study.," J. Prosthet. Dent., vol. 84, no. 4, pp. 413–418, 2000

Walmsley, A. D., Walsh, T. F., Lumley, P., Burke, F. J. T., Shortall, A. C., Hayes-Hall, R., & Pretty, I. (2007). *Restorative Dentistry*. (Elsevier, Ed.) (Second Edi). Churchill Livingstone. Retrieved from https://www.sciencedirect.com/science/book/9780443102462

Advances and Current Trends in Biomechanics – Belinha et al. (Eds)

Joint contact forces occurring during a maximal forward braking and backwards acceleration in elite athletes

R.B. Mateus
Faculty of Human Kinetics, University of Lisbon, Cruz Quebrada, Portugal

V. Ferrer-Roca
High Performance Center (CAR), Sant Cugat del Valles, Spain

F. João & A.P. Veloso
Faculty of Human Kinetics, University of Lisbon, Cruz Quebrada, Portugal

ABSTRACT: Abrupt deceleration is a common practice in several sports, where sudden changes of direction are needed to perform at the highest level. Most of the high-level sports movements are related to accelerations, or decelerations of the body, in order to produce direction changes. The sheer magnitudes of the forces generated by the muscles and acting upon joints and ligaments during such tasks, while considering the amount of times an athlete needs to decelerate and change directions during the course of its career, are hard to fathom. To this end, joint contact forces were estimated via a musculoskeletal modelling approach. This study showed that the highest peak joint contact forces were found at the hip (111.1 ± 47.2 and 99.8 ± 48.7 N/Kg) and knee (122.2 ± 44.6 and 124.9 ± 34.4 N/Kg) joints, along the Anterior/Posterior direction.

1 INTRODUCTION

Nowadays, it is expected on the athlete's fitness levels to be extremely high, to endure the training loads to which they are subjected. To this end, improving joint stabilization during high-impact tasks plays a key factor to be accounted for in training programs that increase muscle coordination and strength (Mazumder *et al.*, 2019). In addition to this, knowledge about how joint kinematics and muscle forces affect joint loading, whilst performing ballistic tasks remains scarce. The use of computational modeling allied with non-invasive experimental techniques to study injury mechanisms is an advantageous approach.

Most of the movements are related to accelerations, or decelerations of the body, to produce direction changes. Several works focus on the former (Schache *et al.*, 2010; Veloso *et al.*, 2015), however deceleration movements are just as important for the success of the athlete (Dalen *et al.*, 2016; Harper, Carling and Kiely, 2019). Not only team sports, such as basketball, volleyball, and soccer, but also individual sports, like tennis or badminton, require several changes of direction, with constant accelerations and decelerations. Concerning the decelerations, they may occur because of the boundary lines that keep players in the game, or as a reaction to other players actions. Hence, deceleration tasks are key to the performance of the athlete in any sport.

To this end, the aim of this study is the estimation of joint contact forces during a maximal forward braking and backward acceleration in elite athletes via a musculoskeletal approach.

DOI: 10.1201/9781003217152-76

2 MATERIALS AND METHODS

Fourteen elite male team-sports injury-free athletes participated in this study (age: 22 ± 4 years, height:185 ± 4 cm, weight: 77 ± 11 kg). The procedures used in this study were approved by the Ethics Committee of the Catalonian Sports Council. All subjects gave their written informed consent after an explanation of the experimental protocol and before the beginning of the study. The volunteers executed a series of five abrupt A/P deceleration trials, in which ground reaction forces and kinematic data were recorded. The best one amongst the set of five was chosen.

Kinematic data was collected at 300 Hz using 8 infrared cameras (Oqus 300, Qualisys AB, Sweden) synchronized in time and space with two force plates (Kistler, Switzerland). 28 reflective markers and semi-rigid marker clusters were used to guide an 8 rigid multibody biomechanical model developed using the Visual 3D platform (C-Motion, Inc.).

Anthropometric measures (body mass, stature) and motion capture tests were performed. The passive markers and four marker clusters were placed based on the calibrated anatomical system technique (Cappozzo *et al.*, 1995) by the same researcher. Specifically, six markers were placed on the trunk, one on top of each acromion, one on the C7 spinous process, two on the sternum area, and one on the spinous apophysis that was aligned with the lower sternum marker (placed so that soft tissue artifact and collinearity was avoided). At the pelvis, two markers were placed on each posterior superior iliac spines and two along each iliac crest. 8 Markers were also placed on the great trochanters head, the lateral and medial femur epicondyles, the lateral and medial ankle malleoli, and on the top of the first and fifth metatarsal heads. Each foot had also one marker on the heel. Finally, the mentioned marker clusters were attached to both thighs and shanks. Kinematic and kinetic variables were low pass filtered using a 4th order Butterworth filter at 8Hz. The biomechanical model built for each participant had 8 segments (feet, shanks, thighs, pelvis, and a trunk segment). The local coordinate systems of each segment were defined per Robertson et al. The hip joint center was computed using the pelvis markers, through a regression equation proposed by Bell et al, the knee joint center was the mid-point of the epicondyles and ankle joint center the mid-point of the malleoli (Robertson *et al.*, 2014).

The Inverse kinematics (IK) problem was solved as a global optimization problem (Lu and O'Connor, 1999). Segment masses were determined according to Dempster, whereas the remaining inertial parameters were computed based on Hanavan.

Muscle and joint contact forces were attained through OpenSim (Delp *et al.*, 2007). A 12 segment, 29 degrees of freedom musculoskeletal model was used to create the simulation. Each lower extremity had five degrees-of-freedom. The hip was modeled as a ball-and-socket joint (3 degrees of freedom), the knee was modeled as a custom joint with 1 degree of freedom (Seth et al., 2010), and the ankle was modeled as a revolute joint (1 degree of freedom). Lumbar motion was modeled as a ball-and-socket joint (3 degrees of freedom). The lower extremity and back joints were actuated by 92 musculotendon actuators. The musculoskeletal model was manually scaled to match each subject's anthropometry. A residual reduction algorithm (RRA) step was used to minimize errors related to kinematic inconsistencies and modelling assumptions. Muscle forces were estimated using a Static Optimization (SO) optimization technique. Joint contact forces were calculated using the JointReaction analysis, available in OpenSim. The forces obtained from SO and the adjusted kinematics from RRA were used for this analysis.

3 RESULTS

This task can be split into two phases: a braking phase, where energy dissipation occurs and deceleration happens, up until around 63 % of the task, and a backward acceleration phase, where the participants change direction. Mean peak joint contact forces, are given in Table 1.

Along the A/P direction, the highest joint contact forces during both the braking and backward acceleration phases were observed at the knee joint of the lower limb in contact with the

Table 1. Mean peak joint contact forces and their respective timing during a maximal single-leg forward braking and backward acceleration task.

	Phase1_X	Phase2_X	Phase1_Y	Phase2_Y	Phase1_Z	Phase2_Z
Hip Peak JCF	111.1 ± 47.2	99.8 ± 48.7	-80.6 ± 27.1	-71.7 ± 24.0	18.4 ± 10.4	12.0 ± 8.6
% Contact	**20.6 ± 21.6**	**64.7 ± 12.1**	**27.4 ± 21.6**	**65.2 ± 11.5**	**18.4 ± 17.3**	**74.6 ± 18.8**
Knee Peak JCF	122.2 ± 44.6	124.9 ± 34.4	-63.3 ± 20.5	-66.2 ± 18.2	-33.1 ± 21.2	-32.0 ± 16.5
% Contact	**25.8 ± 18.5**	**67.8 ± 12.7**	**17.2 ± 19.1**	**70.9 ± 11.4**	**32.1 ± 16.6**	**71.8 ± 12.6**
Ankle Peak JCF	22.3 ± 10.2	12.6 ± 4.4	-49.0 ± 11.7	-49.9 ± 8.5	4.4 ± 2.2	3.9 ± 2.3
% Contact	**5.1 ± 3.2**	**88.1 ± 10.2**	**24.7 ± 17.6**	**75.8 ± 10.8**	**22.4 ± 20.7**	**73.3 ± 12.8**
Sacroiliac Peak JCF	27.3 ± 18.2	38.2 ± 23.5	73.7 ± 27.6	87.2 ± 34.7	12.0 ± 7.9	16.2 ± 10.2
% Contact	**43.9 ± 16.6**	**63.5 ± 8.8**	**29.0 ± 20.6**	**69.9 ± 12.0**	**34.4 ± 20.1**	**68.2 ± 11.9**

Notes: Mean (N=14) ± standard deviation. Peak forces are normalized to the subject's mass and given in N/Kg. Phase 1 corresponds to the braking phase and phase 2 to the backward acceleration phase.

force plate (122.2 ± 44.6 and 124.9 ± 34.4 N/Kg), followed by the forces at the hip joint (111.1 ± 47.2 and 99.8 ± 48.7 N/Kg). Peak joint contact forces at the ankle and sacroiliac joints are significantly lower along this direction.

Compressive forces were observed along the vertical direction at the hip (-80.6 ± 27.1 and -71.7 ± 24.0 N/Kg), knee (-63.3 ± 20.5 and -66.2 ± 18.2 N/Kg), and ankle (-49.0 ± 11.7 and -49.9 ± 8.5 N/Kg) joints, whereas traction forces were recorded at the sacroiliac joint (73.7 ± 27.6 and 87.2 ± 34.7 N/Kg).

Along the mediolateral direction, the largest joint contact forces were observed at the knee joint, laterally directed, reaching approximately 33 N/Kg and 32 N/Kg during the braking and backward acceleration phases, respectively.

4 DISCUSSION

Joint contact forces represent the loads acting at a certain joint to counteract the effects of muscle forces, net joint moments, and ground reaction forces on the same joint.

The ratio of peak joint contact forces to peak ground reaction forces in this study for the hip (4.5-9.2 and 4.8-10.7), knee (3.5-10.1 and 4.4-13.4), ankle (1.8-2.7 and 1.4-3.3), and sacroiliac (2.3-4.1 and 4.1-5.8) joints were significantly different than the ones (2.63-4.63) presented in a study that estimated compressive and shear joint contact forces during a badminton forward lunge (Chen *et al.*, 2020). These differences, which are more prominent across the hip and knee joints, can be explained by the fact that our work included elite-level athletes instead of club-level players. In addition to this, the task studied in our work required the participant to sprint to the force plate, thus reaching greater speeds and resulting in larger joint contact forces.

As expected from this task, the joint reaction forces at the knee and hip are the highest along the A/P direction, with all being applied anteriorly on the tibia and femur, respectively, as they are the main contributors to both the braking and the change of direction stages of the task. The largest forces along the fore-aft direction were recorded at the knee, which may be related to the fact that the muscles that exerted the most force in this task, vasti, are inserted in this joint. In addition to this, the large hip joint contact forces would result in the bending of the distal portion of the femur, applying even more force on the knee joint, to maintain the structural integrity of the rigid bodies in the model. Compressive forces along the vertical directions are applied downwards at hip, knee, and ankle joints. Traction forces along the vertical directions applied at the sacroiliac joint (lower back) arise from the extension of the back while performing this, as the erector spinae and the other lumbar muscles generate forces to resist this extension tendency, thus maintaining balance. These forces may occur as a way for the body to not overload the ankle joint during this task. Following the same line of thought, shear forces along the anterior direction are also observed at the sacroiliac joint.

Along the mediolateral direction, where the lowest magnitudes for the hip, ankle, and sacroiliac joints were observed, there is much more variability in the directions along which these forces are being generated, which might be a by-product of the difference in strategies employed by each athlete in performing the task. At the knee joint, the joint contact forces along the mediolateral direction are also key to the realization of the task, as they help maintaining the knee in a neutral position.

5 CONCLUSIONS

This study showed that the highest peak joint contact forces were found at the hip and knee joints, along the A/P direction.

Since the loadings on the knee ligaments and their contributions are interesting unknowns that may be extremely important, due to their implications on injury prevention and performance-enhancing programs, estimation of these forces may be seen as a stepping stone to better understand the mechanisms behind an abrupt A/P deceleration task.

ACKNOWLEDGEMENTS

This work was supported by CIPER-FCT (I&D unit 447, project reference UIDB/00447/2020).

REFERENCES

Cappozzo, A. *et al.* 1995. 'Position and orientation in space of bones during movement', *Clin. Biomech.*, 10(4), pp. 171–178. doi: 10.1016/0268-0033(95)91394-T.

Chen, T. L. W. *et al.* 2020. 'Joint contact force and movement deceleration among badminton forward lunges: a musculoskeletal modelling study', *Sports Biomechanics*. Routledge, 00(00), pp. 1–13. doi: 10.1080/14763141.2020.1749720.

Dalen, T. *et al.* 2016. 'Player Load, Acceleration, and Deceleration During Forty-Five Competitive Matches of Elite Soccer', *Journal of Strength and Conditioning Research*, 30(2), pp. 351–359.

Delp, S. L. *et al.* 2007. 'OpenSim: Open-source software to create and analyze dynamic simulations of movement', *IEEE Transactions on Biomedical Engineering*, 54(11), pp. 1940–1950. doi: 10.1109/TBME.2007.901024.

Harper, D. J., Carling, C. and Kiely, J. 2019. 'High-Intensity Acceleration and Deceleration Demands in Elite Team Sports Competitive Match Play: A Systematic Review and Meta-Analysis of Observational Studies', *Sports Medicine*. Springer International Publishing, 49(12), pp. 1923–1947. doi: 10.1007/s40279-019-01170-1.

Lu, T. W. and O'Connor, J. J. 1999. 'Bone position estimation from skin marker co-ordinates using global optimisation with joint constraints', *Journal of Biomechanics*, 32(2), pp. 129–134. doi: 10.1016/S0021-9290(98)00158-4.

Mazumder, O. *et al.* 2019. 'Musculoskeletal modeling to predict and reduce Antetrior Cruciate Ligament injury during single leg drop jump activity: Synergistic muscle co-activation approach', *Proceedings of the Annual International Conference of the IEEE Engineering in Medicine and Biology Society, EMBS*. IEEE, pp. 4108–4112. doi: 10.1109/EMBC.2019.8856321.

Robertson, D. G. E. *et al.* (2nd) 2014 *Research Methods in Biomechanics*. Human Kinetics.

Schache, A. G. *et al.* 2010. 'Hamstring muscle forces prior to and immediately following an acute sprinting-related muscle strain injury', *Gait and Posture*. Elsevier B.V., 32(1), pp. 136–140. doi: 10.1016/j.gaitpost.2010.03.006.

Veloso, A. *et al.* 2015. 'Subject-specific musculoskeletal model to identify muscle contribution to the acceleration phase in elite sprinting', *33rd Conference of the International Society of Biomechanics in Sports, Pitiers, France*, 33 (1).

Advances and Current Trends in Biomechanics – Belinha et al. (Eds)

Stress analysis on a PEEK and titanium bar over implants—study with meshless methods

R. Costa
Faculty of Dental Medicine, University of Porto, Portugal

M.H. Figueiral, S.J. Oliveira & J.C. Sampaio-Fernandes
INEGI, Faculty of Dental Medicine, University of Porto, Portugal

D.C. Costa
School of Engineering, University of Minho, Portugal

J. Belinha
ISEP, School of Engineering, Polytechnic of Porto, Portugal

ABSTRACT: Hybrid dentures, consisting of fixed prostheses on implant-supported infrastructures (bars), are an excellent option for oral rehabilitation, especially in extensive tooth loss. The material used for the manufacturing of the bar structure is of fundamental importance, as it influences the structural response and stress distribution during masticatory function. Numerical analysis allows an approximation of the mechanical behavior of oral prostheses and adjacent structures. Using two numerical analysis methods, the finite element analysis and meshless methods (RPIM and NNRPIM), this study aims at analyzing the structural response of two implant-supported bars (without cantilever and with cantilever of 12mm) on 4 dental implants in the mandibular region, made of titanium, PEEK and 30% CFR-PEEK. This investigation concluded that titanium has better structural performance, followed by CFR-PEEK and PEEK. Our data also suggest that both numerical analyses provided similar results and conclusions.

1 INTRODUCTION

Oral rehabilitation is an area of dental medicine in constant evolution due to the introduction of new materials and technologies. Hybrid dentures, consisting of a fixed prosthesis on an implant-supported infrastructure (bar), represent a valuable solution for rehabilitation of edentulous maxilla and mandible, even when alveolar ridge loss is present. It has been suggested that the material used in the manufacturing of the bar is very important for clinical success (dos Santos et al 2015, Bacchi et al. 2013, Abreu et al. 2010). Titanium is currently the most used material for implant-supported bars, mainly due to its biomechanical properties; however, it has aesthetic limitations. Thus, alternative materials that combine good biomechanical properties with better aesthetics have been introduced in dental medicine, as is the case of the polyetheretherketone polymer (PEEK) and its variations (carbon fiber-reinforced PEEK, CFR-PEEK).

The clinical situations are varied and often complex, and numerical analyses are interesting approaches to understand the biomechanical behavior of dental prostheses, providing useful information regarding the bar, implants and bone responses (Bacchi et al. 2013, de la Rosa Castolo et al. 2010). In the field of oral rehabilitation, the finite element analysis method (FEM) has

DOI: 10.1201/9781003217152-77

been widely used. In recent years, however, the significant development of meshless discretization techniques (meshless methods) has made them a valid alternative to FEM-based analyses.

Using meshless methods (Radial Point Interpolation Method – RPIM and Natural Neighbor Radial Point Interpolation Method – NNRPIM), compared to FEM, this study aims at performing a structural analysis of two implant-supported bars (without cantilever and with a cantilever of 12mm) on 4 dental implants in the mandibular region, made of titanium, PEEK and 30% CFR-PEEK.

2 MATERIALS AND METHODS

Two models of implant-supported bars (with and without cantilever) manufactured in titanium, PEEK and CFR-PEEK were studied. FEMAP (student version) and FEMAS (academic software: cmech.webs.com) (Belinha 2016) were the specific software used. FEMAP was used to create the models of the two bars, with the following characteristics: bar width 36mm, bar thickness 3mm, cantilever of 12mm (if present). FEMAS was used to define the boundary conditions and conduct the numerical analyses of the models. As example, the force applied and the boundary conditions of the bar without cantilever are shown in Figures 1 and 2.

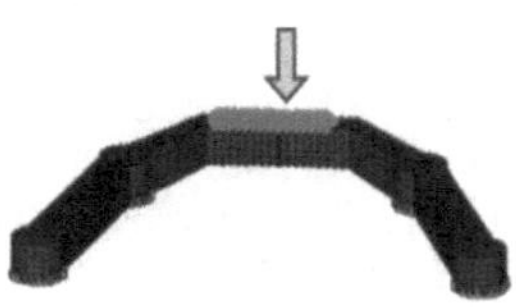

Figure 1. Point of force (f) application.

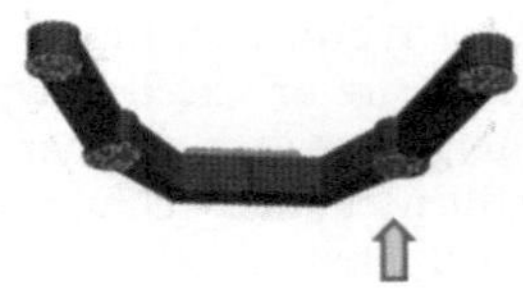

Figure 2. Boundary conditions.

According to the boundary conditions defined, the models were fixed on the base of the 4 pillars corresponding to the implants. Force f1 was applied in the anterior region of the bar without cantilever and force f2 in the posterior region of the bar with cantilever. The magnitude of both f1 and f2 was 1N. In all the numerical analyses performed, the models of the bar, the boundary conditions and the applied forces were maintained, with the only variable being the manufacturing material of the bar and the corresponding properties (elastic modulus and Poisson coefficient). The maximum force that a material can withstand is based on the final stress of the material. Since in all models a force of 1N was applied, the maximum force *Fmax* was obtained with equation 1, where σ1N is the stress obtained with the force of 1N and σU the limit stress of the material (Costa 2019).

$$Fmax = \frac{1N \cdot \sigma U}{\sigma 1N} \tag{1}$$

3 RESULTS

After performing all the simulations, a color map corresponding to each computational analysis, containing the maximum and minimum values and their location within the models, could be obtained. An example of the von Mises stress obtained with FEM for the titanium bar is shown in Figure 3, whereas the displacement field achieved with RPIM for the PEEK bar is displayed in Figure 4.

From the results of each analysis, excel files (.xlsx) were generated with the values of the coordinates of each node, as well as the von Mises stress (Sef) in MPa, strain (Eef) and total displacement (u) in mm. Tables 1 and 2 show the maximum values obtained with the FEM, RPIM and NNRPIM analysis for each pair bar model/manufacturing material. The values of the maximum masticatory force theoretically supported by both bars are summarized in Table 3.

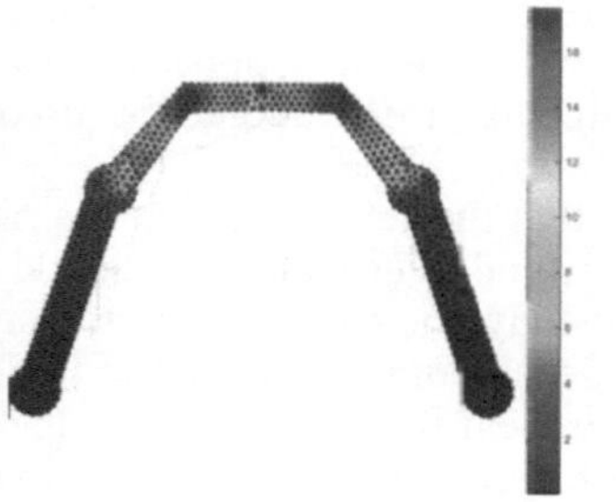

Figure 3. Bar without cantilever.

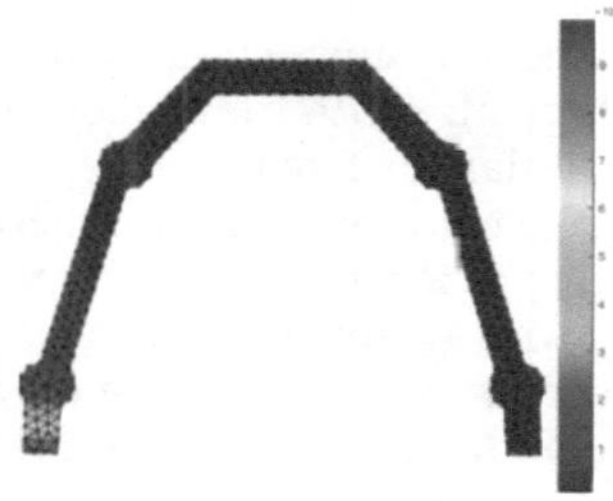

Figure 4. Bar with cantilever.

Table 1. Maximum values obtained from colour map of bar models without cantilever.

	Titanium			PEEK			CFR-PEEK		
Analysis	Sef/MPa	Eef	U/mm	Sef/MPa	Eef	U/mm	Sef/MPa	Eef	U/mm
FEM	17	0.00015	0.0048	17	0.0045	0.14	17	0.0009	0.03
RPIM	20	0.00016	0.0055	20	0.0052	0.16	20	0.001	0.03
NNRPIM	21	0.00017	0.0055	21	0.0054	0.16	21	0.001	0.03

Table 2. Maximum values obtained from colour map of bar models with cantilever.

	Titanium			PEEK			CFR-PEEK		
Analysis	Sef/MPa	Eef	U/mm	Sef/MPa	Eef	U/mm	Sef/MPa	Eef	U/mm
FEM	2.5	0.000025	0.00025	2.5	0.0008	0.008	2.5	0.00014	0.0014
RPIM	4	0.000035	0.0003	3.5	0.001	0.009	3.5	0.0002	0.0018
NNRPIM	3.5	0.00003	0.00025	3	0.0009	0.008	3	0.00018	0.0016

Table 3. Results for the maximum masticatory force supported by both bars.

	Maximum masticatory force	
Material	Bar without cantilever	Bar with cantilever
Titanium	26 N – 57 N	137 N – 384 N
PEEK	4 N – 8 N	23 N – 55 N
CFR-PEEK	5,7 N – 8 N	34 N – 55 N

4 DISCUSSION

In the bar model without cantilever, the values of the von Mises stress were identical in the various materials studied, which was confirmed by the different analyses. Such observation differs from other studies (Abreu et al. 2010), in which it is suggested that the similarity of the stress values would probably result from the approximate values of the elastic modulus of the materials, a condition not verified in our work. Another author attributed the higher von Mises stress of pure PEEK compared to titanium and CFR-PEEK to the simplifications of the FEM model, which considers isotropic linear elastic materials (Sirandoni et al. 2019). The maximum values of von Mises stress and strain were located in the lower region of the bar, close to the mesial portion of the premolar implants. The maximum displacement occurred in the anterior region of the bar, at the same place where the force was applied (f1). According to our data, PEEK is not considered suitable for supporting medium or high masticatory loads, as opposed to other studies (Han et al. 2019, Gallagher et al. 2019) suggesting that PEEK could be used as a reliable material for fixed dental prosthesis in the near future.

Regarding the bar model with cantilever, Gonzalez showed that the deformation of the cantilever is directly proportional to its length and inversely proportional to its width and should not exceed 15mm with 4 implants (Gonzalez 2014). In the current study, the region of greatest stress occurred under the bar, close to the cervical-distal portion of the abutment of the last implant for both the von Mises stress and strain fields. For the displacement field, however, the highest values were found in the distal portion of the cantilever. Titanium bars presented the lowest strain and displacement among the studied materials.

5 CONCLUSION

With the conditions used in the simulations, titanium was the material with the greatest resistance to strain and least displacement, followed by CFR-PEEK and PEEK. The maximum von Mises stress of the different materials were very similar, which was confirmed in the various numerical analyses. For the bar without cantilever, the results obtained with the two discrete methods used, the FEM and the meshless method, confirm that both numerical approaches produce concordant solutions. Although promising, the clinical use of non-metallic materials in extensive oral rehabilitation appears to require further optimizations, to which numerical methods may provide important contributions.

REFERENCES

Abreu, R.T., Spazzin, A.O., Noritomi, P.Y., Consani, R.L., Mesquita, M.F. 2010. Influence of material of overdenture- retaining bar with vertical misfit on three-dimensional stress distribution. *Journal of Prosthodontics* 19(6): 425–431.

Bacchi, A., Consani, R.L.X., Mesquita, M.F., dos Santos, M.B.F. 2013. Effect of framework material and vertical misfit on stress distribution in implant-supported partial prosthesis under load application: 3-D finit element analysis. *Acta Odontologica Scandinavica* 7(5): 1243–1249.

Belinha, J. 2016. Manual for FEMAS M-15.1 finite element and meshless analysis software.

Costa, D.C. 2019. Biomechanical Simulation of Complex Dental Implant Solutions Using Meshless Methods. Master degree thesis, University of Porto, Porto.

de la Rosa Castolo, G., Guevara Perez, S.V., Arnoux, P.J., Badih, L., Bonnet, F., Behr, M. 2019. Implant-supported overdentures with different clinical configurations: Mechanical resistance using a numerical approach. *The Journal of Prosthetic Dentistry* 121(3): 546.e1–546.e10.

dos Santos, M.B., Caldas, R.A., Zen, B.M., Bacchi, A., Correr-Sobrinho, L. 2015. Adaptation of overdenture-bars casted in different metals and their influence on the stress distribution: a laboratory and 3D FEA. *Journal of Biomechanics* 48(1): 8–13.

Gallagher, E.A., Lamoriniere, S., McGarry, P. 2019. Finite element investigation into the use of carbon fibre reinforced PEEK laminated composites for distal radius fracture fixation implants. *Medical Engineering & Physics* 67: 22–32.

Gonzalez, J. 2014. The evolution of dental materials for hybrid prosthesis. *The Open Dentistry Journal* 8:‐85.

Han, X., Yang, D., Yang, C., Spintzyk, S., Scheideler, L., Li, P., Li, D., Geis-Gerstorfer, J., Rupp, F. 2019. Carbon fiber reinforced PEEK composites based on 3D-printing technology for orthopedic and dental applications. *Jornal of Clinical Medicine* 8(2): 240.

Sirandoni, D., Leal, E., Weber, B., Noritomi, P.Y., Fuentes, R., Borie, E. 2019. Effect of Different Framework Materials in Implant-Supported Fixed Mandibular Prostheses: A Finite Element Analysis. *The International Journal of Oral & Maxillofacial Implants* 34(6): 107–14.

Advances and Current Trends in Biomechanics – Belinha et al. (Eds)

Biomechanical evaluation of fixed total rehabilitation on 3 implants

D. Soares, P. Vaz, J. Marques & J.C. Reis Campos
Faculty of Dentistry, University of Porto, Porto, Portugal

M. Vaz & M. Marques
Faculty of Engineering, University of Porto, Porto, Portugal

ABSTRACT: In order to analyze the stresses produced in all the components involved in the chewing process, various techniques can be used such as: holography and finite element analysis, among others. In this study, in a view to assess the distribution and magnitude of bone stress in a total fixed prosthesis on 3 implants, the finite element analysis method was used, which allows the determination of stresses and deformations of complex structures. In effect, 3 trials were carried out, in which loads generated during bilateral chewing, unilateral on the cantilever and posterior to the cantilever, were simulated, based on a model generated from a CT scan of a patient, performed 15 months after loading of the placed implants. The results seem to indicate the area of the peri-implant bone distal to the distal implants as the region of the mandibular bone that suffers the greatest tensions.

1 INTRODUCTION

The search for dental implants is currently one of the main requests of patients in a Dental Medicine consultation. The reasons for this high demand are related not only to the need to replace lost teeth, but also to the replacement of the function associated with them, as well as to the restoration of a dental and facial harmony, increasingly desired by the patients (Lindhe *et al.* 2010).

The rehabilitation of atrophic mandibles is one of the main challenges of modern implantology. Recently, a new treatment option was introduced on the market, using 3 implants placed in the interforaminal region and respective fixed prosthesis.

The Trefoil System (Nobelbiocare®) uses 3 implants and a pre-formed titanium bar in order to support the placement of a permanent acrylic prosthesis screwed to it (Higuchi *et al.* 2019). Achieving primary stability allows immediate loading of the implants upon surgery (Chen *et al.* 2019).

Although it is tempting to use a smaller number of implants, it is necessary to take into account the biomechanical viability of this procedure. This study has as main objective to evaluate the distribution and magnitude of bone stress in a total fixed prosthesis on 3 implants.

2 MATERIAL AND METHODS

In order to assess the magnitude and distribution of bone stress in a fixed prosthesis on 3 implants, computed tomography was performed before the surgical procedure and after a minimum of 6 months under load of the same structure. In this case, it was possible to

DOI: 10.1201/9781003217152-78

conduct the study using a computed tomography scan performed 15 months after placing the prosthesis in the mouth.

The two tomography's were subsequently analyzed using the finite element method which allowed this three-dimensional image to be segmented into simple elements that are grouped in nodes.

The three-dimensional images obtained through computed tomography were used for the design of the mandible, the implants and the bar that serve the study.

These structures were designed and were generated the meshes of finite elements for the purpose of simulate bilateral and unilateral loads on the cantilever and posterior to the cantilever.

Thus, 3 tests were carried out: a) Test 1: 200N vertical bilateral loads in the molar zone; 150N in the premolar zone and 100N in the canine zone.; b) Test 2: 200N unilateral vertical load on the 4th Q implant; and c) Test 3: 200N unilateral vertical load at the distal end of the bar.

Using the Mimics Research 19 software (Materialize, Belgium) together with the Autodesk Fusion 360 software (Autodesk, USA) it was possible to draw the patient's jaw from the computed tomography at 15 months of follow-up and it was possible to isolate it of the remaining elements, namely implants and bar. Subsequently, the implants were designed from the sagittal, axial and coronal sections of the tomography and transferred to the mandible according to their location given by computed tomography. Likewise, the bar was designed according to the images obtained by computed tomography. Autodesk Fusion 360 software (Autodesk, USA) was used for both the bar design and the implant design. Once these structures were designed, they were transferred to the ABAQUS software (Dassault Systèmes, France) - a finite element analysis program.

The implants and bar were defined as titanium and the mandible was divided into cortical and trabecular bone according to the data obtained by computed tomography. These materials were considered to be isotropic and have linear elastic behavior.

Test 1 - Bilateral Chewing – This test aimed to simulate the forces of mastication. For this, bilateral forces of progressively higher loads from anterior to posterior were applied. In particular, loads of 100N were applied in the canine zone, 150N in the premolar zone and 200N in the molar zone (Figure 1).

Figure 1. Illustration of the application force points during the test 1.

Test 2 - Unilateral on one of the distal implants – This test aimed to simulate the absence of cantilever. For this purpose, a force of 200N was applied to a point on the bar above the emergence of the implant, as shown in Figure 2.

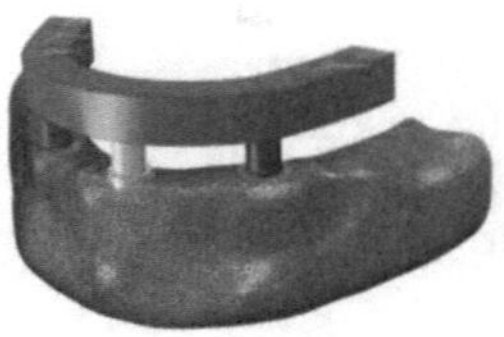

Figure 2. Illustration of the application force points during the test 2.

Test 3 - Unilateral on the cantilever – This test aimed to simulate the presence of forces on the most distal point of the cantilever, in this case applying a force of 200N on it (Figure 3).

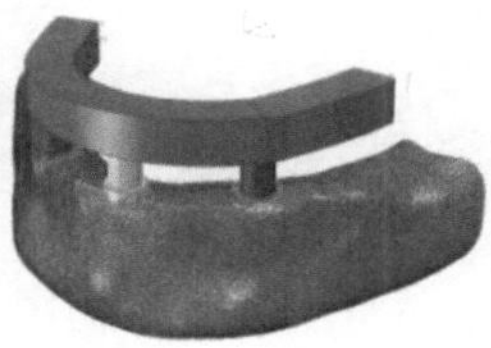

Figure 3. Illustration of the application force points during the test 3.

3 RESULTS

In order to answer the research questions, the maximum stress values of Von Mises were analyzed, as well as the distribution of these stresses. In addition, the maximum displacement in each test was also taken into account, as explained in Table I.

Table 1. Maximum values of Von Mises stresses and maximum displacement over the 3 tests.

	vM maximum [MPa]	Maximum displacement [mm]
Test 1	313.9	0.288
Test 2	177.7	0.159
Test 3	188.9	0.143

The magnitudes of the stresses can be visualized graphically through a color gradient, where blue represents lower tension values and red represents higher tension values.

The highest tension value recorded in test 1 in the mandible was 121.1MPa, in the distal area of the distal implants. (Figure 4)

Figure 4. Distribution and magnitude of stresses in the jaw during test 1.

The highest tension value recorded both in test 2 and 3 in the mandible was 57.1MPa and 36.9MPa, in the distal area of the distal implants. (Figures 5 and 6)

Figure 5. Distribution and magnitude of stresses in the jaw during test 2.

Figure 6. Distribution and magnitude of stresses in the jaw during test 3.

4 DISCUSSION

Considering the tensions in the mandible resulting from the application of loads over the 3 tests, it is clear that it was during bilateral chewing that the greatest tensions were found in the peri-implant bone - 121.1MPa. In short, it appears that the distal region of the peri-implant bone around the distal implants is the point where there is greater tension in the mandible and where bone resorption processes may be more involved (Figures 4, 5 and 6).

It is important to note that a load applied to one point of a prosthesis on implants is transmitted to adjacent implants. The distribution of the load on the implants is influenced by numerous factors such as: the design of the prosthesis - length of the cantilever; the material of the prosthesis; the passivity of the prosthesis on implants; the position, number and distribution of the implants; among others (Powers *et al.* 2006).

It should also be taken into account, that the type of opposing arcade influences the magnitude of the forces produced, with an individual with complete dentition presenting a chewing force 5 to 6 times greater than a completely edentulous subject (Powers *et al.* 2006). In this clinical case, the patient has a removable upper total prosthesis.

In the future, it would be pertinent to assess, whether in fixed total rehabilitation on 3 implants, the angulation of distal implants can improve the distribution of forces in the bone-implant-prosthesis complex.

The reliability and precision of the finite element method depends largely on the precision of the geometric models created. Despite the efforts to reproduce the clinical reality 100%, the simplification of the models is inevitable due to the complexity of the anatomical and prosthetic structures. These simplifications are present in practically all the literature where this method is used, so analysis must take into account these simplifications.

Finally, mean load values were used to simulate a real case of chewing. If average values do not represent the clinical case in question, the use of vertical loads only, also do not represent the real context as well.

The analysis of these results must be careful and rigorous, since direct extrapolation to clinical practice is not advised taking into account the limitations described above. These results are important in the analysis and comparison between test and similar studies, however, they must be contextualized and carefully framed.

5 CONCLUSION

The use of a smaller number of implants to rehabilitate an atrophic mandible has some inherent advantages, such as a potential better cost/benefit for the patient by decreasing the cost of treatment, greater space for cleaning between implants and thus easier to perform the same. However, it is necessary to evaluate biomechanically the viability of this reduction in the number of implants.

The results presented seem to indicate the distal region of the peri-implant bone of the distal implants as the area of the mandible where the greatest tensions were verified, both in a situation of bilateral chewing, and in the simulation of unilateral loads on one of the distal implants and on the most distant region, back of the cantilever.

REFERENCES

Chen S. *et al.* Implant placement and loading protocols in partially edentulous patients: a systematic review. *Clinical Oral Implant Research* 2018;29 \Suppl. 16):106–134.

Higuchi K, *et al.* An innovative implant-supported treatment for the edentulous mandible: case report. *International Journal of Oral & Maxillofacial Implants* 2019;(34)13–16.

Lindhe, J *et al.* (2015) *Clinical periodontology and Implant dentistry*. 6thEdition. Wiley Blackwell.

Powers, JM et al. (2006) Mechanical Properties, in Craig's Restorative Dental Materials. Mosby Elsevier.

Advances and Current Trends in Biomechanics – Belinha et al. (Eds)
© 2022 Copyright the Author(s), ISBN: 978-1-032-10806-3

Experimental studies of animal models with HFPEF

M.M.A. Peyroteo
Institute of Science and Innovation in Mechanical and Industrial Engineering, Porto, Portugal

J. Belinha
School of Engineering Polytechnic of Porto, Porto, Portugal

I. Falcão-Pires & A.F. Leite-Moreira
Faculty of Medicine, University of Porto, Porto, Portugal

R. Natal Jorge
Faculty of Engineering, University of Porto, Porto, Portugal

ABSTRACT: Heart failure with preserved ejection fraction (HFPEF) is considered one of the greatest mysteries in cardiovascular research. Currently, many authors agree that HFPEF is a metabolic disease, with distinct phenotypes according to the comorbidities associated, which hinders the search for a specific and effective therapy for this syndrome. Thus, this review describes the main impact of each comorbidity in HFPEF phenotype along with the main findings of experimental studies using animal models of HFPEF with distinct comorbidities associated. Increased cardiomyocyte passive stiffness and fibrosis of extracellular matrix have been experimentally described, however the mechanisms inducing these alterations are still under debate. The answer could be endothelial dysfunction and reactive fibrosis, but further studies are still required.

1 INTRODCUTION

Heart failure with preserved ejection fraction (HFPEF) is a multifactorial disease resulting in impaired filling of the left ventricle (LV) while preserving ejection fraction (EF) levels above 50% (Singh and Mehta, 2018). Even nowadays, treating HFPEF patients is a challenge, as an effective therapy for this disease remains unknown. One of the reasons is the elevated number of comorbidities associated with HFPEF, increasing the heterogeneity between these patients and the underlying mechanisms of this pathology.

The first step in understanding HFPEF is understanding its precursor pathway – diastolic dysfunction. Diastolic dysfunction is a silent state, in which impaired relaxation is compensated by increasing left atrium (LA) pressure, without any structural disorder of the heart. This abnormal relaxation is attributed to increased passive stiffness and calcium (Ca^{2+}) homeostasis abnormalities, which are strongly related with the comorbidities of HFPEF patients. Therefore, this review will focus on the pathophysiology of HFPEF, followed by several experimental studies with animal models of HFPEF describing the main contributing factors to diastolic dysfunction

DOI: 10.1201/9781003217152-79

2 PATHOPHYSIOLOGY OF HFPEF

Besides impaired diastolic function, HFPEF patients present other disorders such as defective chronotropic response to exercise, LV hypertrophy and increased vascular stiffness (Miranda-Silva et al., 2021). In addition to these cardiac dysfunctions, the vast majority of these patients possess distinct comorbidities that also affect the pathophysiology of HFPEF, such as obesity, diabetes, hypertension, chronic kidney disease (CKD) and chronic obstructive pulmonary disease (COPD). Therefore, several authors believe that therapeutic measures should adapt to the dominant comorbidity of each HFPEF patient (Obokata et al., 2017).

For instance, more than 80% of HFPEF patients are obese (Owan et al., 2006). Thus, obese HFPEF phenotype is highly prevalent and is distinguished by a larger right ventricular size, higher LV filling pressure and greater intolerance to exercise when compared to non-obese HFPEF patients (Obokata et al., 2017). Regarding diabetic HFPEF patients, diabetes presence is associated to higher diastolic LV stiffness and LV mass, along with higher filling and pulse pressures (Mohammed et al., 2012). Moreover, diabetic patients have higher levels of inflammatory markers that lead to a higher risk of hospitalization (Lindman et al., 2014). CKD affects 25% to 50% of HFPEF patients (Ather et al., 2012) and leads to concerning outcomes, aggravating systolic and diastolic dysfunctions (Gori et al., 2014). However, HFPEF patients typically accumulate more than one comorbidity, becoming even more challenging to find a specific therapy suitable to cover multiple conditions (Khalid et al., 2018). Nevertheless, common pathways in these comorbidities have been found such as coronary microvascular endothelial dysfunction, infiltration of activated macrophages and reactive interstitial fibrosis, culminating in increased cardiomyocyte and extracellular matrix (ECM) stiffness, characteristics of ventricular diastolic dysfunction (Campbell and Sorrell, 2015).

3 DIASTOLIC DYSFUNCTION

3.1 *Cardiomyocyte stiffness*

Diastolic dysfunction is mainly characterized by increased LV stiffness, which is often associated with stiffness alterations in the cardiomyocyte or in its ECM. Although there are already studies regarding muscle passive stiffness, few use animal models of HFPEF. One of these studies used ZSF1 rat models of HFPEF with hypertension, diabetes and obesity. When compared to the control group, the skinned cardiomyocytes of ZSF1 HFPEF rats had higher passive stiffness (Hamdani et al., 2013). Another study compared myocardial passive stiffness of a group of patients with hypertension with a group of patients with hypertension and HFPEF (Zile et al., 2015), finding that hypertension is not responsible for altering passive stiffness, since only the group of patients with hypertension and HFEPEF had a significant increase in passive myocardial stiffness. Both studies attributed cardiac titin phosphorylation as the main cause for high myocardial stiffness. Titin is a sarcomeric protein capable of regulating cardyomyocyte's passive stiffness depending on the phosphorylation state of two specific sites in this protein: N2-B unique sequence (N2-Bus) and PEVK region. Studies found that hypophosphorylation at N2-Bus and hyperphosphorylation at PEVK region are responsible for the increased passive stiffness. A distinct study compared the passive stiffness of two distinct patients groups - patients with HFPEF and symptomatic aortic valve stenosis (AS) and another presenting only AS (Borbély et al., 2009). HFPEF patients with AS comorbidity presented significantly higher passive stiffness (6.1 ± 0.4 kN/m^2) when compared with the AS cardiomyocytes group (2.2 ± 0.2 kN/m^2; $P<0.001$). However, besides studying the phosphorylation status of titin, this work also analyzed the isoform shifts of titin. Titin has two isoforms - N2B and N2BA. N2BA isoform is longer and more compliant than the N2B isoform. Although a similar N2BA/N2B ratio was found in this work, the ratio between phosphorylated N2BA (P-N2BA) and phosphorylated N2B (P-N2B) was significantly higher in the HFPEF with AS group (0.77 ± 0.05; $P<0.01$). Since higher levels of P-N2BA has been previously associated with higher passive stiffness, P-N2BA/P-N2B higher ratio can also be a reason for increased passive stiffness in HFPEF.

LV stiffness also depends on the collagen content of ECM. Factors as collagen volume fraction (CVF) and degree of crossbridge cycling can induce myocardial interstitial fibrosis, which ultimately leads to ECM stiffening. A study considering patients diagnosed with acute HFPEF found significantly higher passive stiffness values (6.6 ± 3.0 kN/m^2) supported by significantly higher CVF levels (7.5± 4.0%, P<0.05) (Borbely et al., 2005). Another study previously mentioned (Zile et al., 2015) found that patients with hypertension and HFPEF had an 100% significant increase in insoluble collagen and 130% increase in CVF. High CVF values and insoluble collagen are strongly associated with myocardial fibrosis, thence increased passive stiffness (Rommel et al., 2016; Schelbert et al., 2017). In another study considering hypertensive patients with HFPEF, collagen metabolism has been discussed by analyzing the ratio between metalloproteinases activity (MMPs) and their tissue inhibitors (TIMPs) (Ahmed et al., 2006). A decrease in ECM degradation contributing to increased fibrosis was confirmed by a downregulation of MMPs and an overexpression of TIMPs.

Therefore, both cardiomyocytes and ECM play an important role in the development of HFPEF. In (Hamdani et al., 2013), it was possible to isolate the contribution of titin and collagen to the passive stiffness, finding that at sarcomere lengths (SL) ranging from 2.0 to 2.2 μm, the contribution of titin greatly exceeded the contribution of the ECM (9.3 times higher). However, in (Zile et al., 2015), it is suggested that the increase in passive stiffness is evenly divided between titin and collagen.

3.2 *Calcium homeostasis abnormalities*

Abnormalities in the excitation–contraction coupling (ECC) coordinated by Ca^{2+} handling has also been reported as a cause for impaired relaxation. In a recent study, ZSF1-obese rats with HFPEF syndrome presented, at rest, significantly higher cytosolic free Ca^{2+} concentration (Miranda-Silva et al., 2020). Experimental studies suggested that the sodium-calcium exchanger must be affected, leading to Ca^{2+} extrusion from the sarcolemma to the cytosol (Selby et al., 2011). Moreover, the sarcoendoplasmic reticulum Ca^{2+}-ATPase (SERCA), responsible to transfer Ca^{2+} from the cytosol to the lumen of the sarcoendoplasmic reticulum (SR), is reduced, explaining the elevated levels of cytosolic free Ca^{2+}. Higher mitochondrial Ca^{2+} concentration has also been detected in the same HFPEFE rat model, supporting a direct relation between free Ca^{2+} concentration in the cytosol and the mitochondria (Wüst et al., 2017). It was also detected a slower contraction and relaxation response in this animal model. Ultimately, these higher levels of Ca^{2+} increase the passive stress during diastole, aggravating the diastolic dysfunction.

4 CONCLUSIONS

This work presents a brief description of HFPEF pathophysiology supported by the main experimental findings using HFPEF animal models. Although mechanical and biological characterization of diastolic dysfunction has already been reported, the underlying mechanism leading up to this syndrome is still unknown. Coronary microvascular endothelial dysfunction, infiltration of activated macrophages and reactive interstitial fibrosis may be the answer, however further studies should be conducted to clearly identify their main cellular and molecular pathways. These findings could then contribute to the ongoing search for effective therapeutic options for HFPEF treatment.

ACKNOWLEDGEMENTS

The authors truly acknowledge the funding provided by Ministério da Ciência, Tecnologia e Ensino Superior – Fundação para a Ciência e a Tecnologia (Portugal), under grants: SFRH/

BD/133105/2017. Additionally, the authors gratefully acknowledge the funding provided by LAETA, under project UIDB/50022/2020.

REFERENCES

Ahmed, S.H., Clark, L.L., Pennington, W.R., Webb, C.S., Bonnema, D.D., Leonardi, A.H., McClure, C. D., Spinale, F.G., Zile, M.R., 2006. Matrix metalloproteinases/tissue inhibitors of metalloproteinases: Relationship between changes in proteolytic determinants of matrix composition and structural, functional, and clinical manifestations of hypertensive heart disease. Circulation 113, 2089–2096.

Ather, S., Chan, W., Bozkurt, B., Aguilar, D., Ramasubbu, K., Zachariah, A.A., Wehrens, X.H.T., Deswal, A., 2012. Impact of Noncardiac Comorbidities on Morbidity and Mortality in a Predominantly Male Population With Heart Failure and Preserved Versus Reduced Ejection Fraction. J. Am. Coll. Cardiol. 59, 998–1005.

Borbély, A., Falcao-Pires, I., Van Heerebeek, L., Hamdani, N., Édes, I., Gavina, C., Leite-Moreira, A.F., Bronzwaer, J.G.F., Papp, Z., Van Der Velden, J., Stienen, G.J.M., Paulus, W.J., 2009. Hypophosphorylation of the stiff N2B titin isoform raises cardiomyocyte resting tension in failing human myocardium. Circ. Res. 104, 780–786.

Borbely, A., van der Velden, J., Papp, Z., Bronzwaer, J.G.F., Edes, I., Stienen, G.J.M., Paulus, W.J., 2005. Cardiomyocyte Stiffness in Diastolic Heart Failure. Circulation 111, 774–781.

Campbell, K.S., Sorrell, V.L., 2015. Cell-and molecular-level mechanisms contributing to diastolic dysfunction in HFpEF. J. Appl. Physiol.

Gori, M., Senni, M., Gupta, D.K., Charytan, D.M., Kraigher-Krainer, E., Pieske, B., Claggett, B., Shah, A.M., Santos, A.B.S., Zile, M.R., Voors, A.A., McMurray, J.J. V., Packer, M., Bransford, T., Lefkowitz, M., Solomon, S.D., 2014. Association between renal function and cardiovascular structure and function in heart failure with preserved ejection fraction. Eur. Heart J. 35, 3442–3451.

Hamdani, N., Franssen, C., Lourenco, A., Falcao-Pires, I., Fontoura, D., Leite, S., Plettig, L., Lopez, B., Ottenheijm, C.A., Becher, P.M., Gonzalez, A., Tschope, C., Diez, J., Linke, W.A., Leite-Moreira, A.F., Paulus, W.J., 2013. Myocardial Titin Hypophosphorylation Importantly Contributes to Heart Failure With Preserved Ejection Fraction in a Rat Metabolic Risk Model. Circ. Hear. Fail. 6, 1239–1249.

Khalid, U., Deswal, A., Deswal, A., 2018. Heart Failure With Preserved Ejection Fraction, in: Cardiology Secrets. Elsevier, pp. 209–215.

Lindman, B.R., Dávila-Román, V.G., Mann, D.L., McNulty, S., Semigran, M.J., Lewis, G.D., de las Fuentes, L., Joseph, S.M., Vader, J., Hernandez, A.F., Redfield, M.M., 2014. Cardiovascular Phenotype in HFpEF Patients With or Without Diabetes. J. Am. Coll. Cardiol. 64, 541–549.

Miranda-Silva, D., Lima, T., Rodrigues, P., Leite-Moreira, A., Falcão-Pires, I., 2021. Mechanisms underlying the pathophysiology of heart failure with preserved ejection fraction: the tip of the iceberg. Heart Fail. Rev. 26, 453–478.

Miranda-Silva, D., Wüst, R.C.I., Conceição, G., Gonçalves-Rodrigues, P., Gonçalves, N., Gonçalves, A., Kuster, D.W.D., Leite-Moreira, A.F., van der Velden, J., de Sousa Beleza, J.M., Magalhães, J., Stienen, G.J.M., Falcão-Pires, I., 2020. Disturbed cardiac mitochondrial and cytosolic calcium handling in a metabolic risk-related rat model of heart failure with preserved ejection fraction. Acta Physiol. 228.

Mohammed, S.F., Borlaug, B.A., Roger, V.L., Mirzoyev, S.A., Rodeheffer, R.J., Chirinos, J.A., Redfield, M.M., 2012. Comorbidity and Ventricular and Vascular Structure and Function in Heart Failure With Preserved Ejection Fraction: A Community-Based Study. Circ. Hear. Fail. 5, 710–719.

Obokata, M., Reddy, Y.N.V., Pislaru, S. V., Melenovsky, V., Borlaug, B.A., 2017. Evidence Supporting the Existence of a Distinct Obese Phenotype of Heart Failure With Preserved Ejection FractionClinical Perspective. Circulation 136, 6–19.

Owan, T.E., Hodge, D.O., Herges, R.M., Jacobsen, S.J., Roger, V.L., Redfield, M.M., 2006. Trends in Prevalence and Outcome of Heart Failure with Preserved Ejection Fraction. N. Engl. J. Med. 355, 251–259.

Rommel, K.P., Von Roeder, M., Latuscynski, K., Oberueck, C., Blazek, S., Fengler, K., Besler, C., Sandri, M., Lücke, C., Gutberlet, M., Linke, A., Schuler, G., Lurz, P., 2016. Extracellular Volume Fraction for Characterization of Patients with Heart Failure and Preserved Ejection Fraction. J. Am. Coll. Cardiol. 67, 1815–1825.

Schelbert, E.B., Fridman, Y., Wong, T.C., Abu Daya, H., Piehler, K.M., Kadakkal, A., Miller, C.A., Ugander, M., Maanja, M., Kellman, P., Shah, D.J., Abebe, K.Z., Simon, M.A., Quarta, G., Senni, M., Butler, J., Diez, J., Redfield, M.M., Gheorghiade, M., 2017. Temporal relation between myocardial fibrosis and heart failure with preserved ejection fraction Association with baseline disease severity and subsequent outcome. JAMA Cardiol. 2, 995–1006.

Selby, D.E., Palmer, B.M., LeWinter, M.M., Meyer, M., 2011. Tachycardia-Induced Diastolic Dysfunction and Resting Tone in Myocardium From Patients With a Normal Ejection Fraction. J. Am. Coll. Cardiol. 58, 147–154.

Singh, A., Mehta, Y., 2018. Heart failure with preserved ejection fraction (HFpEF): Implications for the anesthesiologists. J. Anaesthesiol. Clin. Pharmacol.

Wüst, R.C.I., Helmes, M., Martin, J.L., van der Wardt, T.J.T., Musters, R.J.P., van der Velden, J., Stienen, G.J.M., 2017. Rapid frequency-dependent changes in free mitochondrial calcium concentration in rat cardiac myocytes. J. Physiol. 595, 2001–2019.

Zile, M.R., Baicu, C.F., S. Ikonomidis, J., Stroud, R.E., Nietert, P.J., Bradshaw, A.D., Slater, R., Palmer, B.M., Van Buren, P., Meyer, M., M. Redfield, M., A. Bull, D., L. Granzier, H., LeWinter, M. M., 2015. Myocardial Stiffness in Patients With Heart Failure and a Preserved Ejection Fraction: Contributions of Collagen and Titin. Circulation 131, 1247–1259.

Advances and Current Trends in Biomechanics – Belinha et al. (Eds)

Numerical approaches to describe sarcomere functioning

M.M.A. Peyroteo
Institute of Science and Innovation in Mechanical and Industrial Engineering, Porto, Portugal

J. Belinha
School of Engineering, Polytechnic of Porto, Porto, Portugal

I. Falcão-Pires & A.F. Leite-Moreira
Faculty of Medicine, University of Porto, Porto, Portugal

R. Natal Jorge
Faculty of Engineering, University of Porto, Porto, Portugal

ABSTRACT: Sarcomere dysfunction is a common factor in many cardiovascular diseases, hence a thorough understanding of sarcomere functioning is required to identify the main pathologic pathways in each disease. Mathematical modelling is one of the solutions to tackle this problem, since it is a powerful tool capable to describe any phenomenon in an interactive and fast way. Thus, this work will review the main theoretical mechanisms of sarcomere functioning, including its active and passive responses, as well as present the numerical models currently used to explicitly describe these distinct behaviors. Despite their impressive progress throughout the years, numerical models are still lacking on a more detailed description of crossbridge attachment and titin-actin interaction.

1 SARCOMERE FUNCTIONING

1.1 *Active behavior*

Sarcomere is the contractile unit of a cardiac cell, consisting on intercalated myosin filaments and actin filaments disposed in parallel. Each myosin filament is composed of several myosin heads that can attach to actin, forming a crossbridge (XB), due to the interaction between calcium ions (Ca^{2+}) and two proteins composing the actin filament - troponin (Tn) and tropomyosin (Tm) – which together form the troponin–tropomyosin regulatory unit (TTRU) (Galińska-Rakoczy et al., 2008). Sarcomere contraction occurs when Ca^{2+} binds to Tn and induces a conformational change on Tm. Tm achieves a new configuration (i.e. permissive state), exposing the sites where a myosin head can attach to an actin filament, forming a XB (Bremel and Weber, 1972; Trybus and Taylor, 1980). These XB's generate an active force, causing the sliding of actin past myosin, resulting in sarcomere shortening, as explained by the sliding filament theory (Huxley and Niedergerke, 1954; Huxley and Hanson, 1954). When compared to skeletal muscle, cardiac tissue has a higher Ca^{2+} sensitivity, meaning that even low Ca^{2+} concentrations lead to elevated active force. This phenomenon has been explained by cooperative mechanisms in the sarcomere, under three distinct theories (Dupuis et al., 2016). The mechanism XB-TTRU proposes that the attachment of a XB increases the affinity of Tn to Ca^{2+}, suggesting that a bound XB tends to hold Tm in a permissive state (Swartz and Moss, 1992). The mechanism XB-XB states that attached XB's induce the attachment of nearby

DOI: 10.1201/9781003217152-80

XB's, even at low Ca^{2+} concentrations (Fitzsimons et al., 2001). Lastly, the mechanism TTRU-TTRU proposes that a Tm molecule in a permissive state induces conformational changes at nearby Tm's, promoting these molecules to permissive states too (Brandt et al., 1987).

1.2 *Passive behavior*

Sarcomere is also composed by titin filaments that run alongside the myosin filaments. Titin is a protein with extensible and inextensible portions (Labeit and Kolmerer, 1995). Its extensible portion, often known as I-band, unfolds during muscle relaxation, allowing the sarcomere to recover to its resting length (Tskhovrebova and Trinick, 1997). This elastic property is strongly associated with sarcomere's passive behavior as it occurs in the absence of Ca^{2+}. However, titin is a modular polymer, meaning that the I-band can be divided into distinct domains, each one with a different force-extension relationship. These domains are a proximal immunoglobulin region, a middle N2B segment, a long segment rich in proline (P), glutamate (E), valine (V) and lysine (K) residues, named the PEVK region, and a distal immunoglobulin region (Trinick and Tskhovrebova, 2010). It is very important to experimentally describe each of these domains and understand their individual impact on sarcomere's functioning, since titin dysfunctions observed in some pathologies can differ on the impaired domain.

Some studies suggest that titin could also play an important role in sarcomere's active response. When XB attachment is blocked, experimental studies showed that a force is generated through interactions between actin and titin when Ca^{2+} activated (Herzog et al., 2012; Leonard et al., 2010). Therefore, sarcomere functioning depends on this constant interplay between actin, myosin and titin.

2 SARCOMERE NUMERICAL MODELS

Despite being an extremely ambitious task, mathematical models have been able to describe sarcomere's functioning. The first models attempting it were mean field models. Several works have been proposed within this framework presenting acceptable force-length and force-velocity relationships at different Ca^{2+} concentrations (Landesberg and Sideman, 1994; Razumova et al., 1999; Rice et al., 1999). However, as scientists gathered more information on sarcomere dynamics, mean field models were not enough as they were unable to describe the cooperative mechanisms discovered in the meantime. Numerical models were forced to become more complex and refined, as it became clear that the spatial organization of each component of the sarcomere had an impact on the contraction process. Therefore, spatially explicit models were devised. With these models, it was now possible to explicitly incorporate myosin filament and actin filament overlap and the cooperative mechanisms between TTRU's and XB's, which were key factors in sarcomere functioning. Different approaches have been used to describe active and passive sarcomere's behavior. Therefore, a distinction will be made between active and passive numerical models.

2.1 *Active numerical models*

The first spatially explicit model considered a one dimensional array representing the cardiac actin filament discretized with 26 TTRU's (Rice et al., 2003). Using an Ising model approach combined with a Markov chain, each TTRU could shift from 4 distinct states: 0P, 1P, 0N, 1N. Transitions from 0 to 1 represented Ca^{2+} binding to Tn, while transitions from N to P represented the conformational changes of Tm from a non-permissive state to a permissive state, exposing the binding sites between actin and myosin. Assuming distinct transition rates, the model is able to determine the number of TTRU's at a permissive state and correlate it with the active force generated, obtaining accurate Force-pCa relations, with similar shape and steepness to the experimental curves in the literature. However, actin filament functioning was only controlled by Ca^{2+}, neglecting any co-operative mechanism and actin-myosin overlap. A distinct study proposed a Markov model of the actin filament, but included TTRU-TTRU cooperative, allowing a more accurate description of the crossbridges between actin and myosin filaments (Campbell et al., 2010). However, both of these models only considered isometric contractions, due to the elevated computational cost. To overcome this limitation, a following study (Washio et al., 2012) proposed a novel ordinary differential equation (ODE) model, based on the previous model proposed by Rice et. al., but including now actin and myosin filaments and their overlap. The reduced computational cost allowed a wider range of simulations, accounting now for sarcomere length variation. With some mathematical manipulation, the computational cost of this model was further reduced (Regazzoni et al., 2018), and used to study steady-state responses and dynamic responses to distinct Ca^{2+} and sarcomere lengths. Currently, active models are still lacking on a complete description of XB cycling, including energy consumption, as it is often associated with sarcomere dysfunctions.

2.2 *Passive numerical models*

Passive sarcomere behavior has also been mathematically described by several works. A stochastic approach combined with a worm like-chain (WLC) model was able to describe the probability of titin's extension at distinct force levels (Rief et al., 1998). As previously mentioned, titin is a modular polymer, hence the model describes the variation of titin's length according to the number of folded and unfolded domains and the respective folded and unfolded length of each domain. To better characterize titin's elongation, a following study (Li et al., 2002) experimentally described the several domains of the elastic region of titin (i.e. I-band), obtaining more accurate parameters for the WLC model previously proposed, successfully fitting their numerical results into experimental force-extension curves. Following studies (Schappacher-Tilp et al., 2016, 2015) extended these works combining them with active numerical models for a full description of sarcomere functioning. Following a distinct approach, another model (Land et al., 2017) described sarcomere's passive response using spring elements in series and in parallel to reproduce the elastic response shown in experimental data. However, some aspects are still missing, such as Ca^{2+}-induced changes in titin properties and the possible attachment between actin and titin.

3 CONCLUSIONS

This work presents a review on the main numerical models proposed to reproduce sarcomere's functioning. A distinction is made between active and passive responses, as they have distinct main biological pathways. Active models are still lacking on explicit description of XB cycling, while passive models should include titin's interaction with actin. Moreover, it would be important to start combining these responses into a single model, as some pathologies present simultaneous dysfunctions in active and passive response (Tajsharghi and Oldfors, 2013; Tharp et al., 2019). Moreover, in regards to cardiac muscle tissue, sarcomere functioning should be coupled with electrophysiology and hemodynamic models for a full description of muscle response.

ACKNOWLEDGEMENTS

The authors truly acknowledge the funding provided by Ministério da Ciência, Tecnologia e Ensino Superior – Fundação para a Ciência e a Tecnologia (Portugal), under grants: SFRH/BD/133105/2017. Additionally, the authors gratefully acknowledge the funding provided by LAETA, under project UIDB/50022/2020.

REFERENCES

Brandt, P.W., Diamond, M.S., Rutchik, J.S., Schachat, F.H., 1987. Co-operative interactions between troponin-tropomyosin units extend the length of the thin filament in skeletal muscle. J. Mol. Biol. 195, 885–896.

Bremel, R.D., Weber, A., 1972. Cooperation within actin filament in vertebrate skeletal muscle. Nat. New Biol. 238, 97–101.
Campbell, S.G., Lionetti, F. V., Campbell, K.S., McCulloch, A.D., 2010. Coupling of adjacent tropomyosins enhances cross-bridge-mediated cooperative activation in a markov model of the cardiac thin filament. Biophys. J. 98, 2254–2264.
Dupuis, L.J., Lumens, J., Arts, T., Delhaas, T., 2016. Mechano-chemical Interactions in Cardiac Sarcomere Contraction: A Computational Modeling Study. PLOS Comput. Biol. 12, e1005126.
Fitzsimons, D.P., Patel, J.R., Moss, R.L., 2001. Cross-bridge interaction kinetics in rat myocardium are accelerated by strong binding of myosin to the thin filament. J. Physiol. 530, 263–272.
Galińska-Rakoczy, A., Engel, P., Xu, C., Jung, H.S., Craig, R., Tobacman, L.S., Lehman, W., 2008. Structural Basis for the Regulation of Muscle Contraction by Troponin and Tropomyosin. J. Mol. Biol. 379, 929–935.
Herzog, W., Duvall, M., Leonard, T.R., 2012. Molecular mechanisms of muscle force regulation: A role for titin? Exerc. Sport Sci. Rev. 40, 50–57.
Huxley, A.F., Niedergerke, R., 1954. Structural changes in muscle during contraction: Interference microscopy of living muscle fibres. Nature 173, 971–973.
Huxley, H., Hanson, J., 1954. Changes in the Cross-striations of muscle during contraction and stretch and their structural interpretation. Nature 173, 973–976.
Labeit, S., Kolmerer, B., 1995. Titins: Giant proteins in charge of muscle ultrastructure and elasticity. Science (80-.). 270, 293–296.
Land, S., Park-Holohan, S.J., Smith, N.P., dos Remedios, C.G., Kentish, J.C., Niederer, S.A., 2017. A model of cardiac contraction based on novel measurements of tension development in human cardiomyocytes. J. Mol. Cell. Cardiol. 106, 68–83.
Landesberg, A., Sideman, S., 1994. Coupling calcium binding to troponin C and cross-bridge cycling in skinned cardiac cells. Am. J. Physiol. - Hear. Circ. Physiol. 266, 1260–1271.
Leonard, T.R., DuVall, M., Herzog, W., 2010. Force enhancement following stretch in a single sarcomere. Am. J. Physiol. Physiol. 299, C1398–C1401.
Li, H., Linke, W.A., Oberhauser, A.F., Carrion-Vazquez, M., Kerkvliet, J.G., Lu, H., Marszalek, P.E., Fernandez, J.M., 2002. Reverse engineering of the giant muscle protein titin. Nature 418, 998–1002.
Razumova, M. V., Bukatina, A.E., Campbell, K.B., 1999. Stiffness-distortion sarcomere model for muscle simulation. J. Appl. Physiol. 87, 1861–1876.
Regazzoni, F., Dedè, L., Quarteroni, A., 2018. Active contraction of cardiac cells: a reduced model for sarcomere dynamics with cooperative interactions. Biomech. Model. Mechanobiol. 17, 1663–1686.
Rice, J.J., Stolovitzky, G., Tu, Y., De Tombet, P.P., 2003. Ising model of cardiac thin filament activation with nearest-neighbor cooperative interactions. Biophys. J. 84, 897–909.
Rice, J.J., Winslow, R.L., Hunter, W.C., 1999. Comparison of putative cooperative mechanisms in cardiac muscle: Length dependence and dynamic responses. Am. J. Physiol. - Hear. Circ. Physiol. 276, 1734–1754.
Rief, M., Fernandez, J.M., Gaub, H.E., 1998. Elastically coupled two-level systems as a model for biopolymer extensibility. Phys. Rev. Lett. 81, 4764–4767.
Schappacher-Tilp, G., Leonard, T., Desch, G., Herzog, W., 2016. Computing Average Passive Forces in Sarcomeres in Length-Ramp Simulations. PLOS Comput. Biol. 12, e1004904.
Schappacher-Tilp, G., Leonard, T., Desch, G., Herzog, W., 2015. A Novel Three-Filament Model of Force Generation in Eccentric Contraction of Skeletal Muscles. PLoS One 10, e0117634.
Swartz, D.R., Moss, R.L., 1992. Influence of a strong-binding myosin analogue on calcium-sensitive mechanical properties of skinned skeletal muscle fibers. J. Biol. Chem. 267, 20497–20506.
Tajsharghi, H., Oldfors, A., 2013. Myosinopathies: Pathology and mechanisms. Acta Neuropathol.
Tharp, C.A., Haywood, M.E., Sbaizero, O., Taylor, M.R.G., Mestroni, L., 2019. The Giant Protein Titin's Role in Cardiomyopathy: Genetic, Transcriptional, and Post-translational Modifications of TTN and Their Contribution to Cardiac Disease. Front. Physiol. 10, 1436.
Trinick, J., Tskhovrebova, L., 2010. Roles of titin in the structure and elasticity of the sarcomere. J. Biomed. Biotechnol.
Trybus, K.M., Taylor, E.W., 1980. Kinetic studies of the cooperative binding of subfragment 1 to regulated actin. Proc. Natl. Acad. Sci. U. S. A. 77, 7209–7213.
Tskhovrebova, L., Trinick, J., 1997. Direct visualization of extensibility in isolated titin molecules. J. Mol. Biol. 265, 100–106.
Washio, T., Okada, J.I., Sugiura, S., Hisada, T., 2012. Approximation for cooperative interactions of a spatially-detailed cardiac sarcomere model. Cell. Mol. Bioeng. 5, 113–126.

Advances and Current Trends in Biomechanics – Belinha et al. (Eds)
© 2022 Copyright the Author(s), ISBN: 978-1-032-10806-3

Bite force evaluation in patients with removable denture

A. Fernandes
Faculty of Dental Medicine, University of Porto, Porto, Portugal

R.C. Stegun
Faculty of Dentistry, University of São Paulo, São Paulo, Brazil

S.J. Oliveira, M. Sampaio-Fernandes, J.C. Reis-Campos & M.H. Figueiral
INEGI, Faculty of Dental Medicine, University of Porto, Porto, Portugal

ABSTRACT: Bite force suffers changes throughout life, with a physiological decline during aging, due to the loss of muscle mass and teeth. Rehabilitation with removable dentures (RD) allows restoration of the masticatory performance. The current work aims to assess i) the bite force before and after oral rehabilitation with various types of RD, and ii) whether the bite force of treated patients is similar to that observed in a reference group (Gref). The sample consisted of 37 patients rehabilitated with RD: flexible partial RD (PRDflex; n=18), conventional PRD (PRDconv; n=17) and complete denture (CD; n=2). The Gref (n=40) was composed of dentate young adults. The bite force, measured with a gnathodynamometer, was recorded bilaterally with molars in occlusal contact. Statistical analysis was conducted with the GraphPad Prism® software. The bite force increased after oral rehabilitation regardless of the type of denture used, although not reaching the optimal values of the Gref.

1 INTRODUCTION

Bite force consists in the strength of the jaw muscles that determines the amount of energy available to cut or grind food. Its intensity is essentially determined by the muscle capacity and the number of teeth present in occlusion (Braun et al. 1995). The evaluation of bite force has been widely used in dental medicine to explore the mechanisms of mastication, assist in the diagnosis of disorders of the stomatognathic system, assess the effects of prosthodontic rehabilitation, and, to find reference values enabling the understanding of the biomechanics of the different types of dental prostheses (Koc et al. 2010). The bite force changes during life, with an increase until adulthood and a physiological functional decline during aging, resulting from the loss of both muscle mass and teeth. Prosthodontic rehabilitation has been crucial to reestablish the phonetics, aesthetics and masticatory function, namely by recovering the bite force of individuals with tooth loss (De Carvalho Dias et al. 2016). Removable dentures (RD), considered as a non-invasive and low-cost solution (compared to more expensive fixed options), can be divided into two main categories according to the extension: complete dentures (CD) and partial dentures (PRD). Depending on the material in which the skeleton is made, RD can be subdivided into acrylic (polymethylmethacrylate, PPMA), flexible (thermoplastic resins) or cobalt-chromium (Co-Cr).

The current research aims to: a) analyze the bite force before and after oral rehabilitation with RD; b) compare the values of bite force obtained with conventional PRD (Co-Cr and acrylic; PRDconv), flexible (PRDflex) and CD; and c) assess whether after rehabilitation with RD, the bite force values are close to the reference group (dentate young adults, Gref).

DOI: 10.1201/9781003217152-81

2 MATERIAL AND METHODS

The sample consisted of 37 patients that underwent prosthodontic treatment with PRDflex (n=18), PRDconv (n=17) and CD (n=2). These participants were subjected to bite force testing before and after prosthodontic rehabilitation. The Gref group (n=40) was composed of young adults (age between 22 and 29 years) presenting all teeth (excluding third molars).

To record the maximum bite force, a gnathodynamometer (Kratos® model DDK/M) was used. This apparatus is integrated by a load cell that allows measuring the maximum force (in daN) exerted by dental pairs in occlusion. The load cell has two blades each 6 mm thick with 3 mm space between them, being able to detect up to 100 daN of force. Disinfection of the gnathodynamometer was performed with disinfectant wipes for one minute before and after each measurement. Besides, the load cell was protected with a finger from a latex glove to avoid cross contamination. With the participants installed in a chair with a 90° inclination between the backrest and the seat, maintaining the Frankfurt plane parallel to the ground, instructions were given to bite the load cell with as much force as possible without manifesting any type of pain and/or discomfort for 2 to 3 seconds (Figure 1). This procedure was performed in the left and right molar region. Two measurements were made at each point, with the first being used only for calibration. Statistical analysis was performed using the GraphPad Prism® software (version 5.0, GraphPad Software, Inc., 2007) considering $p<0.05$ as the level of statistical significance. Values are expressed as mean±standard deviation).

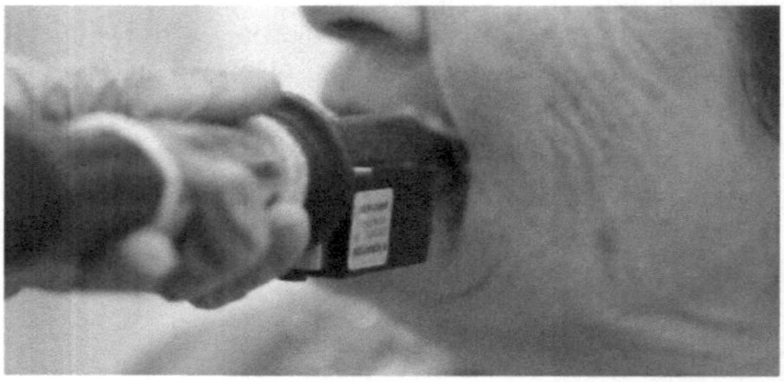

Figure 1. Measurement of maximum bite force with the gnathodynamometer.

3 RESULTS

Analysing the demographic characteristics of the sample, in terms of age, statistically significant differences are only found for the Gref, compared to the other groups ($p<0.0001$ for PRDflex and PRDconv and $p<0.05$ for CD). The female:male ratio is identical in the groups treated with RD (PRDflex, PRDconv and CD); however, a clear predominance of females is observed in the Gref group. Regarding the initial bite force, the mean value is significantly higher in the Gref (16,6 daN±10,1) in comparison with both the PRDflex (3,5 daN±5,7) and PRDconv (1,5 daN ±3,4) groups ($p<0.0001$), with no statistical differences between the PRDflex, PRDconv and CD. The absence of statistical differences between the initial bite force of the Gref and CD (0,0 ±0,0) groups can be attributed to the small number of patients rehabilitated with CD.

The intra-group comparison of the bite force values (Figure 2) reveals that after oral rehabilitation with PRDflex and PRDconv, the mean values are significantly higher in relation to the corresponding initial data ($p=0.0096$ and $p=0.0033$, respectively). This pattern is also observed in the group treated with CD, whose statistical analysis cannot be performed due to the small sample size.

The inter-group analysis of bite force values (Figure 3) shows that there are no significant differences between the mean values of bite force after oral rehabilitation with PRDflex, PRDconv and CD ($p=0.277$). The comparison of bite force values after oral rehabilitation

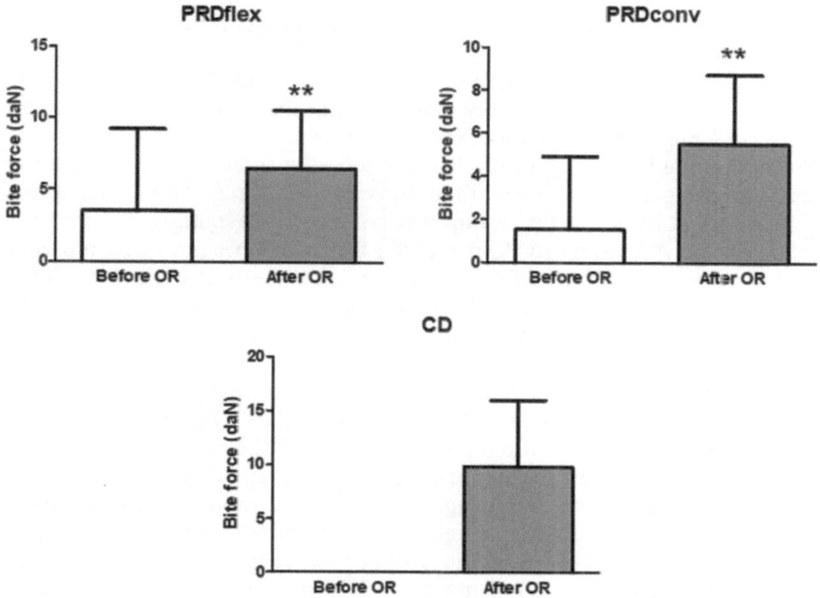

Figure 2. Bite force (daN) before and after oral rehabilitation (OR) with flexible partial removable dentures (PRDflex), conventional partial removable dentures (PRDconv) and complete dentures (CD). Values are expressed as mean±standard deviation. ** p=0.0096 for PRDflex and **p=0.0033 for PRDconv – Wilcoxon test.

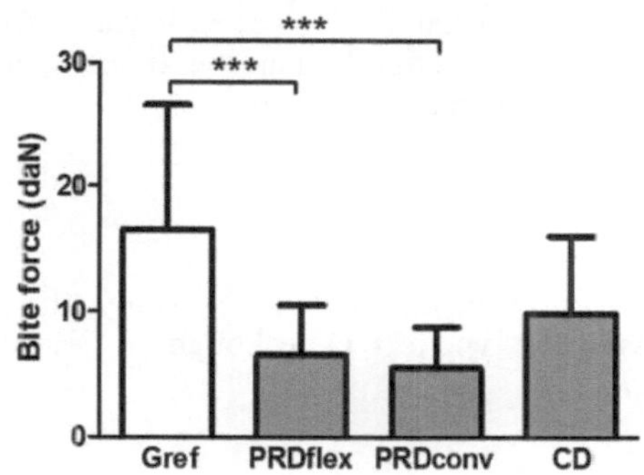

Figure 3. Comparison of the bite force (daN) after oral rehabilitation with flexible partial removable dentures (PRDflex), conventional partial removable dentures (PRDconv), complete dentures (CD) and the reference group (Gref). Values are expressed as mean±standard deviation. ***p<0.0001– Kruskal-Wallis test, followed by Dunn's multiple comparisons test.

with the Gref shows that the mean values obtained after oral rehabilitation with PRDflex and PRDconv remain significantly lower than those of the Gref (p<0.0001). This tendency is also observed in the group rehabilitated with CD, although without statistical significance for the reasons above mentioned.

4 DISCUSSION

The assessment of maximum bite force has contributed to understanding the effects of prosthodontic rehabilitation, providing fundamental insight into the biomechanics of the different types of dental prostheses (Koc et al. 2010). Comparing the intra-group bite force values, it was found that after oral rehabilitation with both PRDflex and PRDconv, the mean values were significantly higher in relation to the corresponding initial data, with a similar pattern being observed in the CD group. Such an increase may be attributed to the reestablishment of occlusal contacts after oral rehabilitation of individuals (Schimmel et al. 2017), with obvious benefits for them. This observation suggests that the type of removable denture does not appear to have a major influence on the patients' bite force.

By comparing the bite force after oral rehabilitation with the standard reference group, the mean values remain significantly lower than those of the Gref. This observation may be explained by the fact that the control group was composed of young participants with complete natural dentition. According to Peyron et al., from 40/50 years of age, muscle mass and strength begin to decrease, likewise affecting the bite force (Peyron et al. 2017) . Thus, despite restoring the number of dental pieces, and even considering the patient is completely adapted to the prosthodontic device, one should not expect the bite force values to be like the maximum scores observed in young adults.

The pilot study herein described had some limitations that are worth mentioning. The first refers to the small size of the sample, mainly in the case of the CD group (n=2), The inclusion of a larger number of participants would have also allowed the creation of subgroups, based on the extension and class of partial edentulism, gender, among others.

At the time of the bite force recording, the tip of gnathodynamometer increases the vertical dimension of occlusion, which might have a negative impact on the measurement (Waltimo & Könönen 1995). However, the distance between the upper and lower incisors caused by the gnathodynamometer did not exceed 20 mm, which is within the acceptable range for these records (Paphangkorakit & Osborn 1997). The bite force measuring was performed unilaterally, which may have led to denture instability in the contralateral side (Koc et al. 2010). To prevent this phenomenon, Schimmel et al. used a metallic piece with the same thickness of the sensor on the opposite side when assessing the bite force in patients with CD (Schimmel et al. 2017). However, it was difficult to obtain the exact value due to activation of the contralateral mandibular elevator muscles, a mechanism that always influences the ipsilateral forces.

5 CONCLUSIONS

In the experimental conditions used in this preliminary study, the bite force increased after oral rehabilitation regardless of the type of RD, although not reaching the optimal values of the standard reference group.

REFERENCES

Braun, S., Bantleon, H. P., Hnat, W. P., Freudenthaler, J. W., Marcotte, M. R. & Johnson, B. E. 1995. A study of bite force, part 1: Relationship to various physical characteristics. *The Angle Orthodontist*, 65, 367–72.

De Carvalho Dias, K., Da Fonte Porto Carreiro, A., Bastos Machado Resende, C. M., Soares Paiva Tôrres, A. C. & Mestriner Júnior, W. 2016. Does a mandibular RDP and new maxillary CD improve masticatory efficiency and quality of life in patients with a mandibular Kennedy class I arch? *Clinical Oral Investigations*, 20, 951–7.

Koc, D., Dogan, A. & Bek, B. 2010. Bite force and influential factors on bite force measurements: a literature review. *European Journal of Dentistry*, 4, 223–32.

Paphangkorakit, J. & Osborn, J. W. 1997. Effect of jaw opening on the direction and magnitude of human incisal bite forces. *Journal of Dental Research*, 76, 561–7.

Peyron, M. A., Woda, A., Bourdiol, P. & Hennequin, M. 2017. Age-related changes in mastication. *Journal of Oral Rehabilitation*, 44, 299–312.

Schimmel, M., Memedi, K., Parga, T., Katsoulis, J. & Müller, F. 2017. Masticatory Performance and Maximum Bite and Lip Force Depend on the Type of Prosthesis. *The International Journal of Prosthodontics*, 30, 565–572.

Waltimo, A. & Könönen, M. 1995. Maximal bite force and its association with signs and symptoms of craniomandibular disorders in young Finnish non-patients. *Acta Odontologica Scandinavica*, 53, 254–8.

Advances and Current Trends in Biomechanics – Belinha et al. (Eds)

Evaluation of masticatory efficiency in patients with removable dentures

M. Correia
Faculty of Dental Medicine, University of Porto, Porto, Portugal

R.C. Stegun
Faculty of Dentistry, University of São Paulo, São Paulo, Brazil

S.J. Oliveira, J.C. Reis-Campos, M. Sampaio-Fernandes & M.H. Figueiral
INEGI, Faculty of Dental Medicine, University of Porto, Porto, Portugal

ABSTRACT: Oral rehabilitation intends to restore masticatory, aesthetic and phonetic functions. In recent years, flexible dentures have been used as alternatives for conventional removable dentures in cobalt-chromium and acrylic resin. This work aims at evaluating the masticatory efficiency in patients treated with removable partial denture (RPD), studying the influence of both the use and the type of RPD (conventional or flexible). Using silicone-based tablets, thirty participants underwent masticatory efficiency tests (18 flexible; 12 conventional), before and after oral rehabilitation. Using a sieve series, the mean geometric diameter (MGD) of samples was calculated with Granucalc1.0 software. After oral rehabilitation a statistically significant decrease in the mean MGD values was found only in the RPD conventional group. The inter-group comparison of the average MGD after oral rehabilitation also showed statistically significant differences. The type of denture appears to influence the masticatory efficiency, with flexible dentures having the lowest performance under our experimental conditions.

1 INTRODUCTION

Chewing is one of the most important functions of the stomatognathic system, being a stimulus for the growth, development and maintenance of this system (Bogucki & Kownacka 2018, Santos et al. 2006, Bessadet et al. 2013). Edentulism causes diverse changes in the craniofacial complex that alter the functional balance of the stomatognathic system, affecting the chewing, swallowing and speech (Bessadet et al. 2013). The main objective of oral rehabilitation in general and of removable dentures in particular, is to restore the masticatory, aesthetic and phonetic functions, thus improving the quality of life of edentulous individuals (Hazari et al. 2015, Torres et al. 2019). Masticatory function can be assessed by both objective and subjective methods.

Over the years, denture base materials have been a constantly evolving issue. Removable dentures in Co-Cr and acrylic resins, known as conventional, are widely used and studied (Torres et al. 2019, Cardoso et al. 2016), but the more recently developed flexible dentures have attracted the interest of many researchers (Bogucki & Kownacka 2018, Fueki et al. 2014).

The present study aims at evaluating the masticatory efficiency in patients treated with RPD, with a focus on the influence of both the use and type of the RPD (conventional or flexible dentures) on this parameter. The null hypotheses were therefore as follows: H01 – There are no

DOI: 10.1201/9781003217152-82

statistically significant differences in masticatory efficiency before and after oral rehabilitation with removable dentures; H02 – The type of denture base material (flexible or conventional) does not have a statistically significant effect on masticatory efficiency.

2 MATERIAL AND METHODS

Masticatory efficiency tests were performed on the 30 participants before and after RPD placement [flexible RPD (RPDflex, n=18) or conventional RPD (RPDconv, n=12)]. The test food, consisting of 12 silicone-based tablets, was chewed for 20 seconds and the content of each test went through a series of sieves with decreasing mesh size. The content weight of each sieve was assessed (Figure 1) and recorded. Based on these values, the mean geometric diameter (MGD) of each sample was calculated with the corresponding mean standard deviation (MDP) using the Granucalc1.0 software. MGDmax was calculated by simulating that all particles were retained in the largest mesh sieve, corresponding to 7998 μm.

Statistical analysis was performed with the GraphPad Prism® software (version 5.0, GraphPad Software, Inc., 2007), with a significance level of 0.05 for a 95% confidence interval, meaning that significant results were considered for $p<0.05$. Values are expressed as mean ±standard deviation.

Figure 1. Sieves' content after the sieving process with decreasing mesh size.

3 RESULTS

In this study, MGD was used as an indicator of the masticatory efficiency, with lower values corresponding to better masticatory efficiencies.

In the group treated with flexible RPD, no statistically significant differences between the mean MGD values before and after denture placement were observed ($p=0.1387$; Figure 2A). On the contrary, for the conventional RPD group, there was a statistically significant decrease in the mean MGD values after denture placement ($p=0.02$; Figure 2B). The inter-group comparison of the mean DGM between the RPDflex and RPDconv groups revealed statistically significant differences ($p=0.0168$; Figure 3), with improved masticatory efficiency being observed in patients with conventional dentures.

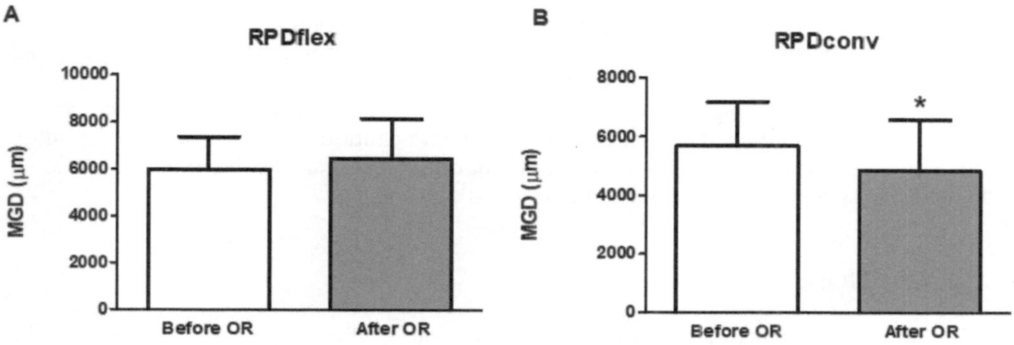

Figure 2. Average MGD values before and after oral rehabilitation with RPD (intra-group comparison). A) Average MGD values of patients treated with flexible RPD (RPDflex) before and after oral rehabilitation (OR); B) Average MGD values of patients treated with conventional RPD (RPDconv) before and after OR. Values are expressed as mean±standard deviation. *p=0.02 – Paired t-test.

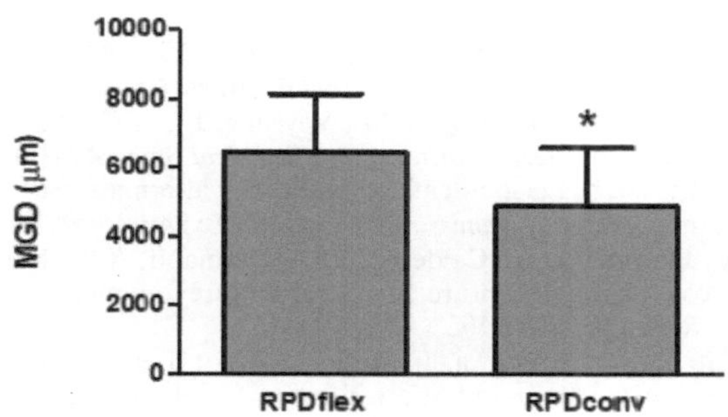

Figure 3. Inter-group comparison of the average MGD values after oral rehabilitation (OR) with flexible RPD (RPDflex) and conventional RPD (RPDconv). Values are expressed as mean±standard deviation. *p=0.0168 – Mann-Whitney test.

4 DISCUSSION

Based on these results, the first null hypothesis (H01) is partially rejected, since after oral rehabilitation with RPDconv the masticatory efficiency was significantly improved, although this was not observed in the RPDflex group. The second null hypothesis (H02) is also rejected as the type of RPD influenced the participants' masticatory efficiency.

According to the literature, masticatory efficiency is impaired in partial edentulous patients, depending on the number and health of the natural remaining teeth, being significantly improved after denture placement (Cardoso et al. 2016, Morita et al. 2018). Concerning the patients treated with conventional RPD, our results lend support to these previous findings. Oral rehabilitation with flexible dentures, however, produced a distinct pattern, with participants showing a subtle reduction of their masticatory efficiency, although not statistically significant. The inter-group analysis corroborates these results, as oral rehabilitation with RPDconv determined enhanced masticatory efficiency in comparison with RPDflex.

Flexible dentures have been presented as interesting alternatives to conventional dentures, particularly in patients with high aesthetic expectations or in case of allergy to acrylic or metal components. In fact, flexible dentures have better aesthetic properties, high flexibility and good fracture resistance, which are compatible with thinner and less porous prosthetic bases. Collectively, these properties improve comfort and satisfaction of denture wearers (Bogucki & Kownacka 2018,

Hazari et al. 2015, Fueki et al. 2014, Singh et al. 2013). On the contrary, conventional dentures provide a firmer prosthetic base, with greater hardness and a minimum deformation capacity and flexibility during function (Bogucki & Kownacka 2018, Akinyamoju et al. 2019).

Although our findings might be related to the type of material used in the manufacturing of the RPD, the results obtained should be interpreted with caution since in this study the adaptation period to the new dentures was not standardized between flexible and conventional RPD, which may represent an important bias.

5 CONCLUSIONS

Despite the methodological limitations of this study, our preliminary results suggest that the use of flexible dentures did not improve the masticatory efficiency, with the opposite being observed with conventional dentures. Overall, these findings suggest that the type of denture influences the masticatory efficiency of patients.

REFERENCES

Akinyamoju, C.A., Dosumu, O.O., Taiwo, J.O., Ogunrinde, T.J., Akinyamoju. A.O. 2019. Oral health-related quality of life: acrylic versus flexible partial dentures. *Ghana Medical Journal* 53(2):163–9.

Bessadet, M., Nicolas, E., Sochat, M., Hennequin, M., Veyrune, J.L. 2013. Impact of removable partial denture prosthesis on chewing efficiency. *Journal of Applied Oral Science* 21(5),392–6.

Bogucki, Z.A., Kownacka, M. 2018. Elastic dental prostheses - alternative solutions for patients using acrylic prostheses: Literature review. *Advances in Clinical and Experimental Medicine* 27(10):1441–5.

Cardoso, R.G., Melo, L.A., Barbosa, G.A., Calderon, P.D., Germano, A.R., Mestriner, W.J. et al. 2016. Impact of mandibular conventional denture and overdenture on quality of life and masticatory efficiency. *Brazilian Oral Research* 30(1):e102.

Fueki, K., Ohkubo, C., Yatabe, M., Arakawa, I., Arita, M., Ino, S. et al. 2014. Clinical application of removable partial dentures using thermoplastic resin-part I: definition and indication of non-metal clasp dentures. *Journal of Prosthodontic Research* 58(1):3–10.

Hazari, P., Bhoyar, A., Mishra, S.K., Yadav, N.S., Mahajan, H. 2015. A Comparison of Masticatory Performance and Efficiency of Complete Dentures Made with High Impact and Flexible Resins: A Pilot Study. *Journal of Clinical and Diagnostic Research* 9(6):ZC29–34.

Morita, K., Tsuka, H., Kato, K., Mori, T., Nishimura, R., Yoshida, M., et al. 2018. Factors related to masticatory performance in healthy elderly individuals. *Journal of Prosthodontic Research* 62(4):432–5.

Santos, C.F., Spadar, C., Mestriner-Júnior, W. 2006. Development of a Colorimetric System for Evaluation of the Masticatory Efficiency. *Brazilian Dental Journal* 17(2):95–99.

Singh, K., Aeran, H., Kumar, N., Gupta, N. 2013. Flexible thermoplastic denture base materials for aesthetical removable partial denture framework. *Journal of Clinical and Diagnostic Research* 7(10):2372–3.

Torres, A., Maciel, A.Q., De Farias, D.B., De Medeiros, A.K.B., Vieira, F., Carreiro, A. 2019. Technical Quality of Complete Dentures: Influence on Masticatory Efficiency and Quality of Life. *Journal of Prosthodontics* 28(1): e21–e6.

Advances and Current Trends in Biomechanics – Belinha et al. (Eds)
© 2022 Copyright the Author(s), ISBN: 978-1-032-10806-3

Incorporation of chlorhexidine in prosthetic base materials

C. Varejão & J.P. Vieira
Faculty of Dental Medicine, University of Porto, Portugal

M. Sampaio-Fernandes, S.J. Oliveira, J.C. Reis-Campos & M.H. Figueiral
INEGI, Faculty of Dental Medicine, University of Porto, Portugal

ABSTRACT: The incorporation of antimicrobial and antifungal agents in denture base materials can be an alternative and/or adjuvant therapy in the treatment of denture-related stomatitis, a pathology frequently associated with the presence of *Candida spp*. This review aims at studying the possibility of incorporating chlorhexidine (CHX) in denture base materials, focusing on its antifungal activity and putative effects on the material properties. An electronic search was performed in the Pubmed database using the keywords: "chlorhexidine", "dental prosthesis", "Candida", "denture stomatitis", "antimicrobal" and "incorporation". Twenty-one out of 155 identified articles were considered relevant for this work, 14 articles were added subsequently, totaling 35 articles. In contrast to acrylic resins, temporary relining materials showed better effectiveness, without compromising their physical properties. CHX incorporation in denture base materials appears to be favorable, since the antimicrobial agent remains at the site of action and lower concentrations of the agent are necessary to obtain the desired therapeutic effects.

1 INTRODUCTION

Denture-related stomatitis is characterized by a generalized inflammatory condition of the oral mucosa under total or partial dentures and is more prevalent in maxilla (Wen et al. 2016). The fungal infection, caused by *Candida albicans,* is an important factor for the development of this pathology (Figueiral et al. 2007).

In order to overcome both the persistence of the fungus on the surface of the prosthesis and on the adjacent mucosa and the recurrence of lesions, antimicrobial and antifungal agents have been incorporated in denture base materials, representing an alternative and/or coadjuvant therapy for the treatment of denture stomatitis (Vieira-Costa et al. 2017).

Chlorhexidine (CHX) is extensively used in Dental Medicine, since it acts against a wide range of Gram-positive and Gram-negative bacteria, yeast, filamentous fungi, and facultative anaerobic and aerobic organisms. This antimicrobial agent, opposed to antifungal drugs, does not create systemic toxicity neither resistance (Vieira-Costa et al. 2017, Bueno et al. 2015, Figueiral et al. 2015).

This work aims at studying the possibility of chlorhexidine incorporation in denture base materials, including tissue conditioners, liners and acrylic resins, also evaluating its effectiveness against *C. albicans* infection, and its putative effects on the physical properties of the prosthetic materials.

DOI: 10.1201/9781003217152-83

2 MATERIALS AND METHODS

In order to address our research question, a narrative literature search was carried out using the online database Pubmed. Various combinations of the following keywords were used: "chlorhexidine", "dental prosthesis", "Candida", "denture stomatitis", e "incorporation". A total of 155 articles were identified in the initial search. 21 articles were searched by title and abstract and full-text articles with no time restriction. Another 14 articles were included for their importance on the theme. Thus, a total of 35 studies were included in the review.

3 RESULTS

The studies addressing the incorporation, release and antifungal efficiency of chlorhexidine were grouped according to denture base material (Tables 1 and 2).

Table 1. Results on incorporation, release and efficacy of chlorhexidine in tissue conditioners and liners.

Tissue conditioners and Liners	Chlorhexidine	Study period	Incorporation	Antifungal activity
PEMA Softone™ - powder+liquid (Bueno et al. 2015)	Chlorhexidine diacetate (powder) 0.016 to 0.384/ml	24h, 48h, 7 and 14 days	On powder	MIC (Minimum Inhibitory Concentration) 0.064 g/ml. Absence of viable fungal cells at 24h and at 14 days.
Visco-gel® - powder+liquid (Radnai et al. 2009)	Chlorhexidine digluconate (gel) 5%,10%,15%, 20%, 25% volume	24h	Gel in the mixture powder/liquid	No inhibitory effect on the fungus (inhibition halos). Chlorhexidine did not diffuse over the material or was inactivated after incorporation.
PMMA Coe-Soft® - powder/liquid (Bertolini et al. 2014)	Chlorhexidine diacetate (powder) 0.5/1/2 wt.% Chlorhexidine Hydrochloride 0.5/1.0/2.0 wt.%	48h	On powder	Effectiveness of chlorhexidine diacetate 1.0 or 2.0%, after 48h.
PEMA Trusoft® - powder/liquid (Bertolini et al. 2014)	Chlorhexidine Diacetate (powder) 0.5/1/2 wt.% Chlorhexidine Hydrochloride 0.5/1.0/2.0 wt.%	48h	On powder	Chlorhexidine Hydrochloride had no inhibitory effect on *C. albicans*. Inhibition halos 1.5x higher than PMMA, after 48h.
PEMA Trusoft® - powder/liquid (Bueno et al. 2015)	Chlorhexidine Diacetate (powder) 0.016 to 0.384/ml	24h, 48h, 7 and 14 days	On powder	Effectiveness during 14 days. MIC=0,064 g/ml
PEM (Lucite) + THFM (Sigma-Aldrich) - powder/liquid (Salim et al. 2012)	Control group: 1g PEM + 0.6mL THFM Test group: 10% (w/w) of PEM replace by Chlorhexidine Diacetate (powder)	28 days	On powder	Effectiveness for 28 days. MIC = 5.03 mg/L.

(Continued)

Table 1. (*Continued*)

Tissue conditioners and Liners	Chlorhexidine	Study period	Incorporation	Antifungal activity
Acrylic gel EMA Visco-gel - powder/ liquid (Addy 1981)	Chlorhexidine Diacetate (powder) 2, 5 and 10%.	100 days	On powder	No Candida cultures. Release during 39 days, in Chlorhexidine Diacetate 5%.

MIC - minimum inhibitory concentration; PMMA - poly(methyl methacrylate); PEM+THFM – poly(ethyl methacrylate) +tetrahydro-furfuryl methacrylate; EMA - ethyl methacrylate.

Table 2. Results on incorporation, release and efficacy of chlorhexidine in acrylic resins.

Acrylic resins	Chlorhexidine	Study period	Incorporation	Antifungal activity
NVP monomer in PMMA = PNVP Lucitone® (Sun et al. 2013)	Chlorhexidine Diacetate (powder) 5 wt.%	14 days	The chlorhexidine incorporation capacity increases with the increase of the % NVP in the sample: - PMMA (samples without NVP) 5.72±0.34μg/cm^2. - 7.92% PNV 98.52 ±2.27μg/cm^2.	Antifungal activity increased with the increase of % PNVP. - 2.4% PNVP (2.41%) sustained release for 15 days (> MIC - (24.63±2.25μg/cm^2); - 7.92% PNVP, allowed a *Candida* reduction of about 63%; after 14 days, reduced for 46%.
Copolymerization of DUDMA with PMAA Sigma-Aldrich (Cao et al. 2010)	Chlorhexidine Diacetate (powder) 10%	14 days	Immersion of samples (aqueous solution during 24h). Binding capacity increased with the % MAA added. Faster release in acidic conditions.	MIC =17.24±0.2 μg/cm^2 Chlorhexidine content in the sample above the MIC value on the 14th day (neutral pH) and reached the MIC on the 7th day (pH 5)
Self-curing PMMA Paladur® (Amin et al. 2009)	Chlorhexidine gluconate (powder) 10% w/w	28 days	On powder.	Effectiveness for 28 days. Prolonged release for 28 days, in distilled water at the oral cavity temperature (high rate of release in the first 4 days; then, slower and gradual).
PMMA resin Lucitone + Coating (Redding et al. 2009)	Chlorhexidine Diacetate 1%	14 days	Immersion in solution.	Coating with chlorhexidine, both high and low porosity, reduced the biofilm by about 98%.
Self-curing Simplex (Lamb & Martin 1983)	Chlorhexidine acetate 2.5, 5, 7.5 e 10% w/w.	3 weeks	On powder. Immersion in water for 21 days.	Measurement of inhibition halos. Positive results for 3 weeks with 7.5% (MIC),

(*Continued*)

Table 2. (*Continued*)

Acrylic resins	Chlorhexidine	Study period	Incorporation	Antifungal activity
				although with alteration of physical properties.
Heat-curing (Trevalon universal) Cold-curing (CroForm) (Addy 1981)	Chlorhexidine acetate (powder) 2, 5 e 10%.	100 days	On powder. The release is proportional to increased concentration.	No *Candida* cultures. Chlorhexidine release: at 5% for 31 (cold) and 32 days (hot); at 10% for 100 days.

MIC - minimum inhibitory concentration; PMMA - poly(methyl methacrylate); PEM+THFM - poly(ethyl methacrylate) +tetrahydro-furfuryl methacrylate; EMA - ethyl methacrylate.

4 DISCUSSION

Taken together, the data collected show that in *in vitro* studies, CHX incorporation in the temporary relining materials, Trusoft and Softone (Bertolini et al. 2014, Bueno et al. 2015), showed better results for efficacy on *C. albicans*, without compromising their physical properties. In acrylic resins, in spite of the benefits of incorporating CHX for 28 days, changes in the physical properties of materials were reported, such as decreased resistance to fracture (Al-Haddad et al. 2014).

The numerous advantages of developing CHX refill systems in acrylic resin is also highlighted by various authors (Cao et al. 2010, Sun et al. 2013), especially in patients with immunosuppression and prosthetic stomatitis recurrence.

5 CONCLUSION

The incorporation of CHX in denture base materials has provided promising results, as the antimicrobial agent remains at the site of action and lower concentrations of the agent are necessary to achieve the desired therapeutic effect. Although valuable, the direct comparison of studies is often difficult because i) different methodologies are used, and ii) there is a great diversity of materials, with important implications in terms of physical properties.

REFERENCES

Addy, M. 1981. In vitro studies into the use of denture base and soft liner materials as carriers for drugs in the mouth. *Journal of Oral Rehabilitation* 8 (2): 131–142.

Al-Haddad, A., Vahid Roudsari, R. & Satterthwaite, J.D. 2014. Fracture toughness of heat cured denture base acrylic resin modified with Chlorhexidine and Fluconazole as bioactive compounds. *Journal of Dentistry*, 42 (2): 180–184.

Amin, W.M., Al-Ali, M.H., Salim, N.A. & Al-Tarawneh, S.K. 2009. A new form of intraoral delivery of antifungal drugs for the treatment of denture-induced oral candidosis. *European Journal of Dentistry* 3 (4): 257–266.

Bertolini, M.M., Portela, M.B., Curvelo, J.A., Soares, R.M., Lourenco, E.J. & Telles, D.M. 2014. Resins-based denture soft lining materials modified by chlorhexidine salt incorporation: an in vitro analysis of antifungal activity, drug release and hardness. *Dental Materials* 30 (8): 793–798.

Bueno, M.G., Urban, V.M., Barberio, G.S., da Silva, W.J., Porto, V.C., Pinto, L., & Neppelenbroek, K. H. 2015. Effect of antimicrobial agents incorporated into resilient denture relines on the Candida albicans biofilm. *Oral Diseases* 21 (1): 57–65.

Cao, Z., Sun, X., Yeh, C.K., Sun, Y. (2010). Rechargeable infection-responsive antifungal denture materials. *Journal of Dental Research* 89 (12): 1517–1521.

Figueiral, M.H., Azul, A.M., Pinto, E., Fonseca, P., Branco, F.M. & Scully, C. 2007. Denture-related stomatitis: identification of aetiological and predisposing factors – a large cohort. *Journal of Oral Rehabilitation* 34 (6): 448–455.

Figueiral, M.H., Fonseca, P., Lopes, M.M., Pinto E., Pereira-Leite, T. & Sampaio-Maia B. 2015. Effect of Denture-Related Stomatitis Fluconazole Treatment on Oral Candida albicans Susceptibility Pro-file and Genotypic Variability. *Open Dentistry Journal* 30 (1): 46–51.

Lamb, D.J. & Martin, M.V. 1983. An in vitro and in vivo study of the effect of incorporation of chlorhexidine into autopolymerizing acrylic resin plates upon the growth of Candida albicans. *Biomaterials* 4 (3): 205–209.

Radnai, M., Whiley, R., Friel, T. & Wright, P.S. 2010. Effect of antifungal gels incorporated into a tissue conditioning material on the growth of Candida albicans. *Gerodontology* 27 (4): 292–296.

Salim, N., Moore, C., Silikas, N., Satterthwaite, J.D. & Rautemaa, R. 2012. Fungicidal amounts of antifungals are released from impregnated denture lining material for up to 28 days. *Journal of Dentistry* 40 (6): 506–512.

Sun, X., Cao, Z., Yeh, C.K. & Sun, Y. 2013. Antifungal activity, biofilm-controlling effect, and biocompatibility of poly(N-vinyl-2-pyrrolidinone)-grafted denture materials. *Colloids and Surfaces B: Biointerfaces* 110: 96–104.

Vieira Costa, J., Alexandre, F., Bettencourt, A., Ribeiro, I., Portugal, J. & Bettencourt Neves, C. 2017. Incorporação de clorexidina em resinas acrílicas – Estudos microbiológicos. *Revista Portuguesa de Estomatologia, Medicina Dentária e Cirurgia Maxilofacial* 58 (S1): 51.

Wen J., Jiang F., Yeh, C. & Sun, Y. 2016. Controlling fungal biofilms with functional drug delivery denture biomaterials. *Colloids and Surfaces B: Biointerfaces* 140: 19–27.

Advances and Current Trends in Biomechanics – Belinha et al. (Eds)
© 2022 Copyright the Author(s), ISBN: 978-1-032-10806-3

Trunk alignment monitoring system in the context of hippotherapy

A.N. Costa, P.R. Gomes, R. Oliveira, F. Santos & J.M. Costa
Universidade Lusíada Norte, Portugal

A. Torrinha, F. Moreira & O. Dantas
Associação Portuguesa de Pais e Amigos do Cidadão Deficiente Mental, Braga, Portugal

A. Lousada
Escola de Equitação – Coudelaria Quinta Oliveira, Póvoa de Varzim, Portugal

C.S. Lima
Departamento de Eletrónica Industrial, Universidade do Minho, Portugal

ABSTRACT: This paper is concerned with a Help-Diagnosis System - Intellectual Disabilities (HDS-ID) that allows professionals in therapy to assess the alignment of the trunk of people with intellectual disabilities in hyppotherapy sessions (HTS). The main objective is to provide therapists a tool that:1) allows the therapist to monitor, in real-time, the trunk alignment autonomously, and 2) create support data to guide the therapists in a specific moment, as well to specify a task schedule, according to previous trunk alignment results. The tool was applied to 26 people with ID, aged between 20 and 46 years old in a therapeutic context. The results showed statistically significant individual differences in the alignment of the static trunk in the sagittal plane, between the biped position and the sitting position on the horse.

1 INTRODUCTION

In the context of rehabilitation programs for people with ID, several researchers have recognized in their studies, the relevance of using the horse's capabilities as a therapeutic tool. Its different attributes and mainly the characteristics of its movements in step (three-dimensional, rhythmic, repetitive and smooth) make it a great therapeutic tool, which improves the quality of life of people with ID in terms of attention, interaction, communication, posture, balance, motor coordination and self-confidence (Atun-Einy et al. 2017), (Espindula et al. 2016).

In addition to the similarity of the horse's movements, step by step, with the human's pelvic movements during a walk (Mikolajczyk et al. 2017), (Moraeset al. 2015), the horse's length and number of steps, per minute, are also similar to the length and number of steps of an adult (Mikolajczyk et al. 2017). These characteristics, together with the horse's ability to cause about 3000 oscillations in the rider's trunk (Mikolajczyk et al. 2017), (Shurtleff et al. 2010), allow the creation of multisensory input in the person, as well as constant dissociation movements of the pelvic girdle and scapular girdle, promoters of the development of adaptive responses to the environment (Atun-Einy et al. 2017), (Espindula et al. 2016), (Mikolajczyk et al. 2017).

In HTS, people with ID are accompanied by a multidisciplinary team composed of a specialist in riding, a therapist and, in some situations, by an auxiliary person. The need for the therapist to position himself laterally and close to people with disabilities, precludes an objective analysis of the rider's trunk, especially in people with a severe diagnosis of ID.

DOI: 10.1201/9781003217152-84

Recent studies refer to the use of three triple-axis accelerometers to compare the rider's movement during a HTS and during a walk (Pálinkás et al. 2013) and Wireless Body Sensor Networks (WBSN) to assess the rider's performance while riding the horse at a trot (Nerino et al. 2011). However, these devices do not allow for a real-time trunk correction.

The solution found consists of a tool that allows monitor the static and dynamic control of the rider's trunk, in real time, and provides a quick and efficient feedback of the possible changes that may occur during a hyppotherapy session, allowing the therapist to make the adjustments and necessary corrections (Gomes et al. 2020).

2 MATERIALS AND METHODS

This study involved 26 people with ID, who attend APPACDM (Portuguese Association of Parents and Friends of the Mentally Handicapped Citizen). The age group is between 20 and 46 years old. According to the Diagnostic and Statistical Manual of Mental Disorders (DSM-V), American Psychiatric Association (2013) 77% of participants have a moderate diagnosis of ID and 23% have a severe diagnosis of ID. The data were obtained in hyppotherapy sessions at Escola de Equitação – Coudelaria Quinta Oliveira, Póvoa de Varzim, Portugal and Clube Hípico do Norte, Esposende, Portugal.

The planning of HTS for people with ID involves the preparation of a set of tasks to evaluate the static and dynamic control of the trunk. One of these tasks for evaluating the dynamic control of the trunk consisted of a turn around the perimeter of an indoor arena (radius: 10m), with the participant looking ahead, Figure 1.

The sensory modules (SM) are based on an intelligent sensor (9-axis absolute orientation), placed on the patient at the level of the shoulder girdle (LSG) SM1, and SM2 fixed in front of the the double handled surcingle (DHS) on the horse. These SM are equipped with wireless communication (Wi-Fi). An application runs on the server that receives data from SM as well as controls and synchronizes wireless communication with these modules. The received data are stored in a Database (DB) created for this purpose. Two dedicated applications (Apps) have been developed: App_1 for data processing so that the therapist can make an assessment of the evolution status of the participants and can plan the therapy sessions, and App_2 to control and synchronize communication with mobile devices so that the therapist has the information, in real time, about the trunk alignment of the participants.

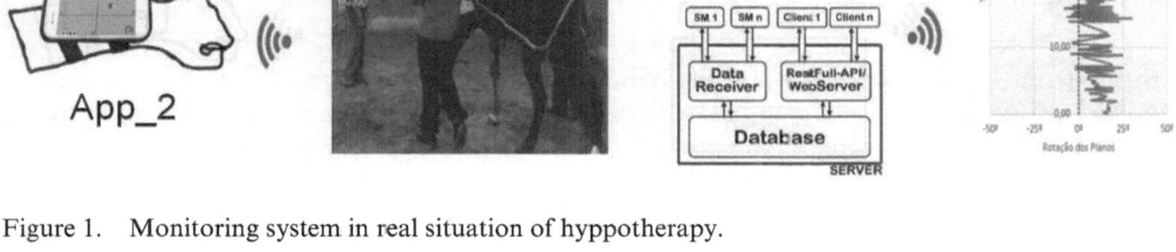

Figure 1. Monitoring system in real situation of hyppotherapy.

Several tasks are performed to validate the tool and the behaviour of all participants at the study, in order to evaluate the importance of the hyppotherapy sessions compared with conventional therapy sessions.

In an institutional context (APPACDM) and accompanied by a therapist, each participant, with a SM1 fixed at the LSG, performed two tasks:

Task 1.1: maintain the biped position, for 60 seconds, without supports, on a flat and hard floor;

Task 1.2: walk about 5 meters, without supports, on a flat and hard floor.

In the context of equestrian centers and accompanied by a riding specialist, a therapist and an assistant, each participant, with a SM1 attached to the LSG, performed four tasks:

Task 2.1: maintain the sitting posture on the horse for 60 seconds, with the horse in a stationary position and with correct alignment of the four limbs;

Task 2.2.1: walk of about 5 meters, with the horse walking, in a straight line, on a flat and hard floor;

Task 2.2.2: walk about 5 meters, in an arena, in a straight line, with the horse in step;

Task 2.3: a lap around the perimeter of an indoor arena (radius: 10 meters), with the participant looking ahead.

3 RESULTS

The study data were analyzed statistically using the Wilcoxon Test of the IBM SPSS Statistics for Windows Software, after confirming the normal distribution. Regarding static trunk control, it was found that all participants showed differences in trunk alignment when sitting on the horse, compared to the bipedal position. Figure 2 shows real-time behaviour of one participant performing Task 2.1, showing that the alignment of the trunk is closer to the midline of the body. There are statistically significant differences for the alignment of the static trunk in the sagittal plane (extension and flexion), in the biped and sitting position on the horse, with sig = 0.042 ($\alpha = 0.05$).

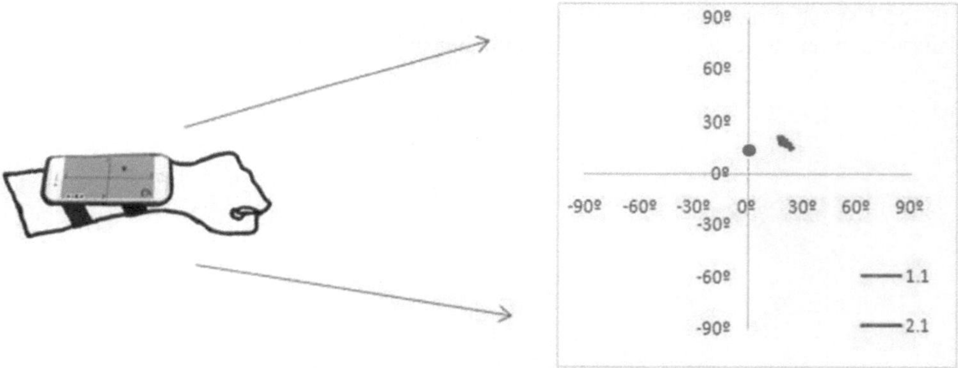

Figure 2. Real-time visualization (red) of Task 2.1 compared with previous data (blue) of Task 1.1.

In the frontal plane, 48% of the participants decreased the degrees of lateral trunk inclination to the right, towards the midline, when sitting on the horse, which represents an individual improvement in the trunk alignment. At the end of sessions of any task the therapist can obtain a scatter plot about the data collect during that session. The tool allows a comparison between the results obtained at institutional context and the hyppotherapy sessions.

4 DISCUSSION

Therapy sessions in an equestrian context have an average duration of about 25 minutes for each patient. However, when a participant has more difficulties, information are sent to the mobile device and therefore the therapist needs to make further adjustments during the session.

The developed system reinforces the versatility of the tool that can be used at different times in a rehabilitation program: initial assessment, follow-up, monitoring and final evaluation, specifically in hyppotherapy.

The system was able to capture the individual variations in the trunk alignment of all participants. Combined with objective feedback from the collected data, the therapist now has the possibility to act more effectively. Figure 3 show two possibilities to combine and compare data: 1) between different tasks or 2) the interaction between the participant and the horse.

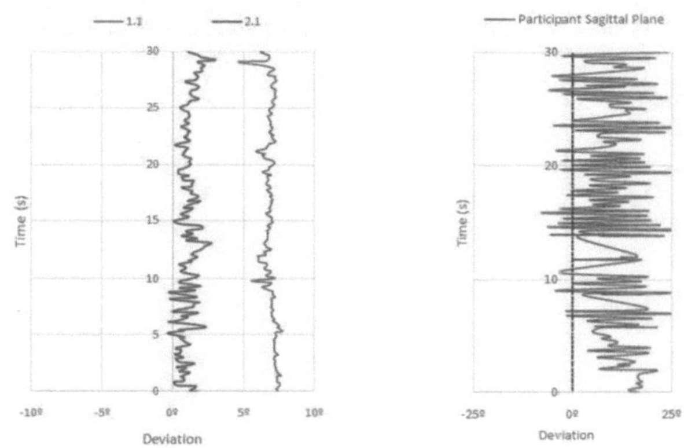

Figure 3. Compare data: 1) Tasks 1.1 and 2.1 (left); 2) Participant sagittal plane and horse (right).

5 CONCLUSIONS

The developed system is a versatile tool that can be used at different times in a rehabilitation program: initial assessment, follow-up, monitoring and final assessment, specifically in hyppotherapy. This tool also allows the assessment of trunk control in static tasks. Combined with an objective feedback of the acquired data, the therapist has the possibility to act in real time in a more effective and objective way.

This system is an effective diagnostic aid tool in real scenarios, which can be adapted to any context. Such a tool makes it possible to analyze the trunk of people with serious difficulties in understanding, difficulties in following guidelines or even collaborative difficulties. When compared to other tools and studies developed in the laboratory context, it has the advantage of being non-evasive and allowing an abstraction of the participants while the sessions are taking place.

ACKNOWLEDGMENT

COMEGI – UI4005, of University Lusíada, supported by national funds through FCT, I.P., Portugal under the project UIDB/04005/2020

REFERENCES

Atun-Einy O.,Lotan M.2017.A Systematic Review of Hippotherapy (horseback riding) for Individuals with Autism is the Cognitive Level Considered? Autism Open Acces, 7(4): 1–11.

EspindulaA.P.,et al.2016.Effects of Hippotherapy on Posture in Individuals with Down Syndrome. Fisioter Mov, 29(3): 497–506.

Gomes P.R., Costa A.N., Torrinha A., et al. 2020. Help-Diagnosis System for Trunk Alignment Evaluation of People with Intellectual Disabilities. IEEE 20th International Conference on BioInformatics and BioEngineering, USA: 715–718.

Mikolajczyk E., Ligeza B., Jankowicz-Szymanska A.2017.The Effect of Hippotherapy on Postural Balance. Eur J Clin Exp Med, 15(1): 45–49.

Moraes A.G., et al.2015.Equoterapia no Controle Postural e Equilíbrio emIndivíduos com Paralisia Cerebral: Revisão Sistemática. RevNeurocienc, 23(4): 546–554.

Nerino R., et al. 2011. WBSN for the Assessment of the Hippotherapy: A Case Study. Int'l Conf. on Body Sensor Networks, IEEE, Dallas: 101–106.

Pálinkás J., et al. 2013. Development and Characterization of a New Measurement Technique for Monitoring Changes in Acceleration during Hippotherapy. Int. Rev. Appl. Sci. Eng., 4: 21–26.

Shurtleff T.L., Engsberg J.R. 2010. Changes in trunk and Head Stability in Children with Cerebral Palsy after Hippotherapy: A Pilot Study. Physical & Occupational Therapy in Pediatrics, 30(2): 150–163.

Advances and Current Trends in Biomechanics – Belinha et al. (Eds)
© 2022 Copyright the Author(s), ISBN: 978-1-032-10806-3

Numerical comparison of the bending properties from three different osteosynthesis plate systems: DCP, LC-DCP and LCP

F.C.P. Sales, F.R.S. Oliveira & J.E. Ribeiro
School of Technology and Management, Polytechnic Institute of Bragança (IPB), Bragança, Portugal

R.R.C. da Costa
Department of Mechanical Engineering, Federal Technological University of Paraná, Cornélio Procópio, Brazil

ABSTRACT: Locking Compression Plates (LCP) are the most recent improvement on osteosynthesis plates. However older plates as DCP are still widely used due to low cost when compared with LCP (about 10-20%). Currently there are few studies that compares the three types of plates to each other. Additionally, due the plates geometry complexity, numerical validations are difficult to be obtained. The objective of this study was to compare the DCP, LC-DCP and LCP plates stiffness with actual and simplified holes using finite element analysis. Numerical models were developed applying appropriated boundary conditions to simulate a 4-point bending test in accordance to ISO 9585. 8 holes DCP models were validated with tests results from reference. Results showed that simplified models had higher properties than non-simplified, besides that, the three plates systems can be used safely, imposing rigid stabilization on fractures site.

1 INTRODUCTION

After a fracture, various medical procedures can be applied to ensure the right alignment and stability between the bone's segments. Conservative management prevents movement in the fractured regions and can be obtained by using casts and splints. On the other hand, operative treatment uses devices as plates, screws, wires, and pins placed internally, near the bone surface, or externally (Bavornratanavech, 2007). Moreover, internal fixation with osteosynthesis plates fulfils mechanical requirements as load transmission from one end of the bone to another and protection against overloads during the healing process (Augat & Rüden, 2018). These procedures had been used since the end of the 19th century and gained recognition mainly after 1958, with AO (Arbeitsgemeinschaft für Osteosynthesefragen) group foundation. AO developed techniques and prosthesis designs as the Dynamic Compression Plate (DCP), an implant able to provide compression on the fracture site due to the screw's displacement. Later, Limited Contact Plates (LC-DCP) were introduced to optimize the blood supply, reducing the effect of the devices contact with the bone surface. The Locking Compression Plates (LCP) are the most recent improvement on osteosynthesis plates, with special holes that can be used as a conventional compression plate or with locking screws, in which, provides angular stability and prevents the device from being pressing towards the bone (Bavornratanavech 2007 & Harasen 2011). However, DCP plates are still being used because of its lower cost, but there are few studies comparing the effectiveness of the three models (Mariolani & Belangero, 2013). Furthermore, many numerical studies use simplified models, due to the high complexity of the plate's geometry, especially the holes shape (Lee et al. 2013 & Baharnezhad et al. 2013). Therefore, this research analyzed the 4-point bending properties of DCP, LCP and LC-DCP plates with simplified and complex 6 screw holes.

DOI: 10.1201/9781003217152-85

2 MATERIALS AND METHODS

Three models were developed based on the AO manuals and commercial devices catalogs (not simplified shapes- N. Spl.). The first geometry was the 4.5mm narrow DCP with 6 oblong holes and 3.8mm thick (Figure 1a). Later, the DCP undersurfaces were "scalloped" away, generating the LC-DCP model (see Figure 1b). Additionally, a new design (Figure 1c) based on the complex LCP minimally invasive plates, including the combined holes with threaded segment and the "sharp" end that allows the plate to slid along the bone surface was considered. All the models were also modified to simpler circular holes (Spl.).

During the modeling, characteristics as thickness, width, number and position of holes were parameterized to enable quick modifications. The length (L) was defined by Equation 1, bellow:

$$L = 2A + E_{cc} + E_{cp}(n - 1) \tag{1}$$

where n is the number of holes, A is the distance between the most externally hole to the plate edge, E_{cc} the distance between the central holes and E_{cp} the distance between the remaining holes. For DCP plates, A is equal to 6.5mm, for LC-DCP it is 8mm and for LCP it is 13mm. For DCP plates, E_{cc} is equal to 26mm and E_{cp} is 16mm; the models LC-DCP and LCP have the same 18mm E_{cc} and E_{cp}.

Besides that, to validate the study, a DCP broad 8 holes were generated with similar dimensions than that used by Baharnezhad et al. (2013) in his experimental (Exp.) and numerical studies (Fem. Lit.).

Assemblies were made to simulate the 4-point bending tests, according to ISO 9585 standard (ISO TC, 2016). The distance between the upper load rollers (k) was 40mm to DCP 8 holes model, 42mm to DCP 6 holes and 36mm for the other plates. The distance between an upper roller and the nearest support roller (h) was 13mm to DCP 8 holes model, 16mm to DCP 6 holes and 18mm for the remaining implants.

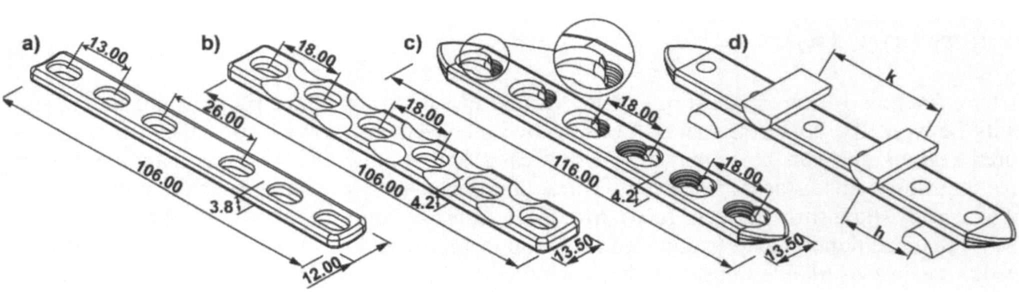

Figure 1. Osteosynthesis plates analyzed (with complex holes geometry): a) DCP b) LC-DCP c) LCP. d) Assembled setup with a simplified LCP (circular holes).

A finite static structural analysis was conducted in Ansys Mechanical Workbench 2019® using S.S. 316L (E= 200GPa, υ=0,28) and plastic strain and stress points defined as: (0, 800MPa), (0.125, 850MPa), (0.250, 960MPa) (Baharnezhad, 2013). To reproduce material plasticity, the multilinear isotropic hardening was applied using the element Solid 187, a 10-node element with 3 degrees of freedom (DOF) at each node. The contact was defined as "bonded" between support rollers and plates, not allowing gaps or sliding between these surfaces (restriction for all DOF). Between the plates and the loading rollers the contact was frictionless, allowing changes on the contact area and with normal direction movements restricted. These contact parameters, the restrictions and the movement imposed are shown on Figure 2.

Boundary conditions were applied so that the displacements were limited for all bodies on the z direction. Bottom supports were fixed and displacement loads were applied in the upper roller, in which, moved 6mm downward, during 30 sub steps. At each step, the reaction loads and displacements from the upper rollers were obtained. Mesh convergence was analyzed until the results for maximum stress and load reached variation less than 5%; starting with a medium quality mesh. Posteriorly, manual meshes were also generated with element size as 1 and 2mm and defeature size as 0.1mm.

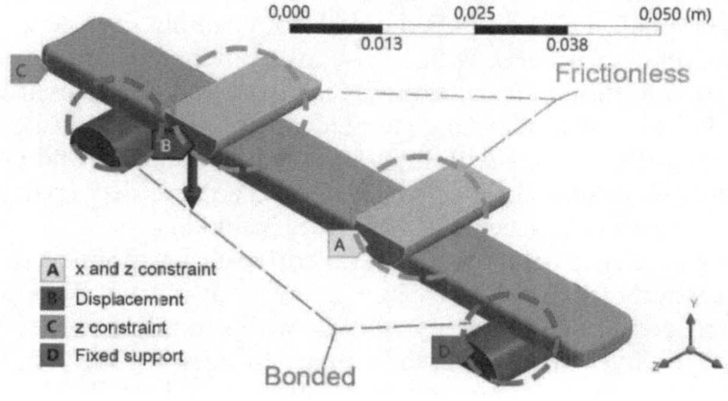

Figure 2. Boundary conditions and contact parameters between rollers and plate.

3 RESULTS AND DISCUSSION

Table 1 shows results obtained during the convergence process for one of the geometries. The first and the second lines were obtained by automatically convergence methods. The other lines present results for the manual process.

The same test was done for all the analyzed models in order to avoid mesh problems influence on the results.

Table 1. The number of officially reported plague cases in the world.

Plate	Elements	Nodes	Max. Stress [MPa]	Max. Load [N]	Stress Variation [%]	Load Variation [%]
LCP (SPL)	5660	10761	9,65e°2	18560		
	20235	30250	9,54e°2	15137	-1.2%	-22.6%
	75057	105427	9,50e°2	14013	-0.4%	-8.0%
	101574	141652	9,51e°2	13746	0.1%	-1.9%

After this process, the results for the Broad 8 holes plate were compared with experimental and numerical data from reference. Values reached for load were 6.9% higher than the experimental results and were considered acceptable, validating the parameters used. The results for all plates are shown in Table 2.

Table 2. Results from 4 point bending numerical analyses.

	DCP Broad 8 holes			DCP 6 holes		LC-DCP 6 holes		LCP 6 holes	
	(exp.)	(Fem)	(Fem. lit)	(Spl.)	(N. Spl.)	(Spl.)	(N. Spl.)	(Spl.)	(N. Spl.)
L [kN]	7.23	7.73	6.88	10.24	7.46	8.87	8.50	9.86	6.37
M [Nm]	65.11	69.55	61.94	81.89	59.66	79.82	76.48	88.77	57.36
EI [Nm²]	15.47	14.50	15.93	30.74	22.88	27.30	23.77	29.70	24.39

4 DISCUSSION

The automatic convergence criterion was unable to obtain satisfactory results for all the analyses done. Since this process compares only the maximum stress obtained in the plate surface, the mesh refinement only happened in the prosthesis geometry. However, the assessed results (output) in 4-points bending tests is the reactive load, obtained in the loading rollers, that did not have their mesh refined. For this reason, stress concentrators were generated, which caused higher reactive load variation even when the mesh had already converged in respect to the stress. In the LCP_SPL plate (LCP, simplified holes), for example, even with automatically generated mesh with high refinement level, with 1.2% convergence error in respect to the stress, still showed about 22.6% of variation in respect to the reactive loading, as presented on Table 1.

The, final mesh selected, was a manual generated, with refinement on both rollers and prothesis surfaces, consequently, the one with higher number of elements and nodes.

All the results from the validated model presented acceptable discrepancy when compared with experimental data from (Baharnezhad, 2013) (see Table 1).

Simplified models had higher bending structural stiffness and moment than non-simplified, due to the smaller screw holes and higher area moment of inertia. This proves the need of caution in the resulting comparison of analysis made with simplified models.

When same hole type plates are compared, similar stiffness is observed. For instance, LCP plates reached a structural bending moment of 24.39 Nm^2 and DCP, 22.88 Nm^2, resulting in a variation of 6%. These results are compatible with the one showed by Mariolani & Belangero (2013), in which proved, in a mechanical point of view, that any of the three plates could be safely applied.

5 CONCLUSION

The adopted analysis procedures showed satisfactory results, comparatively to literature results. Mesh convergence in respect to the reactive load and maximum stress were also obtained, ensuring that the results were not influenced by mesh problems.

These results showed that the holes' representations are the most important factor for the plates model. All evaluated properties reached higher values for circular holes' plates than the complex models, in which implies that non simplified models can be applied when suitable, but simplified should be used carefully. For DCP plates, the structural stiffness variance was 14% and for LCP the values changed 34%. Comparisons between different plate models stiffness showed that all the plates can be used safely, imposing rigid stabilization on fractures site. Results showed that numerical analysis can be validated with test results for DCP 8 holes.

REFERENCES

C.-H. Lee et al., "Simulation-based particle swarm optimization and mechanical validation of screw position and number for the fixation stability of a femoral locking compression plate," *Med. Eng. Phys.*, vol. 36, no. 1, Jan. 2014.

G. Harasen, "Orthopedic hardware and equipment for the beginner. Part 2: plates and screws.," *La Rev. Vet. Can.*, vol. 52, no. 12, pp. 1359–60, Dec. 2011.

G. O. Tong and S. Bavornratanavech, AO Manual of Fracture Management: Minimally Invasive Plate Ot-osynthesis (MIPO), 1st ed. *Clavadelerstrasse: AO Publishing*, 2007.

J. R. L. Mariolani and W. D. Belangero, "Comparing the In Vitro Stiffness of Straight-DCP, Wave-DCP, and LCP Bone Plates for Femoral Osteosynthesis," *ISRN Orthop.*, vol. 2013, pp. 1–6, 2013.

S. Baharnezhad et. al., "Influence of geometry and design parameters on flexural behavior of dynamic compression plates (dcp): experiment and finite element analysis," *J. Mech. Med. Biol.*, vol. 13, Jun. 2013.

ISO 9585: Determination of bending strength and stiffness of bone plates. *ISO, ISO/TC 150/SC 5*, 2016.

P. Augat and C. von Rüden, "Evolution of fracture treatment with bone plates," *Injury, vol. 49*, Jun. 2018.

Advances and Current Trends in Biomechanics – Belinha et al. (Eds)
© 2022 Copyright the Author(s), ISBN: 978-1-032-10806-3

Numerical simulation of blood analogous fluid flow in intracranial aneurysm

M.S. Souza
CMEMS, University of Minho, Guimarães, Portugal

A. Souza
Mechanical Engineering Department, University of Minho, Guimarães, Portugal

R. Lima
MEtRICs, University of Minho, Guimarães, Portugal

J.E. Ribeiro
Polytechnic Institute of Bragança, Bragança, Portugal
CIMO, Polytechnic Institute of Bragança, Bragança, Portugal

ABSTRACT: Intracranial aneurysm (IA) is a pathology that in most cases does not present symptoms in the patient and is still not well understood in relation to the interaction of factors that develop it. Therefore, determining methods to assess risk factors and assist in medical diagnosis is of great importance. With that, the objective of this work was to carry out the hemodynamic study of the flow of fluids inside a real IA. For this, numerical simulations were performed in stationary regime using the Ansys, Fluent software. Therefore, it was possible to show that the use of the CFD tool has great potential in the investigation, diagnosis and definition of treatments, for pathologies such as intracranial aneurysm.

1 INTRODUCTION

Intracranial Aneurysm (IA) is the saccular dilation of the artery that frequently occurs in the arteries of the Willis circle, it is estimated that 2.6% of the population has an aneurysm, which is responsible for 85% of the subarachnoid hemorrhages, with a mortality rate of 50 to 60% after rupture and a dependency rate among survivors of 30 to 40% (Amenta *et al.*, 2012). AI is one of the most feared cerebrovascular pathologies, and the probable causes for its appearance are associated with genetic factors, with increasing age, smoking, hypertension and excessive alcohol use (Tromp *et al.*, 2014). It is known that it is the interactions between biological and physical factors that cause the degradation and weakening of the wall and arise in response to changes in the hemodynamic load and biomechanical stress, however, these interactions are not yet fully understood (Cebral & Raschi, 2013).

Therefore, to understand the pathology of the intracranial aneurysm, it is essential to analyze the different hemodynamic flow regimes within the aneurysm, to determine how they affect the development of endothelial cells and, consequently, the vascular lining. We can relate three approaches to assess the hemodynamic parameters of the vascular system (Souza *et al.*, 2020): *in vivo*, *in vitro* and *in silico*. An integrated approach to the study of flow using computational fluid dynamics (CFD) with experimental validation, has been widely used today for the study of intracranial aneurysms (Roloff *et al.*, 2019). Thus, allowing the analysis

DOI: 10.1201/9781003217152-86

of specific patient geometries, making it possible to establish, as boundary conditions, specific patient properties and, subsequently, obtain results that can quantify hemodynamic parameters (Pinho *et al.*, 2015) The numerical simulations were performed using a CFD tool (Fluent), in a patient-specific IA geometry, obtained from an angiography. For this, a qualitative study was carried out in a stationary regime.

2 MATERIALS AND METHODS

This study consists of a qualitative analysis of the flow behavior in a real intracranial aneurysm through numerical simulations. The behavior of the flow was analyzed through the streamlines. In this section, the IA geometry used will be presented, as well as the properties of the blood analogous fluid and the parameters of the simulations.

2.1 *Geometry*

To obtain the geometry, initially an angiography was used, containing the anatomical area in which the aneurysm is located. Once identified, the image was segmented using an image processing software called ScanIP. This is capable of reading the angiographic images and importing them, translating them into the DICOM format (Digital Imaging and Communications in Medicin). Through the segmentation process, the region of interest was separated, which are the arteries where the aneurysm was located. Finally, the model obtained was converted to STL format.

2.2 *Blood analogous fluid*

Numerical studies need some validation to be relevant to their results. In this study, a comparison was made between the results obtained numerically with the results of experimental tests. Due to the difficulties currently found in the use of blood in in vitro studies, it is necessary to use fluids analogous to blood for experimental blood flow tests (Pinho *et al.*, 2019). These fluids are so called, because they have rheological properties close to those of blood, such as viscosity and density. The fluid used was Dimethyl sulfoxide (DMSO), and the properties of this fluid were inserted in the simulations. Table 1 shows the properties of DMSO and blood.

2.3 *Numerical simulation*

For the numerical study, the governing continuity and Navier-Stokes equations were solved using the commercial software Fluent, Ansys 2017. The equations were solved in a stationary and incompressible regime. The walls were considered rigid and the liquid used is a Newtonian liquid. The mesh obtained has a total of 161491 nodes and 555147 elements. For the boundary conditions, IA input and output regions were defined. The IA has an entrance that is in the internal carotid artery (ICA) and two exits. The left exit is through the anterior cerebral artery (ACA) and the right exit is through the middle cerebral artery (MCA). Inlet and outlet were established according to the hemodynamic behavior of the blood flow. Figure 1 shows the mesh used and the fluid inlet and outlet.

Table 1. Fluid properties used in the simulation.

Fluid	Viscosity (Pa.s)	Density (kg/m^3)	References
DMS0	0.00315	1072	(Campo-Deaño *et al.*, 2013)
Blood	0.00350	1060	(Doutel *et al.*, 2018)

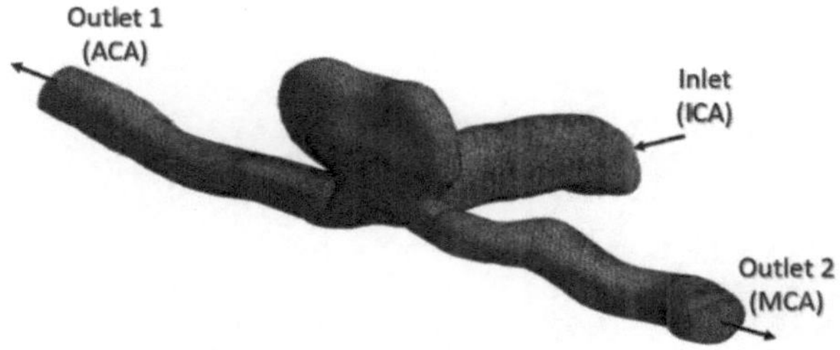

Figure 1. Mesh used in simulations and representation of flow inlet and outlet.

Table 2. Parameters for simulations in steady state.

Velocity (m/s)	Flow rate (mL/s)	Reynolds number	Calculation time (min.)
0.12	1.31	150	45

The flow rate used was 1.31 mL/s, which corresponds to a Reynolds number of 150. To calculate the Reynolds number, the diameter of the aneurysm entry section (Jiang & Strother, 2009), which is 3.78 mm, was used. The parameters of the steady state simulations are shown in Table 2.

The mass flow was entered as an inlet parameter. For outlets, 51% of the outflow was imposed through the anterior cerebral artery and 49% of the outflow through the middle cerebral artery.

The hydrodynamic equations were integrated using the pressure-based solver. gradients in the cell centers were calculated with the Least Squares Cell Based scheme. A second-order and second-order upwind approximations were conducted for the spatial discretization of the pressure and momentum equations, respectively. The SIMPLE procedure was employed to couple the velocity and the pressure.

To validate the mesh used, a mesh was generated with twice as many elements, where it was observed that there was a change in the results of less than 2%, maintaining the same spatial distribution. As a convergence criterion, it was ensured that the residues were sufficiently small and the flow lines were stable with the number of iterations.

3 RESULTS

The result obtained for the simulations made with the flow rate of 1.31 mL/s is shown in Figure 2b. Figure 2a shows experimental results performed under the same conditions as the simulations. Comparing both images, it is observed that the behavior is similar, which validates the simulation performed.

In the Figure 2b and Figure 3, it is possible to observe an area of recirculation in the dome in the aneurysm. It is clear that the fluid loses velocity when it rises in the aneurysm and regains velocity when it leaves the ACA and MCA arteries. The highest speed is observed in the MCA (in red) where there is a narrowing of the artery.

The fluid flow inside the aneurysm can be classified according to its complexity and concentration of entry (Cebral *et al.*, 2011). Thus, it can be said that according to the complexity of the flow, it has a simple flow pattern, that is, it consists of a single recirculation area or vortex structure within the aneurysm. It is also possible to observe that the jet of flow that enters the aneurysm is diffuse, which indicates that it disperses quickly when entering the aneurysm.

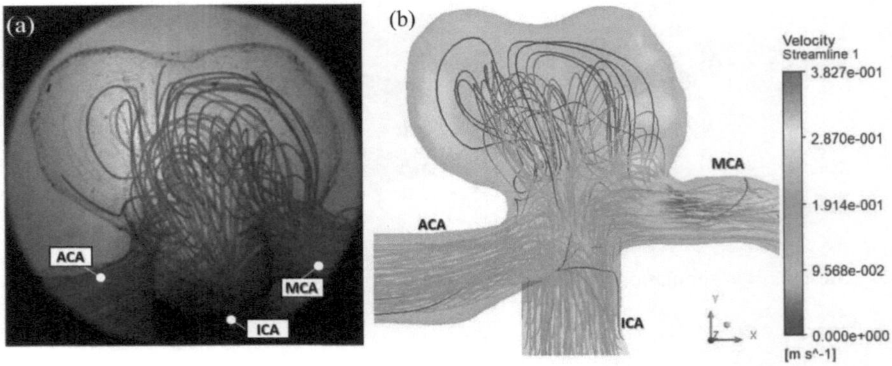

Figure 2. Validation of the numerical method. (a) Experimental test; (b) Numerical simulation.

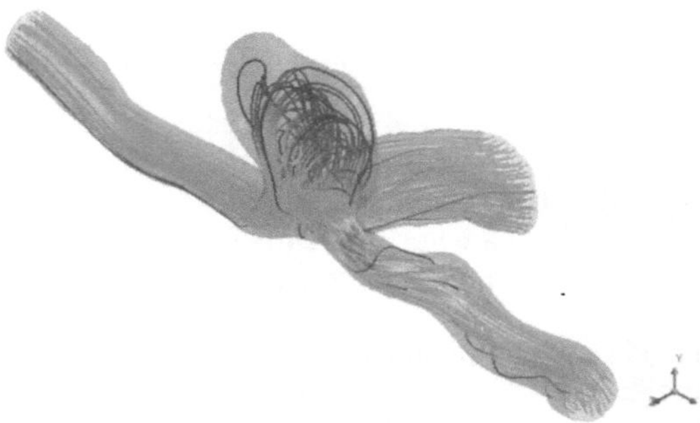

Figure 3. Analysis of streamlines in the IA.

4 CONCLUSION

The developed work had as objective to show the behavior of the flow inside the AI through simulations carried out in stationary regime. It was possible to observe the occurrence of fluid recirculation inside the aneurysm and to classify the flow according to the degree of risk, where it was classified as simple and diffuse. A qualitative comparison of the obtained numerical result was made, with an experimental test under the same conditions, validating the results of the simulations.

REFERENCES

Amenta, P. S. *et al.* (2012) 'Analysis of Nonmodifiable Risk Factors for Intracranial Aneurysm Rupture in a Large, Retrospective Cohort', *Neurosurgery*, 70(3), pp. 693–701. doi: 10.1227/neu.0b013e3182354d68.

Campo-Deaño, L. *et al.* (2013) 'Viscoelasticity of blood and viscoelastic blood analogues for use in polydymethylsiloxane in vitro models of the circulatory system', *Biomicrofluidics*, 7(3). doi: 10.1063/1.4804649.

Cebral, J. R. *et al.* (2011) 'Association of hemodynamic characteristics and cerebral aneurysm rupture', *American Journal of Neuroradiology*, 32(2), pp. 264–270. doi: 10.3174/ajnr.A2274.

Cebral, J. R. & Raschi, M. (2013) 'Suggested connections between risk factors of intracranial aneurysms: A review', *Annals of Biomedical Engineering*, 41(7), pp. 1366–1383. doi: 10.1007/s10439-012-0723-0.

Doutel, E. *et al.* (2018) 'Experimental and numerical methodology to analyze flows in a coronary bifurcation', *European Journal of Mechanics, B/Fluids*. Elsevier Masson SAS., 67, pp. 341–356. doi: 10.1016/j.euromechflu.2017.09.009.

Jiang, J. & Strother, C. (2009) 'Computational fluid dynamics simulations of intracranial aneurysms at varying heart rates: A "patient-specific" study', *Journal of Biomechanical Engineering*, 131(9), pp. 1–11. doi: 10.1115/1.3127251.

Pinho, D. et. al. (2015) 'An in vitro experimental evaluation of the displacement field in an intracranial aneurysm model', Mechanisms and Machine Science, Springer, 24, pp. 261–268. doi: 10.1007/978-3-319-09411-3_28.

Pinho, D. *et al.* (2019) 'Flexible PDMS microparticles to mimic RBCs in blood particulate analogue fluids', *Mechanics Research Communications*. Elsevier Ltd, 100, pp. 18–20. doi: 10.1016/j.mechrescom.2019.103399.

Roloff, C. *et al.* (2019) 'Comparison of intracranial aneurysm flow quantification techniques: Standard PIV vs stereoscopic PIV vs tomographic PIV vs phase-contrast MRI vs CFD', *Journal of NeuroInterventional Surgery*, 11(3), pp. 275–282. doi: 10.1136/neurintsurg-2018-013921.

Souza, A. *et al.* (2020) '3D manufacturing of intracranial aneurysm biomodels for flow visualizations: Low cost fabrication processes', *Mechanics Research Communications*. Elsevier Ltd, 107, p. 103535. doi: 10.1016/j.mechrescom.2020.103535.

Tromp, G. *et al.* (2014) 'Molecular basis and genetic predisposition to intracranial aneurysm', *Annals of Medicine*, 46(8), pp. 597–606. doi: 10.3109/07853890.2014.949299.

Advances and Current Trends in Biomechanics – Belinha et al. (Eds)

Non-parallel implants impressions. Close-tray technique

L. Pimentel
Faculty of Dental Medicine, Universidade do Porto, Oporto, Portugal

M. Gomes
Art and Design School, Oporto, Portugal

J. Portugal
Faculty of Dental Medicine, Universidade de Lisboa, Portugal

P.R. Almeida, J.C. Sampaio-Fernandes & M.H. Figueiral
INEGI, Faculty of Dental Medicine, Universidade do Porto, Oporto, Portugal

ABSTRACT: Non-parallel implant placement allows the use of larger implants, better load distribution and avoids bone regeneration procedures. Despite these benefits, it causes more stress in the impression material during the disinsertion of the tray. The aim of this study is to investigate the influence of the presence of non-parallel implants on the accuracy of three different closed tray impression techniques. A master model was created with two parallel and one non-parallel implants. 27 impressions were carried out (n = 9). The master model and the corresponding impressions were digitalized with an extraoral optical scanner (*Medit Identica SE*) and the angular deviation between them was calculated. Statistical analysis was performed with statistical significance of $\alpha = 0.05$. The normal distribution of the samples was not observed by the Shapiro-Wilk test. According to our results, the impression technique influenced the precision of implant position transfer. The presence of non-parallel implants decreased the accuracy of the impressions.

1 INTRODUCTION

Obtaining accurate working models that register the intraoral position of implants or abutments is essential for the biological and physical long-term stability of implant-supported prostheses. There are several factors influencing the accuracy of implant impressions, including the impression technique used. Although there is already some scientific evidence concerning the accuracy of impression techniques of implants (Lee et al. 2008), there are few studies addressing indirect or closed-tray methods. In the indirect technique, the impression copings are screwed to the implants in the mouth, being later repositioned in the tray after its removal or disinsertion from the mouth. Being easily performed and time-effective, this technique has gain popularity among practitioners, and in most situations, is the most comfortable and safe approach for both patient and clinician, as it is not necessary to unscrew the impression coping with the tray inside the mouth.

Addition silicones (polyvinylsiloxanes) have been increasingly used and accepted as impression material, due to their dimensional stability, accuracy in the reproduction of details and because they are available with various working times and viscosities. They are often referred to as impression materials with better elasticity, better elastic memory, and less permanent deformation than other elastomers (Mandikos 2008).

DOI: 10.1201/9781003217152-87

Some authors (Block et al. 2009, Agliardi et al. 2009) have reported the placement of inclined implants as a useful option on contemporary prosthodontic rehabilitations, with the advantages as follows: i) use of longer implants; ii) improved primary stability; iii) avoidance of extensive cantilevers; iv) improved load distribution on dental arches and v) reduced need of bone grafts procedures. There is no consensual position in the scientific community on the different methods and techniques of implant impressions. According to some authors (Rutkunas et al. 2012, Kurtulmus-Yilmaz et al. 2014), the inclination of dental implants causes a greater distortion in the impression material and a decrease in its accuracy, due to the need of applying greater force, causing higher deformation during tray removal. This effect also increases with the number of implants.

The aim of this study is to investigate the influence of the presence of non-parallel implants on the accuracy of three different closed-tray impression techniques.

2 MATERIALS AND METHODS

A master model, simulating an edentulous mandible, was created using a *Frasaco* mandibular model. Three *Brånemark* hexagon external 4.0 analogs were placed to simulate two parallel implants (B and C, perpendicular to the base of the model) and one non-parallel implant (A) with 15° of inclination, as showed in Figure 1. Three metallic spheres of equal diameter were also included in the model, allowing to overlap the digital images.

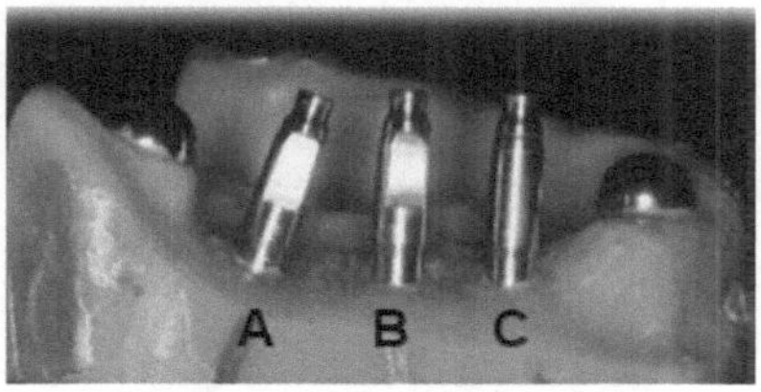

Figure 1. Master model.

Three different closed-tray impression techniques were studied: impression with single addition silicone mono-body viscosity (MB), impression with a mixture of two addition silicone viscosities (DM) and double impression (DI). A metallic structure was built to standardize the impression procedure, since it controls the position of the tray in relation to the master model, the pressure during the introduction and removal of the tray and the amount of the impression material used. A total of 27 impressions of the master model (n=9 per technique) were performed. The addition silicones used were *Coltène/Whaledent monobody medium-body* and *Coltène/Whaledent putty* and *light-body*. The setting time recommended by the manufacturer was doubled (8 minutes), to compensate the temperature difference between the study environment (20° Celsius) and the average temperature of the oral cavity (36° Celsius).

After all impressions, the master model and the samples were digitalized with an extraoral optical scanner (*Medit Identica SE*) and the 3D point clouds resulting from the acquisition of the samples' shape were aligned with the master model (Figure 2 and 3), analyzing the angular deviation of the marked structures using a computer software (*PowerSHAPE 2018, Autodesk*).

The 3-dimensional figure of the impression copings was automatically converted into cylindrical surfaces in order to calculate the angle between the axis of the cylinder and the horizontal reference surface - anatomical base (Figure 4). The angular deviation between the master model and the samples was calculated (*PowerINSPECT 2018, Autodesk plc*).

Statistical analysis was performed with IBM SPSS Statistics for Macintosh, v 21.0. (IBM Corp., Armonk, NY, USA) and a statistical significance of $\alpha=0.05$ was considered. The normal distribution of the samples was not observed (Shapiro-Wilk test). To evaluate the

Figure 2. Digital master model.

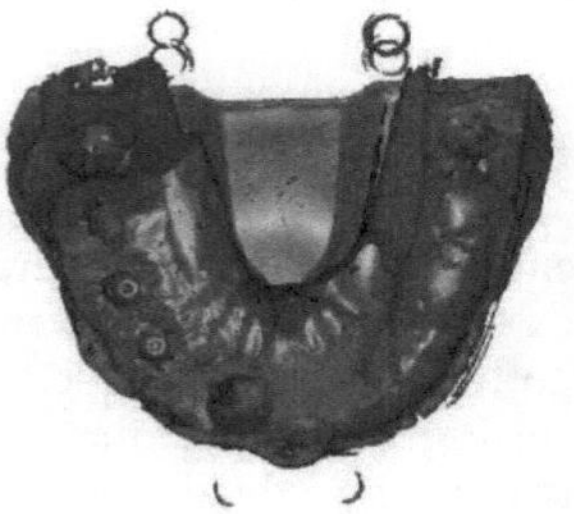

Figure 3. Example of a scanned impression overlapped on digital master model.

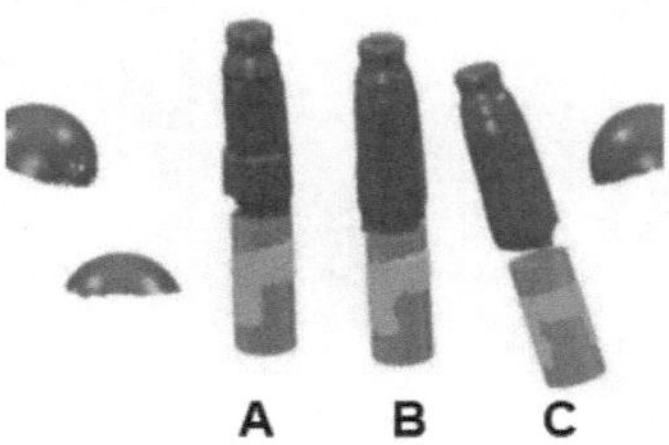

Figure 4. Tridimensional analysis.

effect of the impression methodology on precision, Kruskal-Wallis tests were used followed by multiple comparisons with Mann-Whitney tests with Bonferroni correction. To study the effect of the position of the implants, we performed Friedman Tests.

3 RESULTS

The implant with the highest inclination (A) and the adjacent one (B) presented the highest distortion in all techniques.

The multiple comparisons performed for the deviation of implant A (non-parallel) showed statistically significant differences when we use the DI technique ($p<0.05$). For implant A, there are no statistically significant differences ($p = 1.000$) between the DM and MB techniques. The multiple comparisons performed for implant B only reveal statistically significant differences ($p<0.001$) between the MB and DI techniques. For the parallel implant C, the most distant from the tension zone during tray removal, there are no statistically significant differences between the impression techniques ($p>0.05$).

Comparing the impression copings position regardless of the impression method, no significant differences were observed between them ($p = 0.355$). When we do the same comparison but specifically in the MB group technique, implant B presented significant differences ($p = 0.004$); in the DM technique, there are no statistically significant differences between the different implants ($p = 0.236$); and, in the DI technique there are statistically significant differences, being implants A and B the most deformed ($p<0.001$).

4 DISCUSSION

The accuracy of different techniques, impression materials, implant connection systems and implant positions have been studied elsewhere, but no consistent results have been found (Lee et al. 2008).

According to Assunção et al., the most favorable position of an implant in terms of impression accuracy is perpendicular to the bone surface (Assunção et al. 2004). Other investigation concluded that impressions made in the presence of non-parallel implants were significantly less accurate than those made with parallel implants (Sorrentino et al. 2010). Nevertheless, another study with edentulous patients, reported that the precision of the impression was not affected by the presence of implants with inclinations between 10° and 15° (Papaspyridakos et al. 2016). Other studies show that the inclination has no influence on the precision of the impressions with three or less non-parallel implants with inclinations up to 15° (Conrad et al. 2007).

However, these studies did not simulate clinical situations, as they used non-anatomical master models, not recreating the possible deformation of the impression material. In our investigation, a mandibular simulation model was used in order to emulate the clinical conditions of the impression removal process, trying to reproduce the movements and forces performed in impression tray disinsertion *in vivo*. In this study, the implant with 15° inclination and the adjacent one presented the highest distortion values in all techniques. The parallel implant more distant did not present significant differences between the impression techniques studied. The contradictory results found in the literature may be explained by differences in the methodological approaches, implant connections, techniques, implant inclinations, by the imprecision in the repositioning of the impression copings, and the use or not of plaster in the impression analysis.

The majority of the studies involves the production of plaster models of each impression in which the measurements are directly performed. This procedure increases the number of steps and the possible errors. In this study, we chose to overlap 3D images (*.stl* files), thus taking advantage of the most recent methods to study the accuracy of implant impressions. Such an approach allows to align the files of each impression and the master model, enabling the exact comparison between each technique and implant.

The clinical interpretation of these *in vitro* results should be done with caution, as the model and experimental conditions used do not fully emulate a clinical situation.

5 CONCLUSIONS

Different impression techniques influence the precision and accuracy of the relative position of implants. The double mixing (DM) technique showed the smallest angular deviation and the least deformation for all studied implants. The double impression (DI) technique showed the greatest deformation for the non-parallel implant. The values obtained with all the impression techniques tested do not allow to clinically contraindicate any of them, but further research should be conducted.

REFERENCES

Agliardi E.L., Francetti L., Romeo D., Del Fabbro M. 2009. Immediate rehabilitation of the edentulous maxilla: preliminary results of a single cohort prospective study. *International Journal Oral Maxillofacial Implants* 24 (5): 887–895.

Assunção W.G., Gennari Filho H., Zaniquelli O. 2004. Evaluation of transfer impressions for osseointegrated implants at various angulations. *Implant Dentistry* 13 (4): 358–366.

Block M.S. Haggerty C.J., Fisher G.R. 2009. Nongrafting implant options for restoration of the edentulous maxilla. *Journal Oral Maxillofacial Surgery* 67 (4): 872–881.

Conrad H.J., Pesun I.J., DeLong R., Hodges J.S. 2007. Accuracy of two impression techniques with angulated implants. *Journal of Prosthetic Dentistry* 97 (6): 349–356.

Kurtulmus-Yilmaz S., Ozan O., Ozcelik T.B., Yagiz A. 2014. Digital evaluation of the accuracy of impression techniques and materials in angulated implantes. *Journal of Dentistry* 42 (12): 1551–1559.

Lee H. So J.S., Hochstedler J.L., Ercoli C. 2008. The accuracy of implant impressions: a systematic review. *Journal of Prosthetic Dentistry* 100 (4): 285–291.

Mandikos M.N. 1998. Polyvinyl siloxane impression materials: An update on clinical use. *Australian Dental Journal* 43 (6): 428–434.

Papaspyridakos P., Gallucci G.O., Chen C.J., Hanssen S., Naert I., Vandenberghe B. 2016. Digital versus conventional implant impressions for edentulos patients: accuracy outcomes. *Clinical Oral Implant Research* 27 (4): 465–472.

Rutkunas V., Sveikata K., Savickas R. 2012. Effects of implant angulation, material selection, and impression technique on impression accuracy: A preliminary laboratory study. *International Journal of Prosthodontics* 25 (5): 512–515.

Sorrentino R., Gherlone E.F., Calesini G., Zarone F. 2010. Effect of implant angulation, connection length, and impression material on the dimensional accuracy of implant impressions: An in vitro comparative study. *Clinical Implant Dentistry Research* 12 (Suppl 1): e63–76.

Advances and Current Trends in Biomechanics – Belinha et al. (Eds)
© 2022 Copyright the Author(s), ISBN: 978-1-032-10806-3

Body morphology of biomechanical students from the Polytechnic of Leiria

Henrique A. Almeida & Rita M.T. Ascenso
Computer Science and Communication Research Centre, Polytechnic of Leiria, Portugal
School of Technology and Management, Polytechnic of Leiria, Portugal

Liliana C.R. Ferreira
Center for Mathematics, Fundamental Applications and Operations Research, Faculty of Sciences of the University of Lisbon, Portugal
School of Technology and Management, Polytechnic of Leiria, Portugal

Eunice S.G. Oliveira
Institute for Systems Engineering and Computers, University of Coimbra, Portugal
School of Technology and Management, Polytechnic of Leiria, Portugal

ABSTRACT: Characterization of body morphology focuses on the study of the form and shape of individuals and their specific structural features. Portraiture of body composition is a method of breaking down the body into its core components: fat, protein, minerals, and body water. Body form and shape describe body's individual in a certain moment of time providing insights of overall health condition as well as body performance. Body morphology and composition assessment is used in several areas, from sports to health and is relevant in several age groups, having impact on health status and wellbeing. Considering these two approaches, this research work aims to present the characterization of the body morphology of students from the biomechanics degree of academic years from 2016 to 2020, at the School of Technology and Management of the Polytechnic of Leiria, resulting from data of the Anthropometrics course. Thirteen anthropometric measurements were taken from 66 students to determine parameters such as Body Mass Index, Conicity Index, Anthropometric Ratios (waist-to-height, waist-to-hip) and somatotype which will be presented in the study and will be used to perform statistical analysis of correlations and comparisons between genders.

1 INTRODUCTION

Anthropometry is a research area of great importance which may contribute to the development of studies in other areas, such as ergonomics [Castellucci et al. (2019), Dianat et al. (2018), Hardianto & Hermawan (2021)], health [Gaya et al. (2017), Tallon et al. (2019)], sport [Bayios et al. (2006), Pedretti et al. (2019), Veitía et al. (2018)] and so many other fields. As mentioned in [Dianat et al. 2018], anthropometry is crucial for ergonomic studies, allowing the adjustment of the product and workplace design to a particular end-user population [de Campos et al. (2018), Oyewole et al. (2010), Pentikis et al. (2002)], besides using the anthropometric characterization of the population to design tools and equipment for different occupations [Castellucci et al. (2019)]. The design of ergonomic products, for which anthropometry has played a strong contribute, allows to minimize problems of discomfort, fatigue or even health problems, as shown in several studies [Hardianto & Hermawan (2021), Reis et al.

DOI: 10.1201/9781003217152-88

(2012)]. Anthropometry is also used in sport, namely in high competition sports. There are several studies that make it possible to compare and relate the anthropometric characteristics of sportspeople with their performance, also allowing to assess the anthropometric characteristics that best suit a given sport [Bayios et al. (2006), Pedretti et al. (2019), Veitía et al. (2018)].

Recently, considering the previously mentioned importance of anthropometry, several studies have contributed to fill the lack of anthropometric data in several countries, in order to characterize their populations. In Portugal, as well, anthropometric data has been witnessed to enrich the characterization of the Portuguese population or specific groups of the Portuguese population, for instances, teenagers [Tallon et al. (2019)] and pregnant women [Santana et al. (2018)].

According to Barroso [Barroso et al. (2005)], the anthropometric database in Portugal was frankly scarce. However, in recent years, an increase of studies in this area has been allowed to enrich this database, as well as to relate anthropometry with other areas [Tallon et al. (2019), de Campos et al. (2018), Santana et al. (2018), Barroso et al. (2005), Sousa et al. (2014)].

In this sense, this work aims to contribute to increase the existing anthropometric information in Portugal with the characterization of the body morphology of 66 students based on thirteen anthropometric measurements. Parameters such as Body Mass Index (BMI), Conicity Index (CI), Anthropometric Ratios (waist-to-height (WHtR), Waist-to-Hip (WHR)) and somatotype were determined and are presented in this study. A descriptive statistical analysis of data and comparisons of values between genders are also presented in this work.

2 MATERIALS AND METHODS

For this research it was considered a sample composed by 66 students, contemplating 25 female students and 41 male students, with ages ranging from 18 to 26 years old. The students attending the course of Anthropometrics are from academic years since 2016-2017 until 2019-2020 at the School of Technology and Management of the Polytechnic of Leiria.

All anatomic points, anthropometric measurements and equations needed for obtaining values and credible results are in accordance with the standards from the International Society for the Advancement of Kinanthropometry (ISAK) and the anthropometric measurements were all performed on the dominant side of the evaluated. The used measuring tools were the following: Stadiometer, Balance, Retractable metric measuring tape, Skinfold calliper and Bone callipers, all in accordance with the ISAK standards [Norton & Olds (2005)].

3 RESULTS AND DISCUSSION

Location and dispersion measures were computed for all anthropometric measurements collected and data was plotted to help visualization of the results.

Table 1 presents the mean value and corresponding standard deviation, the maximum and minimum values for the observed sample and they were summarized by gender. It is possible to observe that the general overview of the students present a healthy weight based on the BMI classification (18.5-24.9) [WHO (2019)]. Regarding the CI value, the overall value presents a healthy population, within the limits 1.00-1.73 [Shenoy (2017)]. For the Waist-to-Height ratio, the female and male values are healthy and for the Waist-to-Hip ratio both genders present moderate values.

Boxplots presented in Figure 1 help to identify similarities between genders for Body Mass index with analogous dispersion level and skewness. Conicity index and Waist Height Ratio have similar median values despite dispersion is higher for female students. These two measures present outliers in the higher values in case of male students. More pronounced differences are noted for Waist Hip Ratio for which the value of quartile 1 for male students being almost equal to quartile 3 for female students.

Table 1. Morphological and somatotype values of the students in the sample.

		BMI	CI	WHtR	WHR	Endo	Meso	Ecto
Total	Mean±SD	23.41±3.04	1,17±0.07	0.46±0.05	0.80±0.07	6.07±2.98	4.22±1.24	2.15±1.21
	Maximum	35.00	1.39	0.62	0.99	12.87	7.59	4.63
	Minimum	18.59	0.98	0.37	0.67	1.98	0.80	0.10
Female	Mean±SD	23.40±2.91	1.16±0.07	0.45±0.05	0.76±0.07	8.21±2.43	4.18±1.37	1.75±1.09
	Maximum	30.93	1.29	0.57	0.92	12.87	7.35	4.15
	Minimum	18.59	0.98	0.37	0.67	4.24	0.80	0.10
Male	Mean±SD	23.42±3.11	1.17±0.06	0.46±0.05	0.83±0.06	4.77±2.49	4.24±1.15	2.40±1.21
	Maximum	35.00	1.39	0.62	0.99	12.14	7.59	4.63
	Minimum	18.80	1.06	0.38	0.74	1.98	2.04	0.10

Through a careful analysis of the somatotype biotypes presented in the somatochart (Figure 2) it is possible to observe that the students present a somatotype morphology between the mesomorph-endomorph and the balanced endomorph domains, namely in the meso-morphic endomorph domain. Thus, data shows that the analysed sample presents higher values of adiposity, followed by muscularity with very low values of linearity.

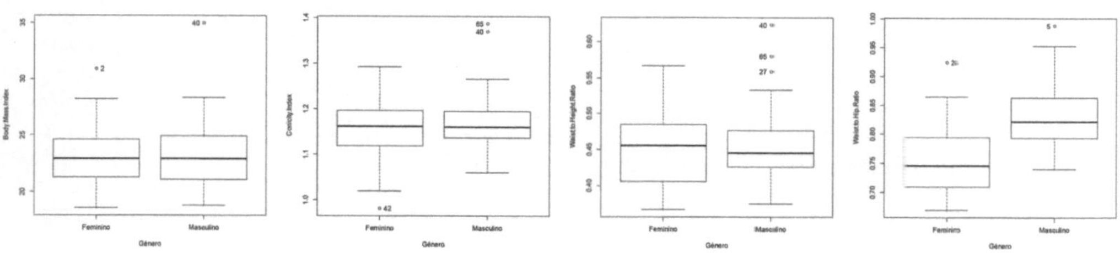

Figure 1. Comparison of morphological values between the female and male students.

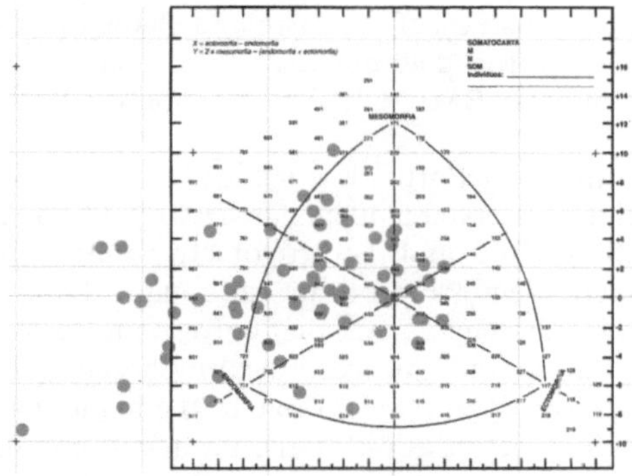

Figure 2. Somatochart of the students.

Considering the somatotype and its biotypes values presented in Table 1, both the female and male students present higher values of endomorphy, above average mesomorphy and low values of ectomorphy. It is possible to observe that both female and male students observed show a mesomorphic endomorph behaviour, meaning that the endomorphy is dominant and the mesomorphy is higher that the ectomorphy. Separately, the female students present a mesomorphic endomorph behaviour while the male students display a mesomorphy-endomorphy behaviour (both the endomorphy and the mesomorphy values are similar with a low value of ectomorphy).

Considering the above data, the analyzed population presents higher values of adiposity, followed by muscularity with very low values of linearity. In Figure 3, it is clear that male students present lower values of endomorphy and present higher values of mesomorphy when compared to female students.

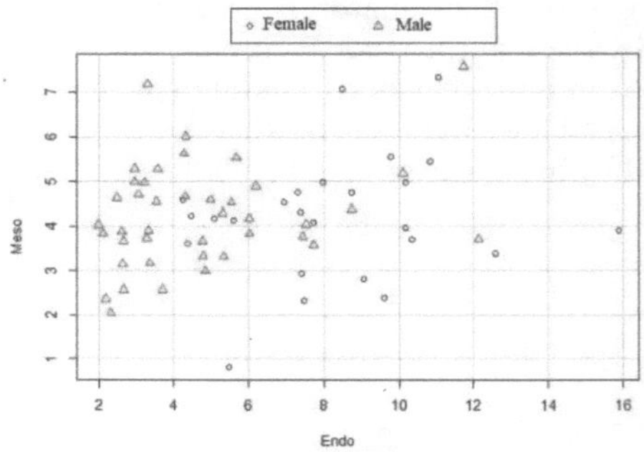

Figure 3. Female and male plot of mesomorphy versus endomorphy.

4 CONCLUSIONS

Body Morphology focuses on the study of the form and shape of living individuals and their specific structural features. For the determination of human body morpho-logy, several parameters may be used, namely, BMI, CI, Anthropometric Ratios (WHtR, WHR) and Somatotype.

Regarding the analysed sample of Portuguese students, it is possible to evaluate the body morphology of the polytechnic observed students as healthy, although some parameters present a high standard deviation. BMI, and Anthropometric Ratios (waist-to-height, waist-to-hip). Somatotype parameters demonstrate that although the average values represent a healthy population, there are extreme values that correspond to overweight people. In conclusion the anthropometric data points towards a healthy population but an in deep analysis shows the existence of some cases of overweight due to the higher values of adiposity parameters. For a more complete characterization of the academic population, in terms of body morphology, more parameters and a more representative population sample will be required in future studies. Then, further analyses could be performed to infer results for Polytechnic students' population, providing insights about anthropometric information and highlighting gender differences. Despite the recognized limitations, this study stresses the usefulness of having update anthropometric database and intends to make a contribution in that sense.

REFERENCES

Barroso, M. P., Arezes, P. M., da Costa, L. G., & Sérgio Miguel, A. 2005. Anthropometric study of Portuguese workers. International Journal of Industrial Ergonomics, 35(5): 401–410.

Bayios, I. A., Bergeles, N. K., Apostolidis, N. G., Noutsos, K. S., & Koskolou, M. D. 2006. Anthropometric, body composition and somatotype differences of Greek elite female basketball, volleyball and handball players. Journal of sports medicine and physical fitness, 46(2): 271.

Castellucci, H. I., Viviani, C. A., Molenbroek, J. F. M., Arezes, P. M., Martínez, M., Aparici, V., & Bragança, S. 2019. Anthropometric characteristics of Chilean workers for ergonomic and design purposes. Ergonomics, 62(3): 459–474.

de Campos R., Carvalho M., Capelassi C., Lopes H., Xu B. 2018. Anthropometric Data Collection of Portuguese Children with Overweight and Obesity. In Di Bucchianico G., Kercher P. (eds) Advances in Design for Inclusion; AHFE 2017. Advances in Intelligent Systems and Computing; vol 587. Springer, Cham.

Dianat, I., Molenbroek, J., & Castellucci, H. I. 2018. A review of the methodology and applications of anthropometry in ergonomics and product design. Ergonomics, 61(12): 1696–1720.

Gaya, A. R., Brand, C., Dias, A. F., Gaya, A. C. A., Lemes, V. B., & Mota, J. 2017. Obesity anthropometric indicators associated with cardiometabolic risk in Portuguese children and adolescents. Preventive medicine reports, 8: 158–162.

Hardianto, A., & Hermawan, D. 2021. Application of Anthropometry Methods in Ergonomic Chair Redesign to Prevent Fatigue A Case Study UKM Lestari Jaya, Tulungagung. In IOP Conference Series: Materials Science and Engineering (Vol. 1071, No. 1, p. 012003). IOP Publishing.

Norton, K., Olds, T. 2005. Anthropometrica: A Textbook of Body Measurement for Sports and Health Courses, CBS Publishers & Distributors.

Oyewole, S. A., Haight, J. M., & Freivalds, A. 2010. The ergonomic design of classroom furniture/computer work station for first graders in the elementary school. International Journal of Industrial Ergonomics, 40(4),437–447.

Pedretti, A., de Oliveira Matta, M., Werneck, F. Z., & Seabra, A. F. T. 2019. Anthropometric characteristics, physical fitness components and technical skills, a comparison between Brazilian and Portuguese young soccer players. Revista andaluza de medicina del deporte, 12(3): 235–238.

Pentikis, J., Lopez, M. S., & Thomas, R. E. 2002. Ergonomics evaluation of a government office building. Work, 18(2): 123–131.

Reis, P. F., Peres, L. S., Tirloni, A. S., Reis, D. C. D., Estrázulas, J. A., Rossato, M., & Moro, A. R. P. 2012. Influence of anthropometry on meat-packing plant workers: an approach to the shoulder joint. Work, 41(Supplement 1): 4612–4617.

Santana, M., Pereira, R., Oliveira, E., Almeida, H. A., & Ascenso, R. M. 2018. Potential defaults of anthropometric assessment parameters during pregnancy. Journal of Mechanical Engineering and Biomechanics, 2(5).

Shenoy, J. U. 2017. Influence of Central Obesity Assessed by Conicity Index on Lung Age in Young Adults. Journal of Clinical & Diagnostic Research 11(4),CC09–CC12.

Sousa, L. S., Pinho, M. E., & Arezes, P. M. 2014. Anthropometric study of the student population of a Portuguese faculty. Occupational Safety and Hygiene II, 447.

Tallon, J.M., Dias, R.S., Silva, A.J., Barros, A., & Costa, A. 2019. Characterization of the anthropometric profile and physical activity levels of Portuguese adolescents. Biometrics & Biostatistics International Journal, 8: 184–193.

Veitía, W. C., Perez, S. L., Revuelta, M. E. G., Campo, Y. D., & García, I. E. 2018. Anthropometrical characteristics of Cuban sporting population: Reference data from a high performance national teams, 1992–2014. Apunts. Medicina de l'Esport, 53(200): 129–137.

WHO. 2019. WHO Mean Body Mass Index (BMI), World Health Organization, Retrieved on 5 February.

Advances and Current Trends in Biomechanics – Belinha et al. (Eds)
© 2022 Copyright the Author(s), ISBN: 978-1-032-10806-3

Friction cone application for assessment of the relation between normal and tangential forces at different maximum vertical jumps

C. Rodrigues & M.V. Correia
C-BER - INESC TEC & FEUP, Porto, Portugal

J.M.C.S. Abrantes
MovLab - CICANT, ULHT, Lisbon, Portugal

M.A.B. Rodrigues
DES, UFPE, Recife, Brazil

J. Nadal
PEB - COPPE, UFRJ, Rio de Janeiro, Brazil

ABSTRACT: This study presents subject specific analysis using friction cone to assess the relation between normal and tangential ground reaction force (GRF) during impulsion phase on maximum vertical jump (MVJ) at countermovement jump (CMJ) and drop jump (DJ) for comparison with squat jump (SJ). Experimental ensemble is composed by six adult male voluntary sports students, performing each 3 SJ, CMJ, and DJ repetitions. During trial tests GRF were acquired with force platform and Simi Motion 3D. Normal and tangential GRF were assessed using friction cone and spherical coordinates for detection of the components relation and achieved performance at SJ, CMJ and DJ. Results conduced to detection of higher friction at DJ with lower impulse time and vertical height, and lower friction at SJ with intermediate impulse time and vertical height, whereas CMJ with intermediate friction presents higher impulse time and vertical height, pointing intermediate friction level association to higher performance.

1 INTRODUCTION

Countermovement is used by human, namely at lower limb on gait, running and jumping as an optimization process of muscle function, with eccentric action preceding muscle concentric action, conducing to more efficient submaximal activities and powerful maximum activities (Komi, 2011). Feet ground contact plays a determinant role on human locomotion with the need for precise control of contact forces with the ground for efficient and stable gait, either on walking, running, or jumping (van Ingen Shenau & van Soest, 1996). Thus, ground reaction forces exerted at the feet contact with the ground, jointly considered applied at the center of pressure (COP), along with weight force applied at the center of gravity (COG) determine the acceleration of COG (Winter, 2005). Since the tangential components of the ground reaction forces (GRF) are responsible for progression and stability in walking, the intensity of the normal GRF component and the relationship between them, associated with the friction of the surfaces in contact, is essential for maintaining balance and longitudinal progression. Although countermovement can be observed in gait, running, and jumping it is at the maximum vertical jump (MVJ) that the countermovement presents its higher expression and

DOI: 10.1201/9781003217152-89

accessibility for evaluation of its contribution. Thus, MVJ with long CM on countermovement jump (CMJ) and short CM on drop jump (DJ) have been used (Asmussen, 1974) to assess CM contribution in relation to no CM condition on squat jump (SJ). On MVJ increasingly used to assess different CM contribution (Bobbert, 1986; Aragón-Vargas, 1997; Bobbert, 2011) normal ground reaction force (GRF_n) has received the largest attention due to the main objective of vertical thrust, with reduced concern on the study of ground reaction tangential forces (GRF_t). Nevertheless, it is GRF_t and its relationship with the friction of the shoes surfaces in contact with the ground and GRF_n, avoiding slipping and maintaining stability to apply GRF_n responsible for vertical impulsion. It is this relation between GRF_t and GRF_n at different MVJ on SJ, CMJ, and DJ we will explore, using the friction cone in conjunction with a spherical coordinate system to assess its relationship with achieved performance at different MVJ, along with the underlying trend contributing for explaining detected relations.

2 MATERIALS AND METHODS

Trial sample is composed by a group of six male voluntary subjects with ages (21,5 ± 1,4) years old, degree students of sports and physical education, without specific sports fitness, training, or injury. After informed consent subjects were weighted (76,7 ± 9,3) kg and their height measured (1,79 ± 0,06) m, with each subject performing after warm-up a total of nine trials corresponding to three repetitions of each MVJ (SJ, CMJ and DJ) movement subphases presented at Figure 1, and 2 to 3 minutes of rest interval between repetitions. During trial tests, ground reaction forces and force moments were acquired with AMTI BP2416-4000 CE force plate at 1000 Hz integrated with Simi Motion 3D Version 7.5.280 software. According to the interest on MVJ, best SJ, CMJ and DJ trial test was selected for each subject, based on longer time interval of flight phase with null value of GRF.

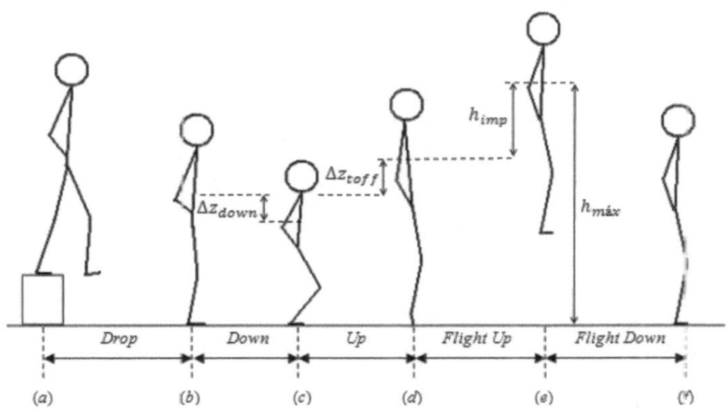

Figure 1. Sequence of movements at DJ (*a* to *f*), CMJ (*b* to *f*) and SJ (*c* to *f*).

Tangential ground reaction force (GRF_t) was obtained from anteroposterior (GRF_x) and medial-lateral (GRF_y) ground reaction force, $GRF_t = (GRF_x^2+GRF_y^2)^{1/2}$, with Coulomb friction (μ) derived from the ratio of the tangential (GRF_t) and the normal ground reaction force ($GRF_n = GRF_z$) corresponding to the vertical ground reaction force, $\mu = GRF_t/GRF_n$. The components of the friction cone were evaluated using spherical coordinates with the azimuth angle ϕ = atan2(GRF_y/GRF_x) between the anteroposterior (GRF_x) and medial-lateral (GRF_y)

ground reaction force, the elevation angle θ = atan(GRF_n/GRF_t) between the normal (GRF_n) and the tangential (GRF_t) ground reaction force, as well as the length corresponding to the spherical distance of the resultant $GRF=(GRF_x^2+GRF_y^2+ GRF_z^2)^{1/2}$ ground reaction force.

3 RESULTS

Representative example of Coulomb friction (μ) time profiles, friction cone azimuth (ϕ) and pitch (θ) angles obtained from ground reaction force time profiles of antero-posterior (GRF_x), medial-lateral (GRF_y) and vertical (GRF_z) components, as well as ground reaction force amplitude (GRF) during impulse phase (IP) are presented at Figures 2 and 3 for one MVJ (CMJ, DJ and SJ) trial experiment of one tested subject. Time reference lines were added to each graph corresponding to critical time instants for association at different physical quantity of the same trial, namely start impulsion (*Imp St*), minimum GRF_z (GRF_z *min*), minimum vertical velocity (v_z *min*), maximum GRF_z (GRF_z *max*), minimum vertical position (Δz min) and maximum vertical velocity (v_z *max*).

According to entire time profile of GRF components, CMJ presented larger time span of impulsion phase (Δt *imp*) and maximum vertical height (*h max*) than SJ, both with higher Δt *imp* and *h max* values than DJ, Figure 2 and Table 1. Average Coulomb friction $<\mu>$ presented during impulsion phase higher value at DJ than CMJ, both with higher values than SJ, whereas CMJ presented higher friction time integral $\int\mu\, dt$ than SJ and DJ, Figure 2 and Table 1, with *h max* presenting negative statistical significative (α=0.023) linear correlation (r=-0.497) with $<\mu>$ and positive statistical significative ($\alpha\leq$0.001) linear correlations (r=0.773) with $\int\mu\, dt$, and (r=0.713) with Δt *imp*. Coulomb friction μ presented different time profiles at CMJ, DJ and SJ with similarities for S1 to S6 subjects at each type of considered MVJ. Thus, while SJ presented several local maximum values with lower amplitude at μ time profiles, CMJ presented one absolute maximum with more pronounced amplitude and DJ presented two maximum with higher amplitude, respectively at the start and the end of impulsion phase, Figure 2.

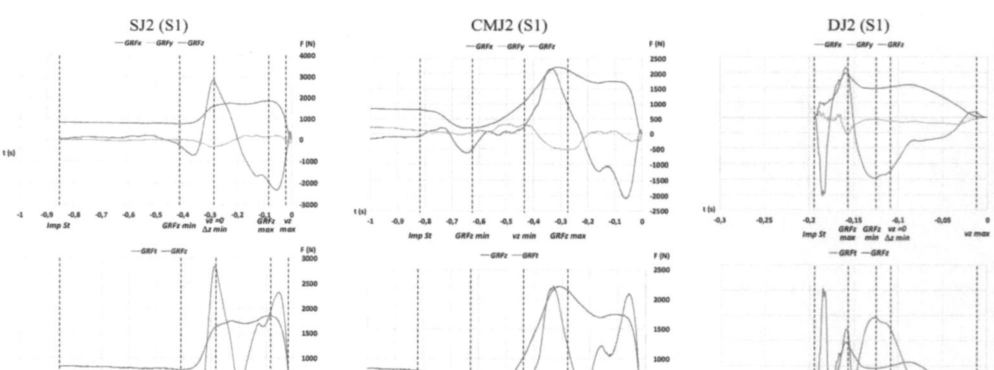

Figure 2. Antero-posterior (GRF_x), medial-lateral (GRF_y), vertical (GRF_z), tangencial (GRF_t) and Coulomb friction (μ) time profiles during impulse phase for one MVJ trial at SJ, CMJ and DJ of one tested subject.

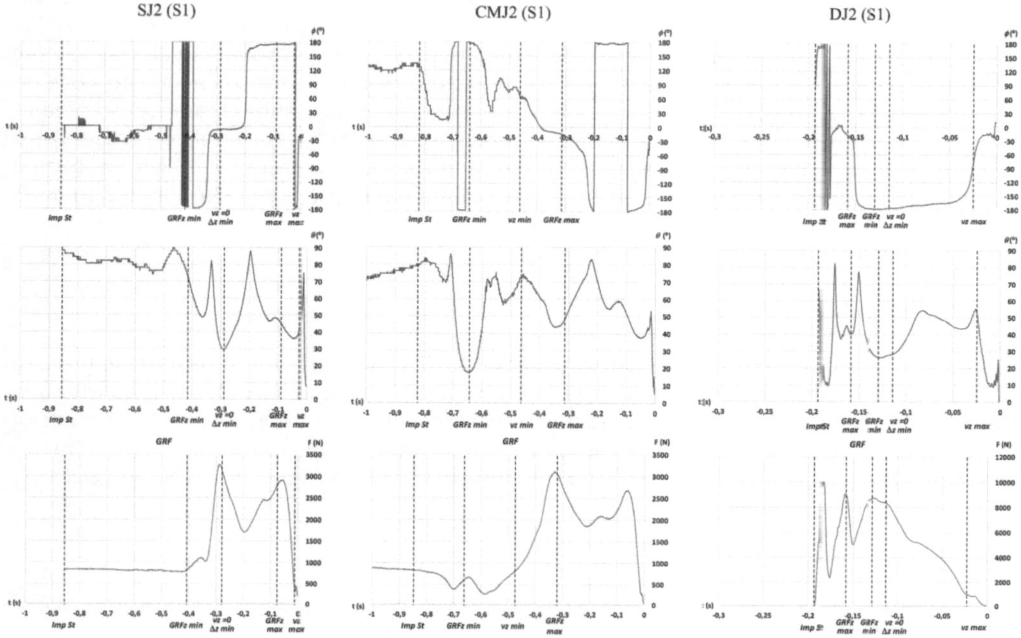

Figure 3. Friction cone azimuth angles ϕ, pitch angles θ and ground reaction force amplitude (*GRF*) time profiles during impulse phase for one MVJ trial at SJ, CMJ and DJ of one tested subject.

Friction cone azimuth ϕ presented different time profiles at SJ, CMJ and DJ with similarities among S1 to S6 subjects for each MVJ type, Figure 3, reflecting variation of antero-posterior and medial-lateral *GRF* components, and determining direction of tangential *GRF* with stability implications at impulsion phase according to magnitudes and variabilities (-25.3 ± 160.7)° on SJ, (-114.1 ± 107.8)° on CMJ and (-112.5 ± 140.8)° at DJ. Friction cone pitch angles θ express the relation of normal and tangential ground reaction force with similar time profilesat each MVJ type, Figure 3, for subjects S1 to S6 and differences between SJ, CMJ and DJ pointing for different stability conditions with higher θ on SJ (61.7 ± 4.0)° and CMJ (62.2 ± 4.3)° than DJ (44.9 ± 5.6)°.

Table 1. Average and st. dev. of Δt_{imp}, h_{max}, $<\mu>$ e $\int \mu\, dt$ at SJ, CMJ and DJ.

	SJ	CMJ	DJ
Δt_{imp} (s)	0.702 ± 0.214	0.804 ± 0.123	0.202 ± 0.023
h_{max} (m)	0.332 ± 0.045	0.364 ± 0.042	0.274 ± 0.032
$<\mu>$	0.513 ± 0.093	0.726 ± 0.218	1.526 ± 0.182
$\int \mu\, dt$	0.349 ± 0.083	0.565 ± 0.102	0.308 ± 0.057

4 DISCUSSION

On one hand subjects need to assure that the ratio between tangential GRF_t and normal GRF_n ground reaction forces is upper limited by the Coulomb friction cone pitch angle θ such that tan $\theta = GRF_t/GRF_n < \mu$, with μ determined by the friction among contact surfaces to prevent slip. On the other hand, due to subject limited capability of force production $GRF = (GRF_x^2 + GRF_y^2 + GRF_z^2)^{1/2}$, if the subject developed force is performed with higher intensity at the tangential plane GRF_t the magnitude of vertical ground reaction force is limited to $GRF_z = (GRF^2 - GRF_t^2)^{1/2}$ at the desired direction on MVJ. For this reason, subjects are confronted at each MVJ type by a trade-off, with the need to balance distribution of applied vertical force GRF_z to achieve higher maximum vertical height, and the need to ensure sufficient tangential ground reaction force *GRF* to prevent slip, while maintaining GRF_t and GRF_z limited to maximum capacity of the subject force production.

5 CONCLUSION

Presented methodology proved adequate extending COP contact behavior at orthostatic position to dynamic conditions on IP at MVJ, as well on complementing normal *GRF* analysis with tangential *GRF* analysis using Coulomb friction and spherical coordinates, allowing detection of the time phases more prone to slip or instability, with possible application at subject strategy correction and relating friction with achieved performance on maximum vertical jump height.

REFERENCES

Aragón-Vargas, L.F. & Gross, M.M. 1997. Kinesiological factors in vertical jump performance: differences among individuals. *Journal of Applied Biomechanics* 13: 24–44.

Asmussen, E. 1974. Storage of Elastic Energy in Skeletal Muscles in Man. *Acta ph. Scand.* 91(3): 385–392.

Bobbert, M.F., Mackay, M., Schinkelshoek, D., Huijing, P.A. & van Ingen Schenau, G.J. 1986. Biomechanical analysis of drop and countermovement jumps. *Europ. J. Appl. Physiol.* 54: 566–573.

Bobbert, M.F. & Richard Casius, L.J. 2011. Spring-like leg behaviour, musculoskeletal mechanics and control in maximum and submaximum height human hopping. *Phil. Trans. R. Soc. B* 366: 1516–1529.

Komi, P.V., Ishikawa, M. & Linnamo, V. 2011. Identification of stretch-shortening cycles in different sports. In J.P. Vilas-Boas, L. Machado, W. Kim, A.P. Veloso (ed.), *Portuguese Journal of Sport Sciences* 11(2): 31–34.

van Ingen Shenau, G.J. & van Soest, A.J. 1996. On the Biomechanical Basis of Dexterity. In M. L. Latash & M.T. Turvey (eds), *Dexterity and Its Development*: 305–338. New Jersey: Lawrence Erlbaum Associates, Inc., Publishers.

Winter, D.A. 2005. *Biomechanics and motor control of human movement*. Hoboken: John Wiley & Sons.

Advances and Current Trends in Biomechanics – Belinha et al. (Eds)
© 2022 Copyright the Author(s), ISBN: 978-1-032-10806-3

Vector coding application for quantification of modified gait

C. Rodrigues & M.V. Correia
C-BER - INESC TEC & FEUP, Porto, Portugal

J.M.C.S. Abrantes
MovLab - CICANT, ULHT, Lisbon, Portugal

M.A.B. Rodrigues
DES, UFPE, Recife, Brazil

J. Nadal
PEB - COPPE, UFRJ, Rio de Janeiro, Brazil

ABSTRACT: This study presents vector coding technique applied to angle-angle joint diagrams to quantify modification of normal gait pattern due to increased knee joint stiffness. Subject specific inverse kinematic analysis was performed during normal gait (NG), stiff knee gait (SKG) and slow running (SR). Case study relates to a healthy 40-year-old male subject weighing 70 kg and 1.86 m tall. Reflective adhesive marks were placed at skin surface of lower limbs selected anatomical points. Images of reflective marks during experimental tests at NG, SKG and SR were captured using camera system. AnyGait software was used for inverse kinematic to obtain lower limb joint angles. Flexion-extension angle-angle diagrams were plotted for the hip, knee, ankle joints and vector coding applied. Results of non-invasive and autonomous vector coding made it possible to objectify and quantify subtle undetectable changes using direct observation of lower limb joint movement at SKG and SR in relation to NG.

1 INTRODUCTION

Normal gait corresponds to the adopted preferred gait by each subject at each movement circumstance according to underlying optimization criteria (Nigg, 2017; Trudeau, 2019; Wilwacher, 2020). As a result of musculoskeletal injury or neurological disease, normal gait subject specific characteristic pattern presents subtle modifications, with the appearance of symptoms that are difficult to detect by direct observation at an early stage (Perry, 2010), pointing for the need of new methods and noninvasive tools for objectivation and quantification of these modifications towards diagnosis and treatment prescription for mitigation and rehabilitation. Stiff knee is characterized by the reduction of the knee joint range of motion as a result of musculoskeletal injury or neurological disease, with impact at the other lower limb joints motion and the change of locomotion with the adoption of the stiff knee gait (SKG) for lower velocities and slow running (SR) for higher velocities in relation to normal gait (NG). Clinical gait analysis has been increasingly applied for diagnosis of musculoskeletal and neurological diseases with the need for assessment metrics contributing for objective and quantified diagnosis independent of the operator. Due to the limitations of discrete measures to represent the variability of a gait pattern, continuous methods using complete time

DOI: 10.1201/9781003217152-90

series are an alternative with high interest in representing and assessing the variability of movement over time. Among these, the representation of relative movement, corresponding to the variation of a variable as a function of another variable of the movement, as is the case of angle-angle diagrams, constitutes a representation of the coordination between the variables involved, as proposed by Sparrow (1987) and Enoka (1994). Regarding the quantification of the relative movement between joints in angle-angle diagrams Hamill (2000) presents the measurement designated as the coupling angle corresponding to the angle with the positive part of the horizontal axis of the vector between two adjacent points in the angle-angle diagram. This is the measure we applied to quantify the modification of the normal pattern of locomotion for the comparison of the intra-joint movement coordination of the lower limbs at NG in relation to SKG and SR.

2 MATERIALS AND METHODS

According to the interest in the analysis of a specific subject and the limitation of the sample mean in the representation of specific subject, a case study of a 40-year-old healthy male volunteer with 70 kg mass and 1.86 m height was assessed. Informed consent was obtained, and adhesive reflective marks were placed on skin surface of palpated anatomical points at the left and right superior anterior and posterior iliac spine, head of the great trochanter, medial and lateral femoral epicondyle, tibial tuberosity, medial and lateral malleolus, and the head of the fifth metatarsus. Trial tests were performed at self-selected walking speed on NG, walk with a stiff knee on SKG, and slow running on SR at the minimum velocity to ensure absence of double contact of both feet with the ground. 3D Cartesian coordinates of the reflecting marks during the tests were obtained by direct linear transformation of the 2D images captured by the calibrated system of 8 Qualysis cameras operating at 1000 Hz. Musculoskeletal modeling was performed using AnyGait v.0.92 software configured for the laboratory's experimental setup, applying the segment model based on a static acquisition and redundant kinematic analysis of dynamic acquisitions, obtaining joint angles through inverse kinematics. The angle-angle diagrams for the knee-hip, knee-ankle and hip-ankle joints were represented during a complete stride at NG, SKG and SR. Vector coding was performed from the coupling angle γ_i and the amplitude of the vector formed by each pair of consecutive points in the angle-angle diagram. According to the directional characteristic of the coupling angle, average direction was calculated based on the mean of the horizontal $\bar{x} = 1/n \sum_{i=1}^{n} \cos \gamma_i$ and vertical $\bar{y} = 1/n \sum_{i=1}^{n} \mathrm{sen} \gamma_i$ projections of the coupling angle along the n iterations, as well as the length of the mean vector $\bar{r} = (\bar{x}^2 + \bar{y}^2)^{1/2}$ as an estimate of the variability associated with its directional concentration.

Figure 1 presents sagittal plane stick-figures with flexion-extension hip, knee and ankle joint angles at selected time instants corresponding to the transitions on each sub-phase on initial double stance (IDS), start of the single support (SS), terminal double stance (TDS), transfer phase (TRF) on NG and SKG, stance (ST), early float (EF), middle swing (MS), late float (LF) on SR.

3 RESULTS

Figure 2 Presents the angle-angle diagrams for the knee-hip, knee-ankle, and hip-ankle joints with differences at sub-phases (*a*)-(*b*) IDS, (*b*)-(*c*) SS, (*c*)-(*d*) TDS, (*d*)-(*e*) TRF and curve shapes on NG and SKG, as well as on (*a*)-(*b*) ST, (*b*)-(*c*) EF, (*c*)-(*d*) MS, (*d*)-(*e*) LF on SR. According to the angle-angle diagrams hip-knee coordination present lower knee angle amplitude on SKG than NG, both lower than SR, with the same occurring

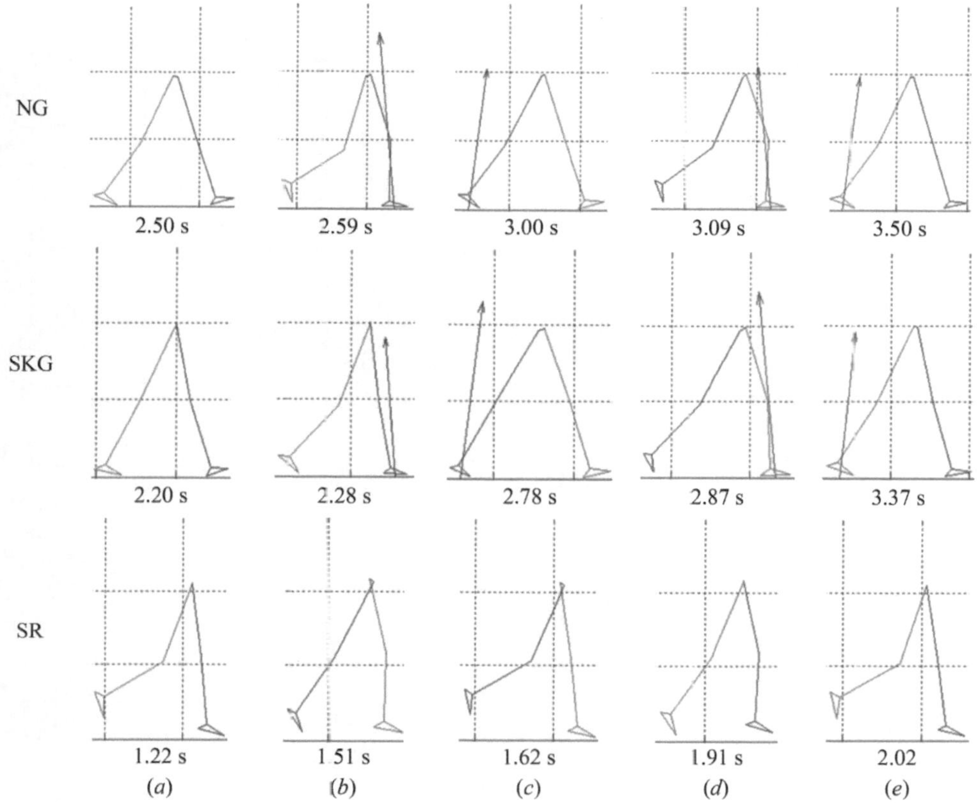

Figure 1. Sagittal plane stick-figures flexion-extension hip, knee, ankle joint angles at (*a*)-(*b*) IDS, (*b*)-(*c*) SS, (*c*)-(*d*) TDS, (*d*)-(*e*) TRF on NG and SKG, and (*a*)-(*b*) ST, (*b*)-(*c*) EF, (*c*)-(*d*) MS, (*d*)-(*e*) LF on SR.

at the knee-ankle angle-angle diagram, and smaller differences on ankle-hip angle-angle diagram between NG, SKG and SR, pointing for greater similarities on A-K coordination at these gait modes. Figure 2 also presents angle-angle corresponding coupling angles and the length of the mean vector, with sub-phases time delimitation where the pattern of γ and r changes with the time sub-phase shift at each gait mode. Table 1 presents gait cycle mean and deviation of the coupling angle γ and length r of the mean vector.

The angle-angle diagrams present at IDS higher variation of the knee angle in relation to the hip and the ankle at NG, with SKG presenting higher variation of the hip and the ankle angles than the knee angle, whereas SR presents at ST higher hip and ankle angle variation than the knee, with the knee and the ankle angle synchronous variation. As regards to SS the hip presents higher angle variation than the knee and the ankle, both at NG and SKG, whereas SR presents on EF higher knee angle variation than the hip and the ankle. At TDS, both NG and SKG present higher knee and ankle angle variation than the hip, while SR presents at MS higher knee and hip angle variation in relation to the ankle. Finally, NG and SKG present at TRF higher knee and hip angle variation, with SR exhibiting at LF reduced angle variation at the hip and the ankle, with moderate knee angle variation.

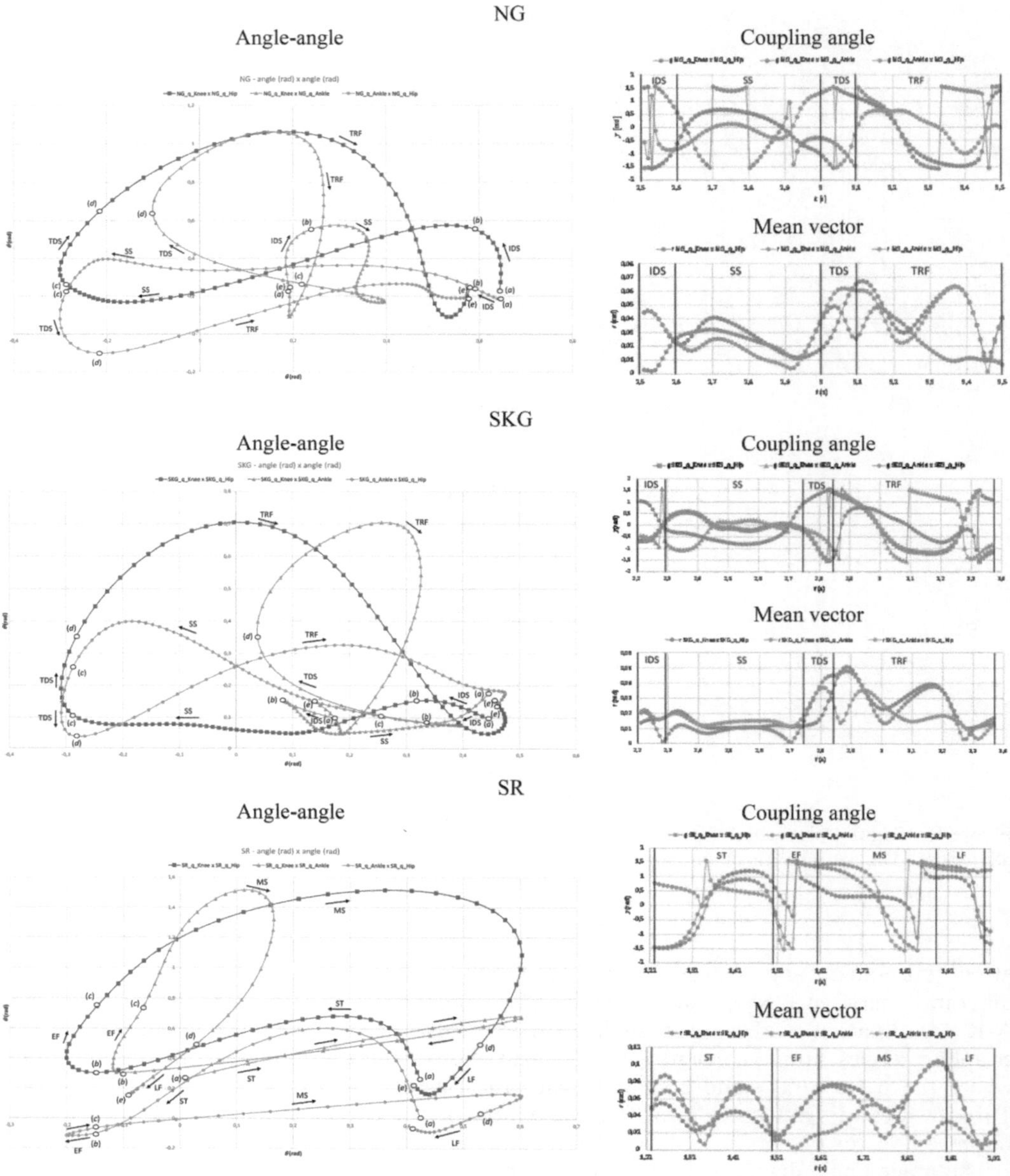

Figure 2. Knee-hip, knee-ankle, and ankle-hip angle-angle diagrams, coupling angle and length of the mean vector at NG, SKG and SR.

Table 1. Gait cycle coupling angle γ and length r of the mean vector.

	NG	SKG	SR
$\bar{\gamma}$ (rad)Knee – Hip(K – H)	-0,098 ± 0,975	-0,163 ± 0,764	0,254 ± 1,073
$\bar{\gamma}$ (rad)Knee – Ankle(K – A)	0,070 ± 1,158	-0,086 ± 0,883	0,775 ± 0,803
$\bar{\gamma}$ (rad)Ankle – Hip(A – H)	-0,025 ± 0,675	-0,027 ± 0,723	0,266 ± 0,827
$\bar{r}$ (rad)Knee – Hip(K – H)	0,036 ± 0,016	0,021 ± 0,012	0,051 ± 0,024
$\bar{r}$ (rad)Knee – Ankle(K – A)	0,033 ± 0,018	0,019 ± 0,013	0,054 ± 0,027
$\bar{r}$ (rad)Ankle – Hip(A – H)	0,024 ± 0,013	0,018 ± 0,007	0,035 ± 0,020

4 DISCUSSION

NG presented higher variability of the gait cycle coupling angle γ between knee-ankle (K-A), and knee-hip (K-H) in relation to ankle-hip (A-H), as well as on SR at the knee-hip (K-H) than the knee-ankle (K-A) and the ankle-hip (A-H), while SKG presents lower γ variability for all the lower limb joints coupling angles. As regards to the directional concentration r, SR presented higher variability than NG for all the lower limb joints coupling angles, both higher than SKG.

5 CONCLUSION

Proposed metrics allowed detection and quantification of the modifications at the lower limb joints relative movement on SKG and SR with respect to NG. Since this is a subject specific study, obtained results are valid for the case study and assessed gait modes at the time of the evaluation. In the case of subjects with injuries or pathologies evaluations must be performed at well-defined time intervals according to the evaluation and rehabilitation plan to allow the detection of changes that may occur gradually. Analyzes must also be specific to each subject to avoid limitation of sample means representation on specific subjects and diversity of case study injury and pathology.

REFERENCES

Enoka, R.M. 1994. *Neuromechanical Basis of Kinesiology*. Champaign: Human Kinetics.

Hamill, J., Haddad, M. & McDermott, W.J. 2000. Issues in quantifying variability from a dynamical system perspective. *Journal of Applied Biomechanics. Sci. Rep*. 10: 1363.

Nigg, B.M., Vienneau, J., Smith, A.C., Trudeau, M.B., Mohr, M. & Nigg, S.R. 2017. The Preferred Movement Path Paradigm: Influence of Running Shoes on Joint Movement. *Med Sci Sports Exerc*. 49(8): 1641–1648.

Perry, J., Burnfield, J.M. 2010. *Gait Analysis: Normal and Pathological Function*. Thorofare: SLACK Inc.

Sparrow, W.A., Donovan, E., van Emmerik, R. & Barry, E.B. 1987. Using relative motion plots to measure changes in intra-limb and inter-limb coordination. *Journal of Motor Behavior* 19: 115–129.

Trudeau, M.B., Willwacher, S., Weir, G., Rohr, E., Ertel, C., Bruggemann, G.P. & Hamill, J. 2019. A novel method for estimating an individual's deviation from their habitual motion path when running. *Footwear Science* 11(3): 135–145.

Willwacher, S., Mählich, D., Trudeau, M.B., Hamill, J., Weir, G., Bruggemann, G.P. & Bratke, G. 2020. The habitual motion path theory: Evidence from cartilage volume reductions in the knee joint after 75 minutes of running. *Sci. Rep*. 10: 1363.

Advances and Current Trends in Biomechanics – Belinha et al. (Eds)

Acetal resin bars for implant-supported overdentures—FEM

V.N. Gomes
Faculty of Dental Medicine, University of Porto, Portugal

D.C. Costa
School of Engineering, University of Minho, Portugal

J. Belinha
School of Engineering, Polytechnic Institute of Porto, Portugal

M. Sampaio-Fernandes, P. Rocha-Almeida, S.J. Oliveira & M.H. Figueiral
INEGI, Faculty of Dental Medicine, University of Porto, Portugal

ABSTRACT: Acetalic resin, or polyoxymethylene (POM), is an alternative material to metal for the manufacturing of bars for implant-supported overdentures. The objective of this work is to study the distribution of von Mises equivalent stresses, equivalent deformations and total displacements when a force is applied on acetalic resin bars with different thicknesses and distal extensions (cantilever) lengths, using the Finite Element Method (FEM). Two Definifit® (Gt-Medical) bars were reproduced virtually in Exocad DentalCAD®, and distal extensions were added. The numerical analysis of the structures was performed using the academic program FEMAS (cmech.webs.com). A vertical force of 1N was applied over the entire bar and, over the abutments, the anterior and lateral arches, and the distal extensions. In both bars, the values of stress, deformation and displacement increase with the length of the distal extension. Bar thickness and cantilever length appear to influence the stress on both acetalic resin bars.

1 INTRODUCTION

An implant-supported overdenture is indicated for patients dissatisfied with the stability and retention provided by conventional removable dentures, although without complaints of pain or discomfort in the mucosa (Barão et al. 2013). A bar is a retentive system that uses supporting implants to anchor overdentures. It consists in an infrastructure between two or more implants, which promotes rigid support and higher quality retention (Burns 2000, Van Kampen et al. 2004). Ferulization of implants offers stability and distribution of occlusal loads throughout the prosthetic structure, as well as the surrounding bone (Stoumpis & Kohal 2011).

With the development of polymers, resinous materials have emerged with differentiating properties from classic metallic materials. The application of acetalic resin or polyoxymethylene (POM) in Dental Medicine have been studied elsewhere.

The objective of this work is to study the distribution of von Mises equivalent stresses, equivalent deformations and total displacements when a force is applied on acetalic resin bars of different thickness and cantilever length, using the Finite Element Method (FEM).

DOI: 10.1201/9781003217152-91

2 MATERIALS AND METHODS

The original Definifit® bars (Gt-Medical) with two different thicknesses (Bar A – 2,0x3,5mm; and Bar B – 2,5x3,5mm), were reproduced virtually in the Exocad DentalCAD® version 2.3 Matera 6990 (64bit) software, and cantilevers of different dimensions (5, 10 and 20mm) and other necessary modifications were done in the .stl file. Both infrastructures were analysed, using the FEM, in the academic program FEMAS (cmech.webs.com). The Poisson's coefficient of 0.2 and the elastic modulus of 2700MPa were the mechanical properties of POM considered for the analysis in each simulation, a vertical force of 1N was applied over the entire bar and individually, over the anterior and posterior abutments, the anterior and lateral arches, and the distal extensions.

3 RESULTS

In the abutments, the highest values of stress are observed at their cervical level (slightly higher on the thicker bar); in the anterior arch, the highest values are found in the midpoint of the arch and close to the anterior abutments (higher on the thicker bar); in the lateral arch, only at the midpoint of the arch (slightly higher on the thicker bar). The maximum values of von Mises stress (MPa), equivalent deformation and total displacement (mm) are higher in the bars with 10 and 20mm cantilevers.

Figure 1 represents von Mises equivalent stresses areas for each bar, when a vertical force is applied over the entire bar.

Hereafter, Figure 2 represents the von Mises equivalent stresses on A (left) and B (right) bars, when a vertical force is applied on the anterior abutment.

The following Figures 3, 4 and 5 show the rising tensions originated, when a vertical force is applied exactly at 5mm, 10mm and 20mm cantilever extensions (of each cantilever bar). The results demonstrated higher stresses in the thicker bar (bar B), for all cantilevers. Due to the larger lever arm of the distal extension, the greater the tension originated by the infrastructure.

Finally, Figure 6 bellow gives an example of a deformation visual effect on 20 mm cantilever.

Figure 1. Representation of von Mises equivalent stress on bar A (left) and bar B (right), when a uniformly force of 1N is applied over the entire bar.

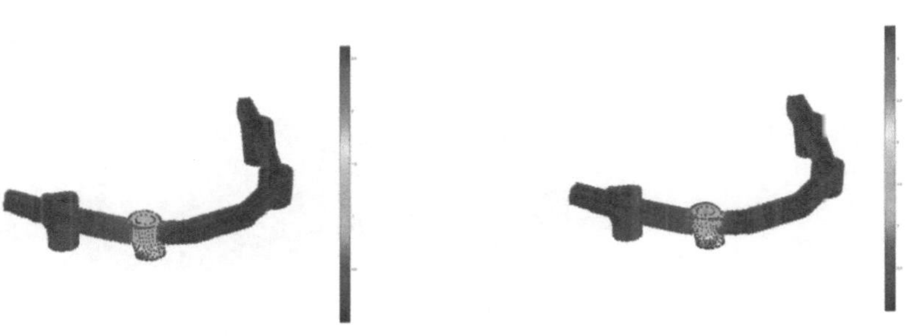

Figure 2. Representation of von Mises equivalent stress on bar A (left) and bar B (right), when a force of 1N is applied over the anterior abutment.

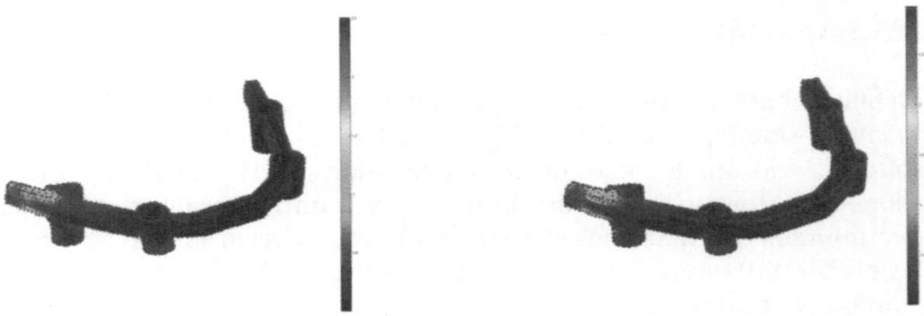

Figure 3. Representation of von Mises equivalent stress on bar A (left) and bar B (right), when a force of 1N is applied on 5mm cantilever.

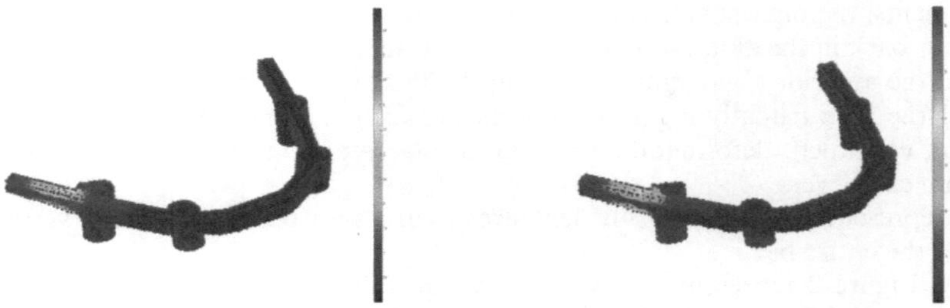

Figure 4. Representation of von Mises equivalent stress on bar A (left) and bar B (right), when a force of 1N is applied on 10 mm cantilever.

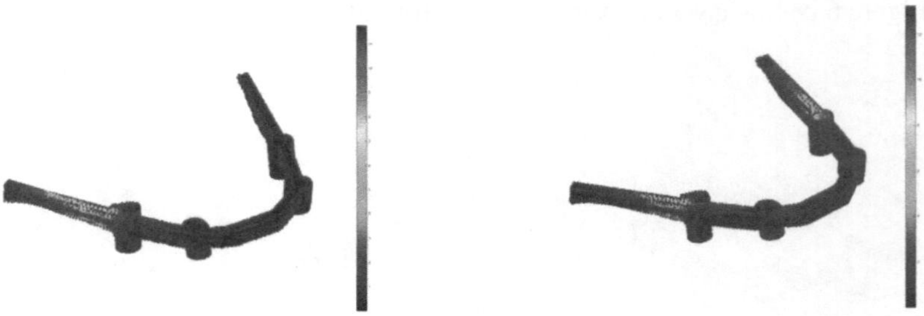

Figure 5. Representation of von Mises equivalent stress on bar A (left) and bar B (right), when a force of 1N is applied on 20 mm cantilever.

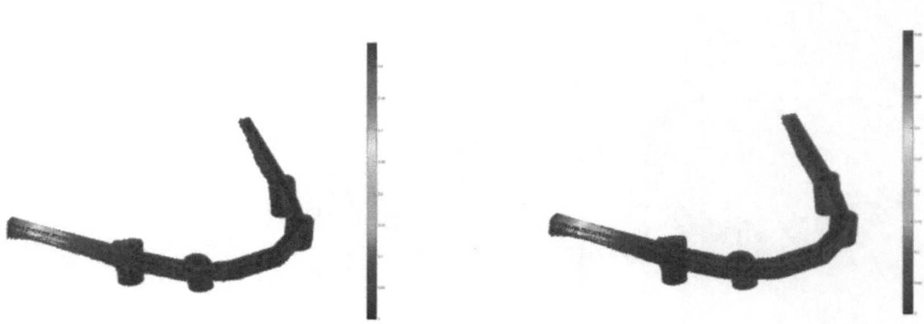

Figure 6. Representation of deformation on bar A (left) and bar B (right), when a force of 1N is applied on 20 mm cantilever.

4 DISCUSSION

Although the bar model herein described has never been investigated before by FEM analysis, the highest stresses observed in the region of the cantilevers and in the center (anterior arch) of the bar are corroborated by other authors in similar studies (Aquino et al. 2018, Waddell et al. 2006). Regardless of the type of infrastructure or means of retention, the stresses are mostly localized around the implant (Barão et al. 2013). In this work, great concentration of von Mises equivalent stresses was observed at the cervical area of each abutment, suggesting high transmission of stresses in the direction of the hypothetical implants.

It is expected that the incidence of forces on the overdenture infrastructure triggers a certain degree of deformation and deflection, depending on the applied force. If the energy incidents on a point is mostly consumed and limited to that same area, less concentration and less energy will be transmitted to other regions of the bar, abutments or implants (Greco et al. 2009).

Cantilevered prostheses must be prepared to withstand the increased stress in the implant closest to the suspended distal extensions (Greco et al. 2009). In the present study, it was found that the values of the stresses double when the force is applied to the 10mm and 20mm cantilever. Likewise, another study corroborates these results, highlighting the influence of the extension of the cantilever on stress distribution along the bar (Rubo & Capello Souza 2009).

Any inaccuracies in the study may arise from the design of the prosthesis (bar and overdenture). This is considered fundamental for a satisfactory biomechanical performance. Thus, some differences of symmetry were found between the physical model and the virtual model of the bar (easily visible in the anterior arch), as a result of possible distortions in the conversion process, namely in the digitalization of the model (CAD) and in the adjustments performed in the software.

5 CONCLUSIONS

The thickness of the bar and the length of the cantilever appear to influence the stress, deformation and displacement of the bar when a vertical force is applied. In both bars, the values of stress, deformation and displacement increase with the increase in the length of the distal extension.

Acetal resin is a promising material for the manufacturing of overdenture supporting bars. However, further studies are needed to define the design and more favorable dimensions, considering the masticatory dynamics of each individual.

REFERENCES

Aquino M.M.O., Kreve S., Ramos E.V., Franco A.B.G., Dias S.C. 2018. Cantilever Protocol Bars in Acrylated Polyetheretherketone (Peek): A Mechanical Compression Assay. *International Journal of Dentistry and Oral Health* 17 (2): 4.

Barão, V.A.R., Delben, J.A., Lima, J., Cabral, T., Assunção W.G. 2013. Comparison of different designs of implant-retained overdentures and fixed full-arch implant-supported prosthesis on stress distribution in edentulous mandible – A computed tomography-based three-dimensional finite element analysis. *Journal of Biomechanics* 46 (7): 1312–1320.

Burns D.R. 2000. Mandibular implant overdenture treatment: Consensus and controversy. *Journal of Prosthodontics* 9 (1): 37–46.

Greco G.D., Jansen W.C., Landre Junior J., Seraidarian P.I. 2009. Stress analysis on the free-end distal extension of an implant-supported mandibular complete denture. *Brazilian Oral Research* 23 (2): 175–181.

Rubo J.H., Capello Souza E.A. 2009. Finite-Element Analysis of Stress on Dental Implant Prosthesis: Finite-Element Analysis of Stress on Dental Implant Prosthesis. *Clinical Implant Dentistry and Related Research* 12 (2): 105–113.

Stoumpis, C., Kohal, R.J. 2011. To splint or not to splint oral implants in the implant-supported overdenture therapy? A systematic literature review: Implant-supported overdenture. *Journal of Oral Rehabilitation* 38 (11): 857–869.

van Kampen, F.M.C., van der Bilt, A., Cune, M.S., Fontijn-Tekamp, F.A., Bosman, F. 2004. Masticatory Function with Implant-supported Overdentures. *Journal of Dental Research* 83 (9): 708–711.

Waddell J.N., Payne A.G.T., Swain M.V. 2006. Physical and metallurgical considerations of failures of soldered bars in bar attachment systems for implant overdentures: A review of the literature. *Journal of Prosthetic Dentistry* 96 (4): 283–288.

Advances and Current Trends in Biomechanics – Belinha et al. (Eds)
© 2022 Copyright the Author(s), ISBN: 978-1-032-10806-3

Neural networks used in a virtual environment to validate a training strategy based on human development

António Ramos Silva
Faculty of Engineering, University of Porto, Porto Portugal
INEGI, Porto, Portugal

Sofia Leite
CINTESIS, University of Porto, Porto, Portugal

Renato Natal & Joaquim Mendes
Faculty of Engineering, University of Porto, Porto Portugal
INEGI, Porto, Portugal

ABSTRACT: Training allows an individual to acquire new skills and improve performance. Maximizing the effect of training mainly requires an adequate training strategy. In the present work, a training rationale based on a theory of human development and skill acquisition, called the Model of Hierarchical Complexity, is set forth for the case of riding a bicycle. Artificial neural network (ANN) models are applied to test the training rationale in a virtual environment developed for the purpose. The goal is to identify the best training sequence of actions, until the complex action of riding a bicycle is achieved. The best training sequence is based on performance metrics and architectural metrics of the ANNs. If this new methodology can be validated, it will be very useful within the scope of digital-twin's new technology

1 INTRODUCTION

Training allows an individual to acquire new skills and improve performance. Generally speaking, training consists of repeating the same action over and over again, until that action is optimally performed and automatized (Curran, 2001; Doyon et al., 2017). However, maximizing the effect of repetitions mainly requires an adequate training strategy: one that is informed by the mechanisms of learning and skill acquisition (Commons and Pekker, 2008; Geijtenbeek, van der Stappen and Panne, 2013; Duran, Sorva and Leite, 2018). This work provides a new methodology based on human development (Commons and Pekker, 2008), virtual environment (VE) and artificial intelligence (Hikosaka et al., 1999; Leite, 2019), for testing the optimality of a given training strategy to the case of riding a bicycle. If this new methodology can be validated in a VE, it can be used within the scope of digital-twin's.

The mechanisms of human development and skill acquisition come from the Model of Hierarchical Complexity (MHC) and will define the training strategy. The MHC is a formal theory of how development and skill acquisition proceeds. It provides sequences of actions of how people learn. These sequences are universal and apply to all domains and all tasks (Giri, Commons and Harringan, 2014). A few concepts are important to define:

i) *Task*: an environmental situation, or problem, that the individual perceives and needs to solve (in this case, the task is how to ride a bicycle);

DOI: 10.1201/9781003217152-92

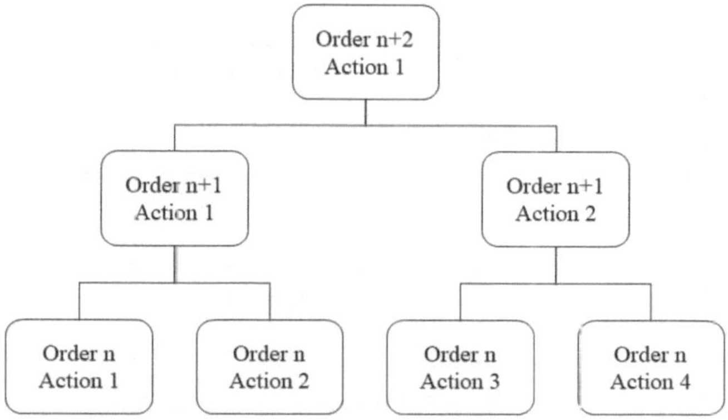

Figure 1. How higher order actions are composed of lower order actions.

ii) *Order of Hierarchical Complexity (OHC)*: the measured difficulty of a given task, where (Figure 1):
   a. The OHC is defined in terms of tasks from the next lower order of hierarchical complexity;
   b. The OHC increases by one if and only if it organizes two or more less complex actions;
   c. The OHC implies that the ordering of the lower task actions is carried out non-arbitrarily.

iii) *Knowledge*: the resources an individual has that allow to perform a given action;
iv) *Action*: the solution that the individual performs to solve a given task;
v) *Task-Action*: a functional pair, where the performed action successfully solves the perceived task;
vi) *Stage*: determined based on the higher OHC task that an individual successfully solves.

Given the axioms above (a., b., and c.), measuring task complexity is a process of recursions, where at each recursion the task is backwards decomposed into sub-tasks. At each stage, increasingly complex tasks are solved by a process of coordinating and integrating lower-order knowledge. When applying this rationale to a training strategy, one first trains less complex sub-tasks and goes on coordinating them to train increasingly complex tasks.

In humans, the MHC rationale has been observed to improve motivation and dedication, as operating in the proximal zone of capacity of the person increases the chance of success and functions as a reward, to improve performance gain and results, to reduce the necessary training time to achieve the same performance proficiency, and to reduce the occurrence of any negative consequence derived from inadequate training.

In artificial settings, ANNs have been recently used for validating this training strategy, too (Leite, 2019; Ramos Silva, 2020). In the first work (Leite, 2019), the first step was to segment the complex task into sub-tasks and sub-actions, then, ANNs were trained to learn each of those sub-actions. Among others, results showed that when comparing this approach to previous work, the final action was learned with more accuracy. Results also showed that there was an economy of resources in architectural terms, since learning a more complex sub-action used the elements of the previous architecture. In the present work, the MHC rationale and the ANN training methodology are applied to training the action of riding a bicycle, in a VE developed for the purpose.

2 MATERIALS AND METHODS

The proposed virtual environment consists of a bicycle moving in space. In order to evaluate different ways of training, the action of riding a bicycle was first segmented into sub-actions (Table 1). ANN models were trained to learn each of these sub-actions (1, 2, and 3) and a control ANN model was trained to learn the action of riding a bicycle from scratch (4). For testing the training strategies, the respective sub-actions ANN were juxtaposed and compared. The list of sub-actions is as follows:

Table 1. List of different training strategies.

Sequence strategy	Sub-actions composing the strategy
A	1→2→3
B	4

1. Learning to balance, assuming the vehicle holds a constant velocity;
2. Learning to guide, assuming the vehicle is in equilibrium;
3. Coordinating 1) and 2);
4. Learning to ride a bicycle from scratch;

To simulate the bicycle dynamics in a 3D numeric model, an implementation in Simscape from MatLab® was used (Figure 2) (Thirumurugan, 2021). This was tested with various properly tuned PID controllers to balance and guide the bicycle through a certain trajectory. The controller (and, thus, the networks) received the horizontal position (X and Y), the turning angle and its derivative (δ and $\dot{\delta}$), the balancing angle and its derivative (ϑ and $\dot{\vartheta}$), the gimbal angle and its derivative (θ and $\dot{\theta}$) and the velocity. The controlled variables were the steering torque, the gimbal torque, rear wheel torque, and the flywheel speed. The referred gimbal is an inertial disc used to create a torque and thus allow the balance of the bicycle. The angle θ is the angle between the bicycle main axis and the disk angle.

To determine the minimum size of the network necessary to perform a certain task, various feedforward ANN were tested, using the Levenberg-Marquardt learning algorithm with no learning rate. For each ANN model, the goal was to find the least complex network architecture capable of learning the proposed sub-action with the best performance, calculated based on the Mean Squared Error (MSE) and required epochs, under a maximum of 100. At each test, a new neuron was added, up to a maximum of 20. After that, a new layer was added with

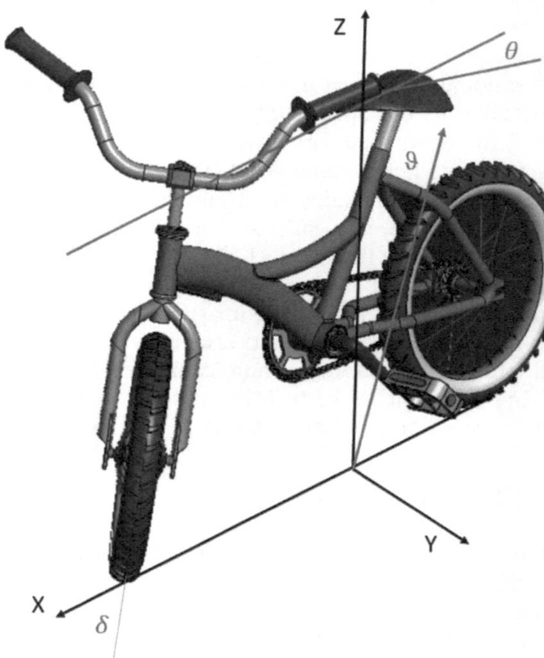

Figure 2. Bicycle diagram and simulation variables.

the number of neurons being increased at each run. The process would stop if a mean squared error (MSE) of 0.001 was reached.

3 RESULTS

For all the sub-actions to be performed, the first network able to obtain a MSE smaller than 0.001 was chosen. This was considered the most efficient network to accomplish the goal. The efficiency was chosen based on architectural metrics (least architectural complexity) and architectural metrics (faster and most accurate learning), in a way that cost increased the least throughout the sequence. The following results were found for each training strategy (Table 2).

Table 2. List of different training strategies.

Sequence designation	Sub-actions	units/layer	connections	MSE	Epochs
A	1	9,1,1	11	0.00078	30
	2	9,2,2	26	0.00087	13
	3	9,3,3	42	0.00099	30
B	4	9,10,3	133	0.00099	8

4 DISCUSSION

Results were illustrative and conclusive about the adequacy of using a training strategy that follows the laws of human development and skill acquisition, as set forth by the Model of Hierarchical Complexity. Segregating the training strategy per sub-action of complexity greatly benefits training in terms of architectural cost, given a similar learning error. In architectural terms, learning to ride a bicycle from scratch requires more than the triple of the number of connections (133) than learning to ride a bicycle from coordinating lower-order actions (42). Considering that this was a simple simulation, these numbers are significant. When scaling up the complexity of the task to solve, these architectural metrics scale up as well. However, it is important to mention that the same economy in training epochs was not verified. Training from scratch required three times less epochs than training from lower-order actions. One explanation is plausible: since this training strategy is based on actual human learning, and since human learning takes the time of a lifetime to stabilize increasingly complex knowledge at the expense of a minimum increase in the neural structure, then, it is expected that training epochs increase.

5 CONCLUSION

This work proposes a methodology based on virtual environment and artificial intelligence for evaluating the optimal sequencing of training. This methodology is based on the rationale of the Model of Hierarchical Complexity, that formalizes human development and skill acquisition. This is the third work in which the presented methodology has been validated, although the first one that validated it using a Virtual Environment, which is very useful for the application to digital twins technology. The creation of digital twins has gained importance in the industry. This methodology enables the evaluation of digital twins of athletes for better developing training procedures and action sequences, for instance.

REFERENCES

Commons, M. L. and Pekker, A. (2008) 'Hierarchical Complexity: A Formal Theory', *World Futures*, 64, pp. 375–382. doi: 10.1080/02604020802301204.

Curran, T. (2001) 'Implicit learning revealed by the method of opposition.', *Trends in Cognitive Sciences*, 5(12), pp. 503–504. doi: 10.1016/S1364-6613(00)01791-5.

Doyon, J. *et al.* (2017) 'Current issues related to motor sequence learning in humans', *Current Opinion in Behavioral Sciences*, 20. doi: 10.1016/j.cobeha.2017.11.012.

Duran, R., Sorva, J. and Leite, S. (2018) 'Towards an analysis of programs complexity from a cognitive perspective', in *Proceedings of the 2018 ACM Conference on International Computing Education Research. International Computing Education Research Conference*, Espoo, pp. 21–30. doi: 10.1145/3105726.3105731.

Geijtenbeek, T., van der Stappen, A. F. and Panne, M. (2013) 'Flexible Muscle-Based Locomotion for Bipedal Creatures', *ACM Transactions on Graphics*, 32, p. 206:1-206:11. doi: 10.1145/2508363.2508399.

Giri, S., Commons, M. L. and Harringan, W. J. (2014) 'There is only one stage domain', *Behavioral developmental Bulletin*, 19(4), pp. 51–61. doi: 10.1037/h0101081.

Hikosaka, O. *et al.* (1999) 'Parallel neural networks for learning sequential procedures.', *Trends in Neurosciences*, 22(10), pp. 464–471. doi: 10.1016/S0166-2236(99)01439-3.

Leite, S. (2019) *Foundation of a Hierarchical Stacked Neural Network model for Simulating Cognitive Development*. Porto.

Ramos Silva, A. (2020) 'Modeling human decision making in chess using the self stacked systems architecture.' *II Chess Researchers International Meeting*, online conference.

Thirumurugan, D. (2021) *Dinesh Thirumurugan (2021). Multi-body Simulator for Bicycle Stability* (https://www.mathworks.com/matlabcentral/fileexchange/69620-multi-body-simulator-for-bicycle-stability*), MATLAB Central File Exchange. Retrieved March 23, 2021*. MATLAB Central File Exchange. Available at: https://www.mathworks.com/matlabcentral/fileexchange/69620-multi-body-simulator-for-bicycle-stability (Accessed: 23 January 2021).

Advances and Current Trends in Biomechanics – Belinha et al. (Eds)

The structural analysis of rib bones using a radial point interpolation meshless method

D. Dias
Faculty of Engineering, University of Porto, Portugal

J. Belinha
School of Engineering, Polytechnic of Porto, Portugal

ABSTRACT: Ribs provide protection to vital inner organs, such as heart and lungs. Yet ribs are highly exposed to impact loads from accidents, leading to fractures. Hence, it is important to understand the structural response mechanics of ribs. In this work, a rib bone is analysed using a radial point interpolation method (RPIM). The most attractive features of RPIM are that it can be combined with finite element method (FEM) and uses its mathematical constructions for more accurate results. Structural analyses of 3D rib bones are performed with six independent load cases as natural boundary conditions and specific essential boundary conditions mimicking the physiological scenario. In this numerical study, the material is considered as linear, elastic, homogenous and isotropic. Displacement, von Mises stress, first and third principal stress fields were obtained with an elasto-static analysis. The results show that the RPIM can produce results similar to FEM with smoother field distributions.

1 INTRODUCTION

Major vital organs, such as the heart and lungs, are protected by the ribs. Under sudden impact loads, ribs can fracture and lead to injuries in the internal organs, causing grave situations. Common in automobile crashes and fallings, rib fractures represent the most common bone fracture, occurring in 10%-20% of all trauma patients (Liebsch *et al.*, 2019). Humans possess 12 pairs of ribs. These curved flattened bones contribute to the thorax wall and consist of cortical and trabecular bones (Netter, 1994).

Multi-dimensional, nonlinear and complex differential equation systems are often encountered in advanced engineering analysis where numerical analysis is essential to obtain approximated solutions. Due to mathematical difficulties, analytical solutions are limited to problems with very simple geometries and external loadings. In the field of biomechanics, the structural response of ribs can be predicted using discretization techniques. The most popular one is the finite element method (FEM), first developed in 1956 for structural analysis of complex aeronautic structures (Turner *et al.*, 1956). The recent developments in computation science and the increasing enhancement of computer hardware, allowed to develop further advanced discretization techniques, such as meshless methods (Belinha, 2014). In meshless methods, the nodes can be randomly distributed since the field function is approximated within a flexible influence domain rather than an element (Belinha, Jorge and Dinis, 2012).

The radial point interpolation meshless method (RPIM) is commonly used for stress analysis of three dimensional (3D) solids, based on the Galerkin weak form formulation using 3D

DOI: 10.1201/9781003217152-93

meshless shape functions constructed using radial basis functions (RBFs) (Liu *et al.*, 2005). The RPIM started with the point interpolation method (PIM), which involves the construction of polynomial interpolants based only on a group of arbitrarily distributed points. Later, the PIM evolved and the RPIM was fully developed (Farahani *et al.*, 2015).

When analysing the impact responses on rib bones, up to date, generally thoracic models or individual ribs have been used. However small-scale refined models are believed to represent an accurate geometric model since large scale models usually present an over-simplified geometry (Li *et al.*, 2010; Subit *et al.*, 2013). Past studies suggest that cortical bone thickness and cross-sectional geometry are important factors of rib impact response analysis, which can be achieved through FE models (Holcombe *et al.*, 2019).

This study aims to compare the numerical performance of the radial point interpolation meshless method (RPIM) and to analyse the impact responses of the first rib when loads are applied to different regions of the rib as well as in distinct directions. The Finite Element and Meshless Analysis Software (FEMAS) academic software (cmech.webs.com) is used to conduct the structural analysis of the ribs.

2 MATERIALS AND METHODS

To prove the flexibility, accuracy and robustness of the RPIM, 3D geometric models of the first rib bone were constructed using an anonymized CT image of the thorax. These models were then discretized in common tetrahedral meshes. Then, the RPIM formulation, running in FEMAS academic software, from the tetrahedral mesh acquires the nodal mesh (the vertices of the tetrahedrons) and assumes the tetrahedral as integration cells, placing inside each tetrahedral four integration points following the Gauss-Legendre quadrature scheme. Afterwards, the influence-domains of each integration point are established, using a simple spherical search and the shape-functions are constructed. Regarding the essential boundary conditions, the ribs are attached to the thoracic vertebrae and the costal cartilage to the sternum, as the white dots in Figure 1(a) indicate. Six independent load cases are analysed, as represented in Figure 1(b). All six load cases are with the same magnitude of 500N and two different directions, loads F1, F2, F3 in the Oz directions and F4, F5, F6 in the Ox directions. The model was simplified and the ribs were assumed as cortical bone. Thus, the material is considered as linear, elastic, isotropic and homogeneous, being the Young's modulus E=13.9GPa and Poisson's ratio υ=0.3 [9]. Afterwards, an elasto-static analysis is performed, and several variable fields are obtained.

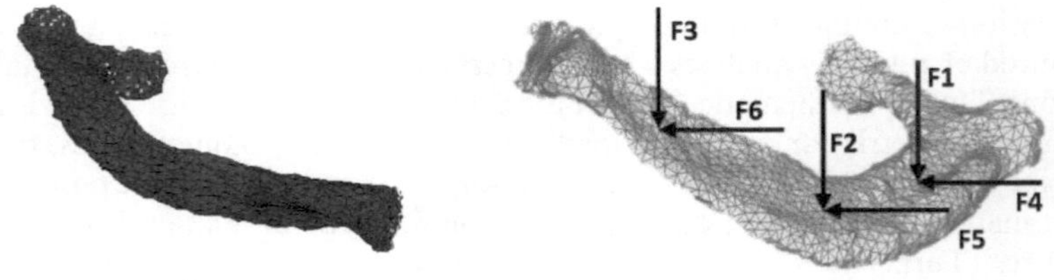

a) b)

Figure 1. A) Essential boundary conditions. b) Natural boundary conditions - 6 independent load cases of 500N.

3 RESULTS AND DISCUSSION

The constructed 3D mesh of the first rib bone was subjected to elasto static analysis using constant strain tetrahedral elements. Absolute displacement, Von Mises equivalent stress, first and third principal stresses were analysed with the 6 different independent load cases.

Obtained absolute displacement colour maps are presented in Figure 2, along with their maximum values. Results show that both RPIM and FEM display similar distributions as well as closer magnitudes. The maximum displacement of 0.79 mm is reached with the F2 load case and the F6 load case produce the minimum of 0.07 mm displacement.

The von Mises failure theory calculates equivalent stress considering the energy absorbed during the distortion. If the equivalent stress exceeds the yield stress, failure occurs. The Figure 3 presents the resultant colour maps obtained from the analysis for six load cases and two methods. While F2 load cause a maximum of 61 MPa, the minimum of 31MPa is caused by the F6 load case. The colour distribution of stress variations shows that higher stresses are obtained where the loads are applied. Furthermore, results from RPIM shows similar and smoother results of FEM.

Principal stresses indicate the maximum and minimum normal stresses in a particular plane. Each material point exhibits three principal stresses. The first principal stress is the highest value in an element and the lowest value is considered as the third principal stress. First and

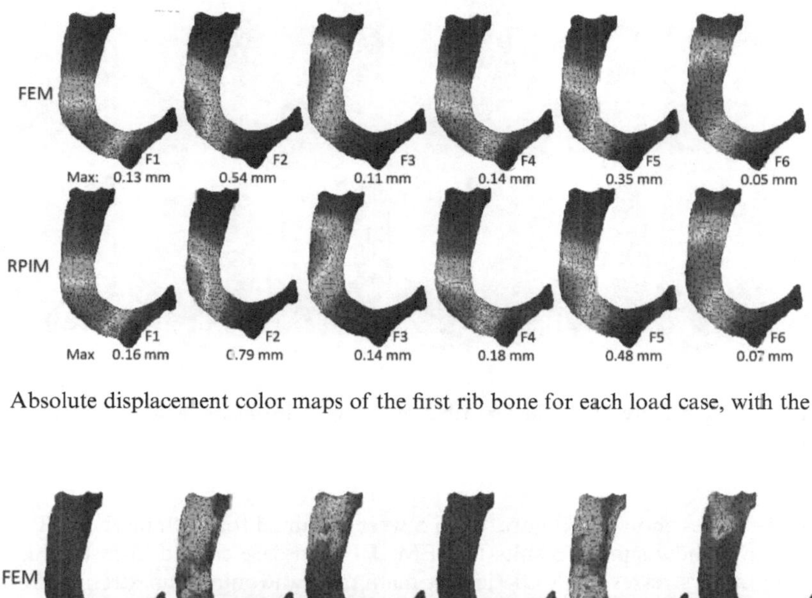

Figure 2. Absolute displacement color maps of the first rib bone for each load case, with the maximum values.

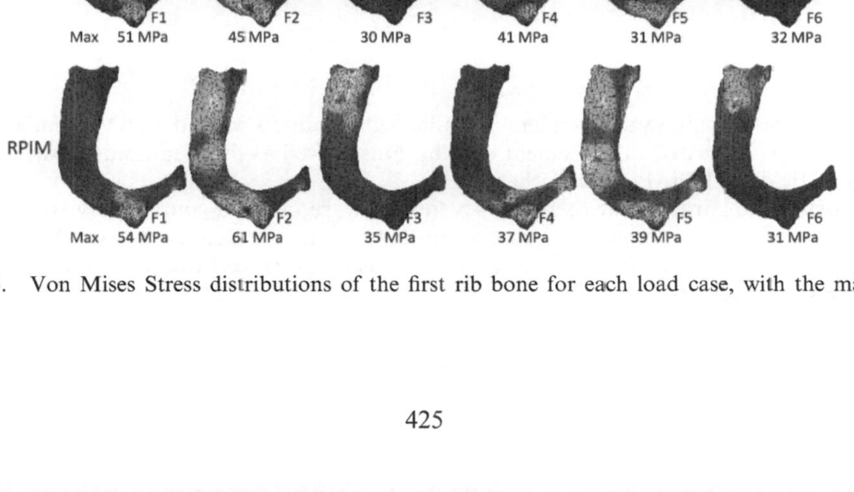

Figure 3. Von Mises Stress distributions of the first rib bone for each load case, with the maximum values.

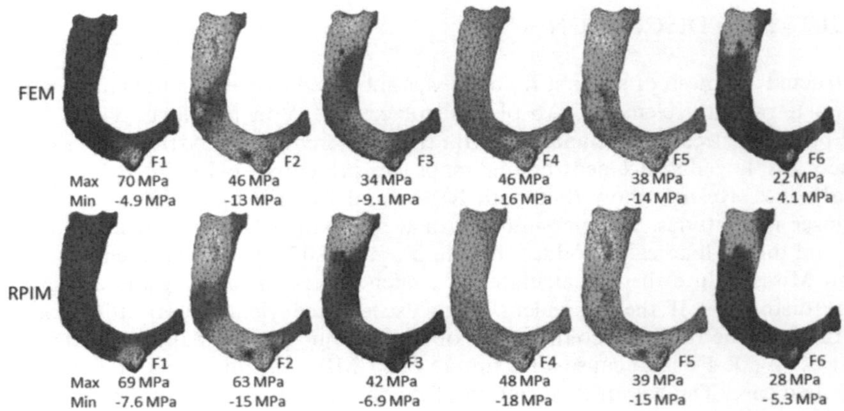

Figure 4. First principal stresses of the first rib bone for each load case, with the maximum and minimum values.

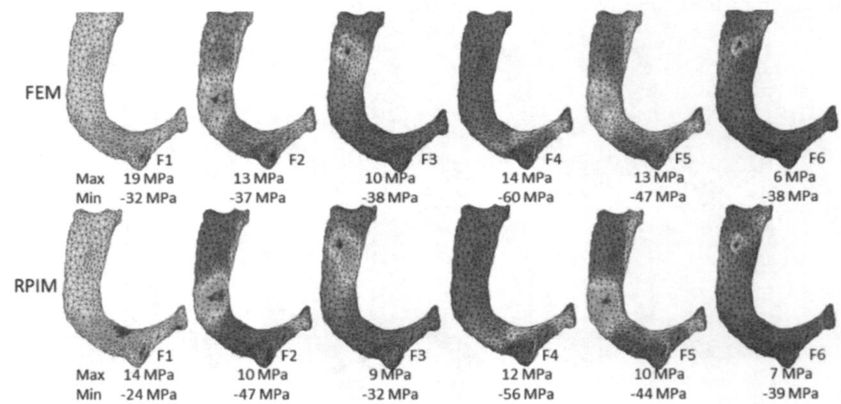

Figure 5. Third principal stresses of the first rib bone for each load case, with the maximum and minimum values.

third principal stresses shown in Figure 4 and 5 were obtained for both methods. Results from the RPIM method show similar results to FEM. F1 load case caused both the highest maximum and minimum stresses while the lowest maximum and minimum stresses are results of load case F4.

4 CONCLUSION

The obtained results allowed to understand that with both FEM and RPIM, similar results can be achieved, stress and displacement distributions as well as the magnitudes. Nevertheless, the variable fields of RPIM are smoother.

This work demonstrates that there are several freeware or open-source software available, capable to enable the complete simulation process, from the acquisition of the discretize model from CT images to the structural analysis using advanced discretization techniques, such as FEM or mesh-less methods.

REFERENCES

Belinha, J. (2014) *Meshless Methods in Biomechanics - Bone Tissue Remodelling Analysis*. Vol.16. Edited by J. M. R. S. Tavares and R. M. Natal Jorge. Lecture Notes in Computational Vision and Biomechanics, Springer Netherlands. doi: 10.1007/978-3-319-06400-0.

Belinha, J., Jorge, R. N. and Dinis, L. M. J. S. (2012) 'Bone tissue remodelling analysis considering a radial point interpolator meshless method', *Engineering Analysis with Boundary Elements*, 36(11), pp. 1660–1670. doi: 10.1016/j.enganabound.2012.05.009.

Farahani, B. V. *et al.* (2015) 'The Axisymmetric Analysis of Circular Plates Using the Radial Point Interpolation Method', *International Journal for Computational Methods in Engineering Science and Mechanics*, 16(6), pp. 336–353. doi: 10.1080/15502287.2015.1103819.

Holcombe, S. A. *et al.* (2019) 'Regional maps of rib cortical bone thickness and cross-sectional geometry', *Journal of Anatomy*, 235(5), pp. 883–891. doi: 10.1111/joa.13045.

Li, Z. *et al.* (2010) 'Influence of mesh density, cortical thickness and material properties on human rib fracture prediction', *Medical Engineering and Physics*, 32(9), pp. 998–1008. doi: 10.1016/j.medengphy.2010.06.015.

Liebsch, C. *et al.* (2019) 'Patterns of serial rib fractures after blunt chest trauma: An analysis of 380 cases', *PLoS ONE*, 14(12). doi: 10.1371/journal.pone.0224105.

Liu, G. R. *et al.* (2005) 'A meshfree radial point interpolation method (RPIM) for three-dimensional solids', *Computational Mechanics*, 36(6), pp. 421–430. doi: 10.1007/s00466-005-0657-6.

Netter, F. H. (1994) *Musculoskeletal system - Netter collection of medical illustrations*. 1 ed. Vol. Saunders.

Subit, D. *et al.* (2013) *Tensile material properties of human rib cortical bone under quasi-static and dynamic failure loading and influence of the bone microstucture on failure characteristics*.

Turner, M. J. *et al.* (1956) 'Stiffness and Deflection Analysis of Complex Structures', *Journal of the Aeronautical Sciences*, 23(9), pp. 805–823. doi: 10.2514/8.3664.

Advances and Current Trends in Biomechanics – Belinha et al. (Eds)
© 2022 Copyright the Author(s), ISBN: 978-1-032-10806-3

Comparing the performance of distinct advanced discretization techniques in the biomechanical analysis of rib bones

D. Dias
Faculty of Engineering, University of Porto, Portugal

J. Belinha
School of Engineering, Polytechnic of Porto, Portugal

ABSTRACT: There are several discretization techniques to analyse the biomechanical response of biostructures, most common technique being the finite element method (FEM). Nevertheless, recently, other advanced discretization techniques have appeared, such as meshless methods. This work aims to provide a comparison between the numerical performances of the FEM and two distinct meshless methods, the radial point interpolation method (RPIM) and the natural neighbour radial point interpolation method (NNRPIM). To compare the performance of these methods, a 3D model of third rib bone is constructed using an anonymized CT image, simulating physiological boundary conditions. The bone material is considered linear-elastic, isotropic and homogeneous. The variable fields, such as displacement, von Mises stress and principal stresses were obtained from the structural analysis. Results show that the NNRPIM is much more computationally expensive than FEM or RPIM. All the solutions obtained with FEM; RPIM and NNRPIM, present similar values and distributions.

1 INTRODUCTION

In the literature, it is possible to find several discretization techniques capable to perform efficient structural analysis of bio-structures. The finite element method (FEM) is the most dominant discretization technique used in structural and biomechanical engineering applications. It was first proposed in 1956 for structural analysis of complex aeronautic structures (Turner *et al.*, 1956) and, since then, it was successfully applied to various scientific fields. However, FEM presents some limitations linked to its mesh-dependent nature. For instance, large mesh distortion in nonlinear geometric analysis can reduce its accuracy and numeric stability. In order to answer some of the FEM limitations, the numerical methods research community started to develop alternative techniques. Meshless methods constitute a class of advanced discretization techniques capable of discretizing the problem domain using only a nodal distribution (Belinha, 2014).

One of the oldest meshless methods being the Smooth Particle Hydrodynamics Method (RKPM). After that more methods such as Diffuse Element Method (DEM), Element Free Galerkin Method (EFGM), meshless local Petrov-Galerkin method (MLPG), Method of the Finite Spheres (MFS), Finite Point Method (FPM), Radial Basis Function Method (RBFM). Each of these methods uses a different (or more evolved version) to approximate a solution to the complex differential equation systems. There are several interpolant meshless methods developed with the ability to overcome the difficulty of imposing essential and natural boundary conditions, such as Point Interpolation Method (PIM), Point Assembly Method and Radial Point Interpolation Method (RPIM). The shape functions of these meshless methods

DOI: 10.1201/9781003217152-94

possess the delta Kronecker property, which allows to directly impose the essential and natural boundary conditions using the same numeric techniques as the FEM. Additionally, interpolating meshless methods can be directly combined with FEM, allowing to discretize part of the domain with FEM and the remaining domain with meshless methods. Compared to FEM, meshless methods possess the advantages of remeshing efficiency and combined with scanning techniques, such as CT and MRI. (Belinha, 2014). Hence, meshless methods are particularly suited to analyse the structural response of biological systems. Since with meshless methods the problem domain can be discretized using only nodes, the 3D geometric model can be defined directly from medical image sets, linking the voxel (or set of voxels) to domain nodes. Moreover, because each voxel possesses its own colour level (associated with the observed biomaterial), it is possible to associate to the corresponding node that material property (Belinha, 2014).

In this work two meshless methods are used, the radial point interpolation method (RPIM) and the natural neighbour radial point interpolation method (NNRPIM). This study aims to compare the performance of FEM with RPIM and NNRPIM. Thus, the third rib bone is analysed using the three discretization techniques.

2 MATERIAL AND METHODS

Rib bones protect major vital organs. However, under sudden impact loads rib bones can fracture and injure those organs. These curved flattened bones contribute to the thorax wall and possess a cortical layer involving trabecular tissue (Netter, 1994). In this study, using an anonymized CT image of the thorax, a 3D geometric model of third rib bone was constructed. The models were discretized using tetrahedrons, which can be directly used by the FEM. Thus, for the FEM analysis constant strain, tetrahedral elements were used. The same 3D discrete model can be used by the RPIM. From the tetrahedral mesh the nodal mesh is acquired (the vertices of the tetrahedrons) and it assumes the tetrahedral as integration cells, placing inside each tetrahedral four integration points following the Gauss-Legendre quadrature scheme. Afterwards, the influence-domains of each integration point are established, using a simple spherical search and the shape-functions are constructed. From the 3D discretized model, the NNRPIM only uses the nodes (the vertices of the tetrahedrons). With the nodal set, the NNRPIM establishes the integration cells and the integration points using the Voronoi diagram and the Delaunay triangulation concepts. The nodal connectivity is enforced using the natural neighbour geometric construction. The shape functions are constructed using the same technique as the RPIM: As Figure 1(a) shows, the ribs are fixed to the thoracic vertebrae and the costal cartilage to the sternum (the white dots). Regarding the applied external load, as represented in Figure 1(b), six independent loads were applied (all with the same magnitude of 500N). Since FEM, RPIM and NNRPIM possess the delta Kronecker property, the essential and natural boundary conditions can be imposed using the direct method in the three numerical methods. The model was simplified and the ribs were assumed as cortical

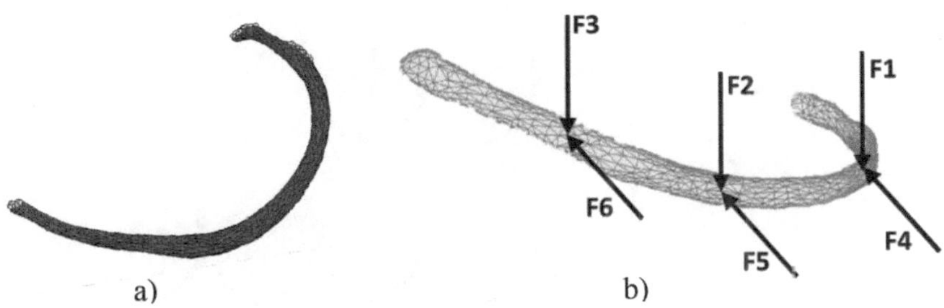

Figure 1. 3D model mesh of the third rib bone. a) Essential boundary conditions in white dots. b) Natural boundary conditions - 6 independent load cases of 500N each.

bone. Thus, the material is considered as linear, elastic, isotropic and heterogeneous, with the Young's modulus $E = 13.9GPa$ and Poisson's ratio $v = 0.3$ (Subit *et al.*, 2013). Afterwards, an elasto-static analysis is performed and several variable fields are obtained. The structural analysis of the ribs is performed with the Finite Element and Meshless Analysis Software (FEMAS), an academic software (cmech.webs.com).

3 RESULTS AND DISCUSSION

The elasto-static analysis, performed using FEMAS academic software, allowed to obtain several variable fields, such as absolute displacement, von Mises equivalent stress, first and third principal stresses.

Figure 2 presents the absolute displacement colour maps of the six load cases obtained from the studied three different methods, with their maximum values. The achieved results suggest that the displacement distributions, as well as their magnitudes of the three methods, are similar.

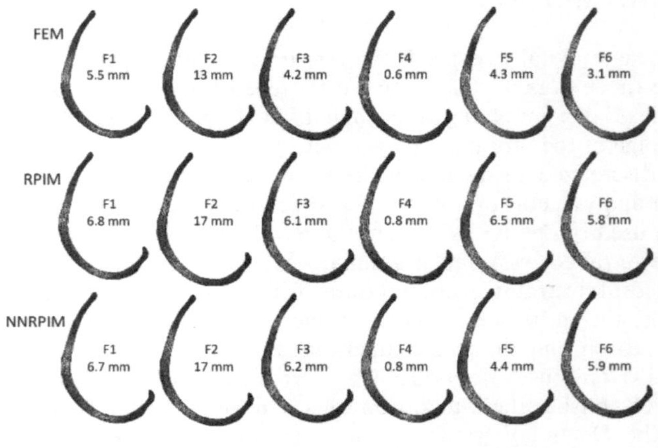

Figure 2. Color maps of absolute displacement with their maximum value.

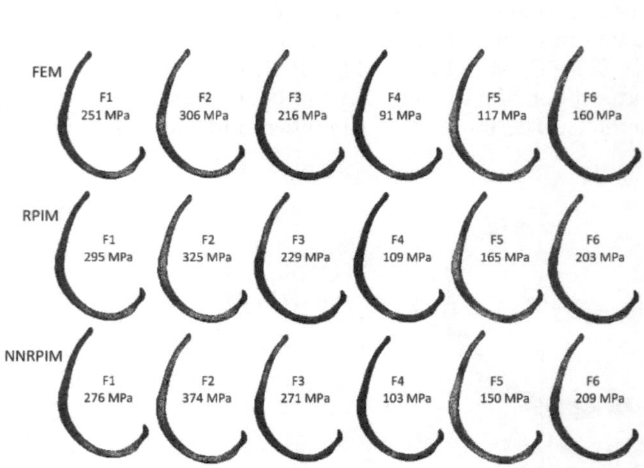

Figure 3. Color maps of von Mises equivalent stress with their maximum value.

Using von Mises failure theory, equivalent stress can be calculated, which indicates the maximum yield stress a point can withstand without failure for a particular load set. The resultant von Mises equivalent stress colour maps of the analysed third rib bone are presented in Figure 3, for FEM, RPIM and NNRPIM methods, respectively. While displacement distributions of the 3 methods appear to be similar, the magnitudes vary slightly in a close range.

The first and third principal stresses are the highest and lowest normal stresses a point can attain in a shear free plane for a given load set. Figures 4 and 5 display the colour maps of the first and third principal stresses in order, with their highest and lowest values. A similar but smoother stress distribution can be observed in the RPIM and NNRPIM colour maps comparing with the FEM. However, the stress magnitudes of RPIM and NNRPIM slightly diverge from FEM.

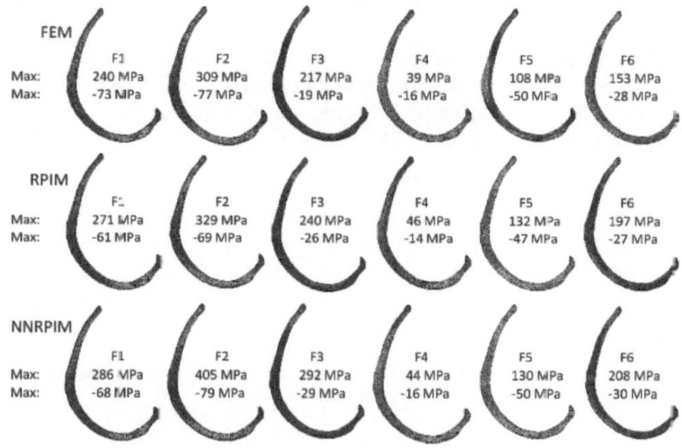

Figure 4. Color maps of first principal stress with their maximum values.

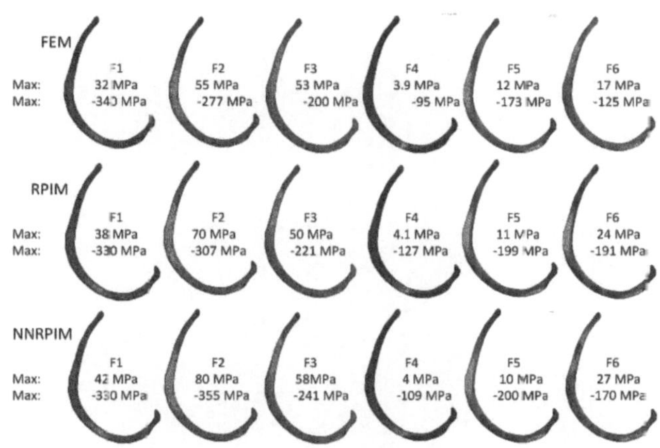

Figure 5. Color maps of third principal stress with their maximum values.

4 CONCLUSION

Even though FEM; RPIM and NNRPIM produce very similar results, their computational cost is quite different. Although the NNRPIM allows obtaining the 3D model directly from the CT images, such meshing flexibility is computationally very expensive. Therefore, when compared with FEM and RPIM, NNRPIM presents the highest computational cost. On the other hand, the processing time of RPIM is very close to the FEM.

REFERENCES

Belinha, J. (2014) *Meshless Methods in Biomechanics - Bone Tissue Remodelling Analysis*. Vol.16. Edited by J. M. R. S. Tavares and R. M. Natal Jorge. Lecture Notes in Computational Vision and Biomechanics, Springer Netherlands. doi: 10.1007/978-3-319-06400-0.

Netter, F. H. (1994) *Musculoskeletal system - Netter collection of medical illustrations*. 1 ed. Vol. Saunders.

Subit, D. *et al*. (2013) *Tensile material properties of human rib cortical bone under quasi-static and dynamic failure loading and influence of the bone microstucture on failure characteristics*.

Turner, M. J. *et al*. (1956) 'Stiffness and Deflection Analysis of Complex Structures', *Journal of the Aeronautical Sciences*, 23(9), pp. 805–823. doi: 10.2514/8.3664.

Advances and Current Trends in Biomechanics – Belinha et al. (Eds)

Finite element analysis of a large scale polylactic acid-based stent for manufacturing using fused deposition modelling

Barbara Malafaya Pinto & Gabriela Caldas
Institute of Science and Innovation in Mechanical and Industrial Engineering (INEGI), Portugal

F.J.L. Alves & R.M.N. Jorge
Department of Mechanical Engineering, Faculty of Engineering, University of Porto, Portugal

Jorge Belinha
Department of Mechanical Engineering, School of Engineering, Polytechnic of Porto, Portugal

ABSTRACT: Additive manufacturing is starting to be common practice for biomedical applications. It allows customized solutions for both patient specific treatments (for therapeutics) and patient specific instruments (for diagnosis or surgical procedures). However, the developed solutions must structurally resist to the mechanical solicitations. Thus, it is important to predict the structural behavior of the designed part/instrument before its production/manufacture. Today, the finite element method (FEM) is the most commonly used discretization technique for structural analysis. FEM's fields of application ranges from large engineering structures to biomechanical analyses at the nanoscale. Thus, in this work, the finite element method (FEM) is used to perform an elasto-static analysis of a large scale stent produced by additive manufacturing - fused deposition modelling (FDM). The objective was to design a robust stent structure capable to withstand the mechanical solicitations at the assemblage phase. After several attempts, a structurally efficient standard modular solution was achieved. The solutions is also capable to be produced by FDM and then assembled respecting the patient's physiological requirements. In the end, the developed solutions was printed by FDM for proof-of-concept.

1 INTRODUCTION

In recent years, consumer demand has shifted from standard products to customized products, so the evolution of additive manufacturing (AM) has become essential. Compared to traditional ones, additive manufacturing consists of layer-by-layer addition of material and provides several advantages, including the development of complex geometries and the customization of products. Currently, the most popular technique is Fused Deposition Modelling (FDM), due to its low cost and the ability to manufacture parts with locally controlled properties. This technique uses thermoplastics, and one of the most common is polylactic acid (PLA), also due to its reduced cost and ease of use. FDM, also known as plastic jet printing (PJP), fused filament modelling (FFM) and thermoplastic extrusion, is included in the material extrusion group, consists of layer-by-layer deposition of fused thermoplastic filaments, from extruder nozzles, adding a layer of material for the construction of the desired physical model. Specifically, the extrusion nozzle has a heat source with temperature management, making it responsible for heating and modifying the physical state of the material (Ligon et al. 2017, Steenhuis & Pretorius 2017, Gkartzou et al. 2017). The extruder is maintained at

DOI: 10.1201/9781003217152-95

a temperature above the material melting temperature to ensure its extrusion and surface adhesion (Ligon et al. 2017, Zhang et al. 2016). Thus, the material is extruded to the printing platform through the extruder, which moves in a predetermined direction. The platform is kept at a temperature lower than the material, so that it cools quickly and consequently solidifies, serving as the basis for the subsequent layer of material (Chua & Leong 2016, Mahamood et al. 2014). After the first layer is completed, the extruder nozzle moves up (or the platform down) and the extrusion nozzle places a second layer, and this procedure is repeated until the model is finished (Ligon et al. 2017).

In this last twenty years, the biomedical field has been increasingly studied, so the AM has rapidly expanded into medical applications, including the development of customized and functional medical devices (such as stents). The stent is a cylindrical endovascular device that enables widening the lumen vessels and arteries, preventing them from narrowing again (restenosis) (Bonsignore 2011, Llorens 2016). This concept is currently used in several parts of the body to intervene in many conditions, such as coronary artery disease (Bonsignore 2011, Spranger et al. 2015). These structures are expandable meshes with a length of 6 to 60 mm and a diameter of 2 to 10 mm (Llorens 2016, Boyer et al. 2019).

This work aims to develop a modular stent solution for FDM additive manufacturing. Due to technical impediments, the developed solution will be designed with a 10:1 scale. Nevertheless, this will allow to test the complete pipeline of the design procedure: (1) model construction; (2) FEM analysis (and corresponding structural optimization); (3) FDM additive manufacturing and (4) the assemblage of the produced modules.

2 MATERIALS AND METHODS

To achieve the proposed goals, a preliminary model was constructed using Solidworks (2018 version), Figure 1. Then, the 3D model was discretized with 4-nodes plain strain tetrahedral elements and the several structural parts forming the prototype part were analyzed using FEM. The designed solutions is innovative, since it proposes a modular assemblage for the stent. Thus, repetitive modules are built and, in the end, those modules are assembled together, allowing to obtain an interconnected flexible structure. For the 3D printing process, a layer height of 0.16 mm, a rectilinear infill with 20% density and an angle of + 45°/-45° were used. The nozzle diameter used was 0.4 mm and the supports structures were always of the same material used for the construction of the model. Since the 3D printer used in this work does not allow printing (with enough precision) the small dimensions required to reproduce at 1:1 scale the stents, all models/prototypes were printed at 10:1 scale. Such preliminary approach allows to test the developed solutions (prototypes) and deliver recommendations for future research works on the topic. In the end, after all the models were printed by an FDM 3D printer, the main observations were highlighted. As indicated in the PLA commercial brochure, the mechanical properties of the PLA assumed in this study were: Young's Modulus of 3.15 GPa, Poisson's coefficient of 0.3 and tensile maximum stress of 105 MPa.

3 RESULTS

The proposed model, Figure 1, consists of six U components connected with double hooks. To bond each wire, a hook has been developed. To analyze the hook mechanical behavior, a linear-elastic finite element analysis was performed in Solidworks using 4-node plain strain hexahedral elements. Firstly, the required opening of the hook (to fit the wire) was measured, as shown in Figure 2(a). Then a total force of 1 N was equally distributed on the inner surface of the hook, fixing its central body, Figure 2(b). Then, the opening of the hook resulting from the force application was calculated and the stress field was obtained with the FEM analysis, as shown in Figure 2(c). After obtaining the produced displacement, the force required to open the hook (in order to attach it to the wire of the cylindrical block) was linearly calculated. Furthermore, the

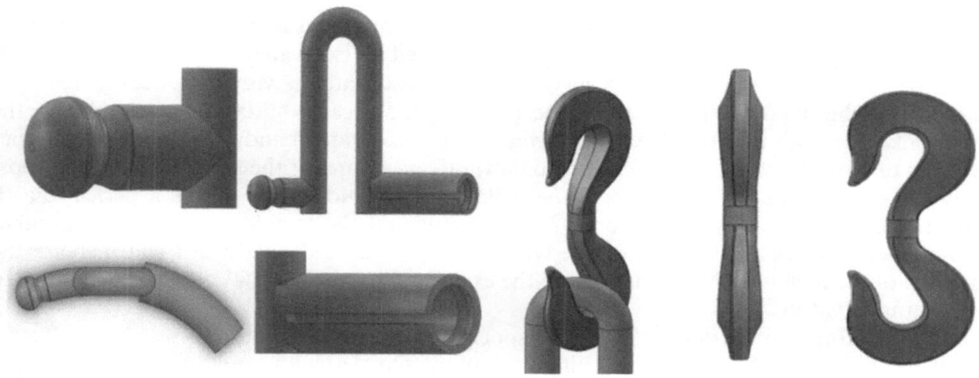

Figure 1. Developed model in Solidworks and several parts composing the model.

corresponding maximum stress was compared with the PLA ultimate stress to verify if the hook would resist to the forces required to fit it into the cylindrical block wires.

The diameter of the wire of the U component (in the zone were the hook attaches) is 6.386 mm. On the other hand, opening of the hook is only 6.100 mm. Therefore, it is necessary to impose a displacement on the hook extremity of 0.286mm (=6.386mm-6.100mm) in order to attach the hook to the U component. The FEM analysis shown that it was necessary a total force of 73.050 N distributed at the inner surface of the hook (Figure 2) in order to obtain a displacement of 0.286mm on the hook extremity and, therefore, deform the hook extremity forcing an opening of 6.386 mm (enough to pass the U component wire). Regarding the stress field in the hook due to the force of 73.050 N (distributed at the inner surface), it was verified that the maximum von Mises stress obtained was of 26.600 MPa, which is lower than the PLA tensile maximum stress (105 MPa). Therefore, it was concluded that the hook would resist to the forces required to fit it into the wire of the U component.

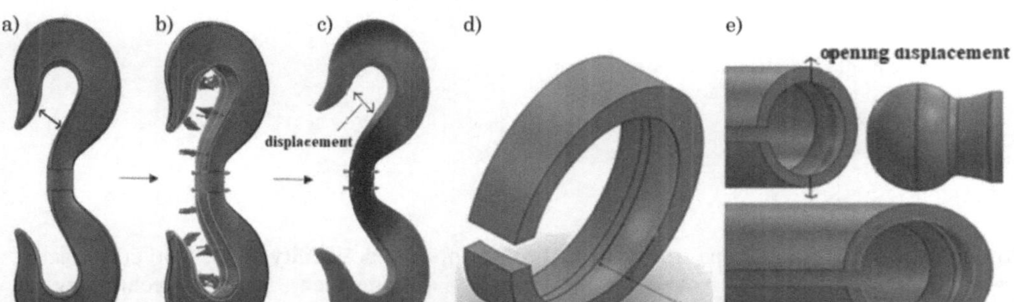

Figure 2. Hook numerical simulation: (a) opening measurement; (b) force application; and (c) displacement obtained. Bandwidth opening: (a) bandwidth size and (b) schematic representation of the opening.

The mechanical behavior of the grappling components of the cylindrical block of wire were also first analyzed using FEM. It is important to understand how the male spherical-end will attach to the female hallow-end of the U components forming the cylindrical block. The male spherical-end of one U component possesses a spherical shape that is inserted in the female hallow-end of another U component. Combining 6 U components allows to build a cylindrical block of wire. Inside the female hallow-end it was created an excavated bandwidth. Figure 2(d) shows the excavated bandwidth. The curvature of the excavated is very large (compared with the width of the bandwidth), therefore, the depth of the excavation is not perceptible in the figure, only the width is noticeable. As the width of the excavated bandwidth increases, so does the depth of the excavation. The excavated bandwidth curvature (of the female hallow-end) possesses the same radius as the sphere of the male spherical-end, which allows a perfect match between the male spherical-end of one U component and the female hallow-end of another U component. Increasing the size of the width of the excavated bandwidth will increase the grappling of the female hallow-end, because the excavated bandwidth will become deeper due to its constant curvature.

In order to force the entrance of the male spherical-end of one U component into the female hallow-end of another U component (Figure 2(e)), the female hallow-end has to be capable to open in the shape of a circumference with a radius, at least, equal to the radius of the sphere of the male spherical-end. Therefore, in the inner surface of the bandwidth it was applied a distributed force. It was found that assuming a bandwidth with 1.880 mm it would be necessary to apply a force of 2.50N (uniformly distributed on the bandwidth) to open the entrance of the female hallow-end enough to allow the entrance of the male spherical-end. With this level of force, it was generated at the female hallow-end a maximum stress of 1.263MPa, which is much lower than PLA tensile maximum stress (105 MPa). The model was printed with a scale of 10:1, possessing a diameter of 100.000 mm and a height of 120.000 mm, as shown in Figure 3.

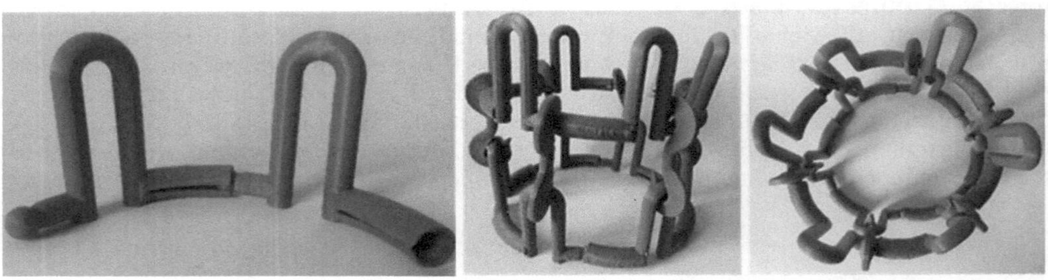

Figure 3. The printed model and its assembly.

4 CONCLUSIONS

Although some limitations, this study allowed to explore the possibility to develop customized biostructures (such as stents) from a low-cost perspective using such innovative technology as 3D printing. In recent years, there has been an exponential development of 3D printing. Thus, this technology is expected to have a promising future and the ability to produce functional products. In the future, it will be possible to use 3D printing for the development of customized stents for commercialization using the FDM technique. Thus, more numerical and biomechanical analyses are required, as well as in vivo tests in order to assess the overall performance of these solutions.

ACKNOWLEDGMENTS

The authors truly acknowledge the funding provided by Ministério da Ciência, Tecnologia e Ensino Superior – Fundação para a Ciência e a Tecnologia (Portugal), under LAETA project: UIDB/50022/2020.

REFERENCES

Bonsignore, C., Open stent design. ed: Fermont, CA: Nitinol Devices and Components, Inc, 2011.

Boyer, C.J., et al., 3D printing for bio-synthetic biliary stents. Bioengineering, 2019. 6(1): p. 16.

Chua, C.K. and K.F. Leong, 3D Printing and Additive Manufacturing: Principles and ApplicationsFifth Edition of Rapid Prototyping Fivth Edition. 2016: World Scientific Publishing Company.

Gkartzou, E., E.P. Koumoulos, and C.A. Charitidis, Production and 3D printing processing of bio-based thermoplastic filament. Manufacturing Review, 2017. 4: p. 1.

Ligon, S.C., et al., Polymers for 3D printing and customized additive manufacturing. Chemical reviews, 2017. 117(15): p. 10212–10290.

Llorens, S.R., 2D functionalized films and 3D direct printed patterned PLLA stents for cardiovascular applications. 2016, Universitat Politècnica de Catalunya.

Mahamood, R.M., et al., Revolutionary additive manufacturing: an overview. Lasers in Engineering,2014, 27(3):161–178.

Spranger, K., et al., Comparison and calibration of a real-time virtual stenting algorithm using finite element analysis and genetic algorithms. Computer methods in applied mechanics and engineering, 2015. 293: p. 462–480.

Steenhuis, H.-J. and L. Pretorius, The additive manufacturing innovation: a range of implications. Journal of Manufacturing Technology Management, 2017. 28(1): p. 122–143.

Zhang, B., et al., 3D printing of high-resolution PLA-based structures by hybrid electrohydrodynamic and fused deposition modeling techniques. Journal of Micromechanics and Microengineering, 2016. 26(2): p. 025015.

Advances and Current Trends in Biomechanics – Belinha et al. (Eds)
© 2022 Copyright the Author(s), ISBN: 978-1-032-10806-3

Review: Prosthetic limb replacement in dogs

M. Teixeira
School of Engineering, Polytechnic of Porto, Portugal

J. Belinha
Department of Mechanical Engineering, School of Engineering, Polytechnic of Porto, Portugal

ABSTRACT: With the growing appreciation and affection for companion animals comes the willingness to help them and give them a good, long life. Technology in veterinary medicine has been making advances over time including the veterinary orthotic and prosthetic industry. Prosthetics are emerging artificial devices that aim to assist those affected by deformities, incomplete limb development, and traumatic or surgical amputation. There are several types of prostheses and different suspension systems, so there is a solution depending on the problem presented. 3D printing is a growing field of great interest in the prosthetics industry as it allows fast, practical, and cheap manufacturing. This review article presents the different prosthetic limb replacements in dogs to provide them a better quality of life.

1 INTRODUCTION

Limb amputation is a procedure often recommended in veterinary medicine as a component of treatment for neoplasms, traumatic injuries, peripheral neuropathies, infections, vascular compromise, ischemic necrosis, and disability resulting from degenerative or congenital processes (Fitzpatrick *et al.*, 2011; Dickerson *et al.*, 2015; Phillips *et al.*, 2017; Contreras *et al.*, 2018). Some contraindications for limb amputation include extreme obesity, coexisting orthopaedic or neurological disease, and diseases that may interfere with the function of the remaining limbs (Fitzpatrick *et al.*, 2011; Dickerson *et al.*, 2015). Prostheses have been rarely used in veterinary medicine due to the ease of adaptation of quadrupeds to move with only three legs, but with the emergence of rehabilitation as a medical specialization field and due to the receptivity and good adaptation of veterinary patients, its use has been growing (Adamson *et al.*, 2005; Mich and Kaufmann, 2018). These devices are intended to allow the animal to move as close to normal as possible and are fixed to incomplete limbs due to deformities, incomplete limb development, traumatic or surgical amputation (Marcellin-Little *et al.*, 2015). Dogs apply a significant level of pressure on their limbs during running and other daily activities, which confers a great challenge in the retention of prostheses, requiring a suspension system to ensure adhesion of the residual limb to the socket. Thus, several types of prostheses are available for application in veterinary medicine, whose suspension systems may include self-suspension of the socket – which makes use of the anatomical shape of the residual limb –, suction suspension – which uses vacuum to keep the prosthesis tightly connected to the residual limb –, and suspension devices – such as harnesses, belts, clips, and sleeves. Transcutaneous prostheses are also possible options for use in dogs (Adamson *et al.*, 2005; Carr *et al.*, 2016; Mich and Kaufmann, 2018; Paternò *et al.*, 2018). Dogs that use prosthetics more often generally have a more positive clinical outcome and higher owner satisfaction, compared to dogs that rarely use these devices (Wendland, Seguin and Duerr, 2019). Rehabilitation during

DOI: 10.1201/9781003217152-96

the adaptation period is crucial for the patient, whether a stump socket prosthesis or a transcutaneous intraosseous prosthesis is used (Mich and Kaufmann, 2018).

2 LIMB AMPUTATION

Most dogs are able to move well even with the absence of a limb, however, if a single limb amputation is required, the quality of life decreases and therefore, multiple limb amputations are rarely performed (Drygas *et al.*, 2008).

Dogs undergoing total limb amputations show significant changes in locomotor biomechanics compared to quadrupedal dogs, which can have damaging effects on long-term musculoskeletal health and lead to decreased quality of life (Wendland, Seguin and Duerr, 2019). Among the concerns owners have about limb amputation are the physical appearance, the perception of reduced quality of life and mobility after surgery and the possibility of overloading the remaining limbs, leading to hypothetical secondary joint pathologies. In addition, there is the guilt associated with limb amputation, the fear that the dog may not tolerate anesthesia, and the perception of a substantial degree of pain associated with surgery. It is, however, important to psychologically support dog owners to take the decision about the amputation as soon as possible, as a delayed decision may lead to further complications (Dickerson *et al.*, 2015; Galindo-Zamora *et al.*, 2016).

It has been reported that dogs with thoracic limb amputations have more difficulty maintaining their balance, and fall more often, being thoracic limb amputations considered more debilitating than pelvic limb amputations (Adamson *et al.*, 2005; Fitzpatrick *et al.*, 2011). Dogs with pelvic limb amputations have more difficulty gaining speed during the early postoperative period. These findings were supported by a recent report documenting more substantial gait changes after thoracic limb amputation compared to pelvic limb amputation (Fitzpatrick *et al.*, 2011).

In a study involving 64 dogs, 91% showed no change in general attitude after amputation, 88% had a complete or near complete recovery of the quality of life they had before amputation, 78% had a better-than-expected recovery and surgical adaptation, and 73% showed no change in their recreational activities after surgery. These data indicated that the overall outcomes for the dogs after amputation were good, and this is extremely important, as some owners may consider euthanasia more compassionate than amputation (Dickerson *et al.*, 2015).

3 STUMP SOCKET PROSTHESIS

Stump socket prostheses are applied after partial limb amputation, they are widely used in human medicine and commercially available for animals (Phillips *et al.*, 2017). These pet devices require precise design and manufacture to provide stability when used. The fur is often left intact because of the protection it gives the skin from abrasive wear and ulcers (Marcellin-Little *et al.*, 2015). Once the animal gets used to using the prosthesis, it is used to the maximum (Phillips *et al.*, 2017).

For the placement of a stump socket prosthesis to be possible, there must be a substantial portion of limb remaining and the larger the stump is the better. Both the manufacture and fitting of pet prosthetic devices must be precise so that the device is stable when worn and does not cause discomfort or injury. There should be no gap or movement of the skin relative to the device when the prosthesis is manipulated. Axial, angular, and rotational stability are critical to successful use of the prosthesis (Marcellin-Little *et al.*, 2015).

Stump socket prostheses can be manufactured by creating a fiberglass impression of the limb that is then used as a mold to create a plaster replica of the animal's limb, or alternatively a handheld 3D scanner or a smart phone with a 3D capture software application can be used to capture a 3D rendering of the skin surface and a polymer replica of the limb can be made using a 3D printer. Thermoplastic polymers, such as polylactic acid, are often used to manufacture prosthetics. Polymer replicas are very precise and follow the contours of the skin surface more accurately than molds, particularly in small patients (Marcellin-Little *et al.*, 2015). Besides this, these plastics are very lightweight and resistant, a simple stump socket prosthesis can weigh about 0.2 kg, which is a benefit for dogs that suffer from arthritis and have to keep their weight down to move comfortably (Phillips *et al.*, 2017; Maniruzzaman, 2019). The prosthesis is printed by modules, so the parts can be re-printed and easily replaced if necessary (Maniruzzaman, 2019). The final prosthesis incorporates padding and foam support for a better adjustment to the limb (Phillips *et al.*, 2017). 3D printing is a trend technology to be applied to animal prosthetics because of the speed of printing and assembly, and the redesign flexibility (which allows easy shape modifications to suit the dog as the body changes) (Maniruzzaman, 2019).

The most common complications resulting from prosthetic use include pressure sores, infection, maladaptation, pain, and loss of mechanical integrity of the prosthesis (Drygas *et al.*, 2008; Phillips *et al.*, 2017; Wendland, Seguin and Duerr, 2019). Gait pattern changes can often occur due to the prosthesis being misaligned/misadjusted or to a limb length discrepancy. These problems should be attended to as soon as possible, and can often be solved by simple alignment adjustments or by changing the thickness of the internal padding of the prosthesis (Jean *et al.*, 2018). The advantages of stump socket prostheses over transcutaneous intraosseous prostheses are: application simplicity; manufacturing low cost; adaptability to different levels of amputation; and lack of a surgical implant, which could come loose or fail (Mich and Kaufmann, 2018).

The prostheses should be routinely checked and cleaned, and any abrasion should be predicted, particularly in the sole part, so that the contact area of the sole can be monitored, and the next sole can be shaped accordingly. The hinges can also wear out, and when this occurs they must be replaced (Marcellin-Little *et al.*, 2015). Despite reduced ability to jump, owners did not perceive reduced function or quality of life (Phillips *et al.*, 2017).

4 TRANSCUTANEOUS INTRAOSSEOUS PROSTHESIS

The limitations of stump socket prostheses associated with the skin-socket interface can be avoided by using transcutaneous prostheses, which constitute an inert implant (or endoprosthesis) usually made of titanium, which is anchored to the limb, and an external prosthesis (exoprosthesis), which is attached to the protruding part of the implant. In this way, the transfer of the ground reaction forces and the mechanical load is done from the prosthesis to the bone and from the bone to the prosthesis at the level of the amputation via the implant (Fitzpatrick *et al.*, 2011; Mich and Kaufmann, 2018). The benefits associated with transcutaneous prostheses include the lack of need for continuous adjustment, better efficiency during gait, improved proprioception, no skin breakdown problems, and no need to constantly put on and take off prostheses. However, there are complications, such as implant infection, pain, failure of the prosthesis to attach to the bone, and periprosthetic fractures. These complications may require the removal of the implant and possibly also segments of the affected limb (Drygas *et al.*, 2008). The transcutaneous intraosseous prosthesis promotes bone and dermal integration at the limb-implant interface, thus achieving a permanent biological seal, and has the mechanical function of coupling the limb to the external prosthesis (Fitzpatrick *et al.*, 2011). Successful clinical results have been reported after the use of transcutaneous prostheses for limb salvage in 4 dogs with malignant neoplasia with good limb function without pain for 8 to 17 months after the surgery (Fitzpatrick *et al.*, 2011).

5 CONCLUSION

Rapid technological advances have allowed the development of prosthetic models that are increasingly sophisticated and adapted to the characteristics and structure of each animal. There are several prostheses for use in veterinary medicine, each with its advantages and disadvantages, and it is up to the owners to contact the veterinarian to clarify the type of prosthesis best suited to the case of their dogs. It is very important to inform the veterinarian of the intention of prosthetics placement at the time of limb amputation of the dog, because for it to be used, the limb cannot be totally amputated.

The stump socket prostheses are not invasive as the intraosseous transcutaneous prostheses nor require surgery and can be removed whenever the owner want, however, it is very important that it is properly aligned and adjusted to the stump so that there are no wounds, discomfort and changes in gait pattern that can lead to musculoskeletal health problems in the long term. Transcutaneous intraosseous prostheses provide greater efficiency during gait, since they consist of an implant and an external prosthesis that is attached to it, giving a better adjustment, and thus providing a gait closer to normal, however, they also have possible disadvantages, especially regarding the implant.

Although 3D printing presents many advantages, and it is of great interest to invest in this mode of manufacture, there are not many studies documented in the literature on prostheses for limb replacement in dogs produced by additive manufacturing.

ACKNOWLEDGMENTS

The authors gratefully acknowledge funding from the Ministry of Science, Technology and Higher Education - Foundation for Science and Technology (Portugal), through the LAETA project: UIDB/50022/2020.

REFERENCES

Adamson, C. *et al.* (2005) 'Assistive devices, orthotics, and prosthetics', *Veterinary Clinics of North America - Small Animal Practice*, 35(6), pp. 1441–1451. doi: 10.1016/j.cvsm.2005.08.009.

Carr, B. J. *et al.* (2016) 'Canine orthopedic devices', *Today's Veterinary Prcatice*, (February), p. 117.

Contreras, E. T. *et al.* (2018) 'Postamputation Orthopedic Surgery in Canine Amputees: Owner Satisfaction and Outcome', *Topics in Companion Animal Medicine*. Elsevier Inc., 33(3), pp. 89–96. doi: 10.1053/j.tcam.2018.07.001.

Dickerson, V. M. *et al.* (2015) 'Outcomes of dogs undergoing limb amputation, owner satisfaction with limb amputation procedures, and owner perceptions regarding postsurgical adaptation: 64 cases (2005–2012)', *Journal of the American Veterinary Medical Association*, 247(7), pp. 786–792. doi: 10.2460/javma.247.7.786.

Drygas, K. A. *et al.* (2008) 'Transcutaneous tibial implants: A surgical procedure for restoring ambulation after amputation of the distal aspect of the tibia in a dog', *Veterinary Surgery*, 37(4), pp. 322–327. doi: 10.1111/j.1532-950X.2008.00384.x.

Fitzpatrick, N. *et al.* (2011) 'Intraosseous Transcutaneous Amputation Prosthesis (ITAP) for Limb Salvage in 4 Dogs', *Veterinary Surgery*, 40(8), pp. 909–925. doi: 10.1111/j.1532-950X.2011.00891.x.

Galindo-Zamora, V. *et al.* (2016) 'Kinetic, kinematic, magnetic resonance and owner evaluation of dogs before and after the amputation of a hind limb', *BMC Veterinary Research*. BMC Veterinary Research, 12(1), pp. 1–14. doi: 10.1186/s12917-016-0644-5.

Jean, B. *et al.* (2018) 'Retrospective Study on External Canine Limb Prosthesis used in 24 patients', *Veterinary Evidence Online*, 3(1), pp. 1–13.

Maniruzzaman, M. (2019) *3D and 4D Printing in Biomedical Applications*. John Wiley & Sons.

Marcellin-Little, D. J. *et al.* (2015) 'Orthoses and Exoprostheses for Companion Animals', *Veterinary Clinics of North America - Small Animal Practice*, 45(1), pp. 167–183. doi: 10.1016/j.cvsm.2014.09.009.

Mich, P. M. and Kaufmann, M. (2018) 'Veterinary Orthotics and Prosthetics', *Canine Sports Medicine and Rehabilitation: Second Edition*, pp. 265–293. doi: 10.1002/9781119380627.

Paternò, L. *et al.* (2018) 'Sockets for limb prostheses: A review of existing technologies and open challenges', *IEEE Transactions on Biomedical Engineering*, 65(9), pp. 1996–2010. doi: 10.1109/TBME.2017.2775100.

Phillips, A. *et al.* (2017) 'Clinical outcome and complications of thoracic and pelvic limb stump and socket prostheses', *Veterinary and Comparative Orthopaedics and Traumatology*, 30(4), pp. 265–271. doi: 10.3415/VCOT-16-09-0127.

Wendland, T. M., Seguin, B. and Duerr, F. M. (2019) 'Retrospective multi-center analysis of canine socket prostheses for partial limbs', *Frontiers in Veterinary Science*, 6(APR). doi: 10.3389/fvets.2019.00100.

Advances and Current Trends in Biomechanics – Belinha et al. (Eds)

Polylactic acid-based stent manufacturing using fused deposition modelling: A large scale prototype

Barbara Malafaya Pinto & Gabriela Caldas
Institute of Science and Innovation in Mechanical and Industrial Engineering (INEGI), Portugal

F.J.L. Alves & R.M.N. Jorge
Department of Mechanical Engineering, Faculty of Engineering, University of Porto, Portugal

Jorge Belinha
Department of Mechanical Engineering, School of Engineering, Polytechnic of Porto, Portugal

ABSTRACT: Cardiovascular complications are one of the major causes of death in the world. As one of the possible treatments, the biopolymeric stents were developed and commercialized through traditional techniques. In comparison to the traditional ones, additive manufacturing provides several advantages, including customization and the development of complex geometries. Nowadays, one of the most popular techniques is the fused deposition modeling, which uses a renewable thermoplastic, the polylactic acid, due to its reduced cost and facility to be used. In recent years, additive manufacturing has been emerging into the medical industry, with the possibility of developing biodevices, including stents, individually adapted to the patient.

1 INTRODUCTION

The biopolymeric stents have received increased attention due to the high efficiency in interventional therapy and their fabrication methods still present significant challenges, including its high-cost. Therefore, this work seeks to complete these gaps by proposing the production of stents in an alternative manufacturing method, known as additive manufacturing, in particular fused deposition modeling (FDM).

The present work aims to develop and test numerically biodevices, such as arterial stents, to be produced with the FDM technique. In the end, large scale prototypes were produced in order to prove that the complete process is feasible. In recent times, the consumer demand is shifting from standardized to customized products. Therefore, it was developed the additive manufacturing, which consists of the model production through layer-by-layer material addition (Mahamood et al. 2014). In opposition to traditional methods, additive manufacturing is simple to use, allows form freedom and customization (Mahamood et al. 2014, Ligon et al. 2017). Thus, one of the most popular techniques is fused deposition modeling, due to its low-cost and facility to be used. This technology is based on the layer-by-layer deposition of extruded thermoplastic filaments (Mahamood et al. 2014, Gkartzou et al. 2017, Zhang et al. 2016, Chacón et al. 2017). Therefore, the polylactic acid is commonly used in FDM, as it is a low-cost thermoplastic biomaterial.

In recent years, 3D printing has been increasing its application in the medical industry for the development of customized biodevices, including stents, for each patient (Ligon et al.

DOI: 10.1201/9781003217152-97

2017). Stents are endovascular structures used for enlarging vessels to increase, for example, the blood flow (Llorens 2016, Bonsignore 2011). These devices are usually made of metals, but recently biopolymeric stents have become popular (Soares et al. 2008).

2 MATERIALS AND METHODS

To achieve the proposed objectives, the practical component was initialized by the model design. For that, four different geometries were developed using Solidworks (version 2018). For one of the geometries, it was necessary to perform a numerical structural simulation to analyze the best dimensions and if it would resist to the assembly. For this step, it was used the Finite Element Method (FEM) in Solidworks (version 2018). Finally, all the models were differently printed by a FDM printer using different types of polylactic acid, and the main conclusions were highlighted.

3 RESULTS AND DISCUSSION

In this work, four different geometries were developed in order to test which one performs better. In Figure 1, it is presented the first model (model 1). This is one of the most popular geometries, characterized by ten wires, each formed in twenty-one zigzags and the wires were connected at the crown points to complete the stent. This geometry was easy to be developed since it was based on the pre-existing stents. This model was vertically printed with supports using a normal PLA and it presented low finishing quality and the high fragility due to its very thin components.

Figure 1. Design and prototype of model 1.

Model 2, presented in Figure 2, is similar to the previous one, but it was composed by ten wires, each one formed in six zigzags. This model is less rigid than the previous one because it has less zigzags. This type is also frequently used for medical use, so the design was also based on pre-existing stents available in literature. This model was printed horizontally, to avoid support material, using a normal PLA and then it was welded at the lateral ends to close the stent. Although it had higher resistance, it was still very fragile at the welded parts. This model was printed again but with flexible PLA, which resulted in higher resistance but also high flexibility, so if a balloon were blown inside of it, the stent would not hold the required shape.

After these geometries, model 3 was developed. Its geometry is presented in Figure 3 and it is an indent connection model, which consisted of wires formed in six zigzags bonded at their indent connectors and, to connect each wire, a hook was developed. This new geometry allow a higher stiffness than the previous solutions. It was also perform a numerical simulation to analyze the stent assembly behavior and the best bandwidth/force correlation to print the stent. This model was also printed with PLA. In the end, it was verified that this geometry had low stability in the indent connectors and the zigzag wires connectors detached easily, disrupting its structural stability.

Last, it was developed model 4, Figure 4, which is a cylindrical grip model composed of wires formed in five zigzag that are bonded by a cylindrical connectors and also, to connect each wire, it was used the hook previous developed. This model is also innovative, and it was developed in order to be more easily fixed and stable to the printing and to the assembly. After printing the model with PLA, it was verified that this simple model was the one with highest stability, resistance and finishing quality.

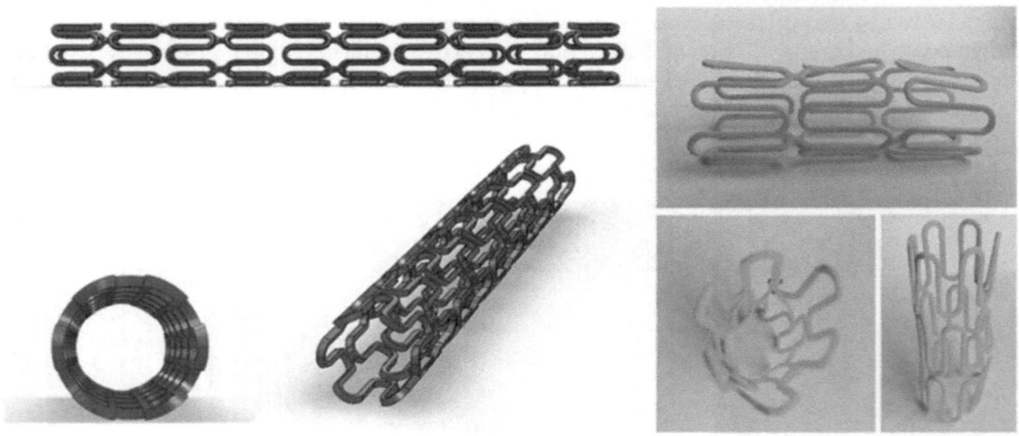

Figure 2. Design and prototype of model 2.

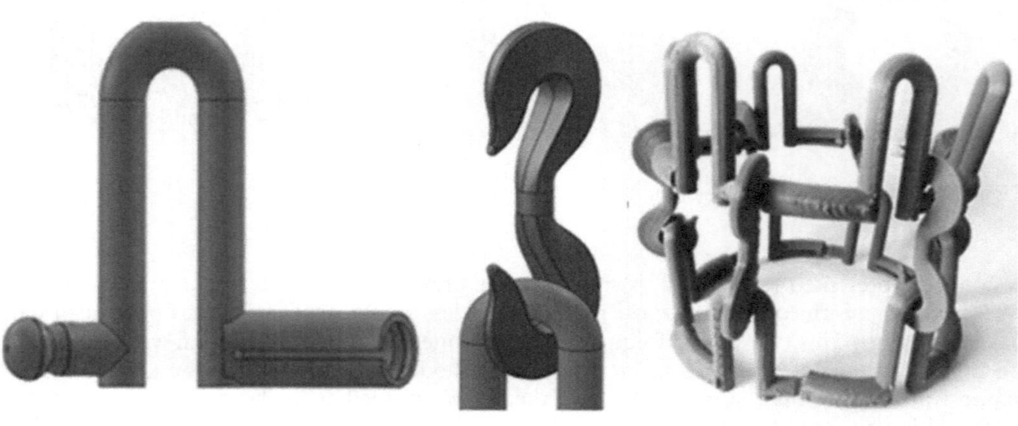

Figure 3. Detail parts and prototype of model 3.

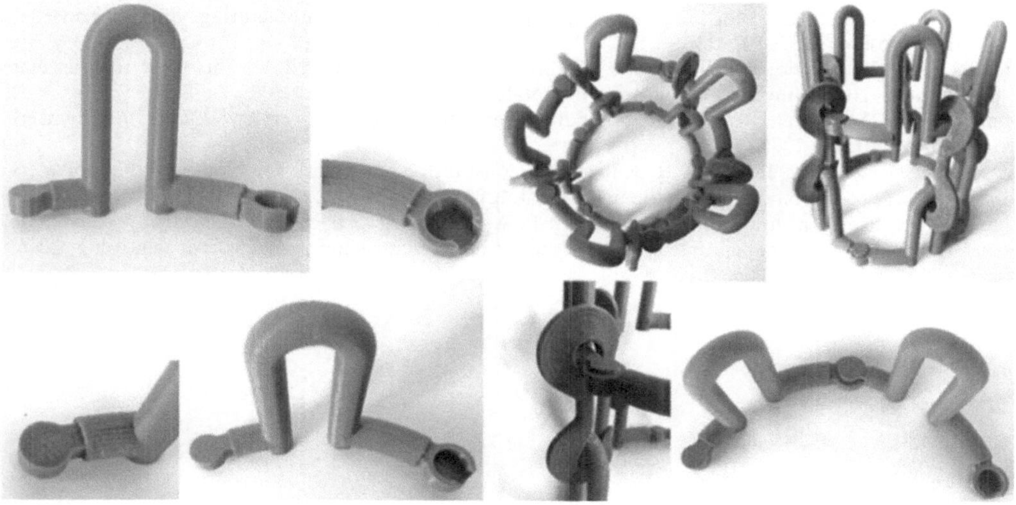

Figure 4. Detail parts and prototype of model 4.

4 CONCLUSIONS

Over the years, there was an increase demand for alternative techniques such as additive manufacturing. There are currently several AM techniques. Fused deposition modelling is the most popular extrusion technique and it has been expanded for broad applications in the field of biomedical engineering, such as the development of personalized and functional medical devices.

In this work, four different geometries were developed and printed. Model 1 and 2, very popular in the medical industry, presented low finishing quality and high fragility. On the other hand, model 3 is an innovative geometry consisting of an indent connection model. But it had low stability in the indent connectors and the zigzag wires connectors detached easily, disrupting its structural stability. Last, model 4 is also a new type of stent consisting of a cylindrical grip model and it showed the highest stability, resistance, finishing quality.

In the future, it should be performed additional numerical analyses to optimize its shape and structural performance. It is also required to produce such models in biocompatible PLA and test them in vivo.

ACKNOWLEDGMENTS

The authors truly acknowledge the funding provided by Ministério da Ciência, Tecnologia e Ensino Superior – Fundação para a Ciência e a Tecnologia (Portugal), under LAETA project: UIDB/50022/2020.

REFERENCES

Bonsignore, C., Open stent design. ed: Fermont, CA: Nitinol Devices and Components, Inc, 2011.

Chacón, J., et al., Additive manufacturing of PLA structures using fused deposition modelling: Effect of process parameters on mechanical properties and their optimal selection. Materials & Design, 2017. 124: p. 143–157.

Gkartzou, E., E.P. Koumoulos, and C.A. Charitidis, Production and 3D printing processing of bio-based thermoplastic filament. Manufacturing Review, 2017. 4: p. 1.

Ligon, S.C., et al., Polymers for 3D printing and customized additive manufacturing. Chemical reviews, 2017. 117(15): p. 10212–10290.

Llorens, S.R., 2D functionalized films and 3D direct printed patterned PLLA stents for cardiovascular applications. 2016, Universitat Politècnica de Catalunya.

Mahamood, R.M., et al., Revolutionary additive manufacturing: an overview. Lasers in Engineering,2014, 27(3):161–178.

Soares, J.S., J.E. Moore Jr, and K.R. Rajagopal, Constitutive framework for biodegradable polymers with applications to biodegradable stents. Asaio Journal, 2008. 54(3): p. 295–301.

Zhang, B., et al., 3D printing of high-resolution PLA-based structures by hybrid electrohydrodynamic and fused deposition modeling techniques. Journal of Micromechanics and Microengineering, 2016. 26(2): p. 025015.

Author index